International Conference on Innovation in Science and Technology (ICIST 2019)

IOP Conference Series: Materials Science and Engineering
Volume 1108

Semarang, Indonesia
17 December 2019

Editors:

A. G. Abdullah
I. Widiaty
A. A. Danuwijaya
C. U. Abdullah
B. Supriyo
Jumi
T. Mulyono

ISBN: 978-1-7138-4601-7
ISSN: 1757-8981

Printed from e-media with permission by:

Curran Associates, Inc.
57 Morehouse Lane
Red Hook, NY 12571

proceedings.com

Some format issues inherent in the e-media version may also appear in this print version.

Institute of Physics
Dirac House, Temple Back
Bristol BS1 6BE UK

Phone: 44 1 17 929 7481
Fax: 44 1 17 920 0979

techtracking@iop.org

Additional copies of this publication are available from:

Curran Associates, Inc.
57 Morehouse Lane
Red Hook, NY 12571 USA
Phone: 845-758-0400
Fax: 845-758-2633
Email: curran@proceedings.com
Web: www.proceedings.com

TABLE OF CONTENTS

Preface

Peer review declaration

External strengthening of reinforced concrete beams with strand without stressing under flexural loading 1
I Iswanto, M Oesman, M Mujiman

Mechanical properties of porous asphalt with gilsonite additive 12
R Rochaeti, J Endawati, T Oktaviani

Reinforcement method of reinforced concrete beam by using non-stressing strand 19
A Susanto, H Kasyanto, S Susilahadi

Deformation and dynamic stability of asphalt concrete by using granite dust as filler 24
M Oesman, A Sundara, J Endawati, D Virnawan

Performance improvement of Asphalt Concrete-Wearing Course with modification of masterbatch SIR20 31
N P Yulianti, S Sulaiman, R Utami

Innovative products of concrete steel bone testing tools in the field of civil engineering building 38
T Hadi, N S Pamungkas, W Wahjoedi, T Mulyono

Monitoring rainfall intensity and moisture water content using soil column experiment 49
M Mukhlisin, B Yunanto, A Suharjono, M Martono, R Apriantoro, R Kusumawardani

Emission comparison of short lived climate forcers over long term greenhouse gases from domestic and transport sector in Semarang city 57
H Huboyo, B Samadiku, O Manullang

Mapping of flood-prone-zone and puddle-prone-zone in Cimahi city, West Java province 65
I Karnisah, Y Astor

Prediction of sediments discharge in watershed with two tank models 77
T Mulyono, J Jumi, A Azizah

Electronic system of rubber belt electro-mechanical continuously variable transmission for motorcycle applications 90
B Supriyo, S Ariyono, S Sihono

Design and development of student attendance information system using QR code in accounting department of Politeknik Negeri Semarang 98
S Sarana, A Sadida, A W Mansyur, A Suwondo

Prediction effective capacity battery of cellular BTS for an effective maintenance 109
M B Harijadi, T Andromeda, M Somantri

Sediments discharge analysis using tank model for disaster mitigation 120
T Mulyono, J Jumi, A Azizah

Speed control system in mobile robot based on Bezier curve trajectory 129
M F Alfatih, M A Riyadi, I Setiawan

VeRO: Smart home assistant for blind with voice recognition 139
L Triyono, T R Yudantoro, S Sukamto, I Hestinigsih

Bandwidth constrained distributed estimation for wireless sensor networks 146
E D Wardihani, A Suharjono, I Sayekti

Online monitoring system for emergency diabetes mellitus patients 152
R Kurniawati, S Suryono, S Ismail

Survey of smart parking application deployment 160
D Nugraha, F Y H Ahmed, M I Abdullah, M G M Johar

Web-based rule-based expert system to screen postpartum depression 174
D Mufidati, M N Widyawati, S Suryono

Web-based rule-based system for early detection of anemia among pregnant mothers 182
S Y Veronica, M N Widyawati, S Suryono

Euclidean distance digital image processing for jaundice detect 189
E P Rahayu, M N Widyawati, S Suryono

GPS-based fire detection system (Global Positioning System) and SMS Gateway 195
A Aryanti, I Mekongga, R S Dewi

Concrete mix design modelling based on variation of hidden layer and neuron of ANN for virtual learning development 202
S Santosa, S Suroso, M B Utomo, M Martono, M Mawardi, Y P Santosa

QoS analysis of WSN (Wireless Sensor Network) using node MCU and accelerometer sensors on bridge monitoring systems 209
D Junesco, E Supriyanto, A Hasan, M Mukhlisin

Accelerometer sensor data analysis of bridge structural health monitoring system 219
A A Hapsari, E Supriyanto, A Hasan, A Suharjono

Double security system with RFID and password based on Arduino Mega 2560 and PLC 227
S Syahid, S Supriyatna, Y Badruzzaman, D Antono, A Santoso

Power saving of IP over WDM: Coherent taxonomy, mechanism and technique: A review 234
L P Pratama, S N Saud, E D Wardihani, N M Lehat

Wind speed monitoring and alert system using sensor and weather forecast 246
M Mardiyono, R E Sari, O N Dini

Victim identification application with face recognition based on android 255
L Triyono, S Handoko, K Y Puspawaty

Performance analysis of shape, color and texture features on tracking information face based on CBIR 264
J Jumi, A Zaenuddin, T Mulyono

Analysis of Deep Water Culture (DWC) hydroponic nutrient solution level control systems 275
A Nursyahid, T A Setyawan, K Sa'diyah, E D Wardihani, H Helmy, A Hasan

Nutrient Film Technique (NFT) hydroponic nutrition controlling system using linear regression method 284
A Nursyahid, H Helmy, A I Karimah, T A Setiawan

Fall detecting clothes in realtime based seniors full body motion capture system using multiple inertial sensors 294
T R Yudantoro, F Pramuditya, R Apriantoro, S Jum'atun

Study method accordance for presstool design learning between POLINES and PT Mekar Armada Jaya in the framework of link & match 299
R M Wattimena, A S Ardjo, H Hartono, A S Alfauzi

Biaxial pressure pack carburizing method to modification local low carbon steel's mechanical properties 305
A Satito, H Hariyanto, S Supandi

Analysis of vibration due to misalignment in the clutch cluster installation of centrifugal pump 312
I G Widodo, A Khoryanton, A Pramono

Model for identification of rice type using combination of shape and color features 320
J Jumi, A Zaenuddin, T Mulyono

Design models of automatic humidity control system in flash off painting line room to reduce defect products at PT. Kubota Indonesia 329
A Khoryanton, G Giyanto, A Satito, S Harmanto, H Haryanto, T Tegar

Effect of Vibrating Fluidized Bed (VFBD) and continuous microwave drying on drying characteristic of green tea 335
S U Handayani, M E Yulianto, S Sutrisno, D Ariwibowo, R Amalia, I Hartati

Improved Savonius double blade performance using modified blade shaped with variations of the wind flow ratios on the blade inlet and outlet side 343
Y D Herlambang, S Supriyo, B Prasetiyo

Smart wind turbine three-layers flat blade with control system Arduino-based for blade angle 350
S Sahid, S Suwarti, B Prasetyo, M Mulyono

Performance of single cylinder diesel engine using triple fuel HSD-LPG-Steam 357
G E Kusuma, S T Sarena, M Santoso

Experimental study of transmission ratio changing mechanism for motorcycle applications 367
B Supriyo, S Ariyono, S Sihono, B Sumiyarso, B Tjahyono

Application of CNC machine router 3-Axis for making of Engravired granite or marble 374
S Suharto, S Suryanto, S Sarana, K Purbono

Maintenance information system development for Powerhouse Department on diesel engine manufacturer 379
A S Ardjo, E D Wardihani, A Slamet, M H Habiburrahman

Android based maintenance information system for machining workshop in polytechnic high education 387
A S Ardjo, E D Wardihani, P Prayitno

Participatory rural appraisal strategy in realizing the standardization of ship components 395
A Khoryanton, P Pratikto, S Suparman, P B Santoso

Hybrid power plant of the photovoltaic-fuel cell 400
Y D Herlambang, T Prasetyo, A Roihatin, Y M Safarudin, F Arifin

Wind energy conversion system with capacity of 450 W in order to support teaching and learning process in the laboratory 409
Y D Herlambang, T H Mulud, B Bono, S Suwarti, G Suwoto

Author Index

Preface

The International Conference on Innovation in Science Technology (ICIST 2019) is an event organized by the Politeknik Negeri Semarang in Semarang - Indonesia. ICIST's collection of articles written by scholars and practitioners in the fields of Electrical Engineering and Computer Science, Mechanical Engineering, and Civil and Environmental Engineering. This conference was organized by the Politeknik Negeri Semarang and was held at Grand Candi Hotel, Jl, Sisingamangaraja No.16, Kaliwiru, Kec. Candisari, Kota Semarang, Jawa Tengah 50232, Indonesia. Advances in Applied Research for Future Innovations, Developing Technologies, and Commercials is a theme based on the fact that over the past few years there has been an impressive increase in the use of technology being a key resource in any information system configuration. Wireless and fixed-line networks complemented by a growing range of mobile devices are having a significant impact on the way we run our lives and our business.

We would like to express our gratitude to all keynote speakers and presenters from around the world who traveled to Semarang to deliver and exchange their ideas. It includes Prof. Dr. Ir. Khisbullah Hudha, M.Sc. (National Defence University of Malaysia (NDUM)), Denni Kurniawan, B.Eng., M.Eng., D.Eng (University Teknologi Brunei), and Dr. Ir. Tedjo Mulyono, MT. (Politeknik Negeri Semarang). We would also like to express our highest appreciation to all those who have supported the success of this conference. To all distinguished authors, attendees, keynote speakers, sponsors, Head of Research and Community Dedication Center Polines and Director of Polines thank you very much. I would also like to express our sincere appreciation to all reviewers for their great support in reviewing all papers.

Finally, I sincerely hope that all participants will get the greatest benefits as well as gaining a pleasant experience during this conference in Semarang, Indonesia. I hope this event contributes to the advancement of science and technology.

The Editors

Ade Gafar Abdullah
Isma Widiaty
Ari Arifin Danuwijaya
Cep Ubad Abdullah
Tedjo Mulyono
Jumi
Bambang Supriyo

IOP Conf. Series: Materials Science and Engineering 1108 (2021) 011001 doi:10.1088/1757-899X/1108/1/011001

LIST OF COMMITTEE

Advisory Board

1. Assoc. Prof. Ir. Supriyadi (Director POLINES, Semarang, Indonesia)
2. Assoc. Prof. Endro Wasito (Deputy Director I POLINES, Semarang, Indonesia)
3. Assoc. Prof. Dr. Budi Prastyo (POLINES, Semarang, Indonesia)
4. Prof. Dr. Ir. Muhammad Mukhlisin (POLINES, Semarang, Indonesia)
5. Prof. Dr. Hishamuddin Jamaluddin (Southern University College, Malaysia)
6. Dr. Amin Suharjono (POLINES, Semarang, Indonesia)
7. Assoc. Prof. Hartono (POLINES, Semarang, Indonesia)
8. ssoc. Prof. Tedjo Mulyono (POLINES, Semarang, Indonesia)
9. Prof. Wang Min-Wen, Ph.D. (National Kaohsiung university of Science and Technology, Taiwan)
10. Prof. Shyu jin-Cherng, Ph.D (National Kaohsiung university of Science and Technology, Taiwan)
11. Assoc. Prof. Dr. Aniza binti Ibrahim (Universiti Pertahanan Nasional Malaysia, Malaysia)
12. Assoc. Prof. Sr. Dr.Khairul Nizam Abdul Maulud (Universiti Kebangsaan Malaysia, Universitas Kebangsaan Malaysia)

Scientific Committee

1. Assoc. Prof. Dr. Totok Prasetyo (POLINES, Semarang, Indonesia)
2. Assoc. Prof. Dr. Ir. Muhammad Mukhlisin (POLINES, Semarang, Indonesia)
3. Dr. Amin Suharjono (POLINES, Semarang, Indonesia)
4. Mr. Sidiq Syamsul Hidayat, Ph.D (POLINES, Semarang, Indonesia)
5. Assoc. Prof. Dr. Sugeng Ariyono (POLINES, Semarang, Indonesia)
6. Assoc. Prof. Dr. Bambang Supriyo (POLINES, Semarang, Indonesia)
7. Assoc. Prof. Dr. Suparni Setyowati Rahayu (POLINES, Semarang, Indonesia)
8. Assoc. Prof. Dr. Yusuf Dewantoro Herlambang (POLINES, Semarang, Indonesia)
9. Dr. Eni Dwi Wardihani (POLINES, Semarang, Indonesia)
10. Dr. Kurnianingsih (POLINES, Semarang, Indonesia)
11. Assoc. Prof. Dr. Ing.Tech. Samuel Beta K (POLINES, Semarang, Indonesia)
12. Dr. Bambang Supriyo (POLINES, Semarang, Indonesia)
13. Assoc. Prof. Dr. Edy Suhartono (POLINES, Semarang, Indonesia)

IOP Conf. Series: Materials Science and Engineering 1108 (2021) 011001 doi:10.1088/1757-899X/1108/1/011001

Organizing Commitee

Conference Chair:

Dr. Jumi, S.Kom., M.Kom. (Politeknik Negeri Semarang, Indonesia)

Member:

1. Assoc. Prof. Dr. Eddy Triyono (Politeknik Negeri Semarang, Indonesia)
2. Mr. Yanuar M.S. (Politeknik Negeri Semarang, Indonesia)
3. Mrs. Anis Roihatin (Politeknik Negeri Semarang, Indonesia)
4. Mr. Helmy (Politeknik Negeri Semarang, Indonesia)
5. Mr. Ari Sriyanto (Politeknik Negeri Semarang, Indonesia)
6. Mrs. Dewi Anggraeni (Politeknik Negeri Semarang, Indonesia)
7. Mr. Mardiyono (Politeknik Negeri Semarang, Indonesia)
8. Mr. M. Anif (Politeknik Negeri Semarang, Indonesia)
9. Mr. Junaidi (Politeknik Negeri Semarang, Indonesia)
10. Mr. Liliek Triyono (Politeknik Negeri Semarang, Indonesia)
11. Mrs. Yuyun (Politeknik Negeri Semarang, Indonesia)
12. Mrs. Ari (Politeknik Negeri Semarang, Indonesia)
13. Assoc. Prof. Dr. Yusuf Dewantoro Herlambang (Politeknik Negeri Semarang, Indonesia)
14. Mr. Sidiq Syamsul Hidayat, Ph.D (Politeknik Negeri Semarang, Indonesia)
15. Dr. Bambang Supriyo (Politeknik Negeri Semarang, Indonesia)

CONFERENCE PHOTOGRAPH

ICIST 2019
International Conference on Innovation in Science and Technology
ADVANCES IN APPLIED RESEARCH FOR FUTURE INNOVATION EMERGING
TECHNOLOGY AND COMMERCIAL
Asst. Prof. Dr. Denni Kurniawan
Dr. Ir. Tedjo Mulyono, M.T.
Politeknik Negeri Semarang

ICIST 2019
International Conference on Innovation in Science and Technology
RISTEKDIKTI
ADVANCES IN APPLIED RESEARCH FOR FUTURE INNOVATION
TECHNOLOGY AND COMMERCIAL

IOP Conf. Series: Materials Science and Engineering 1108 (2021) 011001 doi:10.1088/1757-899X/1108/1/011001

erah Kerja
lah luasan yang dapat
at potong (cutter)

ICIS 2019
nternational Conference o Science and Technology
IN APPLIED RES NNOVATION
TECHNO

ICIST 2019

ICIST 2019

Peer review declaration

All papers published in this volume of *IOP Conference Series: Materials Science and Engineering* have been peer reviewed through processes administered by the Editors. Reviews were conducted by expert referees to the professional and scientific standards expected of a proceedings journal published by IOP Publishing.

- **Type of peer review:**

  A group of reviewers with good reputation of international publication were assigned to review the manuscripts of ICIST 2020 using Konfrenzi Automated Conference System. Prior to the review process, all the reviewers were briefed by Prof. Ade Gafar Abdullah on a virtual meeting regarding the reviewing process. As many as 20 reviewers were then eligible in which each was sent 3 – 4 papers through the submission system. They were given two weeks to finish their review process and given a certain amount of incentive once they finish their task. After the deadline, the reviewed manuscripts were sent to the participants (authors) using the system through their personal account. The authors were given two weeks to revise their manuscripts and send them back to the committee using their account. Papers whose revision was in accordance with the reviewers' comments went into final editing process (paper template layout checking, proofreading, etc.) and those which need further revision went into the second round of review process. All the manuscripts and other required data were submitted to IOP after they passed final editing

- **Conference submission management system:**

  Konfrenzi Automated Conference System

- **Number of submissions received:**

  82

- **Number of submissions sent for review:**

  58

- **Number of submissions accepted:**

  58

- **Acceptance Rate (Number of Submissions Accepted / Number of Submissions Received X 100):**

  70.7%

- **Average number of reviews per paper:**

IOP Conf. Series: Materials Science and Engineering 1108 (2021) 011002 doi:10.1088/1757-899X/1108/1/011002

2

- **Total number of reviewers involved:**

  20

- **Any additional info on review process:**

  -

- **Contact person for queries:**

  ade_gaffar@upi.edu

External strengthening of reinforced concrete beams with strand without stressing under flexural loading

I Iswanto*, M Oesman and M Mujiman

Department of Civil Engineering, Bandung State Polytechnic, Bandung 40012, Indonesia

*iswanto.keto@gmail.com

Abstract. This research was purposed to study the composite behavior and the increased capacity of reinforced concrete beam strength under flexural loading caused by external strand strength without stressing. The specimen is divided into 2 types. Type-I specimen has (150 x 300 x 3200) mm dimension with center point loading, and distance of strand clamp as (Ubars) variable; whereas, the Type-II specimen has (200 x 300 x 3200) mm dimension with third point loading, and the number of strand as variable. The test is carried out by static-monotonic loading, and displacement control. Experimental test results shows that the addition of strand on the lower surface of the beam increase the beam's load capacity. However, the use of U- bars affects the stability of the concrete. The tighter distance of the U-bars decrease the capacity of the strengthening occured. The increased load capacity of Type I specimens, S3-150, S5-150 and S9-150 are 33.2%, 17%, and 12.1% against control beam (BK-150), respectively. Whereas, the increased capacity of the type-II beam strength of test specimens B1-200 and B2-200 are 10% and 17% against control beam (BN-200), respectively. And the increase of initial stiffness of beam Type-I test specimen S3-150, S5-150, and S9-150 are 14%, 18% and 5% against control beam (BK-150), respectively. Meanwhile, the increase of initial stiffness of Type-II beam of test specimens B1-200 and B2-200 are 21% and 24% against control beam (BN-200), respectively. This study shows that concrete and strand bonding is not compatible. The concrete has been reaching the limit strain, as well as the steel reinforcement in the concrete has been in yielded; meanwhile, strand that has fy is much higher than fy of reinforced steel, is still in elastic condition.

1. Introduction

The rapid growth and development of economic establishment and technology in the last decades has led to volume and load increment of heavy vehicle on roads and bridges in Indonesia; thus, constructed roads and bridges mostly receive exceed vehicle's load from the design load. Along with an increase of need for traffic on roads and bridges infrastructure, it is also expected to increase strength, safety, and comfortability.

The preferred strengthening element method of flexural structure is by utilizing FRP (Fiber Reinforced Polymer). However, this method relatively inquires high cost. Whereas, jacketing strengthening method resulted in increased building weight. Besides, retrofitting using stressing strand requires extra scrutiny of the location of an anchor (difficulty of installation) [1]. Based on the several reinforcement alternatives, retrofitting using strand can increase significant capacity strength, since strand has high yield stress and small volume [2]. Research on strengthening of prestressed concrete by

IOP Conf. Series: Materials Science and Engineering 1108 (2021) 012001 doi:10.1088/1757-899X/1108/1/012001

stressing have been conducted. However, such strengthening is complicated implementation, because it requires extra supervision when doing stressing. So, the research on strand strengthening without stressing is needed.

The requirement of the composite structure between the reinforced concrete beam and the external strand strengthening is the compatibility between them. Therefore, the beam with a composite strengthening must be able to work as a monolithic composite structure element. Beam with composite strengthening can work as a monolithic composite beam if the strain on a concrete beam with the strand is equal. To achieve the monolithic composite state is predicted to be done by using the gluing and anchoring method that connects the reinforced concrete beam with the strands [3].

2. Methods

The specimens were divided into 2 (two) types that are described in Table 1. Test method of Type I specimens based on ASTM C293-02 center-point load, displacement control state, while the test method for Type-II specimens based on ASTM-C78 third-point monotonic, displacement control.

Table 1. Test specimen specification.

No	Specimens Code	Longitudinal Bars			Confinement Bars		Cover Concrete (mm)	The Number of Strands	U-Bars	Anchor	Method
		Position	Bar at I-End (Momen-)	Bar at J-End (Momen+)	Bar at I-End (Momen-)	Bar at J-End (Momen+)			Distance (mm)		
Type - I Specimens (150X300X3200) mm											
1	BK-150	Top Bars	2D13	2D13	Ø8-150	Ø8-150	25	-	-	No	*Center point load*
		Bottom Bars	2D13	2D13							
2	S3-150	Top Bars	2D13	2D13				1	750		
		Bottom Bars	2D13	2D13							
3	S5-150	Top Bars	2D13	2D13					375		
		Bottom Bars	2D13	2D13							
4	S9-150	Top Bars	2D13	2D13					187,5		
		Bottom Bars	2D13	2D13							
Type - II Specimens (200X300X3200) mm											
5	BN-200	Top Bars	4D13	2D13	Ø8-150	Ø8-250	40	-	-	No	*Third point load*
		Bottom Bars	2D13	4D13							
6	B1-200	Top Bars	4D13	2D13				1	200	Yes	
		Bottom Bars	2D13	4D13							
7	B2-200	Top Bars	4D13	2D13				2	200	Yes	
		Bottom Bars	2D13	4D13							

Illustration of Type I test design for reinforced concrete beams strengthening by strand is shown in the Figure 1. Strand is attached to the underside of the beam by gluing using epoxy, then the strand is clamped with U-bars steel with distance variations of 750 mm, 375 mm and 187.5 mm implanted into. While the Type-II test specimen, reinforced concrete beam strengthening by is shown in Figure 2. At both ends the strand is carried out by using a steel plate. The strand is attached by gluing with epoxy, then the strand is clamped with U-bars at a distance of 200 mm (this follows the tests conducted by Demir and Tekin [4]) which are stretched into the beam and attached using epoxy.

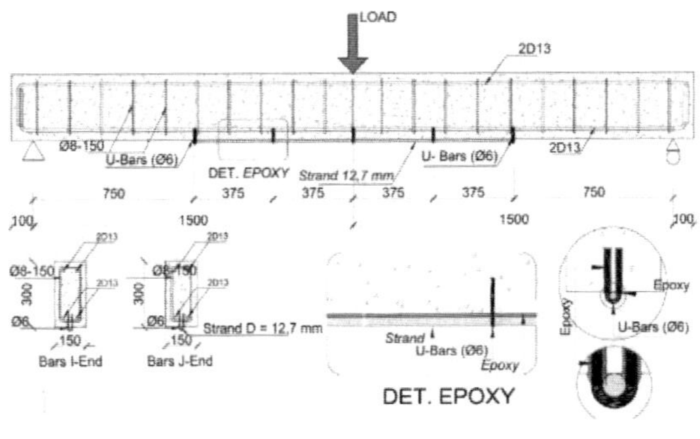

Figure 1. Type I specimens with strengthening.

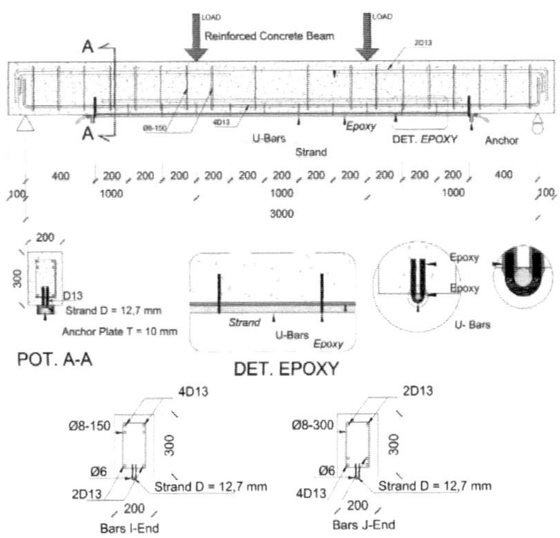

Figure 2. Type II specimens with strengthening.

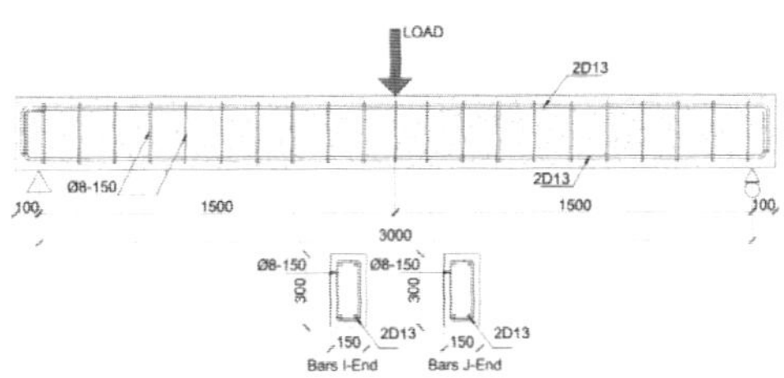

(a) Type I specimens-control test

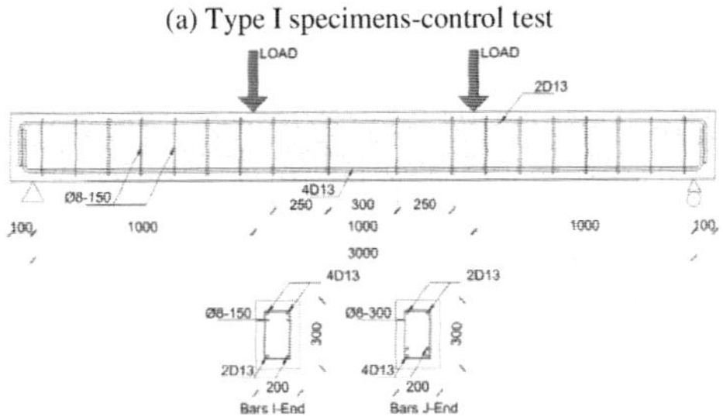

(b) Type II specimens-control test

Figure 3. Control test specimen (a) Type I and (b) Type II.

Strain gauges were installed on steel reinforcement and strands to observe strain; while, the deflection of the concrete beam was measured through LVDT. The strain gauges and LVDT positions can be shown in Figure 4 and Figure 5.

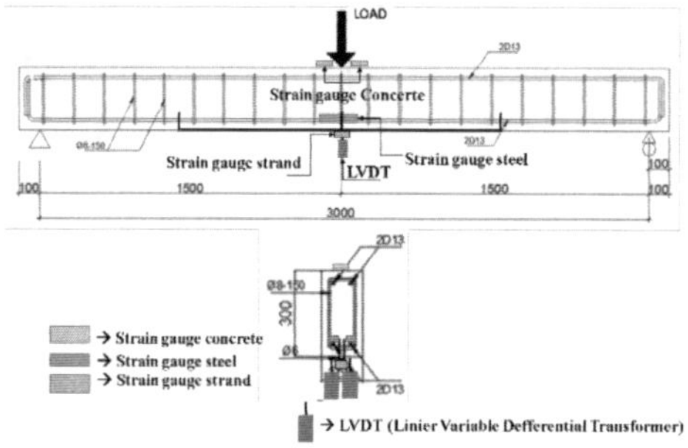

Figure 4. Strain gauge and LVDT positions on Type-I specimens.

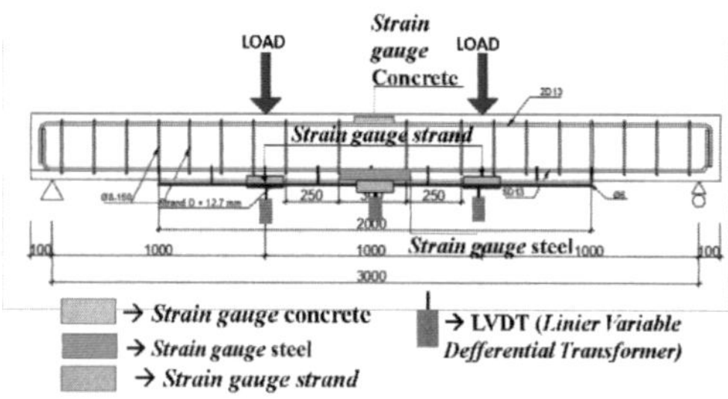

Figure 5. Strain gauge and LVDT positions on Type II test specimens.

2.1. Testing set up

Testing set up, as shown in Figure 6 and Figure 7, the specimen is put on the loading frame with the roll-pin restraint condition at both ends [5]. In the Type-I specimen, the loading was carried out at the center of the beam span or at 1500 mm from the restraint. Meanwhile, the Type II specimen, the loading was done on 1/3 span of reinforced concrete beam at 1000 mm from the restraint.

Figure 6. Set up of type I specimen.

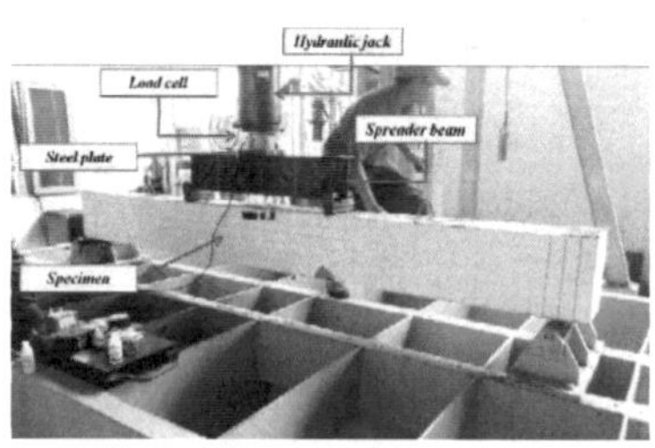

Figure 7. Set up of type II specimen.

3. Test results

3.1. Load-displacement relationship

The result of the loading test is shown in load displacement curves. The curve shows the value of first crack load (Pcr) and maximum load (Pmax). Figure 8 shows the result test of Type-I specimens. The curves have the same behaviour between the composite strand strengthening beams and the control beam (BK-150). It shows that adding a strand compositely on the bottom surface of the reinforced concrete beam increased the load capacity. As shown in the figure 8, with the same displacement, specimen S3-150 produced the highest maximum load, 40.632 kN; whereas, the test specimens S5- 150 and S9-150, respectively were 35.689 N and 34.188 kN. In other words, the addition of strand increased the load capacity by 33.2%, 16.9% and 12.1% respectively to the BK-150 control beam (30.504 kN).

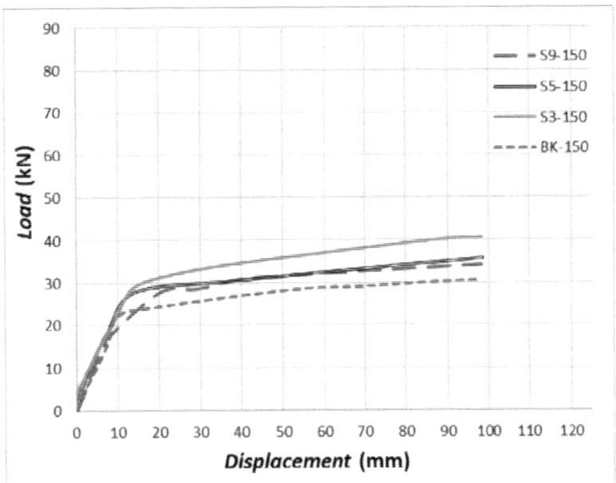

Figure 8. The load-displacement curve of Type I specimens.

The test result of Type-II specimens is shown in Figure 9. As shown in the figure, the load vs displacement curve at the elastic condition had the same behavior between the composite strand strengthening beam and the control beam.

The maximum load of BN-200, B1-200 and B2-200 are 71.557 kN, 78.418 kN and 83.552 kN, respectively. The results show that maximum load of composite strand strengthening beam higher than control beam.

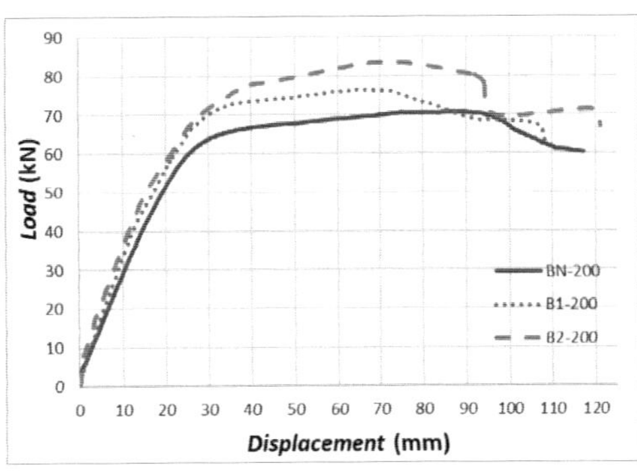

Figure 9. Load-displacement curve Type-II specimens.

Load-displacement curves also show stiffness behavior of the beams during loading tests. The addition of strand as strengthening on the bottom surface of the concrete beam could increase the stiffness of the beam. The results show that, addition of 1 strand on Type I specimens increase the stiffness of S3-150, S5-150 and S9-150 test specimens against control beam by 14%, 18% and 5%, respectively. However, on specimens S9-150, U bars at distance of 187.5 mm (the closest distance), there was a decrease in stiffness. The U bars embedded in the concrete affected the stability of the concrete beam, so that the more closely the bars of the U used, the smaller result of stiffness value and vice versa. Thus, in the use of U-bars should be done restrictions. In this research, the depth of U-bars reinforced concrete was not taken into account, so further studies are needed to determine the depth or length of the U-bars as a strand clamp. The value of the first crack load and the maximum load for each beam, as well as the stiffness value occurring, are listed in Table 2.

Table 2. Stiffness value of type I specimens.

Specimen	First Crack				*Displacement 98 mm*		
	P_{cr}	δ_{cr}	Stiffness		P_{mm}	δ_{mm}	Percent (P_{mm})
	kN	mm	kN/mm	%	kN	mm	%
BK-150	15.75	7.53	2.09	0	30.50	98	0
S3-150	18.83	7.89	2.38	14	40.63	98	33.2
S5-150	18.26	7.39	2.47	18	35.68	98	17
S9-150	17.15	7.82	2.19	5	34.18	98	12.1

As well as the Type I Specimens, the Type II test specimen stiffness analysis was performed during the initial stiffness condition (stiffness in the first crack condition). The first crack in the B1-200 beam decreased when compared to the control test object (BN-200). The cause of this decrease was due to the decrease in cross section of the beam. U-bars (strand clamps) used on Type-II test specimens at 200 mm apart and added with anchoring into the concrete so that the cross-sectional area is reduced more than the Type I test specimens. The large number of cross sectional area reductions due to the U-bars led to a decrease in the first crack load; whereas, strengthening by adding 2 (two) strands (B2-200) may increase the first crack load by 6% against the control beam (BN-200) and increase the first crack loading by 19% due to the influence of the U-bars on the specimen B1-200.

Table 3. Stiffness value of type II specimens.

Specimen	First Crack				Maximum Load		
	P_{cr}	δ_{cr}	Stiffness		P_{mm}	δ_{mm}	Percent (P_{mm})
	kN	mm	kN/mm	%	kN	mm	%
BN-200	44.32	16.46	2.69	0	71.55	86.04	0
B1-200	39.44	12.15	3.24	21	78.41	67.68	10
S9-150	14.12	14.12	3.32	24	83.55	71.17	17

The stiffness analysis of the Type II test specimens showed that the addition of strand can increase the stiffness of test specimens B1-200 and B2-200 to control beam (BN-200) by 21% and 24%, respectively. Meanwhile, the maximum load increased of test objects B1-200 and B2-200 to control beam (BN-200) is 10% and 17%, respectively. Understandably, the difference in the addition of the strand amount gives rise to the stiffness and maximum load value that occurs. However, the increase in the number of strand was not significantly different in stiffness value on test specimens B1-200 (1 strand) with B2-200 (2 strand) only differ 3% while the maximum load that happened was different 7%.

3.2. Pattern of collapse

3.2.1. Type-I specimen test. From the observation of the crack pattern, the control test specimen (BK-150) cracked pattern that occurred in the form of flexural crack and some flexural-shear cracks (the 1/3 spans of middle area). The collapse occurred on the BK-150 specimen began with a crack in the tensile zone under load P and increase in the region which had great moment, the first crack occured. As the given load increased, the crack got longer and wider and new cracks occured in 2/3 spans (1000 mm left-right load). With the addition of the load would cause the increase crack pointing upward toward the neutral line of the beam (there is the flexural collapse occured). And when the load was increased to 76% of the maximum load, the crack direction that occured tends to form a angle of 45° or more to the beam axis (a shear failure occured). The collapse of shear-flexural occured due to the strength of the concrete and the tensile strengthening and the shear strengthening was not sufficient to withstand the value of the load occurred.

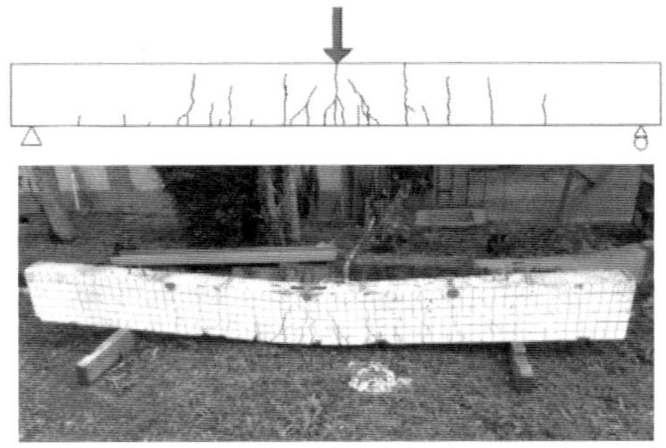

Figure 10. Crack pattern of BK-150 specimen.

Meanwhile, the result of the composite strand strengthening beam specimen observation on crack pattern that occurred was in the form of flexural crack (1/3 spans area of center of load) and dominant shear crack (2/3 spans of load). The collapse occurred was begun with the sound of ruptures in the epoxy lining the strand at the end (the direction of the roll restraint) and the middle of the strand span. Epoxy ruptures then epoxy that attached the strand to the concrete surface slightly apart. The epoxy strand started to detach at the end of the strand in the direction of the roll, then followed by the middle (under loading area). Visually, epoxy was released with a width of ± 1 mm, gradually enlarging in the area. There is a difference in load between the rollers and joints during the process of removing the epoxy. First, the end of the strand in the direction of the roll restraint started to release and followed by the middle (area under load). Due to the epoxy was not applied evenly to the entire surface of the strand during the installation of the reinforcing material, so that the first release of the epoxy is only the end of the strand in the direction of the roll. Simultaneously there was a crack in the layer of concrete blanket, and it increase relatively perpendicular to the neutral line, the first crack occurred.

As the load increased, the epoxy ruptures louder and there was crack in the epoxy then debonding occurred between the strand and the concrete surface. After the specimens cracked on the epoxy or the strand strengthening adhesive, and as the load increases, it showed that the epoxy which is a strand adhesive with concrete loosed gradually until the strand has debonding between strand and concrete. As illustrated in Figure 11 and Figure 12.

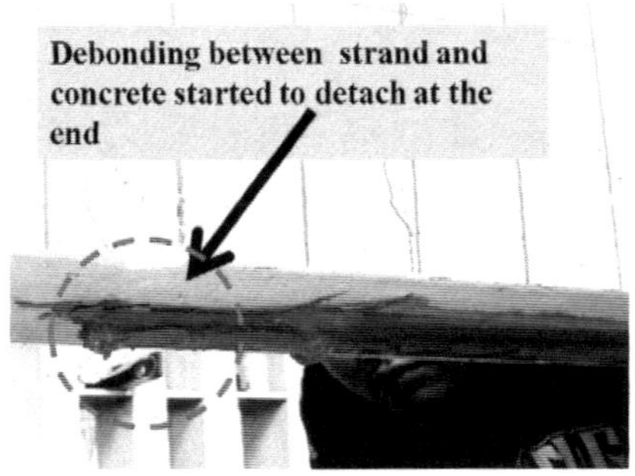

Figure 11. Debonding between strand and concrete at the end of the beam.

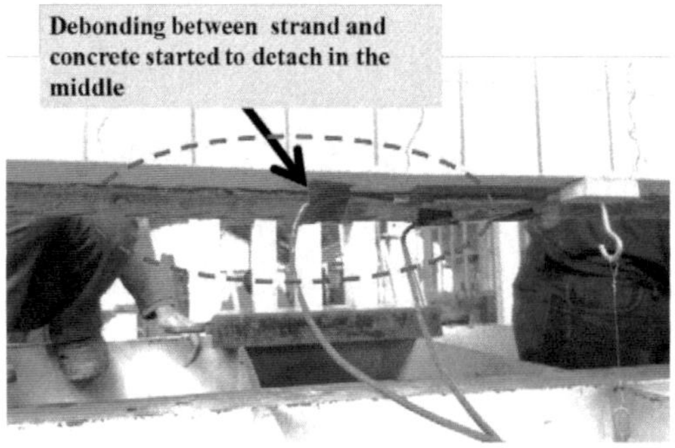

Figure 12. Debonding between strand and concrete in the center of the span.

Simultaneously when debonding occurred the crack width is significantly larger and longer, and new cracks emerged. Then when the load was increased to 71% (S3-150), 66% (S5-150) and 55% (S9-150) of the maximum load, the test specimen had a shear crack at 2/3 span of the source load position (1000 mm) and some diagonal oblique cracks which were a continuation of prior flexural cracks in the 1/3 spans or 500 mm left to right of the load, were flexural-shear cracks. The addition of a strand on the bottom surface of a concrete beam was taped with epoxy and clamped using a U-bars to make the strand did not only increase the pliable capacity as designed. Strand mounted externally on the concrete surface and stretched with U-bars into concrete, to affect the stability of the concrete. As the load increases, the U-bars affected the result of collapse in a shear crack. According to the observation of crack pattern that occurred at strengthening specimen S9-150, it tends to have more cracks than previous strengthening test S3-150 and S5-150. The cracks that emerge generally appeared at a radius of 50 mm from the shape of the U-bars reinforcing strand. Therefore, the U-bars caused weakness in the cross section of the beam that caused crack.

Figure 13. Cracked pattern of specimen S3-150.

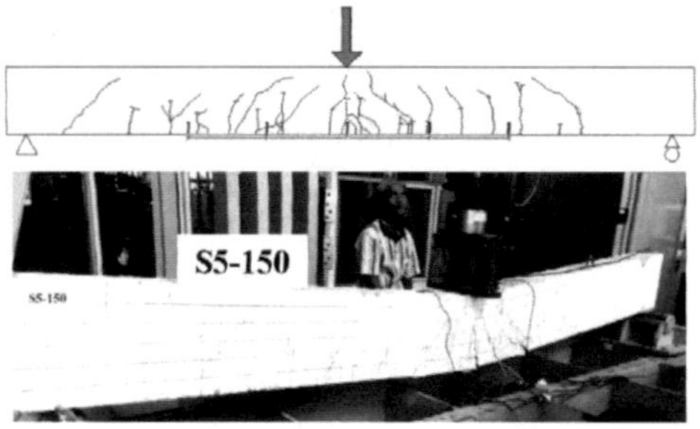

Figure 14. Cracked pattern of specimen S5-150.

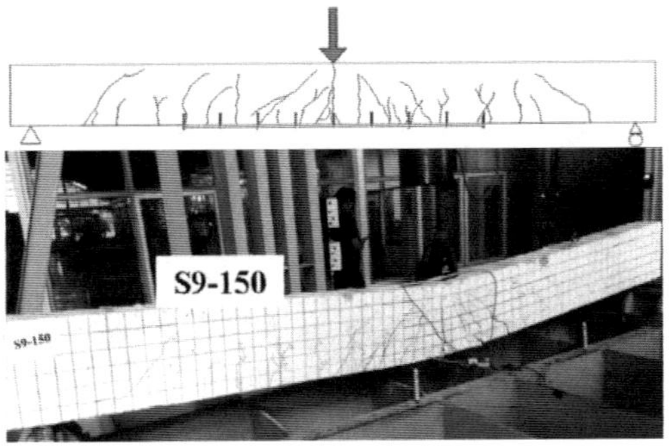

Figure 15. Cracked pattern of specimen S9-150.

3.2.2. Type-II test specimen. According to the observation of the crack pattern, the control test object (BN-200) of the cracking pattern occurred in the form of a flexural crack in the middle of the span between the load or in the area who got a great moment and some flexural cracks in the 2/3 spans of the load against restraint. The collapse occurred in the test specimen BN-200, preceded by a crack in the tensile area between the load and developing in the 2/3 spans of the load, occurred first crack. As the load increased, the crack increased in length and width. With the addition of the load would cause the increase crack pointing upward toward the neutral line of the beam (there was a flexural collapse). And, when the load was increased up to 63% of the maximum load, the crack direction that occured tends to form a 45° angle (shear collapse occured). Also, the collapse of shear-flexural occured due to the strength of the concrete and the tensile strengthening, and the shear strengthening was not sufficient to withstand the value of the load that occured. Cracked failure pattern was seen just before the collapse.

This beam would continue to experience displacement in the absence of a load increase and the crack became more open so that the neutral line continues near to the depressed edge. Eventually it caused the destruction of the concrete. Thus, from the collapsed pattern that occurred was the flexural-shear failure.

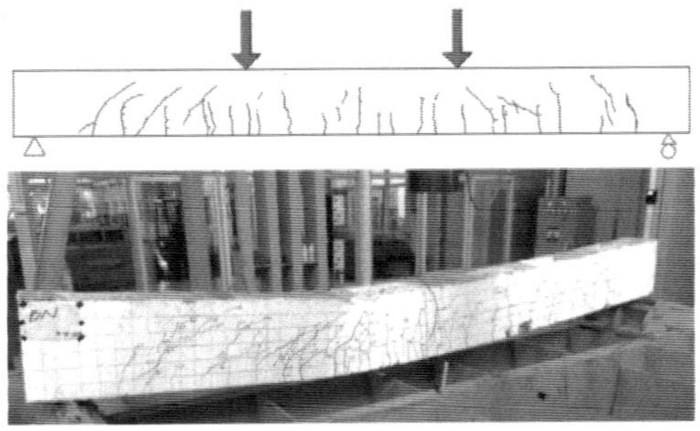

Figure 16. Cracked pattern of test specimen BN-200.

Meanwhile, the specimen of the crack pattern that occurred, begun with the sound of rupture on the epoxy that attaches the strand to the concrete and then epoxy started to detach slightly. The first off epoxy occurred at the end of the strand area close to the position of the anchor. Moreover, the epoxy was cracked and followed by cracked layers of concrete blanket before finally the first crack of the concrete emerge (first crack). The first crack occurred in the tensile area between the load and it increase on the 2/3 span of the load source to the restraint. As the load increased, the crack length and width were increased. With the addition of the load would cause the increase crack pointing upward toward the neutral line of the beam (there was a flexural collapse). And when the load was increased to 58% (B1-200) and 66% (B2-200) of the maximum load, there was a shear crack. The flexural-shear fracture of the reinforcement specimen occurred more often than in the control test object (BN-200).

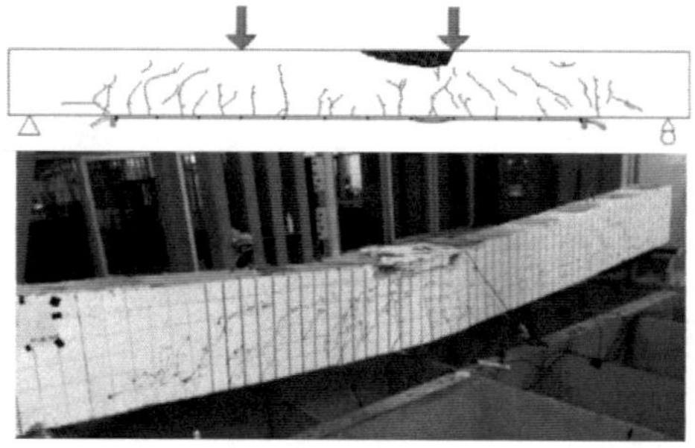

Figure 17. Cracked pattern of specimen B1-200.

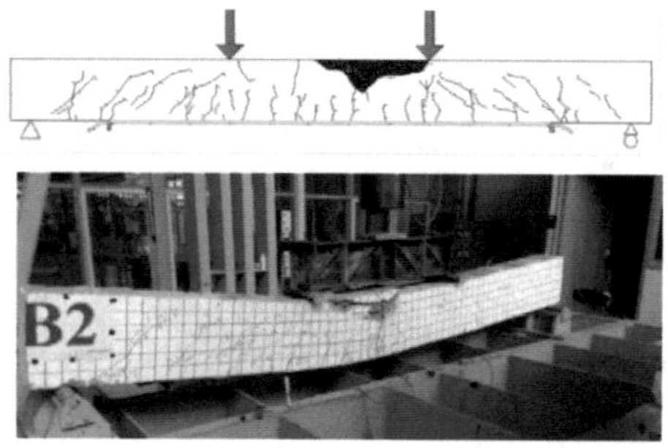

Figure 18. Cracked pattern of specimen B2-200.

Before the maximum load occurred, the curves of load bond and displacement showed the post-serviceability crack area, where the tensile strength stress reached its yield stress. The load-bond curve with displacement is much flatter than the previous area until it reached the maximum load and eventually collapsed. When the beam collapsed marked by a decrease in the load-displacement curve, the crushing occurred in the press region or in other words, the concrete stressing in the compression area reached the limit of (ecu = 0.003), while the strengthening of the strand was assumed to be in an elastic state. Understandly, the bonds between concrete and strand are incompatible, concrete as an element of crushing structure has reached crushing, as well as strengthening that is in the concrete already in yielding condition. Meanwhile, strand that has fy much higher than fy reinforced steel strengthening, when the concrete and reinforced steel strengthening has already collapsed, strand still in elastic condition.

4. Conclusion

The conclusion of this research is as follows:

- The maximum strength capacity of reinforced concrete beams had increased with the addition of strand on the lower surface of the beam. Increased capacity of beam were 12% - 33% to beam control.
- The addition of strands on the reinforced concrete beam was able to increase the initial stiffness, 5%-24% to the control beam.
- The use of U-bars as a strand affected the stability of the concrete; so, the form U-bars needs to be restricted. This study proves that the closer the distance of bars U decreased the capacity of thebeams.
- The control beams had flexural cracks and some shear cracks. Meanwhile, when the composite strand strengthening beam had the dominant shear crack. The strand mounted externally on the concrete surface and anchored with U-bars into concrete, to affect the stability of the concrete. As the load increases, the Ushape rebound affects the collapse pattern which results in a shear crack.

Acknowledgment
We acknowledge with thanks the fund by the DRPM Ditjen Penguatan Risbang and UPPM-Penelitian Terapan POLBAN. So that we can finish this research.

References
[1] Haryanto Y, Satyarno I and Sulistyo D 2012 Strength and ductility of reinforced concrete T-beams strengthened in the negative moment region with wire rope and mortar composite *Journal of the Civil Engineering Forum* **21** 1
[2] Limpaninlachat P, Matsumoto K, Nakamura T, Kono K and Niwa J 2016 Flexural strengthening effect of pre-tensioned UFC panel on reinforced concrete beams *Journal of JSCE* **4**(1) 181-196
[3] Jumaat M Z and Alam M A 2008 Behaviour of U and L shaped end anchored steel plate strengthened reinforced concrete beams *European Journal of Scientific Research* **22**(2) 184-196
[4] Demir A and Tekin M 2011 Strengthening of reinforced concrete (RC) beams with prefabricated reinforced concrete (RC) plates *Scientific Research and Essays* **6**(21) 4577-4586
[5] Krasna W A, Sulistyo D and Supriyadi B 2010 Perilaku Geser pada Keadaan Layan dan Batas Balok Beton Bertulang Berlubang Memanjang *Semesta Teknika* **13**(2) 145-154

Mechanical properties of porous asphalt with gilsonite additive

R Rochaeti*, J Endawati and T Oktaviani

Department of Civil Engineering, Polytechnic State of Bandung, Bandung, Indonesia

*rochaeti054@gmail.com

Abstract. The porous asphalt mixture is an asphalt mixture that consist of high void and a small proportion of fine aggregate. The presence of a high void reduces the value of marshall stability. Many attempts to increase the value of marshall stability by adding substances into the porous asphalt mixture has been done by several researchers. Modified porous asphalt can be achieved by adding gilsonite to the mixture which aims to accomplish the main function of the porous asphalt layer as well as to improve the quality of mixed performance. The method used in this study is an experimental method. The test that conducted in this study using two types of gradations that are Japan and Australia gradation. The research was Marshall characteristic test with 2.5%, 3.5%, 4.5%, 5.5%, 6.5% of gilsonite content and 0% as the control. The results showed that porous asphalt mixture using gilsonite proved to increase the value of Marshall stability compared to non-modified porous asphalt mixture by 24%. The mixture reaches a maximum state at 5.5% gilsonite content. The best performance is produced by the mixture using Japanese gradation.

1. Introduction

The porous asphalt mixture has a high void and a small proportion of fine aggregate. The high airspace caused by the gradation on the porous asphalt mixture is open graded. The air / pore space itself serves as a dual-drainage system that can drain water vertically or horizontally. The porous asphalt mixture is a planned pavement mixture which is laid and compacted to form a surface with a void ratio in the mixture of 22% [1].

The asphalt porous has several advantages as follow: the void contained in the porous asphalt serves to help drain the water vertically and horizontally, providing greater skid resistance, reducing the noise level for users and the surrounding, and it can also improve the safety for the road users [2]. Furthermore, porous asphalt can be used as a reservoir medium that can drain and store rainwater into the soil. In addition to these advantages, the disadvantage which is the main constraint on porous asphalt mixture is the low value of stability caused by the number of void on the porous asphalt [3].

An attempt to increase the value of marshall stability by adding a substance into the porous asphalt mixture has been done by several researchers. There are various supporting materials used, namely; Elvaloy, TAFPACK-Super, Latex, Gilsonite, and so on. The performance research on porous asphalt mix has been done by the Puslitbang Jalan in 1993 by adding gilsonite. The results show that gilsonite can improve the quality of mixed characteristics, such as reducing the value of rutting caused by the loading, reducing sensitivity to temperature and the effect of water immersion, and it can produce good bonds between aggregates. Furthermore, Djakfar et al. has compared the effect of adding gilsonite and latex as an additive to Marshall performance on porous asphalt mixtures [4]. The results obtained that the mixture of porous asphalt with the supporting materials of gilsonite proved to increase the value of marshall stability, compared with latex added ingredients. However, the addition of gilsonite may affect the void in the mixture.

Based on these results, the research will be used as an additional material in the asphalt mixture by reducing the percentage that aims to achieve the function of the asphalt layer and improve performance in parameters.

2. Methods

2.1. Materials

The material (fine and coarse aggregates) used in this study came from Cagak, Subang, West Java. Furthermore, the asphalt used is Pertamina asphalt with penetration of 60/70. The test of aggregate and asphalt is conducted based on *Rancangan 2 Indonesia* for porous asphalt mixture. Furthermore, the additional ingredients used in the asphalt mixture are gilsonite ordered from distributors originating from Bandung as shown in Fig. 1.

Figure 1. Gilsonite before and after milling.

2.2. Machine and specimens

The test specimens used in this study is the middle border of two gradations, which are Japanese gradation and Australian gradation. The variations of asphalt content used are 4.5, 5.0, 5.5, and 6.0%, while the gilsonite content variation is 0, 2.5, 3.5, 4.5, 5.5, and 6.5%. Each level consists of 3 pieces of the specimen. The test is a Marshall characteristic test to obtain stability, VIM, and flow by using the tool as shown in Fig. 2. The results of this test are compared with the specifications of *Rancangan 2 Indonesia*, Australian Asphalt Pavement Association (AAPA) 2004, and Japan Road.

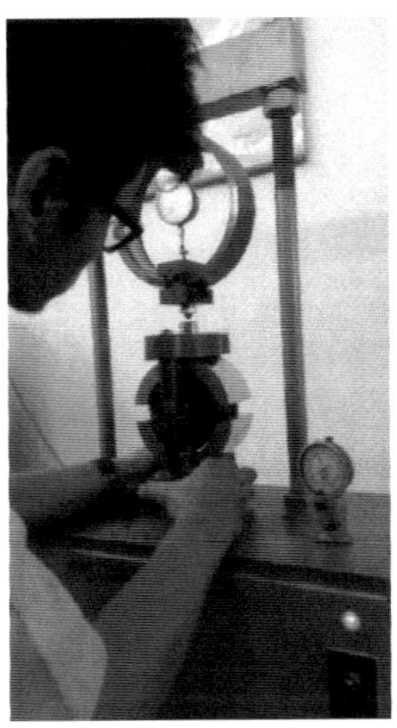

Figure 2. Marshall machine.

3. Results and discussion

In accordance with Rancangan 2 Indonesia, the aggregates and asphalt used in the research have met the characteristic testing requirements.

The test results of Marshall characteristics obtained from this research are in the form of stability, VIM, and flow.

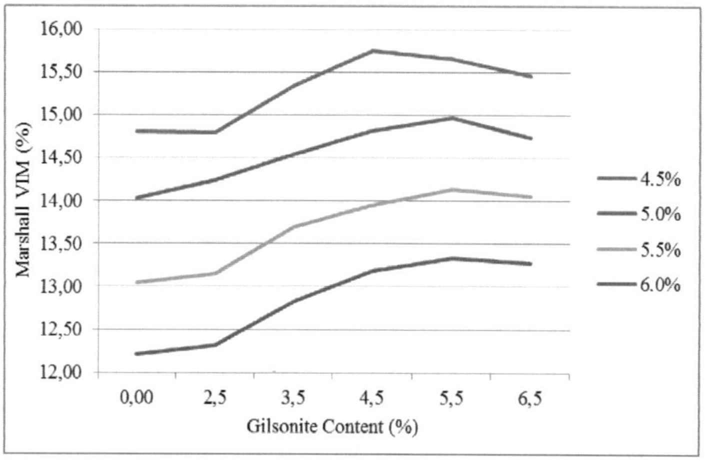

Figure 3. Void In Mixture (VIM) of Japanese gradation.

Fig. 3 the picture shows that the addition of gilsonite causes the VIM values from the Japanese gradations has an improvement. The increase occurred at 5.5% gilsonite.

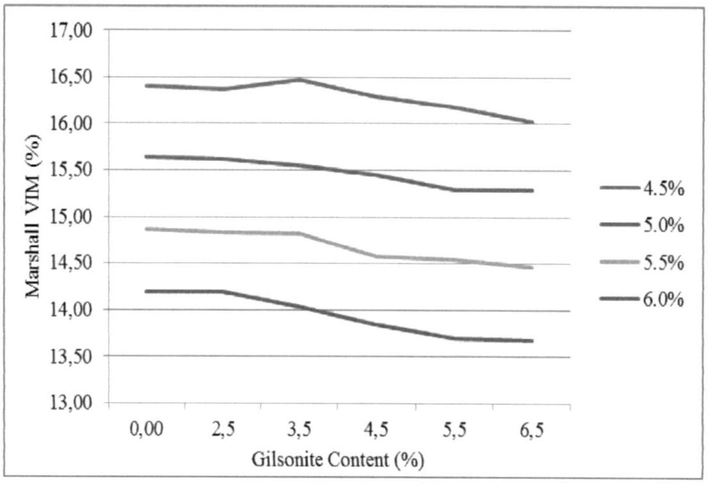

Figure 4. The Australian gradation test mixture.

Fig. 3 the picture shows that the void in the Japanese gradation test mixture inversely proportional to the VIM value of the Australian gradation test mixture in Fig. 4. The VIM values produced by the Japanese graded mixture have a lower value than the Australian gradation test mixture. However, the value does not meet the specification because the VIM value of the two gradations is still below the specifications that have been written in *Rancangan 2 Indonesia* and AAPA 2004. As for Japan Road itself, it has no specification about the void in the mix.

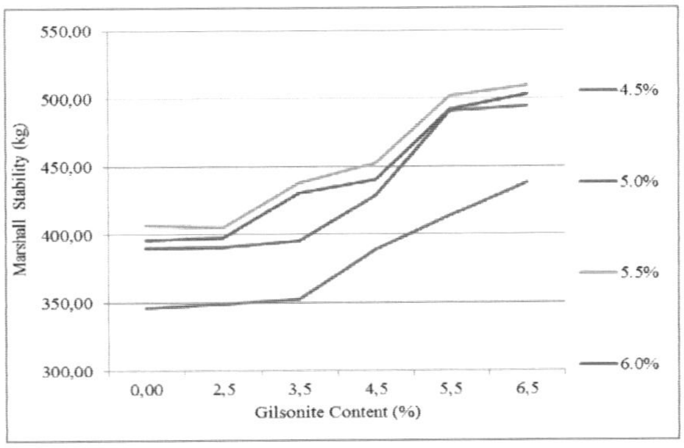

Figure 4. Marshall stability of Japanese gradation.

According to Fig. 5, the Marshall stability value resulted in Japanese gradation shows the same thing as the VIM test results. A stability value that is greater than 500 kg occurs in gilsonite 5.5%.

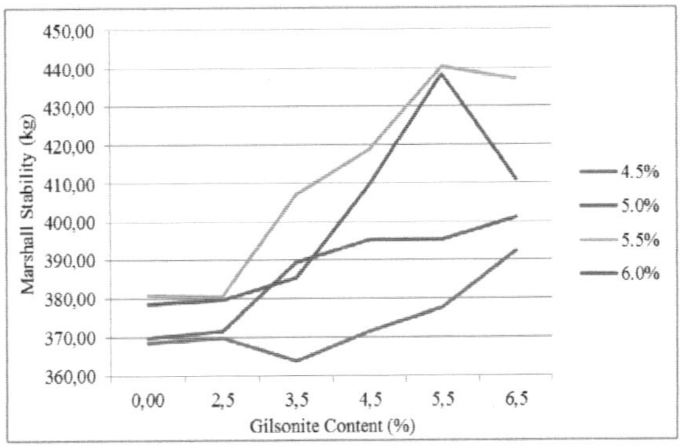

Figure 5. Marshall stability of Australian gradation.

The stability of the Japanese gradation test mixture in Fig. 5 has a higher value than a mixture of Australian gradation test objects in Fig. 6. The stability value of all Japanese gradation test objects meets the Japan Road specification of at least 350 kg [5], while the stability value of the Australian gradation test mixture does not meet the AAPA 2004 specification of at least 500 kg [6]. The value of the stability is related to the void in the mixture of the specimen. It means that, the smaller the value of the void, the better the resulting density, so the better the resulting stability.

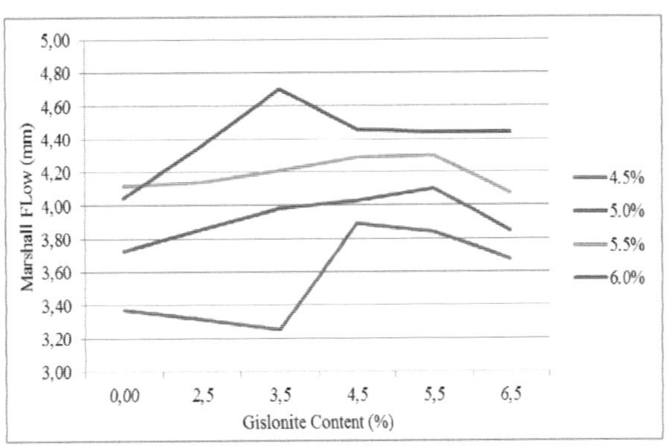

Figure 6. Marshall flow of Japanese gradation.

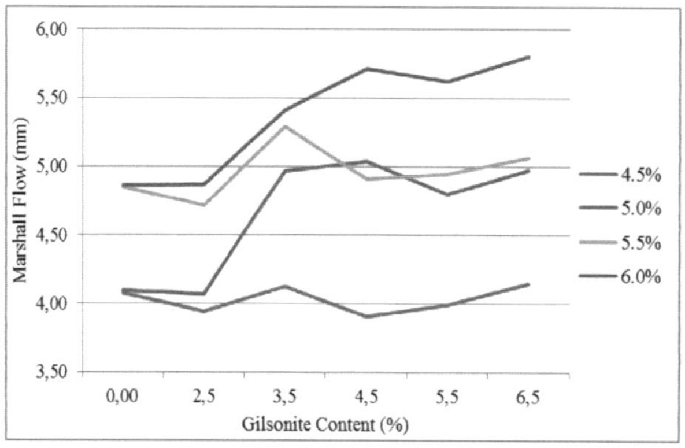

Figure 7. Marshall flow of Australian gradation.

Based on Fig. 7 and Fig. 8, all flow values generated from Japanese and Australian gradation test mixtures meet the specified specifications of AAPA 2004 and Japan Road. The flow value is within the range specified in AAPA 2004 which is 2 - 6 mm [6] and Japan Road is 2 - 4 mm [5]. Using the approach of examination of Marshall characteristics, the optimum gilsonite content obtained was 5.5%. The characteristic quality of the content is then compared with the control specimen as shown in Fig. 9 - Fig. 11.

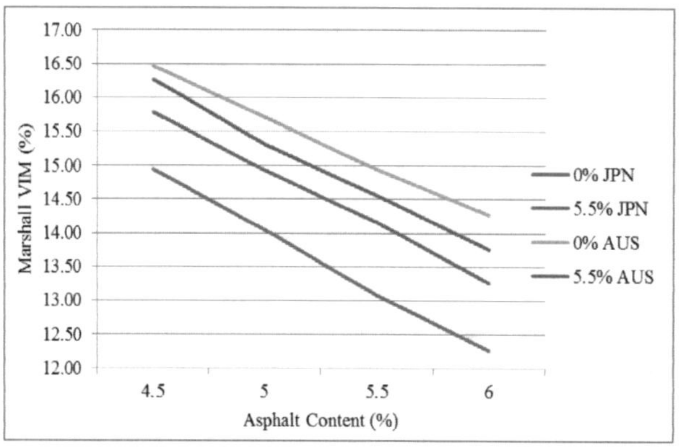

Figure 8. Void In Mixture of porous asphalt.

The results of the void testing in the mixture using the optimum gilsonite content in Fig. 9 shows that the void in the Japanese and Australian gradation test mixture still do not meet the specifications set out in *Rancangan 2 Indonesia* and AAPA 2004, while Japan Road does not have specifications for void in mixture. The void in the Japanese gradation test mixture has a lower than the Australian gradation test object. This happens because Japanese gradation is smoother than Australian gradation, so the void produced by Japanese gradation is less than Australian gradation.

The presence of gilsonite has been shown to affect the void in the mixture. The void in the modified mixture is decreased. This is consistent with the results of the Djakfar et al. study, but the VIM obtained in this study has improved from the Djakfar et al., study.

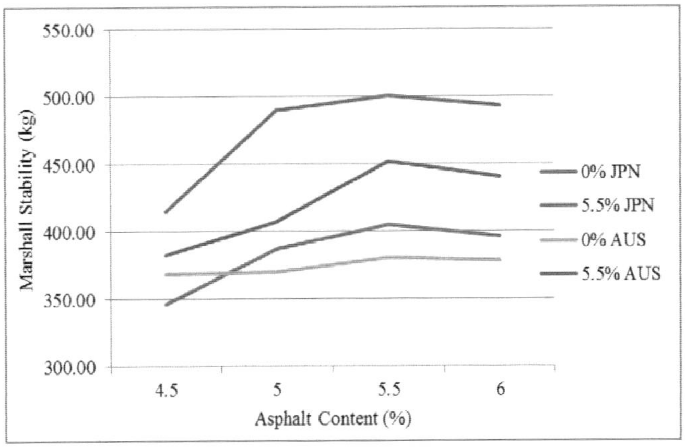

Figure 9. Marshall stability of porous asphalt.

Fig. 10 shows that the stability value from the Japanese gradation test mixture has a higher stability value than the Australian gradation test object. The value of the stability of Australian gradation test objects does not meet the AAPA specification of 2004 which is at least 500 kg, while the stability value of Japanese gradation test objects meets the Japan Road specification of at least 350 kg. However, if the stability value of Japanese and Australian gradation test objects is compared with the specifications of *Rancangan 2 Indonesia*, the Japanese and Australian grading objects have met the minimum of 350 kg [7] specifications.

Marshall stability is a parameter of mixed strength image to the acceptable load. The stability associated with the void in the mixture. The smaller the void, the greater the resulting stability value. The addition of gilsonite in asphalt modification proved to increase the value of stability, this result is in accordance with the study of Djakfar et al.

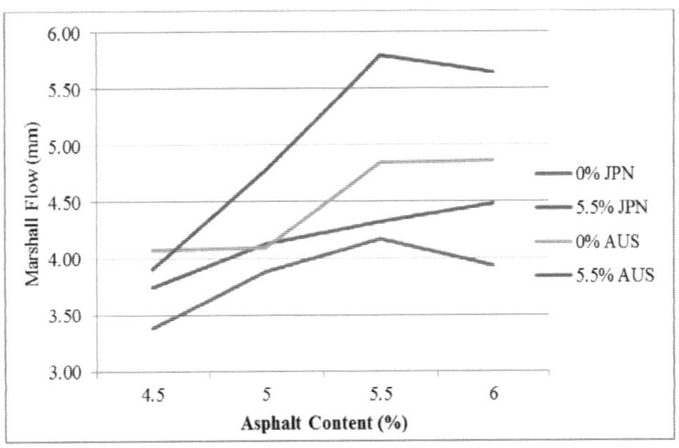

Figure 10. Marshall flow of porous asphalt.

Flow is a parameter that serves to determine the elasticity or change in the plastic form of asphalt mixture due to workload. Low melting values indicate a good resistance to deformation.

The flow value in Fig.11 shows that the Japanese gradation test mixture has a smaller value than the Australian gradation test mixture. However, the mixture of test specimens with the gradations of Japan and Australia has met the specifications determined by Japan Road and AAPA 2004 which are respectively 2 - 4 mm and 2 - 6 mm.

From the overall characteristic testing, Japanese gradation showed better performance than Australian gradation and in summary as shown in Table 1.

Table 1. Resume of Japan characteristic.

Parameter	Specification of Japan Road	Gilsonite Content (%)		Notes
		0	5.5	
VIM (%)	-	13.08	14.16	Increase 8%
Stability (kg)	>350	404.65	500.49	Increase 24%
Flow (mm)	2-4	4.17	4.33	Increase 4%

4. Conclusion

From the result of the research, it can be concluded that:

- The addition of gilsonite in the asphalt has proved to improve the quality of mixed characteristics on some parameters, i.e. stability, the mixture cavity, and flow.
- The mixture reaches a maximum state at 5.5% gilsonite and 5.5% bitumen content.
- The overall parameter of mixed test results using Japanese gradation has a better value than the mixture by using Australian gradation. Thus, the mixture by using Japanese gradation yields better stability than the mixture using Australian gradation.

Acknowledgments

The author acknowledges to UPPM POLBAN that already support the funding for commencement the research, Bogor Rubber Research Center and Bandung Research Center of Roads and Bridges that support the laboratory for the research.

Reference

[1] Nurcahya A 2015 *Analisis Kinerja Campuran Aspal Porus Menggunakan Aspal Pen 60/70 dan Aspal Modifikasi Elvaloy* (Bandung: Program Magister Sistem dan Teknik Jalan Raya, Institut Teknologi)

[2] Kraemer C 1997 *Porous Asphalt: Past and Present* (Madrid: European Conferences On Porous Asphalt)

[3] Falderika 2014 *Evaluasi Modulus Resilien dan Deformasi Permanen Campuran Aspal Porus dengan Bahan Tambah Buton Natural Asphalt (BNA)* (Bandung: Program Studi Magister Sistem dan Teknik Jalan Raya, Institut Teknologi Bandung)

[4] Djakfar L, Bowoputro H and Zaika Y 2015 Pengaruh Penambahan *Additive* terhadap Kinerja Marshall pada Campuran Aspal Porus. *The 18th FSTPT International Symposium,* Unila, Bandar Lampung, *August 28, 2015*

[5] Japan Road Association 1989 *Manual for Asphalt Pavement* Japan Asphalt Association

[6] Australian Asphalt Pavement Association 2004 *National Asphalt Specification*

[7] Departemen Pekerjaan Umum Rancangan 2 2015 *Perancangan dan Pelaksanaan Campuran Aspal Porus* (Indonesia: Departemen Pekerjaan Umum)

IOP Conf. Series: Materials Science and Engineering 1108 (2021) 012003 doi:10.1088/1757-899X/1108/1/012003

Reinforcement method of reinforced concrete beam by using non-stressing strand

A Susanto*, H Kasyanto and S Susilahadi

Civil Engineering Department, Politeknik Negeri Bandung, Bandung, Indonesia

*ambar.susanto@polban.ac.id

Abstract. Many methods can be used in the repair of bridges or other structural buildings. The aim is to increase the existing capacity. In the reinforced concrete beam structure, capacity of structural elements can be carried out by reinforcement method. The research was done by experimental loading test in laboratory. The test objects are consisted of 2 (two) reinforced concrete beams: a normal beam without reinforcement (NB) and a reinforcement beam by using non-stressing srand (RB1). The dimensions of the beams are 150mm x 300mm x 3200 mm. All beams are loaded by static-monotonic load until collapses. From the results of the research, the first yield load for the NB beam and RB1 beam respectively are 22 kN and 29kN. The deformations that occur at the conditions are 9.5mm and 14mm respectively. The collapse occurring on the NB beam is the bending collapse, whereas in the RB1 beam is a bending- shear collapse.

1. Introduction

Some considerations in the repair of bridges, or other buildings, include: age of the building, corrosion of steel reinforcement, defects in construction, transfer of function of the structure, and damage caused by natural disasters.

Reinforcement of structural elements is carried out when the load increases. This reinforcement serves to increase the capacity of these structural elements. Reinforcement can be done, among others, by addition of external prestressed, FRP (fiber reinforced polymer), and steel plate.

The selection of the reinforcement type depends on existing condition, location of the reinforcement, structure capacity, cost, and material. An example is reinforcement by using FRP. The weakness of this material is brittle, not resistant to fire and low durability. For that reason, it is proposed method of strengthening structural elements by using non-stessing strand. The strand, attached to beam by using epoxy, is supported by a U-shaped anchor.

External reinforcement of the beam elements using strand material is one of the methods to restore or increase the capacity of structures that have failed or degraded. On external reinforcement, strand is selected because the material has a high elastic modulus.

2. Literature review

Renata et al., used 15mm x 30mm T-shaped CFRP reinforcement, on reinforced concrete beam. The proposed system successfully combines the advantages of two developed reinforcement techniques i.e. near surface mounted (NSM) and externally bonded (EB) methods. The results of this study can increase the bending capacity and reduce the maximum deformation that occurs [1].

Jain et al., mentioned that the use of FRP for structural reinforcement is very popular. This is because FRP has many advantages, such as on the strength-to-weight ratio and excellent corrosion resistance. The research on reinforced concrete shear reinforcement, Glass Fiber Reinforced Polymer (GFRP) materials are used. Apparently, U-shaped GFRP reinforcement produces better strength compared to reinforcement installed on both sides of the beam [2].

Limpaninlachat et al., utilizes ultra-high strength fiber in prestressed concrete beams. The results showed an increasing in the flexural capacity and structural durability. This study focused on the behavior of bending beams reinforced with strands. The strand is choosed because of the high tensile capacity [3].

Gul et al., examines the reinforcement by adding reinforced steel to the bending area. The results of this study indicate that there are relatively large increasing in flexural capacity, high ductility of the beam, as well as uniform cracks distribution [4].

Hamad et al., conducted research on reinforced beams with steel reinforcement and angle steel. Overall, there is a significant increase in yield. The cross-sectional capacity of the beam increased 3.2 times compared to the normal beam. This means higher shear strength and higher moment resistance [5].

Manisekar et al., examines the use of external prestressing on reinforced concrete beam elements. The beam has decreased strength to 85% of its tensile steel strain. The result, by installing external prestressing can increase the ultimate load by 48% and restore deflection as the original condition [6].

Susanto et al., tested the 2 reinforced concrete beam: Bo (beam without reinforcement) and B1(beam with reinforcement). As a result, the first yield load on B1 is 27% higher than first yield load on Bo. Because of loading test, the Bo specimen has a bending crack on vertical direction. While on the test of B1 specimen, the crack occurs in the diagonal direction. This happens because of the installation of U-shaped anchor. This anchor becomes the support of the strand, whose function is similar to shear reinforcement [7].

3. Methodology

The study was conducted by experimental testing in the laboratory. The specimens consist of 2 (two) reinforced concrete beam, dimension 150mm x 300mm x 3200mm, as shown in Figure 1.

- NB (beam without reinforcement).
- RB1 (beam with strand reinforcement, with 3 pcs anchors).

Anchor is U-shaped reinforcing steel. The placement of 3 pcs of anchors can be seen in Figure 2. The two specimens are simple beam structures. The test is carried out with static-monotonic loads according to ASTM-C293 until collapses by displacement control method [8].

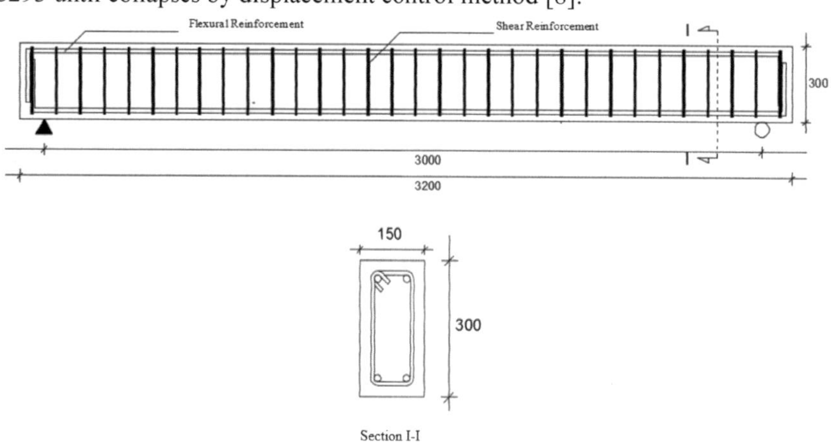

Figure 1. Beam without reinforcement (NB).

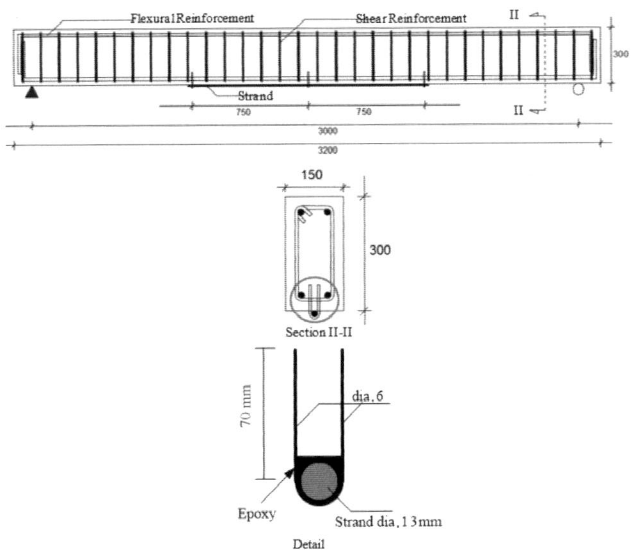

Figure 2. Beam with strand reinforcement (RB1).

4. Analysis and discussions

4.1. Load and deformation

Loading test on 2 specimens result load and deformation values that occur in the first crack condition and the first yield conditions as shown in Table 1. The relationship between load and deformation can be seen in Figure 3.

Table 1. Load ratio and deformation ratio.

Specimen	Load (kN)		Deformation (mm)		Load Ratio to NB		Deformation Ratio to NB	
	$P_{first\ crack}$	P_y	$\delta_{first\ crack}$	δ_y	$P_{first\ crack}$	Py	$\delta_{first\ crack}$	δ_y
NB	15,76	22	7,48	9,5	1	1	1	1
RB1	18,84	29	7,90	14	1,19	1,32	1,06	1,47

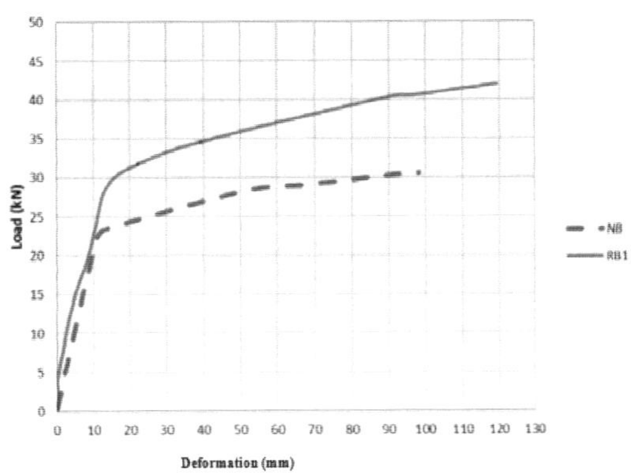

Figure 3. Relationship between load and deformation.

From Table 1 and Figure 3 above, the addition of U-shaped anchor on the bottom side of the beam, ie in the tensile region, influences the magnitude of the force and the deformation that occurs.

4.1.1. The first crack conditions. On the NB beam, the first crack occurs at load P=15.76kN, while deformation = 7.48mm. On the RB1 beam, first crack occurs at the load P=18.84kN, while deformation = 7.9mm. Thus, on first crack conditions, load increases 19% and deformation increases 6%.

4.1.2. The first yield conditions. On the NB beam, the first yield occurs at Py=22kN load, while deformation = 9.5mm. On the RB1 beam, the first yield occurs at Py = 29kN, while deformation = 14.0 mm. It means that on the first yield conditions, yield load increases 32% and deformation increases 47%.

4.2. Pattern cracks and collapse

Beside of to the load and deformation values described above, from the loading test results can also be determined the pattern of cracks and collapse that occur from each samples.

4.2.1. NB beam. Crack pattern that occurs in NB beam is crack bending on vertical direction. Few shear-bending cracks also occur at 1/3 middle span. As the load increases, crack occurs vertically perpendicular to the axis of the beam causing bending failure, as seen in Figure 4.

Figure 4. Crack pattern of NB beam.

4.2.2. RB1 beam. Three (3) pcs U-shaped anchors are installed at a distance of 750 mm along 1500 mm. Anchors are functioned as support of strand that produces bending-shear collapse on the beam. This is because the anchor that inserted into the beam behaves like a shear reinforcement. Many bending-shear, crack occur at the reinforced area and shear crack occur at the outside of reinforced area.

The pattern of crack and collapse of the beams can be seen in Figure 5.

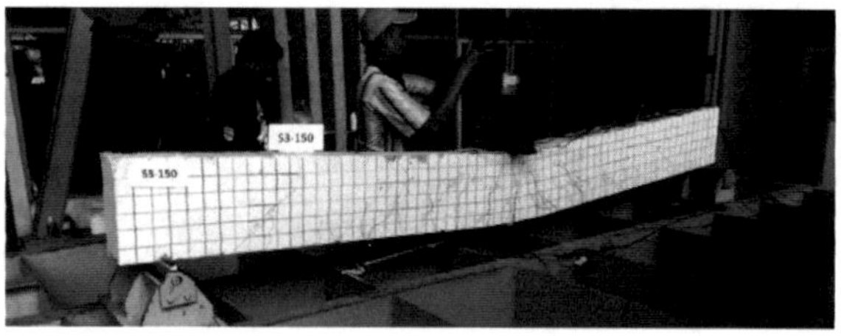

Figure 5. Crack pattern of RB1 beam.

4.3. Composite action

Non-stressing strand for reinforcement contributes to increase bending capacity of beam due to loading system as shown in Table 1. Strands are installed on the bottom side of the beam which glued with epoxy. At a certain distance this strands are supported by U-shaped anchor. Initially, after the first yield, strand on the bottom of beam is detached from concrete and there is not attachment between the concrete and the strand. In this condition between strand and anchor still working together, so there is still a composite action between concrete cross section with strand. As the addition of the load approaches the maximum capacity of the beam, the strand is detached from the anchor. The original anchorage position parallel to the concrete cross-section turned into a tilt because of the tensile force on the strand. At this stage, there is no more composite action between the concrete beam and the strand.

5. Conclusion

From the above discussion it can be concluded that.

- The addition of reinforcement by non-stressing strand can increase the load capacity of the beam until 30% compared to the non-reinforced beam condition (NB).
- Cracks that occur on non-reinforcement beam are bending cracks, whereas the crack pattern that occur on reinforced beam are shear-bending crack. This is because of the addition of U-shaped anchor inserted into the beam as a support of the strand.
- The composite action that occurs between the strand and cross section of the beam only occurs when between the anchor and the strand is still a unity. By the time anchors have begun to tilt, the composite action has been lost.

References

[1] Cholostiakow R K S 2015 Flexural Strengthening of Reinforced Beams Using CFRP T-Shape Profiles *Polymers* 2461-2477

[2] Jain N and Sikka V K 2015 Strengthening Of RC Beams with Externally Bonded gfrps *IOSR Journal of Mechanical and Civil Engineering (IOSR-JMCE) e-ISSN* 2278-1684

[3] Limpaninlachat P, Matsumoto K, Nakamura T, Kono K and Niwa J 2016 Flexural strengthening effect of pre-tensioned UFC panel on reinforced concrete beams *Journal of JSCE* **4**(1) 181-196

[4] Gul A, Alam B, Khan F A, Badrashi Y I and Shahzada K 2015 Strengthening and Evaluation of Reinforced Concrete Beams for Flexure by Using External Steel Reinforcements *International Journal of Scientific Engineering and Technology* **4**(4) 260-263

[5] Hamad B, Masri A, Basha H and Baalbaki O 2011 Behavior of T-shaped reinforced concrete beams partially confined by structural steel *Construction and Building Materials* **25**(2) 1037-1043

[6] Manisekar R, Sivakumar P and Lakshmikandhan K N 2014 Experimental investigations on strengthening of RC beams by external prestressing *Asian Journal Of Civil Engineering (BHRC), India*

[7] Susanto A, Kasyanto H and Susilahadi S 2018 Studi Perkuatan Balok Beton Bertulang Dengan Strand Tanpa Penegangan Dan Tulangan U *Jurnal Poli-Teknologi* **17**(1)

[8] American Society Testing and Material(ASTM)

Deformation and dynamic stability of asphalt concrete by using granite dust as filler

M Oesman, A Sundara, J Endawati and D Virnawan*

Department of Civil Engineering, Polytechnic State of Bandung, Bandung, Indonesia

*dedyvirnawan85@gmail.com

Abstract. Asphalt Concrete – Wearing Course (AC-WC) is a surface layer of flexible pavement which have direct contact with vehicle wheel. One of the surface layer damage categories is the wheel track (rutting). In this research, the experiment was done by using industrial waste of granite stone production in the form of dust as filler in the AC-WC mixture. The variation of filler content used are 3%, 4.5%, 6%, 7.5% and 9% with cement filler as comparator. The test was conducted by using wheel tracking machine on temperature of 60°C, 45°C and 30°C. The test result shows that the optimum filler content using granite dust in the AC-WC mixture is 3%. Meanwhile the optimum filler content on the asphalt concrete using cement filler is 7.5%. Thus, it shows that the higher temperature cause the ability of maintaining the wheel track decreased. Furthermore, if looking at the relation between dynamical stability, deformation depth and deformation velocity on filler concentration is not linear.

1. Introduction

Asphalt concrete - wearing course (AC-WC) is a surface layer of flexible pavement have direct contact with vehicle wheel; wherein, the asphalt concrete mixture comprises coarse aggregate, fine aggregate, filler and asphalt as the binding agent. In general, the material used as a filler in the AC-WC mixture is cement, limestone, dolomite, fly ash and others. The function of the filler in the asphalt concrete mixture is as a filler between the coarser aggregates meaning that the air cavity (void) becomes small, and can produce frictional resistance and locking between the higher bonds.

Chandra states that the AC-WC mixture with granite dust as a filler has better rutting resistance compared to regular stone dust [1]. Asphalt concrete with granite filler has a better fatigue resistance when compared to ordinary stone dust. The asphalt concrete mixture with granite dust as filler shows the linear bitumen content of the filler content. The higher theu filler level will cause the optimum bitumen content becomes lower, because the asphalt will be absorbed by the filler. According to Fasdarsyah, a mixture of AC-WC with granite dust as filler meets the Bina Marga 2010 specification [2].

Furthermore, Widodo states that the more solid AC-WC will cause the ability to maintaining the trace of the wheels better [3]. The higher the temperature will cause a decrease of the AC-WC capability in holding the trace of the wheels.

Based on the description above, the research on the use of granite ash waste as a filler on AC-WC mixture is done with variations of filler content of 3%, 4.5%, 6%, 7.5% and 9%. The determination of variation of filler content based on value of filler level has not been done by previous researcher Chandra [1] and Fasdarsyah [2].

2. Methodology

2.1. Material

The material used in this study is shown in Fig. 1; The coarse aggregate and fine aggregate originated from around West Bandung; whereas, for granite ash filler comes from Ketapang District of West Kalimantan Province, as well as Pertamina asphalt with penetration 60/70. All materials are tested based on the technical specifications of Bina Marga in 2010 revision III.

Figure 1. Materials: (a), (b) coarse aggregate; (c) fine aggregate; (d) granite dust.

2.2. Machine and specimen

This study is begun with testing of all materials based on the 2010 Bina Marga specification of revision III [4]. Subsequently, the test object is made and tested by Marshall method as shown in Fig. 2. Marshall test is done to get the optimum bitumen content which will be used for making Wheel tracking machine (WTM) test specimen [5].

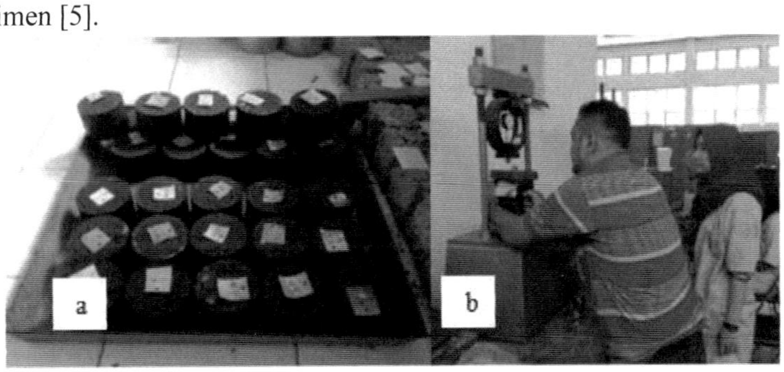

Figure 2. (a) Specimen; (b) Marshall testing.

To obtain deformation behavior and dynamic stability on AC-WC surface layer with granite dust, thus it was tested by using Wheel tracking machine. Square test object with size 30 x 30 x 5 cm, Wheel tracking machine run with speed 21 passing / minute. The tests are carried out at temperatures 60°C, 45°C and 30% °C for 1 hr and tracing observations are performed on the 21st, 105, 210, 315, 945, and 1260 passing [6]. Fig. 3 shows the equipment for the manufacture and testing of deformation and dynamic stability of the AC-WC.

Figure 3. (a) Mixing machine; (b) Compactor machine; (c) Specimen; (d) Wheel tracking machine.

From the test results using the tool Wheel tracking machine shows that the test object is deformed as shown in Fig. 4.

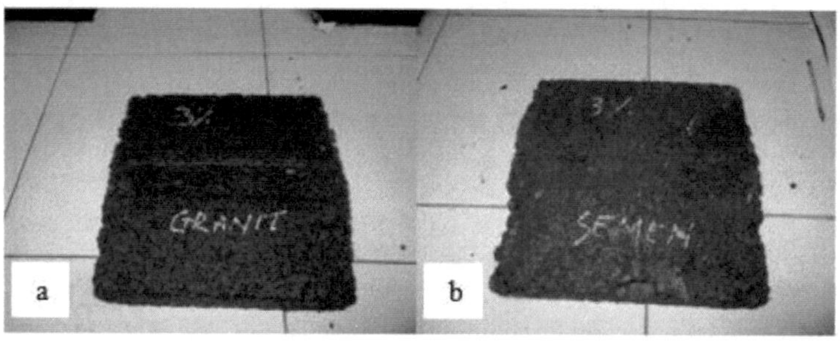

Figure 4. (a) AC-WC with filler granite; (b) AC-WC with filler cement

3. Test result

3.1. Optimum bitumen content

From the results of Marshall testing, we get the optimum bitumen content for each filler content as shown in Table 1.

Table 1. Optimum bitumen content in asphalt concrete.

Filler content	Granit dust	Cement
3%	6.23%	6.20%
4.50%	6.17%	6.18%
6%	6.15%	6.16%
7.50%	6.1%	6.1%
9%	5.9%	6.09%

Table. 1 indicates that the larger the filler content will cause the value of the optimum bitumen content to be smaller because empty cavities on the asphalt concrete have been filled by the filler.

3.2. Structural performance

The results of testing with Wheel tracking machine tool will get the value of dynamic stability, deformation velocity and deformation depth. The deformation occurring on the pavement is one of the parameters in the assessment of pavement performance.

In this research, deformation and dynamic stability testing is done with pavement temperature variation. The temperature of the 60°C test is the most extreme temperature of the pavement, which describes the pavement conditions during the day. Temperature 45°C as a picture of the temperature of the pavement between morning and noon and noon and afternoon. While temperature 30 ° C is a picture of the temperature of pavement in the morning and evening.

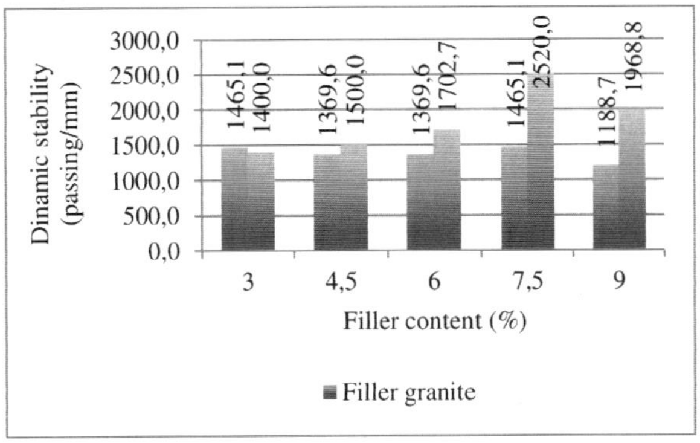

Figure 5. The relationship of dynamic stability with variations of filler content at 60°C.

Fig. 5 shows the relationship of dynamic stability values between asphalt concrete with the content of cement filler and granite filler at 60°C pavement temperature. The picture shows that the relationship of dynamic stability to filler content is not linear; as well as the type of filler. In general, asphalt concrete with cement filler has a greater stability value than asphalt concrete with granite filler. The use of a 7.5% cement filler achieves the highest dynamic stability value, ie 2520.0 passing/mm. Meanwhile, asphalt concrete with granite filler with 3% filler content achieved the highest dynamic stability value, ie 1465.1 passing/mm; which is higher than the achievement of cement filler of the same degree (1400 passing/mm). The addition of filler beyond the optimum level will decrease the value of dynamic stability because it will reduce the contact between coarse aggregate particles [7].

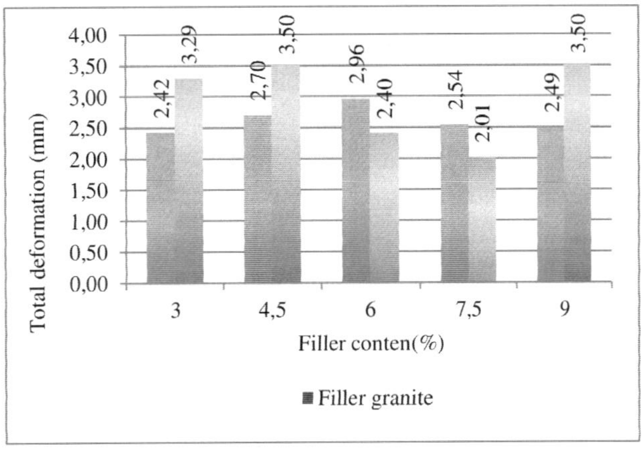

Figure 6. Deformation depth relationship with variation of filler content at temperature 60°C.

Fig. 6 shows the depth of deformation that occurs on the pavement with a temperature of 60°C. The picture shows that the deformation depth that occurs is not linear to the filler content, nor the filler type. For pavement with cement filler, the smallest deformation (2.01 mm) occurs at a 7.5% filler level. While on the granite filler pavement, the smallest deformation (2.42 mm) occurs at 3% filler content.

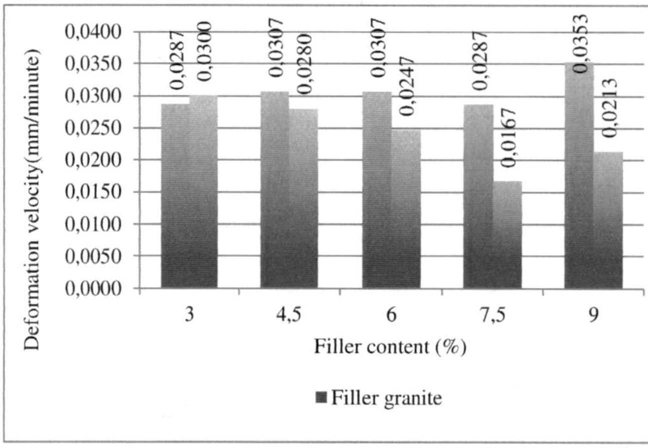

Figure 7. The relationship of deformation velocity with variation of filler content at temperature 60°C.

Fig. 7 shows the relationship between deformation velocity on asphalt concrete with cement filler and granite filler at 60°C pavement temperature. The picture shows that the relation of deformation velocity to filler content is not linear; as well as the type of filler. For the smallest deformation speed (0.0167 min / mm) occurs in asphalt concrete with cement filler, at a 7.5% filler content. While the asphalt concrete with the granite filler variation the smallest deformation velocity (0.0287 min / mm) occurred at 3% filler level. The smaller the speed of deformation will cause the value of dynamic stability which is greater meaning that the depth of deformation becomes small.

Tests of deformation and dynamic stability are carried out with temperature variations of 60°C, 45°C and 30°C as shown in Fig. 8.

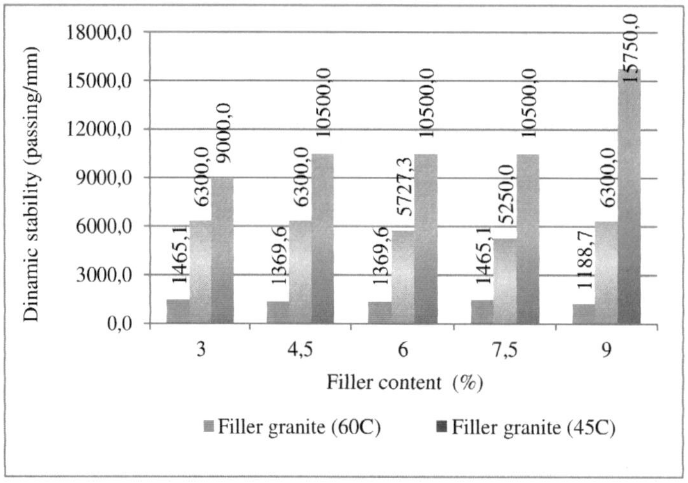

Figure 8. The relationship of dynamic stability with filler content to temperature variation on asphalt concrete with granite filler.

Fig. 8 shows the relationship between dynamic stability and granite filler content to temperature variations. The picture shows that the higher the temperature will cause a decrease of value of dynamic stability on the pavement. The decrease in the value of dynamic stability is due to the highly sensitive nature of the asphalt to the change of the termperature.

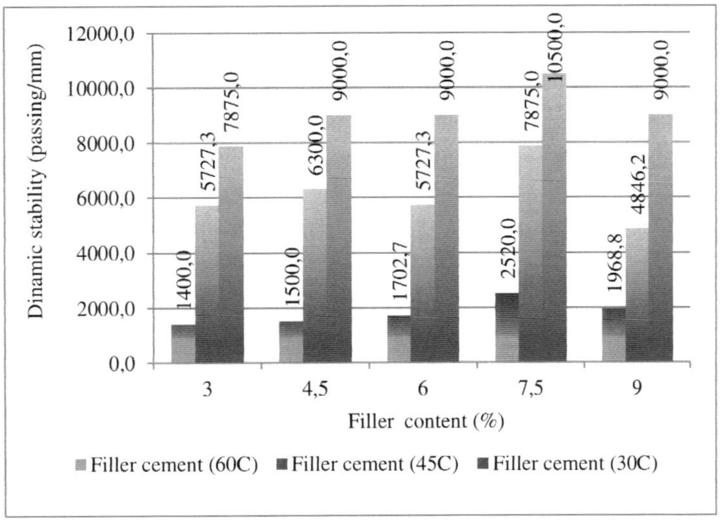

Figure 9. The relationship of dynamic stability with filler content to temperature variation on asphalt concrete with cement filler.

Fig. 9 shows the relationship between dynamic stability and filler levels against temperature variations. With the increase of cement filler content on the pavement, it will increase the value of dynamic stability, but after reaching optimum filler level (7.5%) the value of dynamic stability will decrease again. The higher the temperature will cause the value of dynamic stability to decrease.

4. Conclusion

Based on the results of the research, it can be concluded several things as follows:

- The larger the filler content will cause the value of the optimum bitumen content to be smaller, because the empty cavities in the asphalt concrete mixture are already filled by the filler.
- The relationship of dynamic stability, deformation and deformation velocity to filler content is non-linear.
- The higher the temperature on the pavement, will cause the value of dynamic stability to decrease.
- The use of cement filler with 7.5% content reaches the highest dynamic stability value, ie 2520.0 passing / mm. Meanwhile, asphalt concrete with granite filler with 3% filler content reached 1465.1passing / mm.
- On asphalt concrete with variation of cement filler content, the smallest deformation (2.01 mm) occurs at 7.5% filler level. While on asphalt concrete with variation of granite filler content, the smallest deformation (2.42 mm) occurred at 3% filler content.
- The smallest deformation speed (0.0167 min / mm) occurs in asphalt concrete with cement filler, at 7.5% filler content. While the asphalt concrete with the granite filler variation of the smallest deformation velocity (0.0287 min / mm) at 3% filler content.

Acknowledgment
The author acknowledges to UPPM POLBAN that already support the funding for commencement the research, and Bandung Research Center of Roads and Bridges that support the laboratory for the research.

References

[1] Chandra S and Choudhary R 2013 Performance characteristics of bituminous concrete with industrial wastes as filler *Journal of materials in civil engineering* **25**(11) 1666-1673
[2] Fasdarsyah F and Mukhlis M 2016 Pengaruh Penambahan Filler Granit dan Kramik Pada Campuran Laston AC-WC Terhadap Karakteristik Marshall *Teras Jurnal* **4** 1
[3] Widodo S 2014 Ketahanan Asphalt Concrete Wearing Course Gradasi Halus Terhadap Terjadinya Jejak Roda Kendaraan Pada Berbagai Temperatur Dan Kepadatan *Simposium Nasional RAPI XII, Fakultas Teknik Universitas Muhammadiyah Surakarta*
[4] Dinas Bina Marga 2010 *Spesifikasi Teknis Bina Marga Tahun 2010 Revisi III, Divisi VI. Tentang Perkerasan Aspal* (Kementrian Pekerjaan Umum)
[5] ASTM D6927-15 2015 *Standard Test Method for Marshall Stability and Flow of Asphalt Mixtures* ASTM International, West Conshohocken, PA
[6] JRA 1980 *Manual for Design and Construction of Asphalt Pavement* (Japan Roa d Association)
[7] Mahan H M 2013 Influence of mineral filler-asphalt ratio on asphalt mixture performance *Journal of University of Babylon* **21**(4) 1213-1222

IOP Conf. Series: Materials Science and Engineering 1108 (2021) 012005 doi:10.1088/1757-899X/1108/1/012005

Performance improvement of Asphalt Concrete-Wearing Course with modification of masterbatch SIR20

N P Yulianti*, S Sulaiman and R Utami

Department of Civil Engineering, Polytechnic State of Bandung, Bandung, Indonesia

*nindyaputr@gmail.com

Abstract. *Asphalt Concrete-Wearing Course* is a layer that is located at the top of flexible pavement structure. Types of damage that often occur in the AC-WC layer are rutting and slippery. AC-WC modification with masterbatch SIR20 is a system that improves AC-WC performance in reducing deformation in pavement structure and improving crack and skid resistance. The method that is used in this study is an experimental method that compares and analyzes the performance of the modified AC-WC layer with variations of masterbatch content are 2%, 4%, 6% and 8% and 0% as a control. The 4% of masterbatch content in the AC-WC modification layer is an effective and efficient option to be applied as a surface layer of a pavement structure.

1. Introduction

Nowadays, the use of asphalt penetration 60/ 70 in Indonesia as a binder of a flexible pavement structure, generally in Asphalt Concrete-Wearing Course (AC-WC), is capable of causing various types of damage that can affect the structural and functional performance during the service life, for example, there are many roads that is defective before the service life is reached. It is strengthened by the research of Amal which states that the use of asphalt penetration 60/ 70 in application is less durable where the asphalt easily hardened so the pavement easy to cracked [1]. On the other hand, with the rapid growth of traffic volume and the limited allocation of maintenance funds, it is necessary to develop innovation of AC-WC material and modified AC-WC material mixing technology to minimize the damage of pavement structures.

Asphalt modification with rubber is a system that improves asphalt performance such as reducing deformation in pavement structures and improving crack and skid resistance [2]. However, the solid natural rubber when directly used as an asphalt additive will require a very long mixing time and higher temperature than an asphalt without rubber, because the molecular chains of rubber are still so long so it is less desirable [3].

Furthermore, Yusep suggests an alternative in applying mixing method of modified solid natural rubber from the laboratory scale to the field scale by making the masterbatch [4]. According to Prastanto, the problem of using solid natural rubber as additive of asphalt can be overcome by doing depolymerization technique to natural rubber [5]. Depolymerization is the process of breaking long chains of polymer molecules into shorter molecular chains, otherwise known as the mastication stage of the process of grinding natural rubber in an Open Mill machine with a heat regulated as needed.

Based on the above description, the development research of solid natural rubber as an asphalt mixture modified material of this research will be used a SIR20 solid natural rubber which has been through the process of vulcanization. The selection of 2%, 4%, 6% and 8% natural rubber content of SIR20 was determined based on variations of natural rubber content that have not been done by previous researchers whom are Prastanto [6] and Yusep [4].

2. Methodology

2.1. Material

The material used in this research consists of SIR20 solid natural rubber as the base material of masterbatch polymer; zinc oxide, stearic acid and TMQ anti-oxidant as activators to accelerate the depolymerization of the SIR 20; sulfur as a solid natural rubber vulkanizator as shown in Fig. 1; Asphalt penetration 60/70 production PT. Pertamina; as well as coarse aggregates, fine aggregates and fillers from Cagak, Subang, West Java.

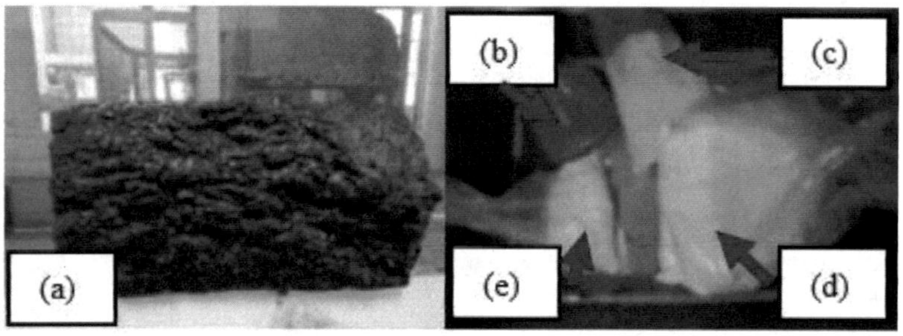

Figure 1. Materials: (a) Solid natural rubber; (b) Zinc oxide; (c) Stearic acid; (d) TMQ anti-oxsidants; (e) Sulfur.

2.2. Machine and specimen

The tool used in this research consists of a mill machine that serves to mix rubber compounds (SIR20 solid rubber, zinc oxide, TMQ anti-oxidant, stearic acid and sulfur) with asphalt penetration 60/ 70 as shown in Fig. 2.

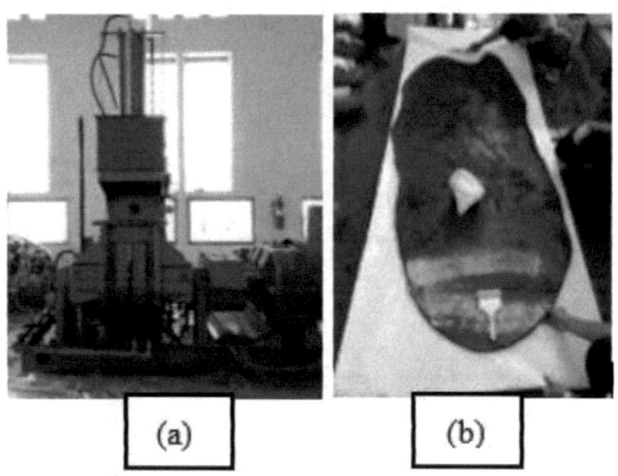

Figure 2. (a) Mill machine; (b) Masterbatch.

Wheel tracking machine serves as a tool of deformation resistance test as shown in Fig. 3a. The test is carried out for 1 hour at 45°C and 60°C temperature. Determination of the temperature of this research is based on the research of Widodo which states that the temperature range of this test from 25°C - 60°C [7]. The specimen in this deformation resistance test is 30 cm x 30 cm x 5 cm as shown in Fig. 3b.

Figure 3. (a) Wheel tracking machine; (b) Deformation resistance specimen.

British pendulum tester serves as a tool of skid resistance test as shown in Fig. 4a. The specimens in this test are 30 cm x 30 cm x 5 cm as shown in Fig. 4b. Prior to testing, the surface of the specimen shall be cleaned and moistened with sufficient water.

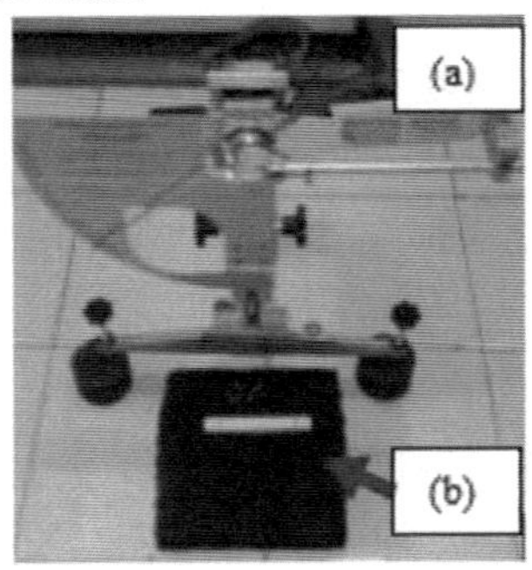

Figure 4. (a) British pendulum tester; (b) Deformation resistance specimen.

3. Test results

3.1. Structural performance

The parameter of structural performance in this research is rutting through dynamic stability test. The deformation test with the Wheel Tracking Machine is performed to simulate the deformation occurring in a pavement structure due to the vehicle path. The following comparison of the three deformation parameters of Wheel Tracking Machine testing results are dynamic stability or deformation resistance, deformation velocity and total deformation.

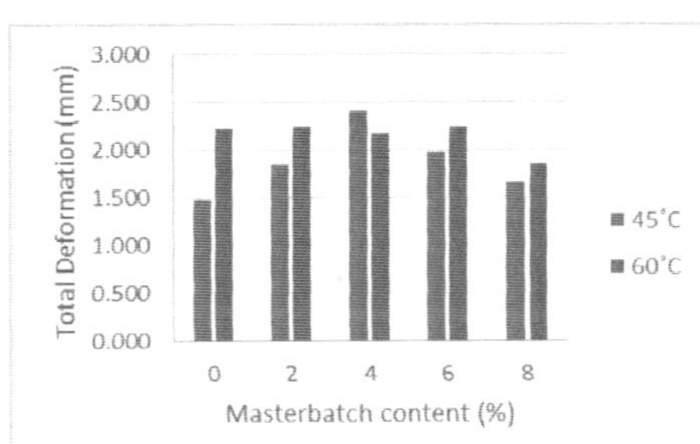

Figure 5. Total deformation.

Based on Fig. 5, it can be concluded that the higher the temperature, the deeper the rutting will occur, this is in accordance with the research of Kurnia [8]. This 60°C temperature describes the extreme pavement temperatures that may occur. At the temperature of 45°C, the lowest total deformation value occurs in the mixture with a 0% masterbatch content. While at a temperature of 60°C, the lowest total deformation value occurs in the mixture with an 8% masterbatch content.

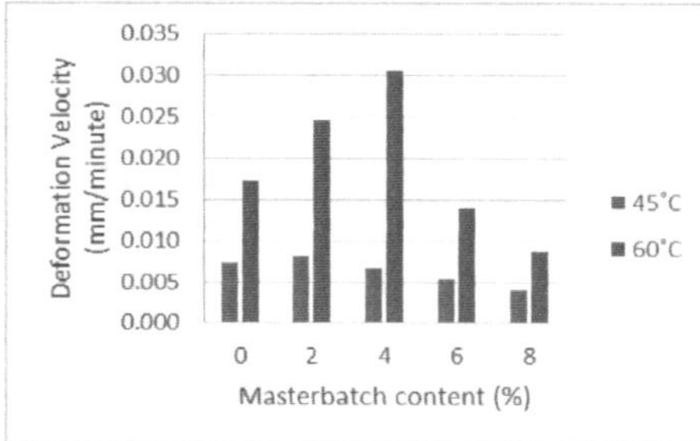

Figure 6. Deformation velocity.

Fig. 6 shows that the more rapid deformation, the deeper the rutting that is produced on the specimen. At this deformation velocity parameter, the mixture with 8% masterbatch content has the lowest deformation velocity value, so the rutting is relative small.

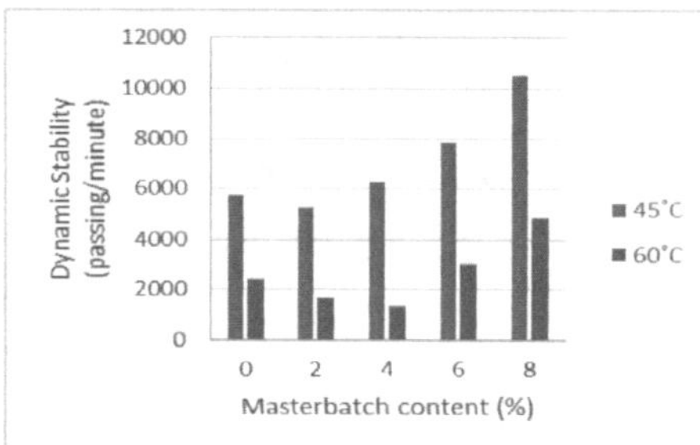

Figure 7. Dynamic stability.

Fig. 7 shows the comparison of dynamic stability or deformation resistance at each mixture with a certain temperature. The highest dynamic stability value at 60°C is indicated by mixture with 8% masterbatch content. This dynamic stability value is influenced by the density and temperature variation of the test [9].

From all of deformation tests performed, AC-WC with 8% masterbatch content had a lower deformation value than the other masterbatch content. It is happening because the rubber contained in the masterbatch made the mixture denser so that the layer has deformation resistance/ higher dynamic stability, as well as the deformation velocity can be retained, then the rutting will be smaller. In this case the rutting or depth of the vehicle's tread trails is referred to as rutting caused by too much repetitive pressure/ loading that impacts damage to the surface layer of a pavement structure.

3.2. Functional performance

The parameter of functional performance in this research is skid resistance. According to Mannering et al., friction resistance is strongly influenced by the season that occurs, such as in dry time all roads have great frictional resistance as well as conversely [10]. The skid resistance is influenced by factors such as surface shape variation and tire conditions, road surface texture, weather conditions and driving conditions [11]. From the result test using the British Pendulum Tester (BPT), the value of British Pendulum Number (BPN) is shown in figure 8 The surface temperature for the test is 27°C.

Num.	Masterbatch content in asphalt penetration 60/ 70 (%)	Distance from each side (cm)	Surface texture	Surface temperature (°C)	Ave. BPN	Correction	BPN
1	0	2	Front	27	82,0	+1	83
2		15	Middle	27	79,2	+1	80
3		2	Back	27	83,4	+1	84
							83
1	2	2	Front	27	92	+1	93
2		15	Middle	27	84	+1	85
3		2	Back	27	88	+1	89
							89
1	4	2	Front	27	87	+1	88
2		15	Middle	27	91	+1	92
3		2	Back	27	89	+1	90
							90
1	6	2	Front	27	92	+1	93
2		15	Middle	27	85	+1	86
3		2	Back	27	90	+1	91
							90
1	8	2	Front	27	80	+1	81
2		15	Middle	27	80	+1	81
3		2	Back	27	80	+1	81
							81

Figure 8. Result of skid resistance test.

Transportation Road Research Laboratory (TRRL) provides minimum standards of result test with BPT tools on motorway freight >2,000 vehicles per day is greater than or equal to 55 BPN [12]. This is also in line with the requirements issued by the TRRL (1989) which can be seen in figure 9.

Category	Location Type	Min. number of skid resistance
A	Tough location such as: Roundabout; radius of bend <150 m on the highway; slope of 1:20 or steeper, with length >100 m; signaled intersection arm on the highway.	65
B	Main/ fast roads, continuous roads and first class roads and heavy urban roads (> 2000 vehicles per day)	55
C	Other locations	45
Note: For Category A and B where the vehicle speed is high (> 95 km / h) additional use is a minimum texture depth of 0.65 mm.		

Figure 9. Minimum number of skid resistance using BPT.

From the result that are obtained after the implementation of skid resistance test, the asphalt mixture with all variations of masterbatch content have meet the requirements. The 2%, 4% and 6% of masterbatch content are increasing BPN values, this indicate that the role of masterbatch provides a good contribution in improving the skid resistance as the surface layer of the flexible pavement structure.

Fig. 8 shows that the value of the skid resistance increasing with the increase of the masterbatch content from 2% to 6%. Nevertheless, at 8% masterbatch content, the skid resistance value decreased significantly from 90 BPN to 81 BPN. Thus it can be concluded that with the addition of masterbatch up to 6%, the value of BPN produced increases, in line with the increasing AC-WC layer and AC-WC modification of skid resistance.

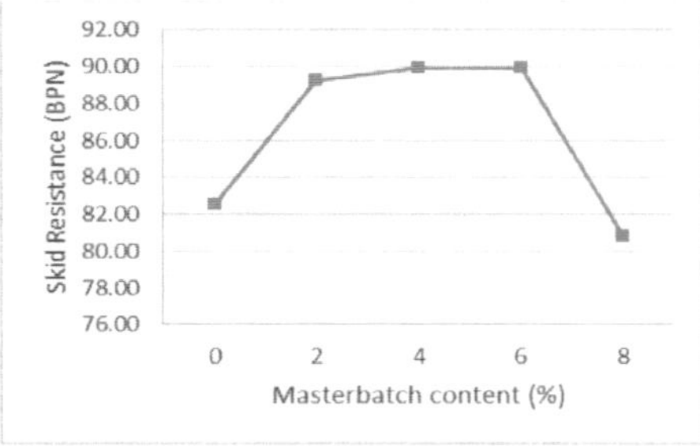

Figure 10. Skid resistance.

From the overall improvement of structural and functional performance of the modified AC-WC layer, it can be seen that the masterbatch content showed the greatest increase at 4% and 8%. In determining the most effective and efficient alternative masterbatch to be applied in Indonesia's highway, a summary of each test parameter is presented in Table 1.

Table 1. Resume.

Parameter	Masterbatch level in asphalt pen. 60/ 70				
	0%	*4%*	*notes*	*8%*	*notes*
Rutting (mm)	2.23	2.18	**increase**	**1.86**	**increase**
Skid resistance (BPN)	83	90	**increase**	**81**	**decrease**

4. Conclusion

From the two parameters that have been tested, it shows that all performance parameters have increased, except for the skid resistance parameter at 8% masterbatch content. Nevertheless, the retaining resistance of 81 BPN still meets the minimum service standard, which is greater than 55 BPN. Therefore, to fulfill the road as one of the strategic National assets, the road should be able to provide the minimum level of service during its service life. This can be accomplished if the performance of the pavement structure can be increased, especially on the surface layer of AC-WC.

Thus, the selection of 4% masterbatch content is an effective alternative of performance improvement and efficient alternative of added cost compared with 8% masterbatch. However, the techno economy needs further research. In general, it can be applied to improve the performance of pavement structures in Indonesia both from structural and functional performance.

Acknowledgment

The author acknowledges to UPPM POLBAN that already support the funding for commencement the research, Bogor Rubber Research Center and Bandung Research Center of Roads and Bridges that support the laboratory for the research.

References

[1] Amal A S 2011 Pemanfaatan Getah Karet Pada Aspal AC 60/70 Terhadap Stabilitas Marshall Pada Asphalt Treated Base (ATB) *Media Teknik Sipil* **9** 1 8 – 16
[2] Suroso T W 2007 Peningkatan Kinerja Campuran Beraspal dengan Karet Alam dan Karet Sintetis *Jurnal Jalan-Jembatan* **24**(1) 14 - 25
[3] Vichitcholchai N, Panmai J and Na-Ranong 2006 Modification of Asphalt Cement by Natural Rubber for Pavement Construction *Rubber Thai Journal* **79**(3) 170
[4] Yusep 2011 *Aspal Karet* (Pusat Penelitian Bangunan Jalan dan Jembatan, DSIP F-PJS-02-32-08)
[5] Prastanto H 2014 Depolimerisasi Karet Alam Secara Mekanis untuk Bahan Aditif Aspal *Jurnal Penelitian Karet* **32**(1) 81 - 87

[6] Prastanto H, Cifriadi A and Ramadhan A 2015 Karakteristik dan Hasil Uji Marshall Aspal Termodifikasi dengan Karet Alam Terdepolimerisasi sebagai Aditif *Jurnal Penelitian Karet* **33**(1) 75 - 82

[7] Widodo S 2013 Ketahanan Asphalt Concrete Wearing Course Gradasi Halus Terhadap Terjadinya Jejak Roda Kendaraan Pada Berbagai Temperature Dan Kepadatan *Simposium Nasional RAPI XII, Fakultas Teknik Universitas Muhammadiyah Surakarta*

[8] Kurnia R 2014 *Kinerja Campuran Beraspal Lapis Pengikat Menggunakan Aspal Pen 40/50 Dengan Variasi Gradasi Bina Marga Dan Asphal Institute (Bagian Dari Studi Perpetual Pavement)* (Bandung: Program Studi Sistem dan Teknik Jalan Raya, Institut Teknologi Bandung)

[9] Prery A A, Subagio B S, Kosasih D and Hendarto S 2016 Evaluasi Karakteristik Modulus Resilien Dan Deformasi Permanen Campuran Beton Beraspal (AC-Binder Course) Menggunakan Campuran Agregat Berabrasi Tinggi *Jurnal Teknik Sipil, Jurnal Teoretis Dan Terapan Bidang Rekayasa Sipil* **23** (3) 223-232

[10] Mannering F, Kilareski W and Washburn S 2007 *Principles of highway engineering and traffic analysis* (John Wiley & Sons)

[11] Suwardo 2014 *Investigasi Kekesatan Perkerasan Jalan Menggunakan Wessex Skid Tester* Teknik (Yogyakarta: Sipil Fakultas Teknik UGM)

[12] Transportation Road Research Laboratory 2014 *Instructions for Using the Portable Skid Resistance Tester* (London: Road Note 27)

IOP Conf. Series: Materials Science and Engineering 1108 (2021) 012006 doi:10.1088/1757-899X/1108/1/012006

Innovative products of concrete steel bone testing tools in the field of civil engineering building

T Hadi*, N S Pamungkas, W Wahjoedi and T Mulyono

Civil Engineering, Politeknik Negeri Semarang, Semarang, Indonesia

*tjokro.hd@polines.ac.id

Abstract. Bending test is one type of material testing conducted to determine the mechanical properties of a technical material. In designing this bending test required a standard design method that uses standard machine elements as well as commonly found in the market. In practice in the field, there are only a few materials testing practitioners who pay attention to aspects and effects of variations in the dimensions of the test specimen on the data from the test results. The test is carried out with the intention to obtain information about fulfilling the quality requirements and the feasibility of the material to be used as construction material for buildings, bridges and other physical buildings. As for one application that can be done is to look at aspects of the problems that are around that are associated with practicum activities in the laboratory and in the work industry. Therefore it is necessary to adjust between making the final report with the needs needed in the industrial world. For this reason, the bending test is made to study the effect of dimensional variations on test specimens in arcing testing. For this reason, this study with the title "Innovative Products for Concrete Reinforcement Steel Curved Test Equipment in the Field of Building Materials in the Department of Civil Engineering" is very important to do. The research was conducted to improve the performance of tools in the building materials laboratory. The final results of the study are expected to function optimally and easily used for laboratory work.

1. Preliminary

1.1. Background

In this modern era many technologies have been developed to make it easier to support development. With the rapid development in Indonesia, the existing facilities such as the construction of buildings, hotels, supermarkets, homes and other buildings, so that more and more that require the use of concrete iron as the foundation framework of the building.

One way to strengthen buildings, especially concrete construction buildings that are widely known is to use reinforcing steel as reinforcement. Because the use of reinforcing steel is very closely related to the security and safety of buildings and especially the people around it, the government has determined that reinforcing steel produced and used in Indonesia must meet the concrete steel quality standards based on the Indonesian National standard, SNI 2052- 2014.

Reinforcing steel is a very important material in construction, especially in reinforced concrete construction is unavoidable considering reinforcing steel is one of the main determinants in the strength or failure of construction.

IOP Conf. Series: Materials Science and Engineering 1108 (2021) 012006 doi:10.1088/1757-899X/1108/1/012006

The bending and modulus of elasticity test on a material is carried out using a load where the main stress is in the form of bending. The modulus of elasticity in bending and tensile or compressive tests will experience a slight difference even though the specimens are the same. That is because the modulus of elasticity in the tensile or compressive test is in one direction, namely the tensile or compressive direction. Whereas in the bending test, the modulus of elasticity is in two directions, namely tensile and compressive.

So the research team wants to create a tool for testing that is the bending test, so the research team hopes that with this bending test tool we can find out the curvilinear strength of the products on the market even though it may not be fully. This curvilinear test tool (bending) can test products such as concrete reinforcement steel.

1.2. Research purposes
The proposed research has the aim to make the design of concrete reinforcing steel arch test equipment to support practicum equipment in the building materials laboratory. For specific purposes is:

- Making concrete reinforcing steel arch test equipment tangible and ready to use for the benefit of practicum in order to function optimally.
- Design of curved test equipment that has good performance.

In this overall stage it provides three measurable outcomes as contributions, namely:

- Prototype modification of concrete reinforcing steel arch test equipment.
- SOP (Standard Operating Procedure) or steps in operating the equipment.
- Publications in scientific journals and drafts for the submission of IPR.

In this overall stage it provides three measurable outcomes as contributions, namely:

- Prototype modification of concrete reinforcing steel arch test equipment.
- SOP (Standard Operating Procedure) or steps in operating the equipment.
- Publications in scientific journals and drafts for the submission of IPR.

1.3. Innovation
This research is an additional development as a support tool for building materials laboratory practicum test in Semarang State Polytechnic Department of Civil Engineering.

1.4. Urgency
With this research, it is expected that urgency can be realized in the final results of this study, namely:

- Improving the quality of practical learning in the building materials laboratory of the Semarang State Polytechnic Civil Engineering Department.
- Adding a collection of tools in the Laboratory of Building Materials, Civil Engineering Department, Semarang State Polytechnic.

2. Literature review

2.1. Bending test definition
Curvature test is one form of testing to determine the mechanical ability of a material that can be seen visually [1]. In principle, this test is done to measure the strength of the material due to axial loading that can be seen from the shape and deflection that occurs in the specimen at a certain unit of length between the support / support The bending test (bending test) is a process of testing the material by pressing to get results in the form of data about the bending strength (bending test) of a material being tested.

2.2. Basic theory of bending
Bending is a process of plastic deformation of a metal against a linear axis with little or no change in surface area [2]. A piece of iron can be bent using a press machine. This bending process usually uses a die in the form of V, U, W or other patterns. The bending process causes the metal which is on the outside of the neutral axis to pull, while on the other side experiences pressure.

2.3. Bending test process
Bending process (Bending) is a work by putting pressure on a particular part so that plastic deformation occurs on the part under pressure [3]. While the bending process is the process of bending or bending using a manual bending tool or using a bending machine.

The bending process that works on the design of this tool, which is adopting techniques or bending processes by means of rotary or rotation contained in the pipe bending machine. Then the types of bending machines that will be used when designing and building will be explained at the next point.

2.4. Definition of concrete reinforcement steel
Concrete reinforcement steel is steel in the form of a circular rod that is used for reinforcing concrete, which is produced from billet raw material by hot rolling [4]. Based on its shape, concrete reinforcing steel can be divided into 2 (two) types, namely plain concrete reinforcing steel and fin / threaded reinforcing steel.

Figure 1. Plain concrete and fin steel reinforcement.

2.5. Physical-mechanical characteristics of reinforcing steel
The mechanical properties of a material including reinforcing steel become information parameters in the basic design of a material's strength and as supporting data for material specifications [5]. The mechanical properties of a material obtained from a tensile test result give a bigger picture than other mechanical tests, including the parameters of yield stress (deformation), tensile strength, breaking stress, strain, modulus of elasticity, percentage increase in length, percentage reduction in size, toughness , plasticity, elasticity, plasticity, yield type and fracture shape experienced by material.

Other material properties can be obtained from impact tests that produce toughness, hardness tests that provide information about the resistance of a material when penetrated by other objects and have a correlation with tensile testing, and bending tests which provide information about material plasticity.

The mechanical properties of a material need to be well known, because the material is used for various purposes in various circumstances. Mechanical properties are determined by the type and comparison of the atoms that compose the material which includes the types of elements and their composition, and the environment at the time of testing in obtaining mechanical properties [6].

2.6. Research road map
Research road map in this research can be seen in figure 2.

Has been done		Current research	Further research
2017	**2018**	**2019**	**next year**
design and build the pendrol fixing device	**Design of flexural test equipment for railroad concrete bearings as a supporting tool for building materials practical tests**	**Modification of the railroad concrete bearing flexural test equipment as a supporting tool for building materials practicum tests**	**Improvements if there are or proposed registration of Copyright / Patents**

Figure 2. Research road map.

3. Research methods

In accordance with the objectives of the study, namely, to make a pendrol plug, carried out by the following methods. The research begins with the preparation of materials and design of work drawings. Construction of a foundation, support pillars, permanent support and foundation. Continuing the initial hydraulic push jack assembly and hydraulic pressure booster system.

Manufacture of tanks and olie pipe installations, pressure regulating faucets and curved test object mounting aids. Furthermore, the arrangement of electric motors, pressure gauge dial and other electrical accessories. The final stage after everything is installed is tested by the tool. The research flow process can be seen in the following flowchart.

3.1. Flowchart

Flowchart in this research can be seen in figure 3.

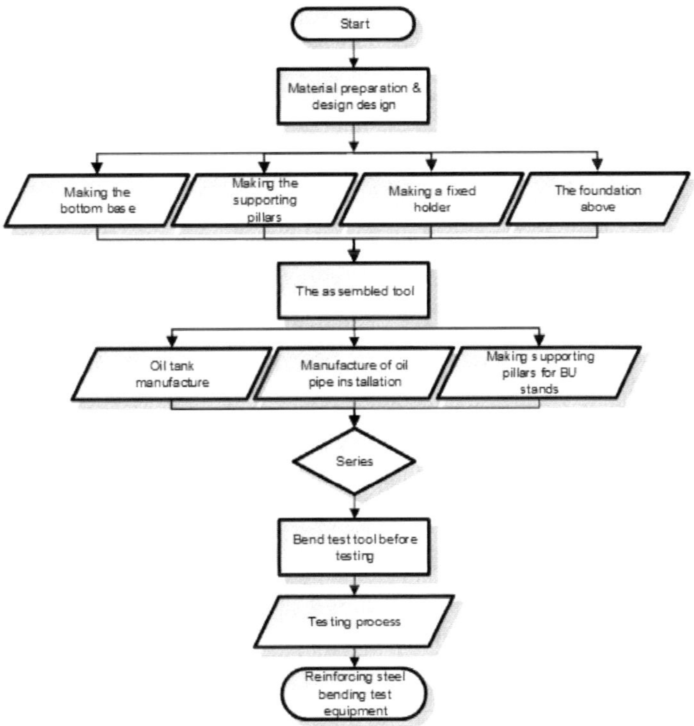

Figure 3. Flowchart research implementation flow.

3.2. Place and time of research

The study was conducted in fiscal year 2019 in the laboratory and workshop majoring in civil engineering Semarang State Polytechnic which is located at JL. Prof. Soedarto, SH. Tembalang, Semarang.

3.3. Equipment material

The materials needed in this study include: plain concrete iron, screw concrete, bottom foundation, support pipe, hydraulic jack, hydraulic pressure strengthening system, top runway, electric motor, pressure speed regulating valve, load gauge manometer.

3.4. Research equipment

Equipment used in this study include: welding machines, lathe equipment, drilling machines, screw making equipment, painting equipment.

IOP Conf. Series: Materials Science and Engineering 1108 (2021) 012006 doi:10.1088/1757-899X/1108/1/012006

3.5. Drawing of proposed tool design
Proposed tool design in this research can be seen in figure 4.

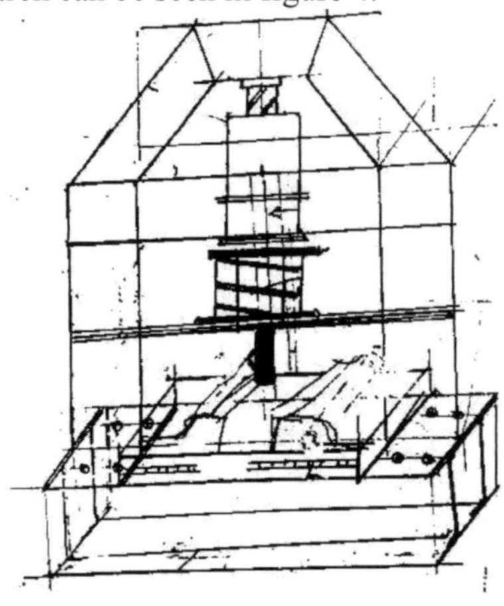

Figure 4. Proposed tool.

3.6. Data analysis
Data of laboratory test results in some time needed in one test compared to the conditions before the concrete bearing bending test equipment is shorter and the calculation is more optimal.

3.7. Stages in research
The steps in testing include:
- Prepare a bending test device according to the size settings on the diameter
- The material being tested is buckled according to the diameter size
- Prepare the bending needle according to the size of the diameter of the steel bar
- Begin the buckling testing process with the condition that the sample being tested results bend at an angle of 180 degrees (U-shaped)
- Start the buckling test process.
- After finishing the test, bending the results of the test diameter is seen visually with a magnifying glass. Can be seen in the print out of the test results the results are okay or broken.

4. Results and discussion

4.1. Research results
The result in this research can be seen in figure 5 - figure 20.

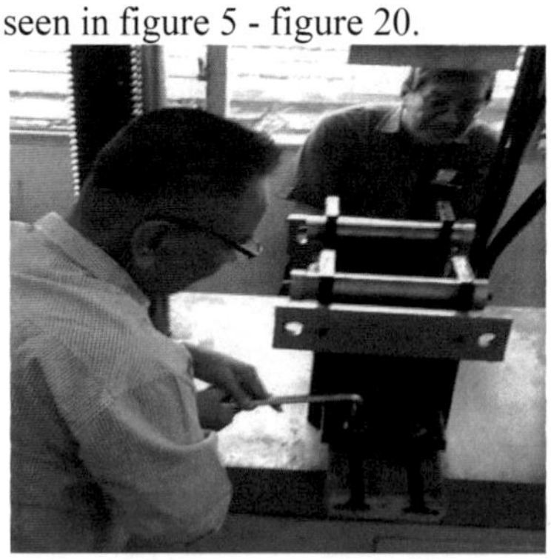

Figure 5. The team is installing the test equipment that has been modified.

Figure 6. The test object is attached to the modified test equipment.

Figure 7. Installation of test specimens before testing.

Figure 8. Test object results after testing.

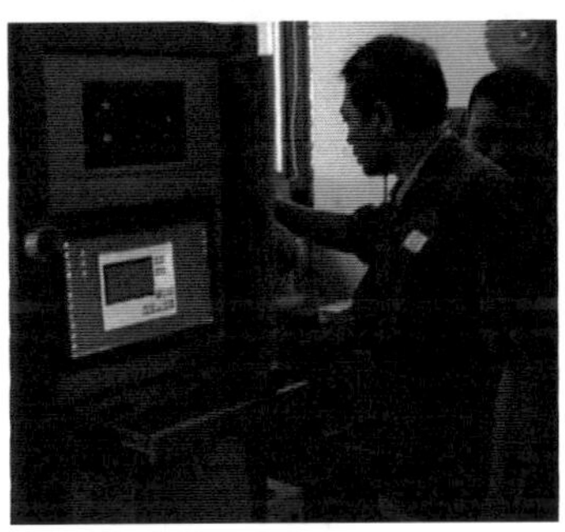

Figure 9. Output results from testing.

Figure 10. The test object is the result of testing.

REINFORCING STEEL
Baja Tulangan

Ex. PT. HANIL JAYA STEEL (HIJ)

Total Sample.
Size Ø : 8, 10, 13, 16, 19, 22, 25, 32
Vol. : 3 unit x 1 meter
Code : TBL-HIJ
Project : Tambak Lorok CCPP Block 3 Project
Test : 2 unit = Yield streng & 2 unit =Bending

Figure 11. Specifications of objects to be tested.

LABORATORIUM PENGUJI JURUSAN TEKNIK SIPIL
POLITEKNIK NEGERI SEMARANG
Jl. Prof. Sudharto, SH Tembalang, Semarang, Jawa Tengah 50275
Telp. 024 7473417, Fax 024 7472396

M-16 → BERAT = 171 GRM
M-24 → B = 460
M18 = 370 GRM

UJI TARIK DAN LENGKUNG BAJA TULANGAN BETON
SNI 2052:2014

JOB . 584

Tanggal Datang : 8 - 11 - 2017
Pelanggan : PT. HUTAMA KARYA
Proyek :
Lokasi Pekerjaan :
Nomor Job : 584/PJ-CI/KI/017
Jenis Benda Uji :
Tanggal Uji :
Dikerjakan : TJOKRO HADI, SST. MT

EX . PT. HANIL JAYA STEEL (HIJ)

No	Benda Uji	Panjang	Berat (B)	Berat/M	Ø	L_0	L_1	$Ø_1$	Ø dalam (mm)	Model Sirip	Jarak Sirip Melintang (mm)	Lebar Sirip Membujur (mm)		Tinggi Sirip (mm)	Hasil Uji Lengkung
1	PØ8 - 1	100	380	380	7,9	64	88,2	4,7							OKE
2	PØ8 - 2	100	366	366	7,9	64	83,3	4,9							OKE
3	SØ8 - 1										6,7	2,2	2,1	0,5	
4	SØ8 - 2										6,7	2,1	2,2	0,5	
5	SØ10 - 1	100	596	0,596	9,84	80	95,3	6,6	9,3	ST.IKAN	6,8	3,3	3,4	0,5	OKE
6	SØ10 - 2	100	588	0,588	9,77	80	94,4	7,3	9,3	"	6,7	3,4	3,3	0,4	OKE
7															
8	SØ13 - 1	100	1018	1,018	12,85	104	125,4	9,8	12,1	ST.IKAN	7,3	3,4	3,6	0,8	OKE
9	SØ13 - 2	100	1005	1,005	12,77	104	128,0	7,0	12,1	"	7,2	3,6	3,4	0,7	OKE
10			1539												
11	SØ16 - 1	100	~~1085~~	1,539	15,80	128	151,6	9,1	15,0	ST.IKAN	10,9	4,6	4,6	0,8	OKE
12	SØ16 - 2	100	1541	1,541	15,82	128	158,4	10,4	15,0	"	10,9	4,6	4,6	0,8	OKE
13							6X								
14	SØ19 - 1	100	2165	2,165	18,75	114	138	14,4	17,8	ST.IKAN	12,3	5,2	5,3	1,0	OKE
15	SØ19 - 2	100	2168	2,168	18,76	114	138,4	13,3	17,8	"	12,2	5,3	5,2	1,0	OKE
							6X								
2	SØ22 - 1	100	2948	2,948	21,84	132	159,6	16,0	21,5	ST.IKAN	14,3	5,5	5,2	1,5	OKE
3	SØ22 - 2	100	2920	2,920	21,77	132	161,8	16,0	21,5	"	14,3	5,6	5,3	1,6	OKE
4															
5	SØ25 - 1	100	3738	3,738	24,63	150	189	16,2	23,4	ST.IKAN	15,2	6,4	6,7	1,6	OKE
6	SØ25 - 2	100	3740	3,740	24,64	150	184,8	15,0	23,4	"	15,2	6,5	6,6	1,7	OKE
7															
8															
9	SØ29 - 1														
10	SØ29 - 2														
11															
12															
13	SØ32 - 1	100	6228	6,228	31,79	192	241	24,3	30,0	ST.IKAN	19,9	7,8	8,1	2,4	OKE
14	SØ32 - 2	100	6200	6,200	31,72	192	232	21,9	30,0	"	19,9	8,2	7,8	2,5	OKE
15															

Figure 12. Initial test table.

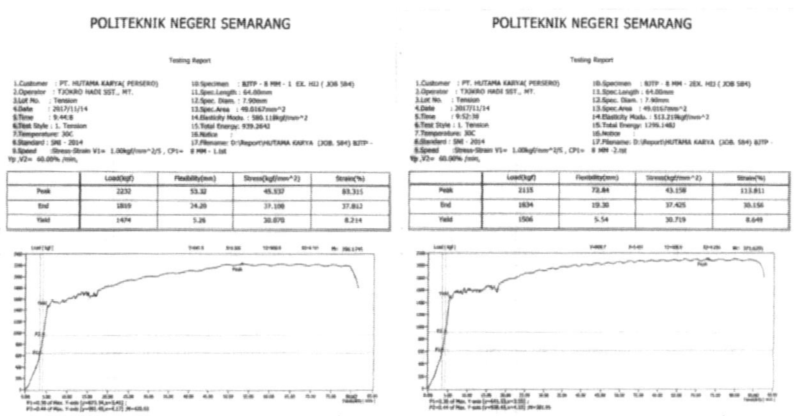

Figure 13. Test results with a diameter of 8 mm.

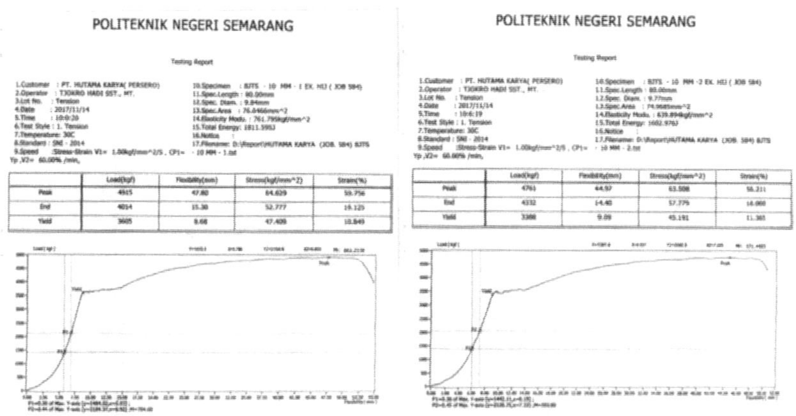

Figure 14. Test results with a diameter of 10 mm.

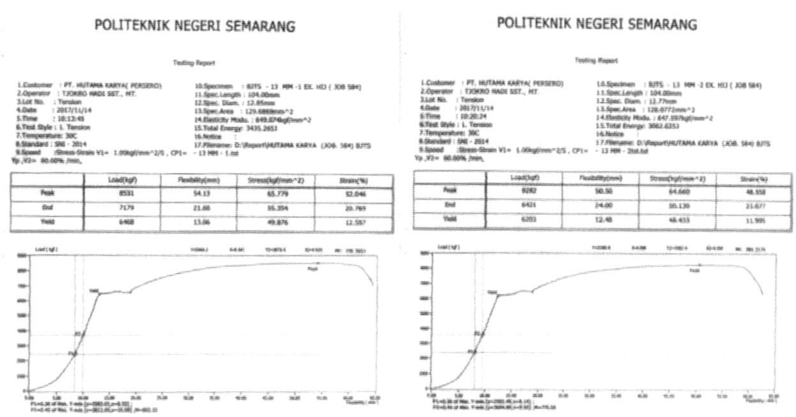

Figure 15. Test results with a diameter of 13 mm.

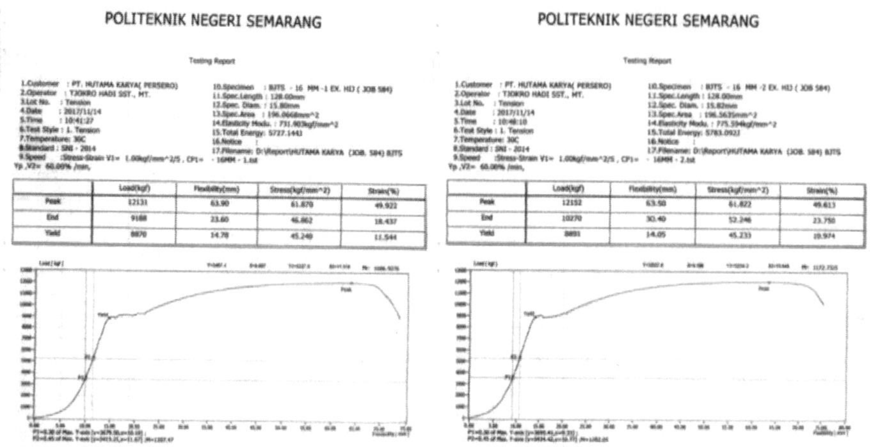

Figure 16. Test results with a diameter of 16 mm.

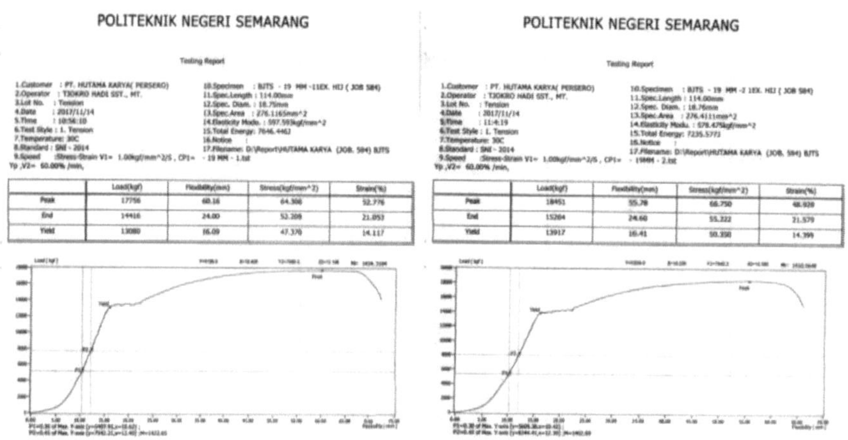

Figure 17. Test results with a diameter of 19 mm.

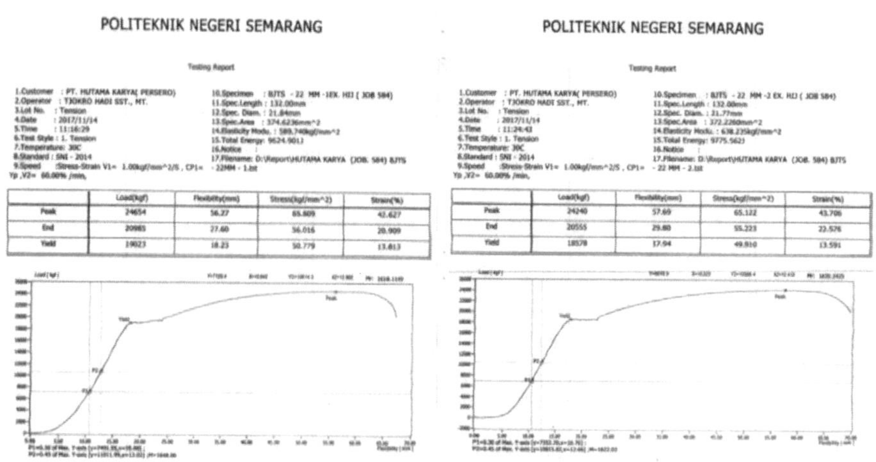

Figure 18. Test results with a diameter of 22 mm.

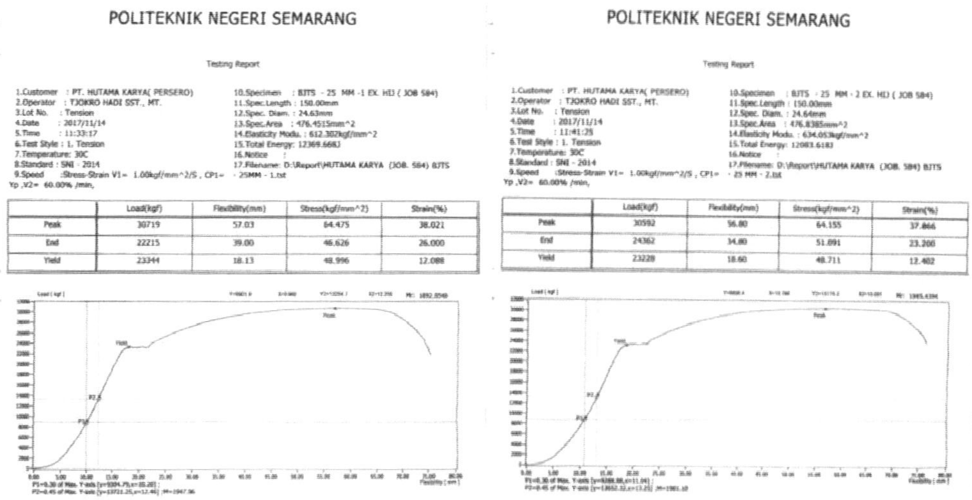

POLITEKNIK NEGERI SEMARANG

Testing Report

1.Customer : PT. HUTAMA KARYA(PERSERO)
2.Operator : TJOKRO HADI SST., MT.
3.Lot No. : Tension
4.Date : 2017/11/14
5.Time : 11:33:17
6.Test Style : 1. Tension
7.Temperature: 30C
8.Standard : SNI - 2014
9.Speed :Stress-Strain V1= 1.00kgf/mm^2/S , CP1=
Yp ,V2= 60.00% /min,
10.Specimen : BJTS - 25 MM -1 EX. HD (JOB 584)
11.Spec.Length : 150.00mm
12.Spec. Diam. : 24.63mm
13.Spec.Area : 476.4515mm^2
14.Elasticity Modu. : 612.302kgf/mm^2
15.Total Energy: 12369.668J
16.Notice :
17.Filename: D:\Report\HUTAMA KARYA (JOB. 584) BJTS - 25MM - 1.tst

	Load(kgf)	Flexibility(mm)	Stress(kgf/mm^2)	Strain(%)
Peak	30719	57.03	64.475	38.021
End	22215	39.00	46.626	26.000
Yield	23344	18.13	48.996	12.088

POLITEKNIK NEGERI SEMARANG

Testing Report

1.Customer : PT. HUTAMA KARYA(PERSERO)
2.Operator : TJOKRO HADI SST., MT.
3.Lot No. : Tension
4.Date : 2017/11/14
5.Time : 11:41:25
6.Test Style : 1. Tension
7.Temperature: 30C
8.Standard : SNI - 2014
9.Speed :Stress-Strain V1= 1.00kgf/mm^2/S , CP1=
Yp ,V2= 60.00% /min,
10.Specimen : BJTS - 25 MM - 2 EX. HD (JOB 584)
11.Spec.Length : 150.00mm
12.Spec. Diam. : 24.64mm
13.Spec.Area : 476.8385mm^2
14.Elasticity Modu. : 634.053kgf/mm^2
15.Total Energy: 12083.618J
16.Notice :
17.Filename: D:\Report\HUTAMA KARYA (JOB. 584) BJTS - 25 MM - 2.tst

	Load(kgf)	Flexibility(mm)	Stress(kgf/mm^2)	Strain(%)
Peak	30592	56.80	64.155	37.866
End	24362	34.80	51.091	23.200
Yield	23228	18.60	48.711	12.402

Figure 19. Test results with a diameter of 25 mm.

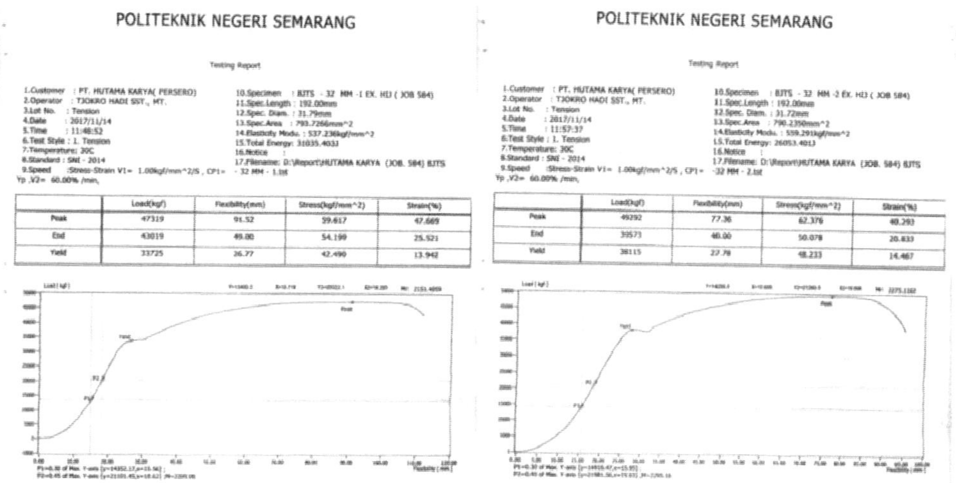

POLITEKNIK NEGERI SEMARANG

Testing Report

1.Customer : PT. HUTAMA KARYA(PERSERO)
2.Operator : TJOKRO HADI SST., MT.
3.Lot No. : Tension
4.Date : 2017/11/14
5.Time : 11:48:52
6.Test Style : 1. Tension
7.Temperature: 30C
8.Standard : SNI - 2014
9.Speed :Stress-Strain V1= 1.00kgf/mm^2/S , CP1=
Yp ,V2= 60.00% /min,
10.Specimen : BJTS - 32 MM -1 EX. HD (JOB 584)
11.Spec.Length : 192.00mm
12.Spec. Diam. : 31.79mm
13.Spec.Area : 793.7266mm^2
14.Elasticity Modu. : 537.236kgf/mm^2
15.Total Energy: 31035.403J
16.Notice :
17.Filename: D:\Report\HUTAMA KARYA (JOB. 584) BJTS - 32 MM - 1.tst

	Load(kgf)	Flexibility(mm)	Stress(kgf/mm^2)	Strain(%)
Peak	47319	91.52	59.617	47.665
End	43019	49.00	54.199	25.521
Yield	33725	26.77	42.490	13.942

POLITEKNIK NEGERI SEMARANG

Testing Report

1.Customer : PT. HUTAMA KARYA(PERSERO)
2.Operator : TJOKRO HADI SST., MT.
3.Lot No. : Tension
4.Date : 2017/11/14
5.Time : 11:57:37
6.Test Style : 1. Tension
7.Temperature: 30C
8.Standard : SNI - 2014
9.Speed :Stress-Strain V1= 1.00kgf/mm^2/S , CP1=
Yp ,V2= 60.00% /min,
10.Specimen : BJTS - 32 MM -2 EX. HD (JOB 584)
11.Spec.Length : 192.00mm
12.Spec. Diam. : 31.72mm
13.Spec.Area : 790.2350mm^2
14.Elasticity Modu. : 559.291kgf/mm^2
15.Total Energy: 26053.401J
16.Notice :
17.Filename: D:\Report\HUTAMA KARYA (JOB. 584) BJTS -32 MM - 2.tst

	Load(kgf)	Flexibility(mm)	Stress(kgf/mm^2)	Strain(%)
Peak	49292	77.36	62.376	40.293
End	39573	40.00	50.078	20.833
Yield	38115	27.78	48.233	14.467

Figure 20. Test results with a diameter of 32 mm.

4.2. Discussion

- All the reinforcing iron that will be bent / bending is tested its tensile strength in order to obtain the Stress yield in units of Kgf / mm2.
- Prepare a bend test according to the size settings on the diameter
- The material being tested is buckled according to the diameter size
- Prepare the bending needle according to the size of the diameter of the steel bar
- Begin the buckling testing process with the condition that the sample being tested results bend at an angle of 180 degrees (U-shaped)
- Start the buckling test process.
- After finishing the test, bending the results of the test diameter is seen visually with a magnifying glass. Can be seen in the print out of the test results the results are okay or broken.
- If it is observed getting cracks and even broken results, the reinforcement is recommended not to be used in buildings.

5. Conclusion

- BJTP reinforcement 8 mm in diameter. Yield Stress results are 30,070 kgf / mm^2. With conclusions OK / good buckling results.
- BJTP reinforcement iron 10 mm in diameter. Yield Stress results are 47,409 kgf / mm^2. With conclusions OK / good buckling results.
- BJTP reinforcement diameter of 13 mm. Yield Stress results are 49,876 kgf
- / mm^2. With conclusions OK / good buckling results.
- BJTP reinforcement diameter 16 mm. Yield Stress results are 45,240 kgf / mm^2. With conclusions OK / good buckling results.
- BJTP reinforcement diameter 19 mm. Yield Stress results are 47,370 kgf / mm^2. With conclusions OK / good buckling results.
- BJTP reinforcement 22 mm in diameter. Yield Stress results are 50,779 kgf
- / mm^2. With conclusions OK / good buckling results.
- BJTP reinforcement 25 mm in diameter. Yield Stress results are 48,996 kgf
- / mm^2. With conclusions OK / good buckling results.
- BJTP reinforcement 32 mm in diameter. Yield Stress yields 42,490 kgf / mm^2. With conclusions OK / good buckling results.

From the conclusions of the test, for all concrete reinforcement diameters, if the tensile test results where Yield Stress <55,000 kgf / mm2, it will get a good buckling test result.

References

[1] Tsauri S 2018 *Perancangan dan Pembuatan Alat Uji Bending Untuk Pengujian Material Logam Yang Akan Digunakan di Lab Metalurgi Fisik Universitas Islam As-Syafi'iyah* (Jakarta: Universitas Islam As-Syafi'iyah)
[2] Son L and Satria F 2014 Rancang bangun alat bending sengkang pada kolom skala laboratorium *Teknika* **21**(3)
[3] Kurnia M 2016 *Rancang Bangun Mesin Bending Otomatis Untuk Begel Diameter 8 Mm (Proses Pembuatan)* (Palembang: Politeknik Negeri Sriwijaya)
[4] Gazal M 2014 *Rancang Bangun Alat Bantu Penekuk Begel Cincin Segiempat Untuk Konstruksi Beton (Biaya Produksi)* (Palembang: Politeknik Negeri Sriwijaya)
[5] Prayitno A, Dalil M D M and Yanuar Y 2013 Evaluasi Mutu Produk dari Produk-produk Baja Tulangan Domestik Berdasarkan Konsistensi Kekuatannya *Prosiding SEMIRATA 2013* **1**(1)
[6] Dieter G E and Bacon D J 1986 *Mechanical metallurgy* (New York: McGraw-hill)

IOP Conf. Series: Materials Science and Engineering 1108 (2021) 012007 doi:10.1088/1757-899X/1108/1/012007

Monitoring rainfall intensity and moisture water content using soil column experiment

M Mukhlisin[1]*, B Yunanto[2], A Suharjono[2], M Martono[1], R Apriantoro[2] and R Kusumawardani[3]

[1] Civil Engineering, Politeknik Negeri Semarang, Semarang, Indonesia
[2] Electrical Engineering, Politeknik Negeri Semarang, Semarang, Indonesia
[3] Civil Engineering, Universitas Negeri Semarang, Semarang, Indonesia

*mmukhlis2@gmail.com

Abstract. Indonesia is a tropical country that has two seasons, rainy season and dry season. The wetting and drying of soil due to changing of the seasons caused change on soil properties such as moisture soil water content, effective soil porosity, soil suction etc. During rainy season, rainfall infiltration caused increasing of moisture soil water content in a soil slope, tends to reduce soil shear strength and followed by slope instability. In this soil column experiment was set up with rain gauge station and YL-69 sensor and capacitive soil moisture sensor to monitor rainfall infiltration and soil moisture water content in red clay soil. A wireless sensor network with low power wide area network technology integrated with the cloud to observe rainfall intensity and soil water content in soil column with realtime monitoring. The focused on improving the accuracy of rainfall intensity and soil water content sensors readings using regression methods. The experiment results showed that rainfall intensity was observed accurately while for the YL-69 sensor, the error reading was 5.15% and the error reading capacitive soil moisture sensor was 2.99%.

1. Introduction

Indonesia is a tropical country located in the Ring of Fire region. This makes Indonesia often suffer from disasters such as earthquakes, tsunami, eruptions, floods and landslides. Based on the National Disaster Management Agency data, 1586 disasters have occurred in Indonesia from January to April 2019. Hydrometeorological disaster is a dominant disaster with a percentage of 98%, while the other 2% is caused by geological disasters. Floods and landslides are the most dominant disasters occurred [1].

Natural disasters are unpredictable events that have a large impact. One example of the impact of natural disaster flooding and landslide occurred in Sentani, Papua, March 2019 caused 112 people died, 82 people were lost 965 injured and losses and damages were estimated at 668 billion rupiah [1].

Slope failure or landslides are usually caused by several factors including soil materials, human activities on slope, topography, hillslide development and climatic conditions, vegetation and rainfall. But, in tropical countries, rainfall becomes a major factor in the cause of slope failure [2,3]. During and after rainfall, parts of the rainwater reach the surface of the slope infiltrate to the ground, while others flow on the surface. When water infiltrating into the slope, metric suction is decreasing due to increased soil humidity. Then, due to this, the soil structure becomes transformed and reduces the frictional and cohesive strength between particles [4]. It gives a negative effect on slope stability because of negative pore water pressure between the soil particles and shear strength is decreasing [5–7].

Soil Moisture water content can be measured using several techniques such as Thermo gravimetric, Neutron moisture meter, Time Domain Reflectometry (TDR), Capacitance technique and Frequency Domain Reflectometry (FDR), Resistive sensor, Thermal Dissipation Block and Tensiometer [8]. Before measurement in the field, water infiltration can be simulated using soil column experiment to analyze soil moisture content [9] using tomography techniques to measure soil water content utilizing Electrical Capacitance Volume Tomography (ECVT). Meanwhile, Ibrahim et al., uses a TDR sensor and a tensiometers to measure soil moisture content [10]. Both methods have good accuracy and non-destructive. However, both of them utilize cables for the transmission of data to the local computer used for data acquisition/data logger. If it wants to be applied to real conditions, it will certainly require expensive costs and monitoring cannot be done remotely.

Internet of Thing (IoT) is a breakthrough that allows to conduct rainfall monitoring and soil moisture water content remotely and realtime. Vani and Rao develops prototype to monitor soil moisture using the YL-69 sensor (resistive soil moisture sensor) [11]. Data that is read is sent through a Wifi-based Wireless Sensor Network (WSN) integrated with a cloud server. Similar research is also conducted by Juca utilizing NodeMCU ESP8266 and YL-69 sensors to monitor the soil moisture content that is integrated with Thingspeak Cloud [12]. Meanwhile, utilize a capacitive soil moisture sensor to read soil moisture and NodeMCU ESP8266 is integrated with Thingspeak Cloud for data acquisition [13]. In addition to the development on the WSN and IoT sides, Sudha and Sinha also conducted an evaluation to calculate the accuracy of the YL-69 sensor on clay soil medium with an average measurement error is 53.14% [14].

Devices capable of realtime and cloud-integrated monitoring is still insufficient if not supported with energy-efficient technology. When the sensor is installed in a disaster prone area where the area is difficult to reach by humans and battery replacement is very rare, of course it is a new problem. In addition, the accuracy of the sensor needs to be improved so that slope stability analysis can be done to the maximum. In this study, we proposed device to monitor rainfall and soil moisture water content with IoT Cloud-based Low Power Wide Area Network (LPWAN) technology using LoRa as WSN.

2. Methodology

2.1. System design

The proposed system of this study is a prototype of soil moisture using low cost soil column experiment consisting of node sensors, LoRa Gateway and Cloud Server. Sensor node consists of rain gauge sensor, soil column made of PVC with resistive soil moisture sensor (YL-69) and capacitive soil moisture sensor, data acquisition device and LoRa transceiver (LoRa MiniDev). The sensor node uses energy from a rechargeable battery using a solar cell. Voltage Input used in the sensor node is 5 volt. Meanwhile, the LoRa gateway consists of NodeMCU and LoRa Bee radio transceiver devices connected to the Wireless Router (WLAN / IEEE 802.11 N). System architecture design can be seen on Fig. 1 and Soil column experiment can be seen on Fig. 2.

The network that was built on this study based on research conducted by Suharjono utilizes IEEE 802.11 N network as network backbone with 4G network as redundant on Gateway side [15]. Meanwhile, on the side WSN based LPWAN, LoRa chosen because of the wide coverage area and energy consumption is more efficient. Soil moisture sensors measure soil moisture water content based on the voltage value caused by resistance or capacitance changes in red clay soils. Water is a good electrical conductor. When water is absorbed into the soil, the sensor's soil moisture electrode will be interconnected so that the electric current can flow on the sensor. Then the voltage read by the sensor is sent to the ATMega328p microcontroller embeded on LoRa MiniDev. ATMega328p reads the input voltage value from the sensor using Analog to Digital Converter (ADC) with 8 bit resolution. Then the ADC value is converted to voltage again using (1). Then the value of the voltage is mapped to the scale of 1-10 adjusts to the scale on the 3 way meter to measure soil moisture water content.

$$Voltage = \left(\frac{3.3}{1023}\right) x\ ADC\ Value \tag{1}$$

Data that is read by the rain gauge sensor and soil moisture sensor is then transmitted using LoRa module to LoRa Gateway. Then LoRa Gateway performs a parsing to separate the data. After that, the data is encapsulated into IEEE 802.11 N format and sent to the cloud using the Message Queuing Telemetry Transport (MQTT) protocol over a Wifi network. The MQTT protocol is used on the uplink side because MQTT is a type of data-agnostic protocol where data that is capable of sending is not limited to certain types of data, but all types of data can be transmitted such as binary, text, XML and JSON data. This protocol implements data compression and data reduction techniques to perform energy efficiency [16,17].

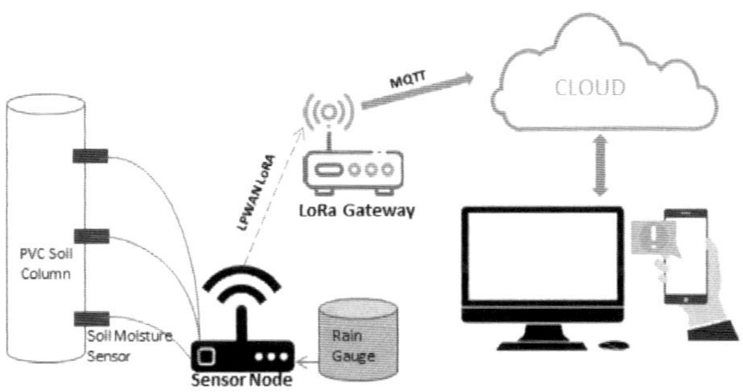

Figure 1. System architecture.

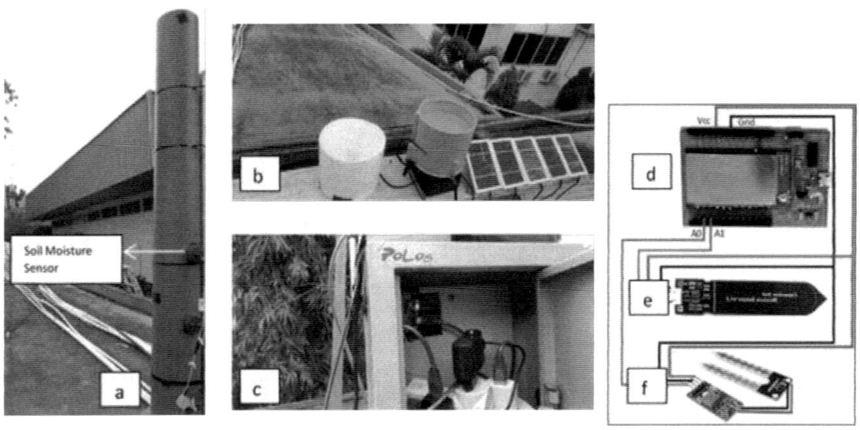

Figure 2. Infiltration column apparatus and instrumentations: (a) soil column, (b) rain gauge sensor, (c) data logger and lora transceiver (sensor node), (d) Lora MiniDev, (e) capacitive soil moisture sensor and (f) resistive soil moisture sensor.

2.2. Testing method

Testing conducted on laboratory environments using artificial rain from showers with constant water discharge. In this study, the measured soil moisture water content is the top layer in the soil column. Resistive and capacitive soil moisture sensor inserted into the soil column along with 3 way meter. Then the soil column was given a rain-made up to 3 way meter indicates a reading ranging from 1 to 10 scales. Each scale reads at 3 way meter, the value of the read voltage of the node sensor will be recorded every 5 second interval. Tests on each scale are done for 5 minutes. The results of the sensor measurements are then compared to the conventional measurements of the 3 way meters using linear regression methods. Error measurement on sensors can be calculated using (2).

$$Error(\%) = \left|\frac{exact - approximate}{exact}\right| x100 \quad (2)$$

3. Results and discussion

3.1. Resistive soil moisture sensor

The soil moisture sensor consists of two probe's pads which are used to measure the Volumetric content of water. The two probe's pads allow the current to pass through the soil and then it gets the resistance value to measure the moisture value. When there is water, the soil will conduct more electricity which means that there will be less resistance. Therefore, the moisture level will be higher, and the output voltage will be lower. Dry soil conducts electricity poorly, so when there is less water, then the soil will conduct less electricity which means that there will be more resistance. Therefore, the moisture level will be lower output voltage will be higher. The Correlation between output voltage (V) and soil water content is shown in Fig. 3. From Fig. 3 a linear equation of average voltage as a function of soil water content by using resistive sensor-based can be obtain as (3),

$$y = -0{,}2789x + 3{,}4412 \quad (3)$$

Where y is average voltage value and x are soil water content. Variable y as the result of conversion from ADC value to a voltage value, and x as soil water content at real conditions (%). From (3), the value of x can be represented as (4). This equation which is used to determine the soil moisture water content in the Arduino program is (4).

$$x = \frac{y - 3{,}4412}{-0{,}2789} \quad (4)$$

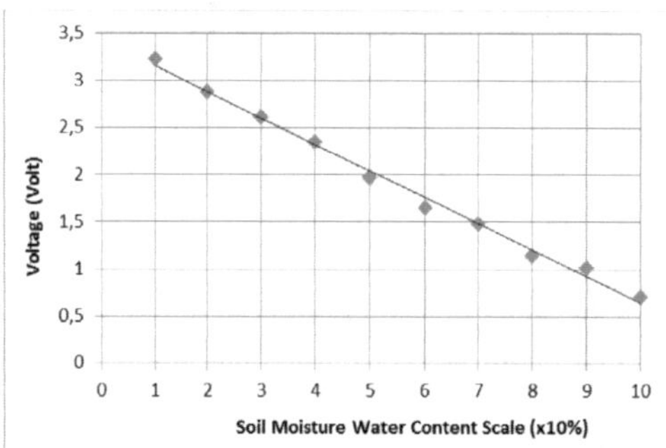

Figure 3. Correlation between output voltage and soil water content in resistive soil moisture sensor.

3.2. Capacitive soil moisture sensor

The electrical component known as a capacitor consist of three pieces. A positive plate, a negative plate and the space in-between the plates, known as the dielectric. The physical form and construction of practical capacitors vary widely, and many capacitor types are in common use. Most capacitors contain at least two electrical conductors often in the form of metallic plates or surfaces separated by a dielectric medium.

A capacitive moisture sensor works by measuring the changes in capacitance caused by the changes in the dielectric. It doesn't measure moisture directly (pure water doesn't conduct electricity well), instead it measures the ions that are dissolved in the moisture. These ions and their concentration can be affected by a number of factors, for example adding fertilizer for instance will decrease the resistance of the soil. Capacitive measuring basically measures the dielectric that is formed by the soil and the water is the most important factor that affects the dielectric. The correlation between output voltage (V) and soil water content is shown in Fig. 4. From Fig. 4 a linear equation of average voltage as a function of soil water content by using capacitive sensor-based can be obtain as (5),

$$y = -0{,}241x + 3{,}5398 \tag{5}$$

Where y is average voltage value and x are soil water content. Variable y as the result of conversion from ADC value to a voltage value, and x as soil water content at real conditions (%). From (5), the value of x can be represented as (6). This equation which is used to determine the soil moisture water content in the Arduino program is (6).

$$x = \frac{y - 3{,}5398}{-0{,}241} \tag{6}$$

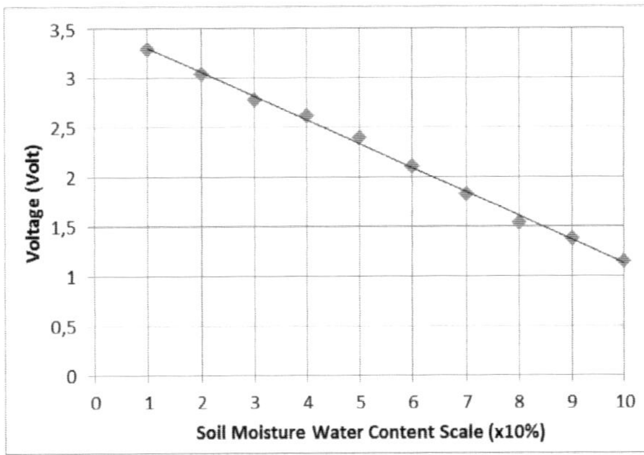

Figure 4. Correlation between output voltage and soil water content in capacitive soil moisture sensor.

Based on the data and equation obtained, the result is then compared to see the level of linearity between resistive-based sensors with capacitive-based sensors. It can be seen in Fig. 5. The comparison results between soil moisture sensor and conventional moisture measurement can be shown in Table 1. From Table 1, the measurement using (4) obtains the average error of measurement of 5.15% and measurement using (6) obtains the average error of measurement of 2.99 %.

Capacitive-based sensor has some advantages. It not only avoids corrosion of the probe but also gives a better reading of the moisture content of the soil as opposed to using a resistive soil moisture sensor. Since the contacts (the plus plate and the minus plate of the capacitor) aren't exposed to the soil, there is no corrosion of the sensor itself. That's why capacitive-based sensors have better accuracy compared to resistive-based sensors. In addition, capacitive-based sensors have more lifespan than resistive-based sensors. Testing errors can be minimized by setting more precise ADC values and soil wetting the red clay on the soil column is done more evenly.

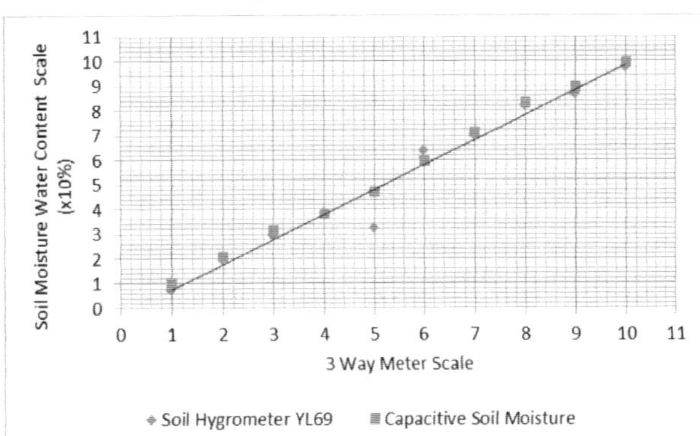

Figure 5. Measurement comparision using (4) on YL-69 and (6) on capacitive soil moisture sensor.

Table 1. Measurement of soil water content using (4) and (6).

3 Way Meter (x10%)	Resistive Soil Moisture (x10%)	Error (%)	Average Error (%)	Capacitive Soil Moisture (x10%)	Error (%)	Average Error (%)
1	0.75013	24.98		1.03473	3.47	
2	2.01836	0.92		2.0694	3.47	
3	2.94404	1.87		3.14734	4.91	
4	3.88996	2.75		3.81169	4.71	
5	3.25207	5.04	5.15	4.71028	5.79	2.99
6	6.41216	6.87		5.94706	0.88	
7	7.03307	0.47		7.10063	1.43	
8	8.23807	2.98		8.3093	3.87	
9	8.68549	3.5		8.95982	0.45	
10	9.78273	2.17		9.90673	0.94	

3.3. Website-based and Android application monitoring system

Soil moisture water content data that is read by the sensor is then transmitted to the cloud server and it's displayed using widgets on the website and the android application, so that it's easily readable by users. It can be seen in Fig. 6. In addition to a website-based monitoring system, data can also be monitored using an android application as shown in Fig. 7. It aims to facilitate the user when they are in a difficult place to use a laptop. By using Android application, monitoring can be done anywhere quickly and efficiently.

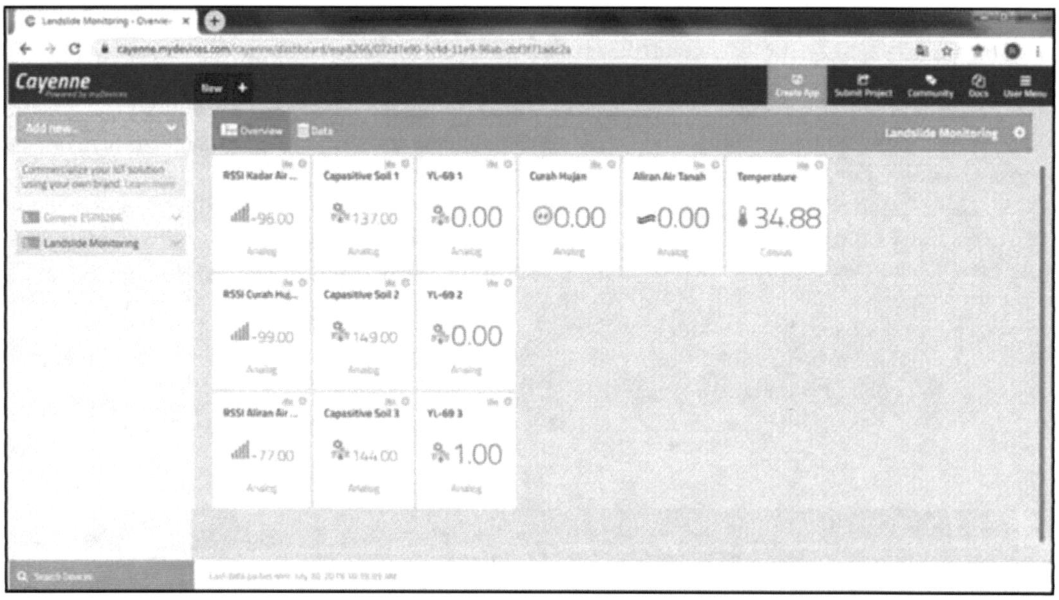

Figure 6. Web-based monitoring system.

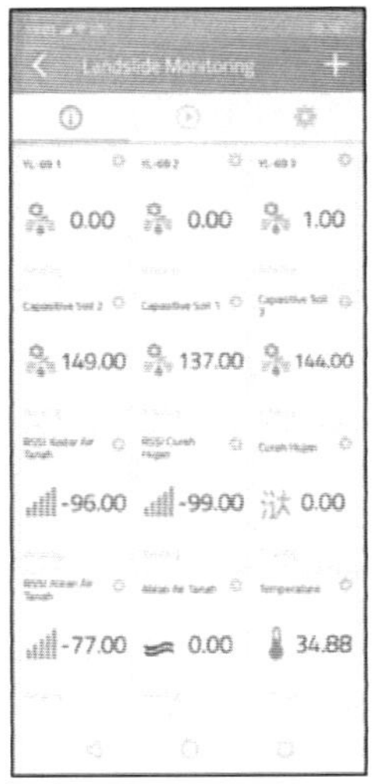

Figure 7. Android application monitoring system.

4. Conclusion

In this paper, the accuracy of resistive dan capacitive soil moisture sensor have been successfully improved. Both of data were compared and sent to the cloud server. Based on data that has been obtained from experiments, the capacitive-based sensor has a better level of accuracy in the soil test medium of red clay using soil column experiment. Corrosion of sensor probe greatly affects the level of accuracy of soil moisture water content measurement. Data's transmitted using LPWAN and Cloud IoT Network can run well. Likewise, with website-based and android application monitoring system, both of them can displayed information according to the data's that has been transmitted.

References

[1] CNN Indoinesia 2019 *Bencana Januari-April 2019 lebih banyak dibanding tahun lalu* [Online] Retrieved from: www.cnnindonesia.com

[2] Bujang H B, Ali F H and Low T 2006 Water infiltration characteristics of unsaturated soil slope and its effect on suction and stability *Geotech. Geol. Eng.* **24** 1293--1306

[3] Gofar N, Lee L M and Kassim A 2008 Instrumented soil column model for rainfall infiltration study *Geotropika, International Conference of Geotechnical & Highway Engineering*

[4] Gofar N, Lee M L and Asof M 2006 Transient seepage and slope stability analysis for rainfall-induced landslide: A case study *Malaysian J. Civ. Eng.* **18** 1–13

[5] Gasmo J, Rahardjo H and Leong E C 2000 Infiltration effects on stability of a residual soil slope *Comput. Geotech.* **26** 145--165

[6] Mukhlisin M, Ramadhan B M and El-Shafie A 2011 Effect of soil water retention model on slope stability analysis *Int. J. Phys. Sci.* **6** 4629–35

[7] Mukhlisin M and Taha M R 2012 Numerical model of antecedent rainfall effect on slope stability at a hillslope of weathered granitic soil formation *J. Geol. Soc. India* **79** 525--531

[8] Lekshmi S, Singh D N, Baghini M S and Shojaei M 2014 A critical review of soil moisture measurement. Measurement *Measurement* **54** 92–105

[9] Mukhlisin M, Saputra A, El-Shafie A and Taha M R 2012 Measurement of dynamic soil water content based on electrochemical capacitance tomography *Int. J. Electrochem. Sci.* **7** 5457–66

[10] Ibrahim A, Mukhlisin M and Jaafar O 2014 Effect of rainfall infiltration into unsaturated soil using soil column Soil Column *Sains Malaysiana* **43** 1477–84

[11] Vani P D and Rao K R 2016 Measurement and monitoring of soil moisture using cloud IoT and android system *Indian J. Sci. Technol.* **9** 1–8

[12] Juca S C S, Rabelo S L, Gonçalves D L C, Silva V F, Pereira R I S and da Silva S A 2018 Construction of soil moisture and irrigation IoT monitoring system using Project Based Learning *Int. J. Innov. Educ. Res.* **6** 99–111

[13] Nagahage E A A D, Nagahage, I S P and Fujino T 2019 Calibration and Validation of a Low-Cost Capacitive Moisture Sensor to Integrate the Automated Soil Moisture Monitoring System *Agriculture* **9** 141

[14] Sudha L K and Sinha A 2015 Effect of moisture content in the sandy, clay and loamy soil using humidity sensor YL-69 *Int. J. Sci. Res.* **6** 107–10

[15] Suharjono A, Apriantoro R, Mukhlisin M, Anif M, Hidayat S S, Setiawan T A and Kadiran S A 2018 Network Design and Performance Evaluation of MQTT Based HetNet Using LoRa Network and IEEE 802.11 for Internet of Things *6th International Technical Conference*

[16] Pal S and Ghosh, Sourav Bhattacharya S 2017 Study and implementation of environment monitoring system based on MQTT *Environ. Earth Sci. Res. J.* **4** 23–8

[17] Ali A, Shah G A and Arshad J 2016 Energy efficient techniques for M2M communication: A survey *J. Netw. Comput. Appl.* **68** 42--55

Emission comparison of short lived climate forcers over long term greenhouse gases from domestic and transport sector in Semarang city

H Huboyo*, B Samadiku and O Manullang

Environmental Engineering Department, Diponegoro University, Semarang, Indonesia

*huboyo@undip.ac.id

Abstract. Short-lived climate forcers (SLCF) controls are important because of their shorter impact and more immediate benefits in the short term than carbon dioxide controls that have a 100-year time horizon. In this study, we estimated the emission of SLCF from domestic and transport sector in Semarang city using emission factor from IPCC and CORINAIR. We also predicted the emission dispersion using AEMOD View. The total consumption of LPG domestic sector in Semarang city is 48,532 tons in 2017 and produces total SCLF emissions 0.537 tons/year and 11,538 tons/year for BC and CH_4 respectively. This corresponds to CO_2 emissions of 145,611 tons/year and 230.76 tons/year of N_2O emissions. While from transport sector, the motorcycle has the largest VKT i.e 18.282.061.600 km/year in 2017. The emission, based on VKT data combined with active vehicles data, reach 178.23 tons/year, 802.52 tons/year for BC and CH_4 respectively. Transport sector produces total CO_2 emissions almost 15 times to that of domestic sector. Based on emission model, the emission distribution was concentrated in the districts located in the city center. The pollutant was accumulated in the Northwest, city center and Southeast of the city.

1. Introduction

Anthropogenic climate change is largely driven by human induced changes in the composition of the atmosphere, including long-lived GHGs (that have lifetimes of approximately eight years or more) and short-lived climate pollutants removed from the atmosphere or transformed to CO_2 relatively quickly (i.e. a few weeks for BC to about 12 years for methane) [1]. Some short-lived climate pollutants such as black carbon (BC), tropospheric ozone (O_3), and methane (CH_4), are defined as gases and particles that have relatively short atmospheric lifetimes with significant warming impacts on the climate. short-lived climate forcers (SLCFs) have detrimental impacts on air quality, directly or via formation of secondary pollutants [2]. BC fluctuation, for example, could effect to chemical evolution and lead to disturbing Aerosol Optical Depth beyond the sources [3].

Climate change control through the control of greenhouse gas emissions included in the Indonesia's priorities in the next few years. Reductions in SLCFs have the potential to decrease the rate and degree of warming in the next few decades, with SLCF mitigation having a rapid effect on temperature [4]. To reduce SLCF precursor i.e. tropospheric ozon by reducing NOx and NMVOC, while to reduce the global radiative forcing, lessen CH4 emission could be an efficient way [5]. The relation of SLCF and CO2 could be approached by analyzing global-mean surface temperature with CO2, CH4, HFC, related-BC and SO2 mitigation [6]. The BC concentration also correlate well with

PM2.5, The ratio of BC/PM2.5 generally has consistent average of 5.29% during observation in China [7]. This observation indicates that PM2.5 measurement is important to know the BC concentration.

Many important sources emit more than one short-lived greenhouse pollutant, therefore, control of this SLCF is difficult e.g., black carbon and organic carbon from combustion of biomass, coal, and diesel fuel. The needs on emission inventory as stated by the Ministry of Environment and Forestry of Indonesia was to determine the comprehensive amount of pollutants from the source within areas and periods of time, with the ultimate objective is to enable the Municipal Government in Indonesia to make emission inventory result as a baseline for policy making and urban air pollution control strategy.

In this study we estimate the emission load of SLCFs from domestic and transport sector in Semarang city. We also compare them with CO2 and N2O emission from the same sources. The SLCFs we investigate comprise of methane (CH4) and black carbon (BC). We predicted the hypothetical dispersion of these emissions using AERMOD View.

2. Methodology

2.1. Estimate the emission load from domestic sector

To calculate the emission load from household activities, we estimate from LPG consumption in the household. Calculation using Tier 1 method, adopting from IPCC using following equation [8]:

$$\text{Emission} = \sum_a [Fuel\ a\ x\ EF\ a]$$

where, Emission = Specific pollutant emission load (kg), Fuel a = fuel consumption of type a (GJ), EF a = emission factor (g / GJ).

We did survey (randomly) to 100 households consist of poor households, medium class households and affluent households to know the average LPG consumption based on household categories.

2.2. Estimate the emission load from transport sector

To calculate the emission from transportation, we used Tier-2 method i.e. using vehicle kilometer traveled-VKT (vehicle kilometer traveled-VKT). The term VKT is intended to know the mileage of the vehicle in a given time (day/ week/month/year). The emissions are treated as area sources based on district. Total VKT of mobile source is the number of multiplication of vehicle population and average mileage per year for each vehicle category/sub-category.

The equations for estimating the total VKT of a mobile source are:

$$\mathbf{VKTb,c} = \mathit{VKTb,c}\ \text{odo}\ .\ \text{Nb,c}$$

Where: *VKTb*,c odo = VKT vehicle category b and fuel c based on odometer survey (km/year); **VKTb,c** = VKT of all motor vehicles category b using fuel c (km/year); Nb,c = number of motor vehicle category b using fuel c

2.3. Emission factor

To calculate the emission from domestic sector, we used emission factor from CORINAIR [9] for each parameter as follows: BC: 11 mg/kg, CO_2: 63.1 g/MJ, CH_4: 5 g/MJ, N_2O: 0.1 g/MJ.

The emission factor for transport sector we use and adopt from Ministry of Environment and Forestry and EPA guideline as follows:

Table 1. Emission factor used in this study.

Vehicle Types	Parameters			
	CO2 (g/kg)	CH4	N2O	BC (g/kg)
Motorcycle	3.180	0,26	0,002	0.0510
Gasoline Personal Car	3.180	0,07	0,005	0.2800
Diesel Personal Car	3.172	0,01	0,014	0.7900
Bus	3.172	0,06	0,031	0.4000
Truck	3.172	0,01	0,031	0.6400

3. Results and discussion

Based on statistical data in Semarang City in 2017, the population of the city reaches 1,701,172 inhabitants. The population growth rate during 2010-2016 was 10.85% due to rapid urbanization. The average city density of Semarang was 4,628 residents/sqkm. Based on government classification, approximately 9% of the population of Semarang is categorized poor people, the middle class is around 17-43% while the affluent people reaches 11%. Our survey confirmed the poor household generally has less than 1 million rupiahs/month income, while affluent households collect more than 4 million rupiahs/month.

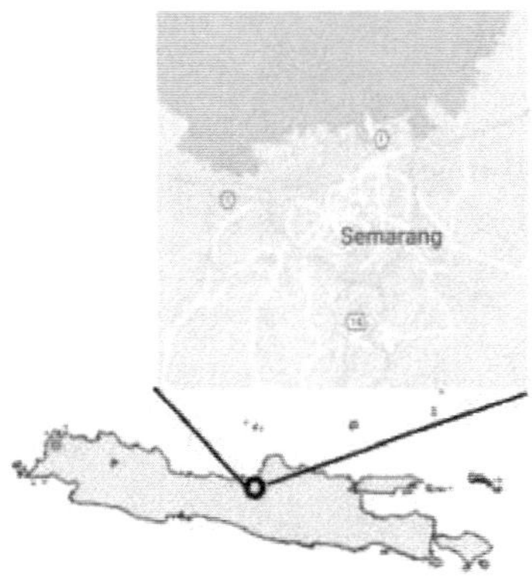

Figure 1. Semarang city map.

3.1. Domestic activities and emission load

Based on our survey, the average consumption of domestic LPG for affluent households is 9.92 kg/month/households, while for middle class household is about is 9.34 kg/month/household. Last, for poor people is 5.83 kg/month/household. This shows that for the poor, they less consume domestic LPG. However, the number of household member also affects the LPG consumption. The average cooking duration was dominantly at 2-3 hours/day for all household categories. The affluent households usually use 12 kg LPG cylinder, while for the poor generally use 3 kg LPG cylinder. While the middle class shows combination of using LPG 12 kg and 3 kg cylinder.

Based on data from the Ministry of Energy and Mineral Resources (2018), the city of Semarang has 52 LPG cylinder distributors and based on our observation it was found that approximately 200 cylinders were distributed daily. If we assume that each distributor distributes the same number of LPG, then based on top-down calculation, the amount of LPG distributed in 2017 was 3,796,000 cylinders. It was quite different if we calculate based on bottom-up approach which shows 2,734,172 cylinders. Several distributors sold the LPG to the Semarang periphery inhabitants that might be live administratively to neighboring cities such as Demak, Kendal and Semarang Regency.

Our estimation, the total consumption of LPG domestic sector in Semarang City is 48,532 tons in 2017 and produces total SCLF emissions 0.537 tons/year and 11,538 tons/year and for BC and CH_4

respectively. This corresponds to CO_2 emissions of 145,611 tons/year and N_2O emission of 230.76 tons/year. The distribution of emission loads for each district (represented by CO_2) is shown in Figure 2.

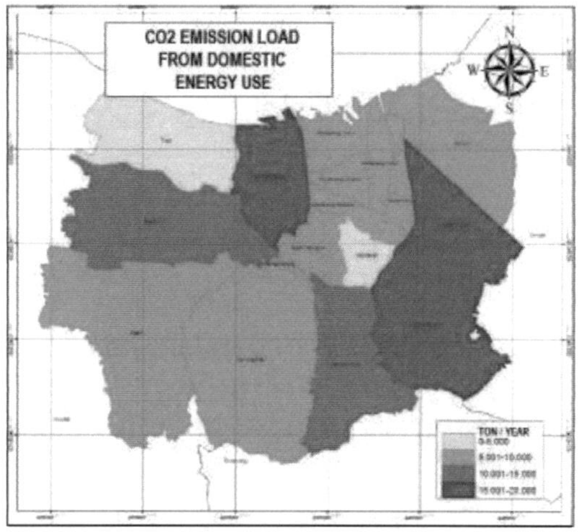

Figure 2. CO_2 emission load from domestic sector.

3.2. Domestic emission dispersion

Emission dispersion estimation was simulated using AERMOD View program with ISCST3 modes. We simulated the model input derived from standardized gas stove parameters (emission rate, height of emissions) for each household using area sources category. The average wind velocity based on meteorological data provided by Meteorological and Geophysics Agency in 2017 is 1.81 m/s with the highest wind speed is 11.10 m / s and the lowest wind speed is less than 0.5 m/s. The dominant wind direction of Semarang city in 2017 comes from the northwest and southeast direction as depicted in the Figure 3.

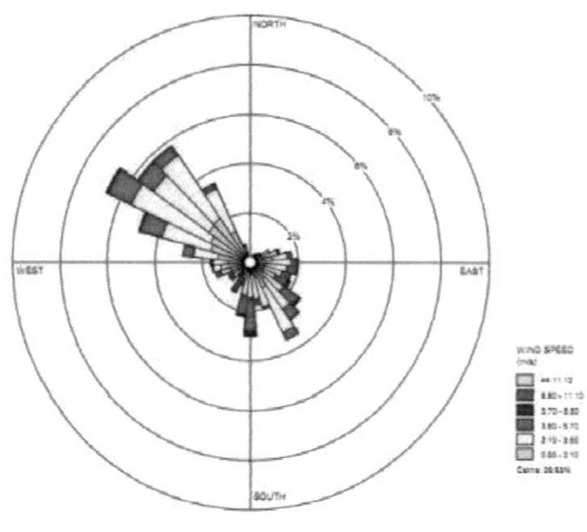

Figure 3. Prevailing wind at Semarang city throughout the year.

The simulation results of the distribution emission from domestic sector (represented by CO_2 emission) is shown in the Figure 4 below.

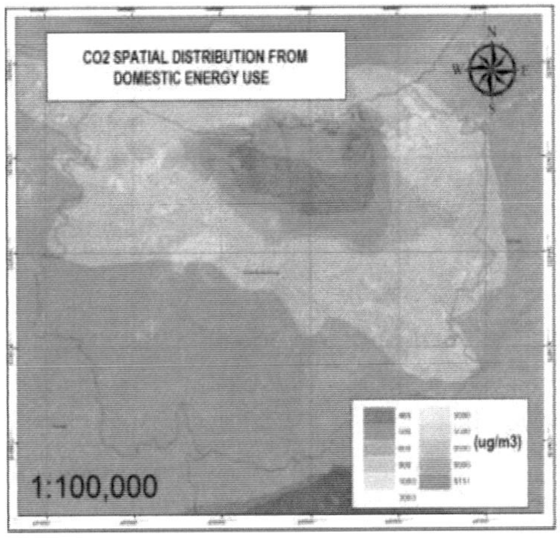

Figure 4. CO_2 emission dispersion from domestic sector.

From the above figure, the emission distribution was concentrated in the districts lied in the city center. The pollutants were accumulated in the northwest, at the city center and southeast of the city. Thus, if we compare Figure 2 and Figure 4, it be concluded that higher emission in the Northwestern and Southeastern of the city will accumulate in the city center. This might be caused, not only by the prevailing wind which come from Northwest and Southeast, also caused by the land terrain in the southern of the city which has higher elevation than in the city center.

3.3. Transport activities and emission load

In this study, we focus on transportation on road transportation i.e. from highways and road networks. We collected current vehicles data from Local Income and Management Agency at province level. These vehicles data reflect the active vehicles that still in use. Active vehicles here are vehicles that are still registered by paying taxes therefore are considered still actively producing emissions on the road.

Table 2. Number of active vehicles in Semarang city.

Year	Vehicle types			
	PC	*Truck*	*Bus*	*MC*
2012	156,614	56,666	2,345	822,700
2013	136,191	52,289	1,917	696,086
2014	141,669	49,748	1,675	661,180
2015	178,542	65,825	2,394	1,067,815
2016	176,964	53,788	1,728	728,824
2017	226,073	78,448	3,271	1,252,196

PC: Personal Car, MC: Motorcycle

Based on table above, the vehicle in Semarang City is dominated by two-wheeled vehicles or motorcycles that are 80% in the proportion followed by passenger cars by 15%.

Transport sector emissions calculations are calculated based on the mileage of each vehicle type or Vehicle Kilometer Traveled (VKT). In this study the type of vehicle is divided into trucks, buses, motorcycles, passenger cars fueled by gasoline, and diesel-fueled passenger cars. The fraction between gasoline and diesel passenger cars in Semarang City based on previous research was equal to 97% and 3% [10].

In this study the VKT data was derived from odometer readings observation for each vehicle type using the results of previous research that can be seen in following table 3.

Table 3. Vehicles VKT based on vehicles types.

Vehicles Types	Number (units)	Average of VKT (km/year)
Bus	3,271	43,435 [10]
Truck	78,448	10,115 [11]
Gasoline Personal Car	219,291	15,330 [10]
Diesel Personal Car	6,782	15,330 [10]
Motorcycle	1,252,196	14,600 [10]

Based on above table, the motorcycle has the largest VKT i.e. 18.282.061.600 km/year in 2017. The emission, based on VKT data combined with active vehicles data, reach 178.23 tons/year, 802.52 tons/year and 88.78 tons/year for BC, CH_4 and N_2O respectively. This transport sector produces total CO_2 emissions of 2,123,302 tons/year (almost 15 times to that of domestic sector). The distribution of emission loads for each district (represented by CO2) is shown in Figure 5.

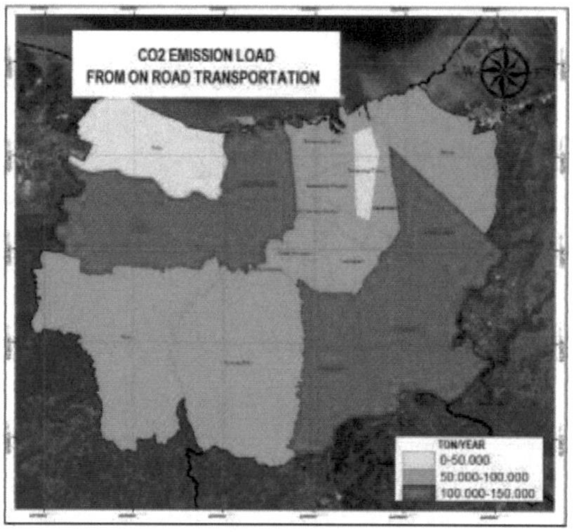

Figure 5. CO_2 emission load from on road transport sector.

3.4. Transport emission dispersion

Likewise, in domestic emission, the estimation of emission dispersion from transport sector was simulated using AERMOD View program with ISCST3 modes. We modified the model input to accommodate for transport emission. We still use historical meteorological data collected from Meteorological and Geophysics Agency in 2017. Likewise, from domestic sector, the emission distribution was concentrated in the districts located in the city center. The pollutants spread from transport sector was more outward rather than those from domestic sector. However, it was still accumulated in the Northwest, city center and Southeast of the city.

The simulation results of the distribution emission from domestic sector (represented by CO_2 emission) is shown in the Figure 6.

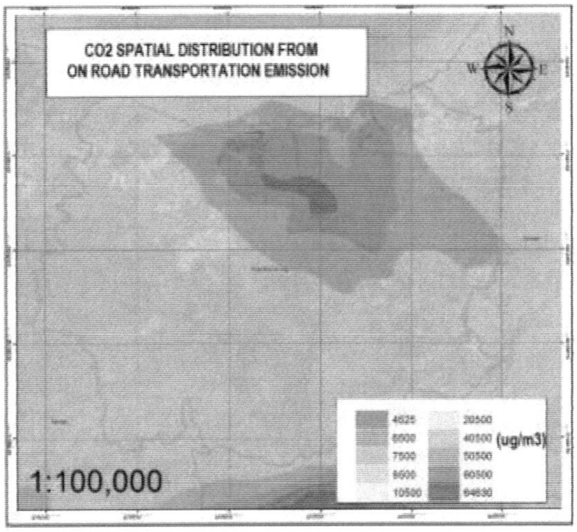

Figure 6. CO_2 emission dispersion from on road transport sector.

3.5. Comparing with other cities

We compare the transport emission (represented by CO_2) in Semarang city with those in other cities. Based on Fig. 7, the population is directly proportional to the amount of GHG's produced. Denpasar City which has a population of 880,600 in 2015 produces CO_2 pollutants as much as 984,280 tons/year. While the city of Semarang which has a population of 1,729,428 people in 2017 produces CO2 emission as much as 2,123,302 tons/year. The Jakarta city which has the largest number of city population i.e. 9,057,632 inhabitants in the year 2009 produces CO_2 emission of 11,744,480 tons /year.

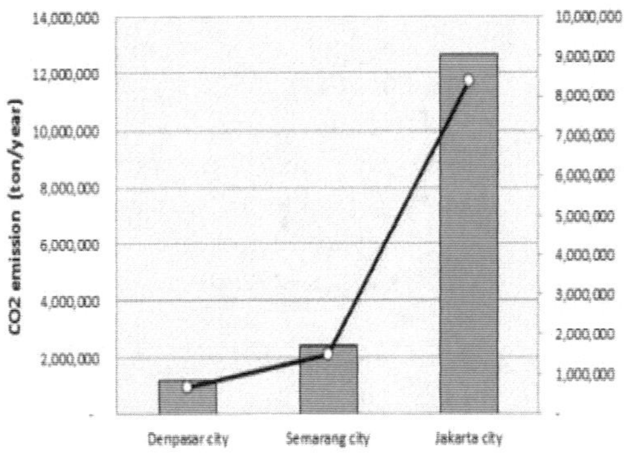

Figure 7. Comparing CO_2 emission load from on road transport sector in other cities.

4. Conclusion

Short-lived climate forcers (SLCF) controls are important because of their shorter impact and has immediate benefits in the short-term period. Based on this study results, the total consumption of LPG domestic sector in Semarang City is 48,532 tons in 2017 and produces total SCLF emissions 0.537 tons/year, 11,538 tons/year and 230.76 tons/year for BC, CH_4 and N_2O respectively. This is corresponding to CO_2 emissions of 145,611 tons/year. While from transport sector, the motorcycle has the largest VKT i.e. 18.282.061.600 km/year in 2017. The emission, based on VKT data combined with active vehicles data, reach 178.23 tons/year, 802.52 tons/year and 88.78 tons/year for BC, CH_4 and N_2O respectively. Transport sector produces total CO2 emissions almost 15 times to that of domestic sector. Based on emission model, the emission distribution was concentrated in the districts located in the city center. The pollutants were accumulated in the Northwest, city center and Southeast of the city. Comparing to the other cities, the emission of CO_2 from transport sector is directly proportional to the number of the population.

Acknowledgment
This study was supported by Directorate General of Research and Development Reinforcement, Ministry of Research, Technology and Higher Education, Republic of Indonesia (Grant Number.: 101-75/UN7.P4.3/PP/2018). We also appreciate the data collection by Indri Marlena, and manuscript preparation by Dian Pertiwi. We also thanks to PT. Stadia Reka for their permission for using the AERMOD View software.

References

[1] IPIECA.org. 2014 *Understanding short-lived climate forcers* (United Kingdom: Blackfriars Road London SE1 8NL)
[2] Kirtman B, Power S B, Adedoyin A J, Boer G J, Bojariu R, Camilloni I and Prather M 2013 *Near-term climate change: projections and predictability*
[3] Sadiq M, Tao W, Liu J and Tao S 2015 Air quality and climate responses to anthropogenic black carbon emission changes from East Asia, North America and Europe *Atmospheric Environment* **120** 262-276
[4] Myhre G, Samset B H, Schulz M, Balkanski Y, Bauer S, Berntsen T K and Easter R C 2013 Radiative forcing of the direct aerosol effect from AeroCom Phase II simulations *Atmospheric Chemistry and Physics* **13** 1853-1877
[5] Akimoto H, Kurokawa J I, Sudo K, Nagashima T, Takemura T, Klimont Z and Suzuki K 2015 SLCP co-control approach in East Asia: Tropospheric ozone reduction strategy by simultaneous reduction of NOx/NMVOC and methane *Atmospheric Environment* **122** 588-595
[6] Rogelj J, Schaeffer M, Meinshausen M, Shindell D T, Hare W, Klimont Z and Schellnhuber H J 2014 Disentangling the effects of CO2 and short-lived climate forcer mitigation *Proceedings of the National Academy of Sciences* **111**(46) 16325-16330
[7] Zha S, Cheng T, Tao J, Zhang R, Chen J, Zhang Y and Du J 2014 Characteristics and relevant remote sources of black carbon aerosol in Shanghai *Atmospheric research* **135** 159-171
[8] IPCC (Intergovernmental Panel on Climate Change) 2006 Guidelines for national greenhouse gas inventories *Task force on national greenhouse gas inventories (TFI)*
[9] CORINAIR 2009 *Atmospheric Emission Inventory Guidebook 3th Edition* (European Environment Agency)
[10] Buanawati T 2017 *Estimation of Conventional Air Pollutants (SOx, NOx, CO, AND PM) From Personal Vehicles Based On International Vehicle Emission (IVE) Method in Semarang City Road* (Semarang: Diponegoro University)
[11] Huboyo H S, Sutrisno E, Sutrisno A M and Tiarani V L 2018 Spatial Distribution of Conventional Air Pollutant and GHGs from Land Transportation in Two Developing Cities and Main Co-Benefit Actions For Reducing It In *E3S Web of Conferences* **31** 09005

IOP Conf. Series: Materials Science and Engineering 1108 (2021) 012009 doi:10.1088/1757-899X/1108/1/012009

Mapping of flood-prone-zone and puddle-prone-zone in Cimahi city, West Java province

I Karnisah* and Y Astor

Politeknik Negeri Bandung (POLBAN), Bandung, Indonesia

*iinkarnisah@polban.ac.id

Abstract. Climate change and advanced urbanization level followed by the changing function of an area affect to the amount of water absorption area. It causes puddle and flood in city, include Cimahi, West Java, which is in the last 10 years has been surrounded by flood and puddle. To avoid the flood case in Cimahi, the re-identification and flood-factor inventory are needed. The identification and inventory are using the simplest method (literary study from any source), start from discussing session with Cimahi government, interviewing and filling questioner to the citizen who live in flood and puddle area, measuring river and drainage profile, counting the capacity of river and drainage, until simulating the 2-dimension and 3-dimension of flood and puddle. The result of the identification and inventory is shown in Flood-Prone-Zone Map and Puddle-Prone-Zone Map. According to Cimahi's Flood-Prone-Zone Map, it shows that the flood-safe area is about 110,747 ha (2,70%), the not-flood-prone area is about 494,233 ha (12,05%), the flood-prone area is about 2016,474ha (49,18%), and the very-flood-prone area is about 1478,799 ha (36,07%). According to Cimahi's Puddle-Prone-Zone Map, it shows that the puddle-safe area is about 128,616 ha (3,14%), the not-puddle-prone area is about 551,246 ha (13,44%), the puddle prone area is about 2702,826 ha (65,92%), and the very-puddle prone area is about 717,556 ha (17,50%). The Flood-Prone-Zone Map and the Puddle-Prone-Zone Map can be used as a reference for the Cimahi Government in formulating the policy of regional development.

1. Introduction

Cimahi City West Java Province is one of the urban areas that often occur flood. Geographically Cimahi City is basin valley that slopes to the south with altitude in the north 900-1025 meters above sea level (Cipageran Urban District North Cimahi). This valley is the slope of Mount Burangrang and Mount Tangkuban Perahu, while the altitude in the south is about 850-900 meters above sea level (partly in Cibeber and Leuwi Gajah Urban District South Cimahi) leading to the Citarum River [1]. This topography condition that causes South Cimahi more often affected by floods compared with North Cimahi.

Based on the location of administration in Figure 1, Cimahi City is a buffer city of Bandung. Cimahi city is bordered directly with West Bandung regency and Bandung City [2]. The flood that occurred in South Cimahi is a more complex problem (Figure 2) than in North and Central Cimahi, it is caused by the flood location located in the border area so to solve the flood problem it is need to cross local government involving West Bandung Regency, Bandung and West Java Provincial Government [3].

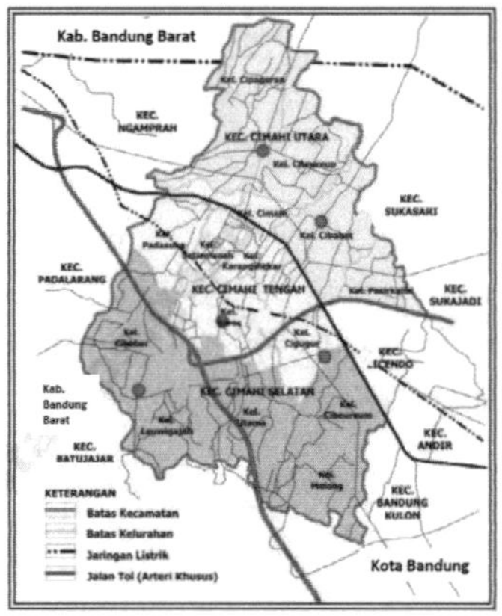

Figure 1. Administrative boundaries of Cimahi City.

In addition, the population in South Cimahi is higher than the population in North and Central Cimahi [4] (Table 1) causing the need for residential and garbage produced increase. This condition causes the Watershed (DAS) in South Cimahi shallow and no longer functioning properly.

Figure 2. Flood in Melong Urban district, South Cimahi.

Table 1. Population of Cimahi city year of 2010-2014.

District	Population (Male + Female)				
	2010	**2011**	**2012**	**2013**	**2014**
South Cimahi	230.623	236.778	241.374	245.989	250.337
Central Cimahi	163.070	165.653	167.090	168.494	169.677
North Cimahi	147.484	151.162	153.833	156.508	159.001
Total	541.177	553.593	562.297	570.991	579.015

Many efforts have been done by the Cimahi City Government independently or in cooperation with the Government of Bandung, West Bandung Regency and West Java Provincial to handle the problem of this flood, which is by doing short and medium term solutions, including the normalization of the river and non-structural solution through community empowerment. Meanwhile, to overcome the problem of flood in the long term, it is necessary to conduct a comprehensive study related to the management of water resources in an integrated manner through flood engineering and management.

The definition of city flood engineering and management is the application of scientific and mathematics principles for the purpose of practical (engineering) integrated process to achieve the goal of overcoming flood problem systematically, effectively and efficiently in cities or urban areas [5]. Referring to the definition, that to conduct a comprehensive and integrated process is needed an accurate and actual data and information as input to build a flood management model in accordance with the characteristics of Cimahi City. Therefore, it is necessary a flood-prone-zone and puddle-prone-zone mapping model that can display the data as a whole and actual, which able to visualize flood phenomenon that occurred in Cimahi City, West Java Province.

This research builds zonation map model of flood prone-zone and puddle-prone-zone in an effort to overcome the problem of flood and puddle in Cimahi City, West Java Province. Zonation map model of flood-prone-zone and puddle-prone-zone is built based on a comprehensive study on the problem of flood and puddle in the City of Cimahi using textual and geographical data conditions of the Cimahi City.

2. Research methodology

Figure 3 is the research methodology that used in building a Model of flood-prone-zone and puddle-prone zone in Cimahi City:

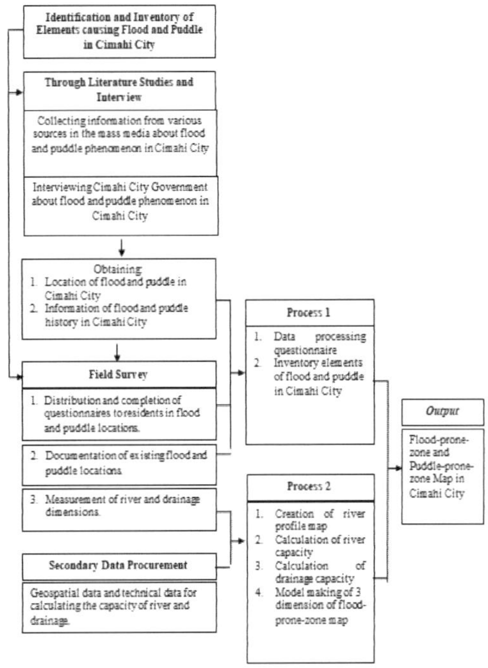

Figure 3. Flowchart of research methodology.

Identify the elements that causing flood and puddle, also impacts and losses caused by flood and puddle that occur in the City of Cimahi. Identification is done through literature study to find information of flood and puddle that ever happened in Cimahi City so that obtained information of location of flood and puddle, time of incident, and reason of flood and puddle occur. Information searching is done through printed mass media, electronic, and other references such as official government accounts and agencies in Cimahi City, as well as amateur photos and videos from Cimahi City residents displayed through social media. In addition to the literature study, also conducted presentations and discussions with the Cimahi City Government discussing the definition of flood and puddle, as well as effort to overcome the flood and puddle that has been done by the Cimahi City Government. Flood is defined as the condition of the increasing water volume in the river that causes the water which is not accommodated and then it flows into the surrounding land. Meanwhile the puddle is a condition of the water that is not accommodated in the drainage channel so it flows out the drainage channel or wets the surrounding area. Based on the results of identification from various sources, it obtains 6 points of flood and 27 points of puddles in Cimahi City.

IOP Conf. Series: Materials Science and Engineering 1108 (2021) 012009 doi:10.1088/1757-899X/1108/1/012009

Inventory of necessary data, including prevailing laws and local government policies related to flood prevention in Cimahi City. Primary data obtain directly from the field such as measurement data and survey data. Secondary data are obtained from related agencies such as rainfall data, Satellite Imagery, Earth Rising Map, Digital Elevation Model (DEM), Watershed Map, River Network Map, Land Use Map, and other data.

Survey at flood-prone-zone and puddle-prone-zone. Measure the situation and cross-section of the river to obtain information of existing conditions (Figure 4). Observe the high frequent of flood and puddle that often occur (Figure 5) and direct interviews and questionnaires that has been given to flood victims and local government.

Figure 4. Measuring the cross-section of Cilember river.

Figure 5. Measurement of drainage channel dimension.

Based on observation and measurement of the 4 rivers in Cimahi, which are Cilember River, Cisangkan River, Cibereum River and Cimahi River, the cause of flood in Cimahi City is dominant due to the narrowing and shallowing of river channel [6]. Based on the result of observation and measurement of drainage channel cross-section in the location of puddle in Cimahi City, the cause of puddle in Cimahi City is dominant due to the condition of drainage channel that covered by garbage and the drainage channel dimension that is not in accordance with guidelines and planning.

Calculation of river capacity (Figure 6) and drainage (Figure 7) at flood and puddle area in Cimahi City. Calculations are performed using HEC RAS software for modelling the flow.

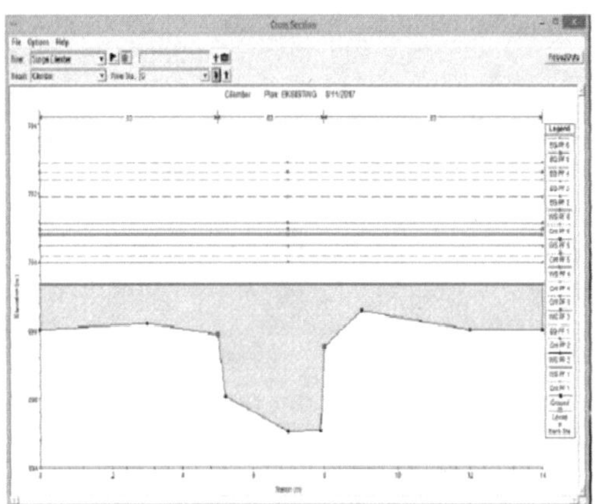

Figure 6. Calculation of Cilember river capacity.

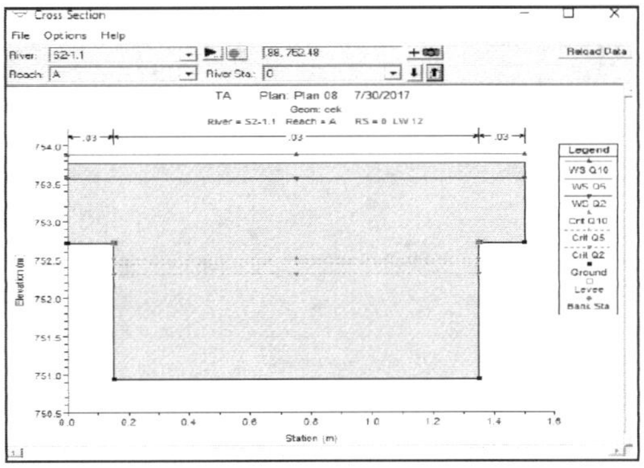

Figure 7. Calculation of drainage channel capacity.

Flood modeling in 3 dimensions (Figure 8). Conducted to get a visualization of the flood and puddle phenomenon in the Cimahi City, so the location that affected by flood is known and the area of flood can be calculated. The modeling results are obtained information that the change of flood and puddle high in Cimahi City does not significantly affect the area of flood and puddle, it because Cimahi City is basin valley.

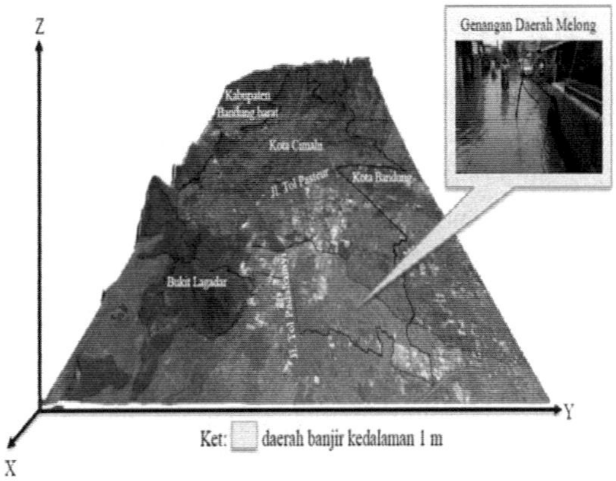

Figure 8. Dimension visualization of flood area.

Making the Map of Flood-prone-zone and Puddle prone-zone in Cimahi City. The parameters, that is used in the manufacture, are slope, land use, rainfall, soil type, river buffer and puddle point. Each parameter will be given the weight, the greater the effect of parameters on flood and puddle the higher weights used. Each parameter consists of classes, each class has given a score based on the class's influence on the flood and puddle. Flow chart that shows the Map of Flood-prone zone and Puddle-prone-zone making process in Cimahi City is showed in attachment section.

2.1. Making of slope map

The slope is a factor that has a great effect on the flood prone. The more sloping the slope the runoff flow will be slower that can cause flood prone. Conversely, the steeper the slope the runoff will be faster, so the water that falls to the ground will flow quickly. Based on RTRW Cimahi City Year of 2012-2032, the most slope in Cimahi City is at 0-8% (Table 2) [1].

Table 2. Slope in Cimahi city based on districts.

District	Slope (Ha)				Total
	0-8%	8-15%	15-25%	25-40%	
South Cimahi	1483,07	13,60	68,69	68,69	1.654,30
Central Cimahi	977,44	-	6,44	5,74	989,62
North Cimahi	1288,71	22,30	80,11	17,84	1.408,96
Total	3749,22	35,90	175,49	92,26	4.052,88

Source: RTWR Cimahi city year of 2012-2032

Making of Slope Map using Digital Elevation Model (DEM) from Shuttle Radar Topography Mission (SRTM) 30 m accuracy. Scores used in each class show in Table 3.

Table 3. Score of land slope classes.

Num.	Class	Score
1	Flat (0-3%)	9
2	Ramps (3-8%)	7
3	Slant (8-15%)	5
4	Steep (15-30%)	3
5	Very steep (30-45%)	1

2.2. Making of land use map

Land use is classified into two which are built and non-built land use. Land use consists of industrial facilities, health, commercial, sports, education, communications, offices and community services, defense and security, places of worship, tourist attractions and habitation. The non-constructed land use consists of rice fields, parks, shrubs, vacant lots, gardens, cemeteries, landfills, water bodies, and fields [7]. Based on the Land Use Data in Cimahi City in 2012 the largest land use is habitation, followed by the use of rice fields [1]. The correlation of land cover types to flood-prone and puddle-prone level is that the land use can reduce rainwater absorption.

Making of Slope Map using Digital Elevation Model (DEM) from Shuttle Radar Topography Mission (SRTM) 30 m accuracy. Scores used in each class show in Table 3. Making of Land Use Map is using Cimahi City Satellite Imagery data, visual interpretation method and digitization process. Land use in Cimahi City consists of settlements, factories, rice fields, vegetation, plantations and open land. Scores used in each class are (Table 4):

Table 4. Score of land use classes.

Num.	Class	Score
1	Factory	5
2	Habitation	3,5
3	Open field	3
4	Vegetation	2
5	Plantation	1,5
6	Rice field	1

2.3. Making of rainfall map

Rainfall is the most influencing factor of flood and puddle. Rainfall data that is used is annual rainfall data from BBWS Citarum and Irrigation Agency in 2004-2016. The stations that are used are Cidadap-Montaya station, Dago Pakar-Bengkok, Cisondari, Cipaku-Ciparay and Cicalengka. Based on RTRW Cimahi City Year of 2012- 2032, Cimahi City has low rainfall. Meanwhile, according BMKG July and November Year of 2014, the Cimahi City has a medium rainfall that is 151-200mm but sometimes it raises into 201-300 mm. Making of rainfall maps includes calculating the average annual rainfall for each station. Interpolation of Inverse Distance Weighted (IDW) method is performed. Scores used in each class show in Table 5:

Table 5. Score of rainfall classes.

Num	Class	Score
1	>3000mm (Very wet)	9
2	2501-3000mm (Wet)	7
3	2001mm-2500mm (Moist)	5
4	1501mm-2000mm (Dry)	3
5	<1500mm (Very Dry)	1

2.4. Making of soil type map

Soil type is very influential on infiltration process (soil absorption). If the capacity of the rainwater that goes down directly to the ground does not exceed or less than the infiltration capacity, the infiltration rate can be said to be equal to the rain intensity. Based on data from the Center of Groundwater and Environmental Geology year of 1996, the extent of land type in Cimahi City is mostly Tufaan clay and silty-clay which has a very low rate of infiltration compared with other soil types (Table 6) [8].

Table 6. Soil type in Cimahi city.

District	Soil type (Ha)				Total
	1	2	3	4	
South Cimahi	290,44	3,79	0	1.360,07	1.654,30
Central Cimahi	24,72	0	0	964,90	989,62
North Cimahi	0	0	93,74	1.315,22	1.408,96
Total	315,16	3,79	93,74	3.640,19	4.052,88

Description: 1 = Andesite, 2 = Clay and sandy-clay, 3 = Sandstone, conglomerate and tufaan sandstone, 4 = Tufaan clay and silty-clay. Scores used in each class show in Table 7:

Table 7. Score of soil type classes.

Num	Class	Score
1	Tufaan clay and silt clay	4
2	Clay and sandy-clay	3
3	Andesite	2
4	Sandstone, conglomerate and tufaan sandstone	1

2.5. Making of puddle and river buffer map

Distance from the river and drainage is one of the factors that influence the level of flood and puddle. The closer to the river and drainage channels the more likely to be affected by flood and puddle. The distance from the river and drainage is determined based on the existing condition in the field. The puddle and river buffer maps are made using the Buffer method and the Multiple Ring Buffer tools. Scores used in each river buffer class show in Table 8.

Table 8. Score of river buffer classes.

Num	Class	Distance of buffer	Score
1	Very prone	0-25m	7
2	Prone	>25m-100m	5
3	Somewhat prone	>100m-250m	3

Table 9. Score of puddle buffer classes.

Num	Class	Distance of buffer	Score
1	Very prone	0-100m	7
2	Prone	>100m-300m	5
3	Somewhat prone	>300m	1

After the scoring for each parameter is done, the next thing is to determine the score of flood-prone-zone and the puddle-prone-zone parameters (Table 10 and Table 11), as follows:

Table 10. Score of flood-prone-zone.

Num	Parameter	Score
1	Slope	0,3
2	Soil type	0,25
3	River buffer	0,2
4	Land use	0,15
5	Rainfall	0,1

Table 11. Score of puddle-prone-zone.

Num	Parameter	Score
1	Slope	0,3
2	Land use	0,25
3	Soil type	0,2
4	Puddle buffer	0,15
5	Rainfall	0,1

Flood and puddle values are determined from the total sum of scores of all parameters affecting floods and puddles. The prone value is determined by the following equation [9]:

$$K = \sum_{i=1}^{n} (W_i \times X_i)$$

Information:
K = Value of Prone
W_i = Score for the i-th parameter
X_i = Score Class for the i-th parameter

The interval width of each class is done by dividing the same number of values obtained by the number of class intervals specified by the following equation:

$$i = \frac{R}{n}$$

i = Width of Interval
R = Difference in maximum and minimum scores
n = Number of flood prone classes

Table 12. Score of flood prone level.

Num	Flood prone level	Score
1	Very prone	6.75-5.487
2	Prone	5.487-4.225
3	Not prone	4225-2962
4	Secure	2.962-1.7

Each class of flood and puddle prone has characteristics based on frequency, duration and depth.

Table 13. Score of prone characteristic.

Num	Class of prone	Characteristic		
		Frequency (year)	Duration (days)	Depth (m)
1	Secure	Never	-	-
2	Not prone	1-2	-	-
3	Prone	1-2	1-2	0,5-1,0
4	Very prone	Annually	2-15	0,5-3,0

3. Result and discussion

There are 2 main outputs of this research which are Flood-prone-zone and Puddle-prone-zone Map. In contrast to previous similar research, the Flood-prone-zone and Puddle-prone-zone Map from this research has more value which lies in better data reliability, because the process of identification and inventory of flood and puddle elements is done using several methods, ranging from the simplest method (literature study from various sources), discussions with Cimahi City government, interviews and questionnaires to residents in flood and puddle locations, up to river and drainage cross-sectional measurement, calculation of river capacity and drainage, and simulation of flood and puddle 2 dimensions and 3 dimensions.

3.1. Flood-prone-zone map

Flood-prone-zone map is displayed in 2 formats which are vector and raster (figure 9 and figure 10), as follows:

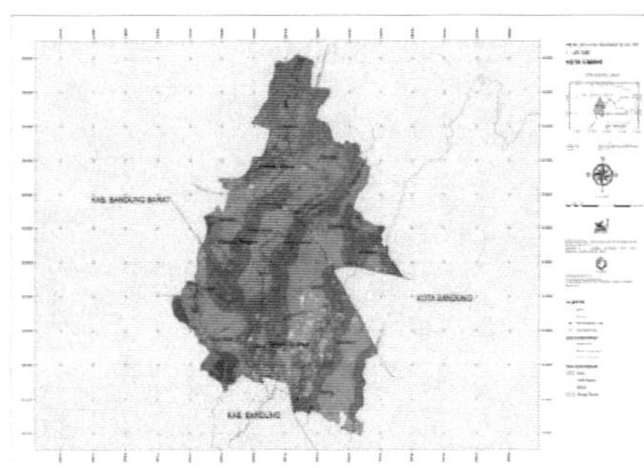

Figure 9. Flood-prone-zone in vector and raster format.

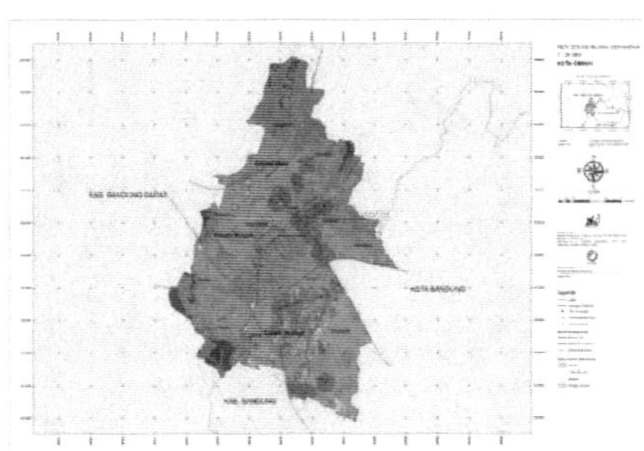

Figure 10. Puddle-prone-zone in vector and raster format.

Based on flood-prone-zone map above, calculate the area of each flood-prone-zone of Cimahi City area. In this study there are differences of the area of Cimahi City, based on Map of Administrative Boundary of the Earth Rising Map of Indonesia which is downloaded from Geospatial Information Agency Portal it obtains the area of 4,100,244 ha. Meanwhile, according to RTRW Cimahi City Year of 2012-2031 it obtains information that the area is 4,052.88 ha, and according to Book of Cimahi in Year of 2015 it obtains information that the area is 40.2 km2.

The result of calculating flood-prone-zone of Cimahi City area taken from Rupa Bumi Indonesia map, obtained information that the secure area of flood is 110,747 ha (2,70%), not flooding is 494,233 ha (12,05%), prone to flood is 2016,474 ha (49.18%), and very vulnerable is 1478,799 ha (36.07%).

3.2. *Puddle-prone-zone map*

Puddle-prone-zone map is displayed in 2 formats which are vector and raster, as follows: The results of the calculating puddle-prone-zone of Cimahi City taken from the Rupa Bumi Indonesia Map, obtained information that the secure area of puddle is 128.616 ha (3.14%), not puddle is 551,246 ha (13.44%), puddle is 2702,826 ha (65,92%), and very vulnerable to puddle is 717,556 ha (17,50%). After the location of flood and puddle along with its extent, place each flood and puddle location in Cimahi City Spatial Pattern Map sourced from Spatial Plan of Cimahi City Region year of 2012-20132 (Figure 11). The compatibility of spatial pattern on flood-prone-zone and puddle-prone-zone is showed in figure 12 and table 14.

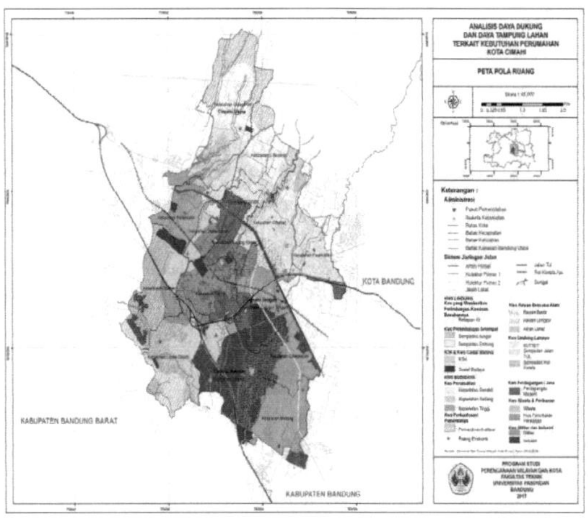

Figure 11. Cimahi city spatial pattern map year of 2012-2032.

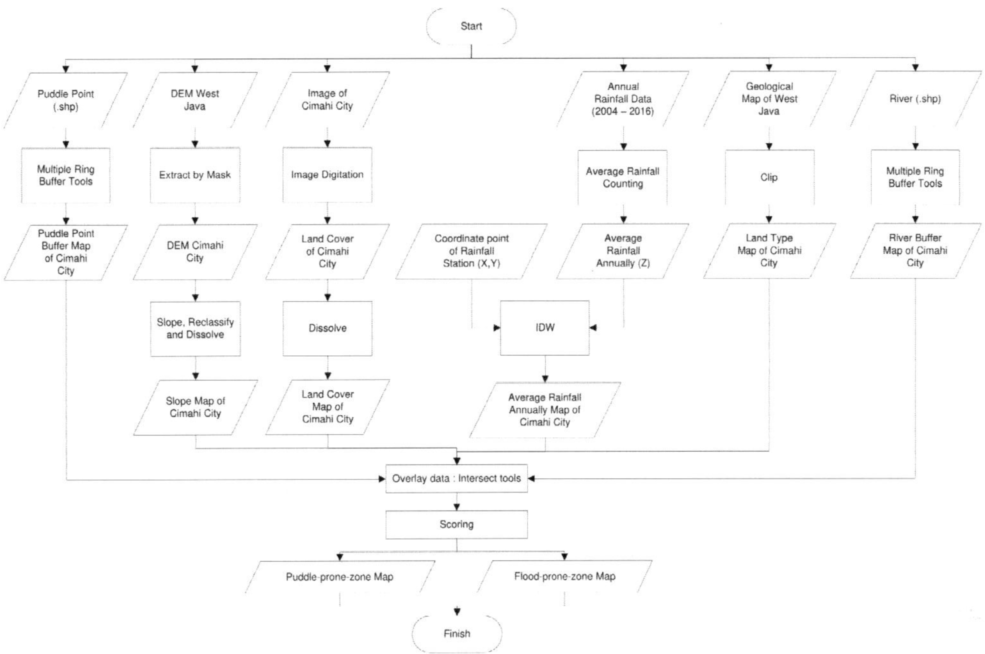

Figure 12. Flowchart of flood-prone-zone and puddle-prone-zone map making process.

Table 14. Compatibility of spatial pattern on flood-prone-zone and inundation-prone-zone.

Num	Spatial Pattern	Location	Flood-prone-zone Map				Inundation-prone-zone Map			
			VP	P	NP	S	VP	P	NP	S
1	RTH	Kel. Melong	1	1			1	1		
		Kel. Cibereum	1	1				1		
		Kel. Cigugur	1	1			1	1		
		Kel. Padasuka	1	1	1	1	1	1	1	1
2	Socio-cultural	Kel. Padasuka			1	1	1	1		
		Kel. Cibeber		1	1		1	1		
		Kel. Baros	1	1				1		
		Kel. Leuwigajah				1				
		Kel. Cipageran			1				1	
3	Low Density Habitation	Kel. Cipageran	1	1				1		
		Kel. Citereup		1			1	1		
		Kel. Cimahi	1	1			1	1		
		Kel. Cibabat	1	1			1	1		
		Kel. Pasirkaliki	1	1				1		
		Kel. Setiamanah	1					1		
		Kel. Cibeber				1	1			
		Kel. Leuwigajah			1	1	1	1		
4	Medium Density Habitation	Kel. Cibeber	1	1				1	1	
		Kel. Leuwigajah	1	1	1			1	1	
		Kel. Utama	1	1				1		
5	Office or Institution	Kel. Leuwigajah	1				1	1		
		Kel. Cibeber	1				1			
		Kel. Baros	1	1				1		
		Kel. Setiamanah		1				1		
		Kel. Karang Mekar	1	1				1		
		Kel. Cigugur	1					1		
6	Modern Trade	Kel. Baros	1					1		
		Kel. Cigugur	1				1	1		
		Kel. Cimahi	1	1			1			
		Kel. Setiamanah	1				1	1		
		Kel. Karang Mekar		1			1	1		
		Kel. Cibabat		1			1			
7	Tours	Kel. Cipageran		1	1	1			1	1
		Kel. Citereup	1	1	1			1	1	
		Kel. Cibabat	1	1				1	1	1
8	Military	Kel. Utama		1				1		
9	Industry	Kel. Utama	1	1			1	1		
		Kel. Cigugur		1			1	1		
		kel. Cibeureum	1					1		
		Kel. Melong	1	1			1	1		
		Kel. Cibeber			1	1			1	1

Description:
VP= Very Prone,
P= Prone,
NP= Not Prone,
S= Safe.

4. Conclusion

- It provides 6 points of floods and 27 points of puddles in Cimahi City.
- The cause of floods in Cimahi City is dominant due to the narrowing and shallowing of river channels. Meanwhile, the cause of puddles in Cimahi City is dominant due to the condition of drainage channels that is covered with garbage and drainage channel dimensions that are not in accordance with guidelines and planning.
- Change of the flood water and puddle height in Cimahi City did not significantly affect the area of flood and puddle.
- It provides information that the area of secure flood is 110,747 ha (2,70%), not flooding is 494,233 ha (12,05%), prone is 2016,474ha (49,18%), and very prone is 1478,799 ha (36.07%). As for the puddle, it was found that the secure area of puddle is 128,616 ha (3,14%), not prone is 551,246 ha (13,44%), prone is 2702,826 ha (65,92%), and very prone is 717,556 ha (17.50%).
- Nine types of spatial patterns in the City of Cimahi mostly located in the zone that is very prone and prone to flooding and puddles.

Acknowledgment
The author acknowledges to Ministry of Research, Technology and Higher Education for the opportunities that have been given, to the Director of Bandung State Polytechnic. Ir. Rachmad Imbang Tritjahjono, MT, Head of UPPM Ir. Ediana Sutjiredjeki, M.Sc, and Team of Bandung State Polytechnic. Also to the Assistant Economy and Development of the Regional Secretariat of Cimahi City Government Drs. Benny Bachtiar, M.Si, the expert staff from Bappeda Cimahi City, BNPD Cimahi City, and Cimahi City Public Works Department.

References

[1] Regional Development Planning Board Cimahi 2012 *RTRW Cimahi City Year of 2012-2032* (Cimahi: Regional Development Planning Agency Cimahi)
[2] Central Bureau of Statistics Cimahi City 2015 *Limits of Cimahi City Administration. Cimahi: Central Bureau of Statistics Cimahi City* [Online] Retrieved from: https: //cimahikota.bps.go.id
[3] People's Mind 2016 *Flooding in Melong Cimahi* [Online] Retrieved from: http://www.pikiran-rakyat.com
[4] Central Bureau of Statistics Cimahi City 2015 *Total Population Cimahi City 2010-2014* (Cimahi: Central Bureau of Statistics Cimahi City)
[5] Kodoatie R J 2014 *Flood Engineering and Management of the City* (Yogyakarta: Andi Publisher)
[6] Cimahi City Government 2014 *Cross section of Cilember River* (Cimahi: Cimahi Municipal Government)
[7] University of Pasundan 2017 *Analysis of Support Capacity and Capacity of Land Related to Cimahi City Housing Needs* (Bandung: University of Pasundan)
[8] Center for Ground Water and Environmental Geology 1996 *Type of Land in Cimahi City*
[9] Purnama A 2008 *Mapping of Flood-prone Areas in Cisadane River Basin Using Geographic Information System* (Bogor: Bogor Agricultural University)

IOP Conf. Series: Materials Science and Engineering 1108 (2021) 012010 doi:10.1088/1757-899X/1108/1/012010

Prediction of sediments discharge in watershed with two tank models

T Mulyono*, J Jumi and A Azizah

Politeknik Negeri Semarang, Semarang, Indonesia

*teted64@gmail.com

Abstract. Uncontrolled erosion would cause considerable damages, such as soil fertility decline, water structures damage and reservoirs sedimentation. As the data for the sedimentation rate are limited, several models have been developed to predict the surface erosion and the rate of sedimentation. However, the availability of sufficient, diverse and extensive data is needed for the implementation of the models, both for the model calibration and the verification. The result of the analysis shows that both of the Water Tank Models that represent the erosion-sedimentation rate process, in which Water Tank 1 being the three-tank cascade system and Water Tank 2 being two-tank cascade system, are not optimum. This can be observed from the values of volume error (VE), relative error (RE), root-mean-square error (RMSE) and correlation coefficient (R) that show the effect of 1.5 hours of rain period in the sedimentation rate. The field condition shows considerable sedimentation, on the other hand, the models' simulations show decreasing sedimentation rates. The optimum model's parameters for Water Tank 1 and Water Tank 2 are 924.51%-1049.26% for the relative error, 50.81% - 121.42% for the volume error, 0.9 for the correlation coefficient and 6703.59-17,297.85 for the root-mean-square error. The parameters and constant's values of the models are different relative to the drainage basins' condition.

1. Introduction

Water resource problems in Indonesia are originated by several causes, but the most prominent one is the environmental degradation as a result of deforestation in the watershed, thus resulting in its decreasing ability to store water. This irresponsible deforestation had first resulted in 58 critical watersheds in Indonesia [1], the number then increased to 62 critical watersheds [2], and currently, there are 108 watersheds (DAS) in critical condition [3]. The criteria used to determine whether the watersheds are in critical conditions are the low percentage of land cover, high annual erosion rate, the ratio of maximum and minimum river discharge rate, and excessive mud deposit (sediment load). The indicators of critical watersheds provide the averments that the environmental system supporting the hydrological cycle has been damaged, e.g. reduced forest area surrounding the river basins and higher levels of erosion and sedimentation, resulting in floods, landslides, and soil fertility decrease.

Uncontrolled erosion would cause considerable damages, such as soil fertility decline, water structures damage and reservoirs sedimentation [4]. Estimating the rate of sedimentation is difficult due to the variability of the data. Moreover, despite the variability, as the mechanism of erosion to sedimentation rate is quite complex, the availability of the data is limited. The analysis of sedimentation rate is usually carried out by measuring the sediment rating curve which requires sufficient data covering various range of small to large sedimentation rate which need extra time, cost and energy. Therefore, for estimating or predicting the erosion, sedimentation rate at the watershed is needed for dam and reservoir planning, soil conservation design, land usage planning, water quality management and effective control strategies to reduce the risk of water runoff, and for protection against erosion.

A number of models and approaches for predicting the erosion and sedimentation rate have been developed, such as the WEPP model [5], the KINEROS model [6], the ANSWERS model [7], the AGNPS model [8,9] and the SWAT model [10,11]. Those models have included the erosion analysis and sedimentation rate in the area of rivers that are caused by rainfall and water runoff [12]. Applying the Time-Area method for predicting sediment deposit adjusted with time [13-15] with lumped sediment runoff models analyzing sediment flow covering areas of steep land and in rivers around the watersheds.

The flow of sediment in the river is mainly caused by erosion that occurs in the rivers' areas [16]. Erosion and sediment displacements are included in the hydrological process that occurs in the watersheds [17]. The erosion - sediment discharge at the in the watershed occurs in steep land and in the channels or rivers.

Starting with the occurrence of rain, the process of sediment discharge in steep land is affected by rain and surface runoff, the process would result in sediment deposit in the steep land, and the process of sedimentation rate in the channel or river in which the sediment is originated from steep land, erosion and riverbanks. The process would result in sedimentary reservoir in a river or channel. Therefore, the concept of reservoir models will represent the prediction of sedimentation rate in the watershed. One of these models is the water tank model.

Based on the portrayals of the existing models, a hydrological model in the form of a tank model can be developed to predict sediment flow in the watershed that can represent the process of erosion and sedimentation in the watershed. Several models for sediment flow analyses are the rain-flow model which incorporate sediment elements with lumped methods based on existing distributions, but this model has not been applied, thus just acceptable in theory.

2. Related works

Optimization is an activity to get results repeatedly and influence each other. The best results that will be obtained are minimal or maximum values. If there is no function of several variables, it can be used as a result to provide conditions that provide minimum results or maximum results [18], as in Figure I.

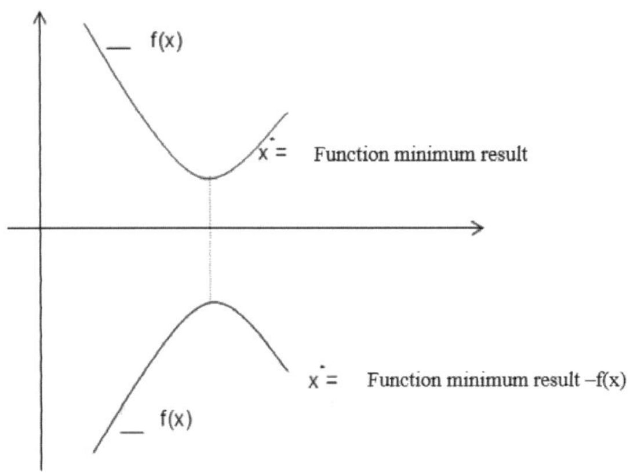

Figure 1. Minimum and maximum results.

In recent years, there have been several types of variations of global optimization methods applied in automatic tank model calibration, Cooper et al [19] applying the Shuffle Complex Evolution (SCE) method, simulated annealing (SA) method and Genetic Algorithm (GA) for parameter optimization of tank models in the arrangement of two tanks, [20] applied the Shuffle Complex Evolution (SCE) method and Powell multistart for calibration of tank models in Taiwan, then Setiawan et al [21] and Setiawan [22], verifying with the Marquardt algorithm that showed efficiency and effectiveness in determining the parameter optimization of the tank model for two watersheds in Japan and in Indonesia, and Kuok et al [23] apply global optimization methods (GOMs) namely the Shuffle Complex Evolution (SCE) method, the Particle Swarm Optimization (PSO) method and Genetic Algorithm (GA) for calibration and parameter optimization of tank models. From the above statement, there is no general agreement among researchers which method is most suitable for calibrating the tank model. Therefore, in this study parameter optimization of the tank model was carried out to predict sediment discharge in the watershed using the Genetic Algorithm (GA) method with Matlab.

3. Research method

The step in the research method for tank model development to predict sediment discharge in the watershed is data collection and processing. Data collection includes secondary data such as watershed maps (DAS), land use maps, rainfall station maps. Then the primary data or observation data in the Kreo watershed include observation and measurement of rainfall data, data on elevated sediment discharge, discharge data with a period of time must be the same and carried out at the alleged post or river flow observation station (SPAS). Furthermore, data processing includes data recording, analysis and interpretation of result

3.1. Data processing analysis

Data processing analysis in this research will be an analysis of the results of primary data measurement in the form of rain data, debit data, sediment discharge data and analysis of sample data testing in the laboratory.

Rain data is measured and recorded using the help of a telemetry device, automatic rainfall raingauge (ARR) rain gauge which is connected to a computer via the internet, so that rain data is read directly on the computer.

The discharge data begins with measuring the water level based on the automatic water level tool known as Automatic Water Level Recorder (AWLR). Once known the water level is measured by the wet cross-sectional area of the river. Then measured the velocity of the river current with the Current meter then it can be calculated by multiplying the cross-sectional area with the current velocity, as in Equation (1).

$$Q = A \times V \tag{1}$$

Where Q is the discharge (m^3 / sec), A is the wet cross-sectional area (m^2), and V is the current velocity (m / sec). While the sediment discharge data, in this study the method of measuring the elevated sediment discharge using the Equal Discharge Increment (EDI) method, in which a cross section is divided into several sub-sections, where each sub-section must have the same discharge. Then the sediment measurement in this way is carried out in the middle of each sub-section [24]. The basic equation for sediment rate (elevated sediment discharge) uses Equations (2) and (3).

$$Q_s = 0{,}0864 \times C \times Q_w \tag{2}$$

$$C = aQ_w^b \tag{3}$$

Where, Qs is daily sediment discharge (ton / day), C is the concentration of elevated sediment (mg / liter), Qw is discharge (m3/sec), a and b are constants obtained from regression.

The potential discharge of serious errors that arise in the use of curved sediment rate to calculate sediment load is the use of average daily debit data, especially if the average daily debit is used from the average of three measurements (06.00 am, 12:00 noon and 17.00 pm). This is due to heavy rain occurring at midnight night, so that the peak of the floods that occur will occur at night [24]. Meanwhile, according to Simons and references number [24-26], the equations of sediment transport can be classified in the basic form as in Equation (4).

$$Qs = A\ (B\text{-}Bc)\ D \tag{4}$$

Where, Qs is a sediment discharge, A is a parameter that relates to flow and sediment characteristics, B is a parameter that can be either Q discharge, average flow velocity u, slope of Sw water level, slope of Sf energy, slope of So river base, shear stress τ, current strength τu, uS unit current strength etc., Bc is a critical condition parameter associated with B in the initial movement (incipient motion), D is a parameter related to flow and sediment characteristics.

Sediment concentration from a river section is a comparison between sediment discharge and river flow discharge [27]. This value can be formulated as in Equation (5).

$$Q_S = C \times Q \tag{5}$$

Where, Qs is a sediment discharge, C is the sediment concentration, Q is the flow discharge. For the measurement of this EDI method Equation (5) can be changed to Equation (6). by entering Equation (7).

$$Q_S = \sum_{1}^{n} \overline{C}_i q_i \Delta x_i \tag{6}$$

$$\text{if } Q = \sum_{1}^{n} q_i \Delta x_i \tag{7}$$

Then the average concentration is as in Equation (8).

$$\overline{C_i} = \frac{\sum_{1}^{n} \overline{C}_i q_i \Delta x_i}{\sum_{1}^{n} q_i \Delta x_i} \tag{8}$$

Where, is the average concentration of sediment on a river section, Ci is the concentration of sediment in the sub-section to i, is the width of the sub-section of the river to i, qi is the debit per width of the sub-section and n is the number of vertical measurements. Because in the way of EDI qi value, q2 = ... = qn = Q / n then Equation 8 can be changed to Equation (9) or (10).

$$\overline{\mathrm{C}} = \frac{\frac{Q}{n} \sum_{1}^{n} C_1}{Q} \tag{9}$$

$$\overline{\mathrm{C}} = \frac{\sum_{1}^{n} C_1}{n} \tag{10}$$

Equation (10) is the average sediment concentration in the measured river section. Analysis of testing sample data in the laboratory is an analysis of the weight of the sediment and can be done as follows:

Samples in the form of water and soil erosion particles, which have been included in the bottle are labeled: sequence number and location and date of occurrence, first drained ± 24 hours so that the soil particles settle, after which the water is disposed of. Soil particles and residual water are then put into porcelain dishes that are known to be of weight and are labeled accordingly (date and number). At this stage, recording is done: the cup code number, the date of the rain event, the serial number, and the weight of the cup used. From a sample the weight of the cup used was used, as in Equation (11).

$$Wm = \ldots\ldots\ gram \tag{11}$$

Then the cup and its contents are put in the oven at 110ºC for ± 24 hours, so the soil becomes dry (absolute). After that the cup containing dry soil (eroded soil) is removed from the oven, drained ± 15 minutes after it is put into the desiccator and after the cold is weighed. From a sample, the weight of the Wm plate, and weight (plate + eroded soil) was recorded in grams, as in Equation (12).

$$W(m + s) = ...\text{grm} \tag{12}$$

W (m + s) = weight (cup + eroded soil) ... gram

Then Ws can state the weight of dry soil or eroded soil per sample per location per rain event, the amount can be calculated: Ws = [W (m + s) - Wm] gram for 500 CC water volume. Then the amount of eroded soil per location of rain events, can be calculated as in Equation (13).

$$\text{Wtot} = (\text{Vtot}/500) \times \text{Ws} = ...\text{gram} \tag{13}$$

3.2. Model accuracy test criteria

The model compiled to simulate processes in nature is a model that must be able to approach the actual process. In this connection, whatever the form of the model, whatever approach is used, the output of a model must be able to approach the process that actually occurs in nature. However, it is almost impossible for the natural processes that occur to be matched correctly.

Therefore, there will always be a deviation between the measured output (observed) and the calculated output (simulated) described in Figure 2. To see the extent of the results of the optimization approach to the model parameter values in the Sub-watershed, the calibration count output by means of optimization needs compared to the results obtained in a measured way.

The accuracy of the model depends on three factors: the accuracy of input data, the effectiveness of parameter assessment and errors in the model itself [28]. Testing the accuracy of the model can be done by testing the measured data and simulation results data. The model is said to be careful if there is a high correlation between measured data and simulation results data and has the smallest possible value deviation.

In this study used correlation coefficient (R), volume error (VE), mean relative error (RE) and the least squares average error (RMSE) as a test model accuracy criterion, and calculated in the auxiliary program. Limitation of criteria for the accuracy of the test model in this study is VE <% 5, -10% <RE <- 10%, R> 0.7 and RMSE close to 0 (zero).

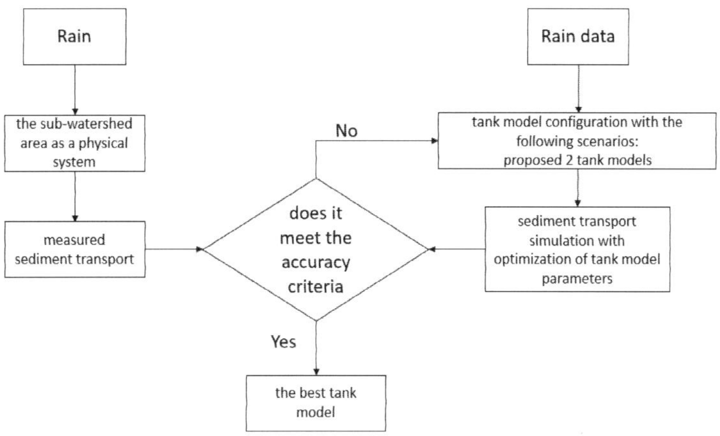

Figure 2. Model calibration scheme.

The correlation coefficient (R) is the price that shows the magnitude of the attachment between the measured value and the simulation value. If the price of R is worth 1, then it can be said that the correlation between the two is very close, but if R is zero then the two values are said to have no relationship at all. The correlation coefficient is formulated as in Equation (14) [29].

$$R=\sqrt{\frac{Dt^2-D^2}{Dt^2}} \tag{14}$$

$$Dt^2=\sum_{i=1}^{N}\left(Sed_{obs}^{i}-\overline{Sed}\right)^2$$

$$D^2=\sum_{i=1}^{N}\left(Sed_{obs}^{i}-Sed_{sim}^{i}\right)$$

Where, Sed^i_{sim} is simulation sedimen debit in the i (ton / day) period, Sed^i_{obs} is the measured sediment discharge in the i (ton / day) period, N is the amount of data.

Volume error (volume error, VE) is a value that shows the difference in the volume of simulation results and measured during the simulation period. If the VE value is very small, it means that the total volume of the simulation value and measured is almost the same. Conversely, if VE is very large, there is a deviation from the simulation results and is measured. Volume errors are formulated as in Equation (15) [30].

$$VE=\frac{\sum_{i=1}^{N}Sed_{obs}^{i}-\sum_{i=1}^{N}Sed_{sim}^{i}}{\sum_{i=1}^{N}Sed_{obs}^{i}}x100 \tag{15}$$

Where, VE is the difference in the volume of sediment concentration, Sed^i_{sim} is simulation sedimen debit in the i (ton / day) period, Sed^i_{obs} is the measured sediment discharge in the i (ton / day) period, N is the amount of data.

Relative mean error (Relative Error, RE) serves to determine the mean relative deviation of simulated sediment discharge to measured sediment discharge. The advantage of using RE is the ease of checking simulation results, because if the value is close to zero it means that the deviation of simulated sediment discharge to measured sediment discharge is very small. The relative mean error is formulated as in Equation (16) [31].

$$RE=\frac{1}{N}\sum_{i=1}^{N}\frac{\left|Sed_{sim}^{i}-Sed_{obs}^{i}\right|}{Sed_{obs}^{i}} \tag{16}$$

Where, RE is a relative error, Sed^i_{sim} is simulation sedimen debit in the i (ton / day) period, Sed^i_{obs} is the measured sediment discharge in the i (ton / day) period, N is the amount of data.
The average root of the squared error (Root Mean Squared Error, RMSE) is a measure of the difference between the simulated sediment discharge value and the measured sediment discharge value. The RMSE is formulated as in Equation (17) [29].

$$RMSE=\sqrt{\frac{\left(\sum Sed_{sim}^{i}-Sed_{obs}^{i}\right)^2}{N}} \tag{17}$$

Whereas, the RMSE is the root of the average number of squared errors, Sed^{i}_{sim} is simulation sedimen debit in the i (ton / day) period, Sed^{i}_{obs} is the measured sediment discharge in the i (ton / day) period, N is the amount of data.

3.3. Sensitivity analysis
Sensitivity analysis on the tank model for predicting sediment discharge is intended to determine the parameters that most influence the model output, which will be used as a guide to predicting sediment discharge in the watershed and to find the most sensitive parameters for sediment discharge, it is necessary step is done by adding and subtracting the value of each parameter by 10%, then running, then see the output of the model.

To get the sensitivity index (S), a formula is used as in Equation (18).

$$S = \left[\frac{P_{10} - M_{10}}{Baseline} \right] \tag{18}$$

Where, S is the sensitivity index with% unit, P_{10} is the simulation result by making 10% addition, M_{10} is the simulation result by making a 10% reduction, BASELINE is the initial (basic) simulation result. Each sensitivity index (S) occurs, then compared with the value of the largest sensitivity index and then the results are made in percent form. To test the sensitivity of the tank model for prediction of sediment discharge is applied to the Kreo River Basin.

4. Results and analysis

4.1. Development of a tank model with hypothetic data
The results of the analysis of tank model development with hypothetical sediment discharge data which includes hypothetical data of 1 hydrograph, hypothetical 2 hydrograph data, hypothetical delay 1 data, hypothetical delay 2 data and ideal hypothetical data, indicate that the tank model program can work well. This is indicated by the value of sediment discharge resulting from the tank model simulation and the model accuracy value (model performance) of the 2 proposed tank models (Model Tank 1 in Figure 3, Model Tank 2 in Figure 4).

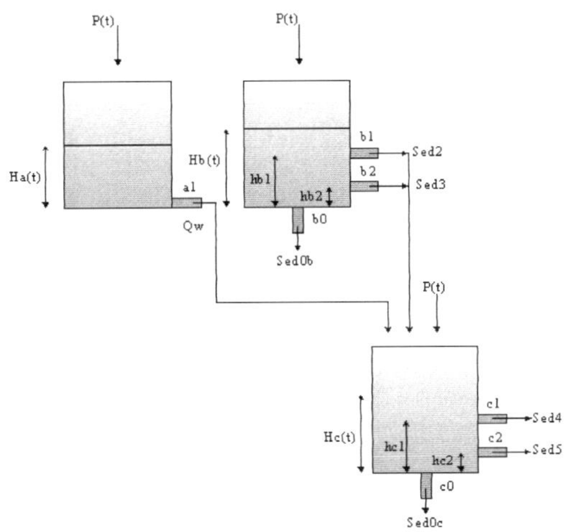

Figure 3. Configuration or model structure of tank 1.

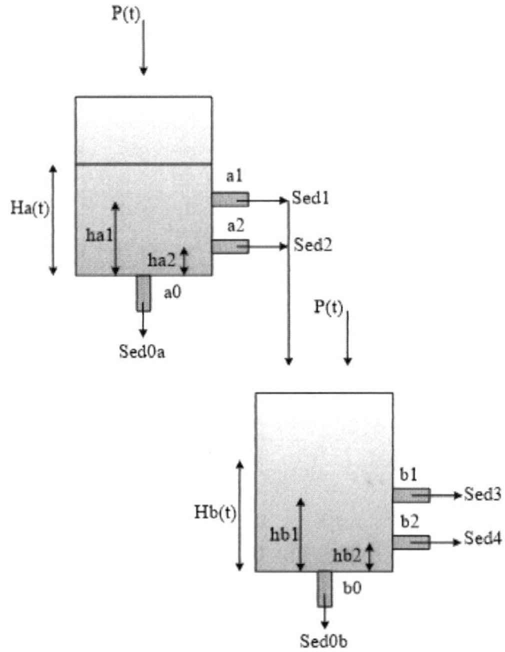

Figure 4. Configuration or model structure of tank 2.

4.2. Development of tank models with observation data

The results of the research on the development of tank models for the prediction of sediment discharge in the watershed using observation data in the Kreo Sub-watershed in 4 (four) rain events on 4 (four) days, namely on 6, 10, 26 and 28 January 2014, for 2 (two) The proposed tank model (Model Tank 1 and Tank Model 2), shows that the Tank Model 1 and Tank 2 model are not good enough to apply this based on the results of the model accuracy criteria which are the relative error value (RE), the volume error (VE) value (Table I), correlation coefficient value (R) and root mean square error (RMSE) value (Table II).

Table 1. Value of accuracy criteria 2 tank model for sediment debit prediction in the Kreo sub-watershed for Relative Errors (RE) and Volume Errors (VE).

Rain	RE		VE	
	Tank Model 1	**Tank Model 2**	**Tank Model 1**	**Tank Model i 2**
06/01/2014	413,63	277,67	119,73	57,08
10/01/2014	154,41	300,64	62,63	203,70
26/01/2014	433,74	375,67	165,54	129,69
28/01/2014	799,42	624,85	51,65	21,55

Table 2. Value of accuracy criteria 2 tank model for sediment debit prediction in the Kreo sub-watershed for Correlation Coefficient (R) and Average Root Amount of Error Squares (RMSE).

Rain	RE		VE	
	Tank Model 1	**Tank Model 2**	**Tank Model 1**	**Tank Model 2**
06/01/2014	0,52	0,52	275,4	183,11
10/01/2014	0,65	0,65	742,6	1851,7
26/01/2014	0,89	0,89	81,71	63,64
28/01/2014	0,83	0,83	1223,5	1006,3

The results of the tank model analysis on January 6 and 26, 2014, for the Tank Model 1 and Tank Model 2 of the sediment hydrograph graph on the rising limb, the simulation sediment discharge value was greater (above) than the measured sediment discharge value, this was bulk rain tends to be high, of course the amount of surface flow increases so that the output value of sediment discharge in the tank model becomes larger and the recession limb value of the simulated sediment discharge is smaller (below) than the measured sediment discharge value, this is due to rainfall tends to decrease flow the surface also decreases so that the output value of sediment discharge in the tank model becomes small, and the condition of the field of rainfall decreases, the value of sediment discharge decreases but tends to remain high, while the peak value of simulated sediment discharge tends to be lower than the measured sediment discharge value. The simulation results of sediment discharge value of Tank Model 1 and 2 do not approach the measured sediment discharge value, see Figure 5.

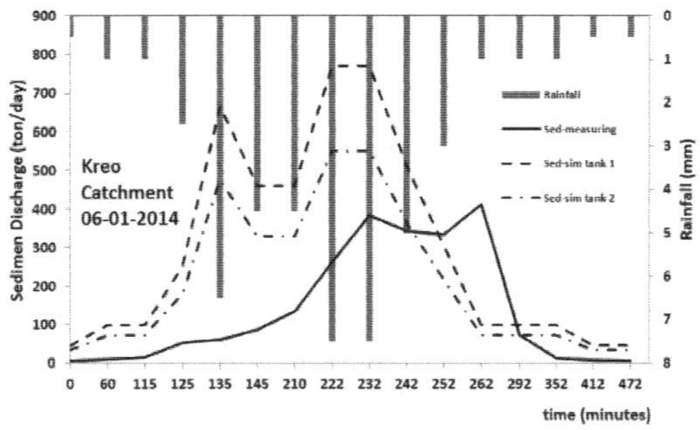

Figure 5. Comparison of sediment discharge observation and simulation of tank models in the Kreo sub-watershed on January 6, 2014.

In the Tank Model 1 and Tank Model 2 the simulation results show the tendency of the results of simulation sediment discharge values to be higher than the value of measured sediment discharge, the possibility of accumulation of sediment discharge values from these tanks. However, the results of the tank model have a wide range of differences. The deviation of the sediment discharge value from the simulation results on observations is likely caused by several things including:

- The synchronization factor when measuring the initial occurrence of rain and measuring sediment discharge (sediment concentration), at least must be precise, which usually uses concentration time, and the process of measuring sediment discharge (sediment concentration) and careful and correct sampling is very difficult so it is likely to contain element of error.
- Activities of humans and / or livestock on land or in rivers, cause high enough sediment even though rain is not so heavy (normal rain).
- Data length, if the data used for calibration is too short or is less likely to affect the model output.
- The tank model is a lumped model, using a single value on input parameters without spatial variability and producing a single output and assuming steady state hydraulic conditions

However, the series of observable sediment discharge values and simulated sediment discharge have a pretty good resemblance.

4.3. Development of other das data input tank models

4.3.1. Lesti river sub-watershed, East Java. The results of the analysis of the tank model development for the prediction of sediment discharge in the Lesti River sub-watershed, with the time series rainfall data on October 3-6, 2003 and applying 2 tank models, shows that the Tank Model 1 and Tank Model 2 are not good enough. to apply this based on the results of the model accuracy criteria are the RE value, VE value, R value and RMSE value (Table III).

Table 3. Accuracy criteria of 2 tank models and lumped method for prediction of sediment discharge in the Lesti river sub watershed for RE, VE, R and RMSE.

Research Criteria	Lumped Method (Apip)	Distributed Method (Apip)	Tank Model 1	Tank Model 2
RE	**231,73**	214,41	1049,26	924,90
VE	6,06	3,28	98,18	50,81
R	0,90	0,80	0,90	0,90
RMSE	3179,23	3545,76	12371,71	6703,59

For all 2 tank models the value of R = 0.8, quite well this has similar shape, only the value of sediment discharge has a difference range can be seen in Figure 6, while 2 other tank models have a large enough VE value away from the requirements between 50,81-98,18. But the RE and RMSE values are still quite large.

Lumped Method by references number [32-35] is an analysis of the Lumped method physically based on a distribution model (distributed model) for the prediction of sediment flow on the scale of the Watershed. In a physical-based analysis based on a distribution model for sediment flow, the watershed is divided into segments or grids, and in its analysis segment or grid uses 12 parameters, including the depth of capillary soil layer (dm); depth of capillary soil layer + depth of non-capillary soil layer (da); hydraulic conductivity of non-capillary (ka) soil layer; non-linear exponent constant for capillary (β) soil layer; roughness coefficient of manning (n); median grain size (D50); off ground capability (k); release and settling efficiency (α); sediment density (ρs); water density (ρw); gravity (g) and kinematic viscosity of water (NU). Furthermore, with a lumped procedure estimation method, the estimated sediment flow can be determined.

The results of the analysis of the tank model development for the prediction of sediment discharge in the Lesti River Sub-watershed, with rainfall data of 2 hours on September 19, 2005 and applying 2 tank models, Tank 1 and Model Tank 2, are not good enough to apply this based on the accuracy criteria the model is the RE value, VE value, R value and RMSE value (Table V).

For all 2 tank models the value of R = 0.9, quite well this has similar shape, only the value of sediment discharge has a difference range can be seen in Figure 6. RE, VE and RMSE values for Tank Model 1 and Tank Model 2 tend to be large, because the simulation value of sediment discharge is greater than the value of measured sediment discharge, to start from rising limb, peak and falling limb, because the upper hole parameter factor of the tank for the lowest tank has a sensitivity between 11-18%, quite influential against simulation results.

Table 4. Tank Model Accuracy Criteria, Lumped Method, Distributed Method for Prediction of Sediment Discharge in the Lesti River Sub watershed with 2 hours of Rain Data for RE, VE, R and RMSE.

Research Criteria	Lumped Method (Apip)	Tank Model 1	Tank Model 2
RE	441,85	168,27	212,91
VE	117,53	58,59	73,29
R	0,60	0,80	0,80
RMSE	2260,94	2036,33	2191,57

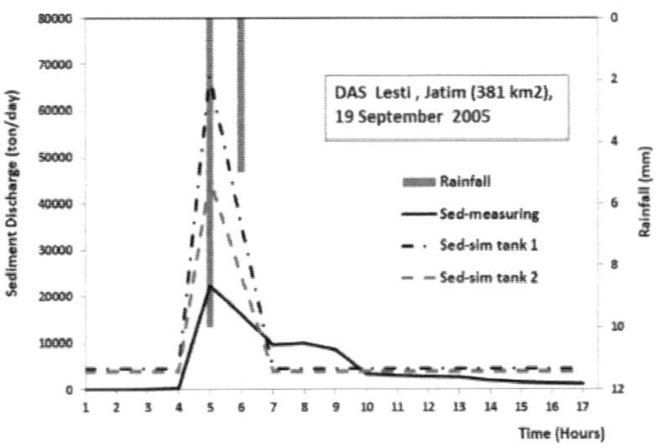

Figure 6. Sediment discharge simulation results 2 tank models with 2-hour rainfall data in the Lesti river sub-watershed.

5. Conclusion

In this study the tank model was developed with its parameters for predicting sediment discharge in the watershed, using observation data input data, other watershed data in the form of rainfall data, and sediment discharge data, this can be summarized as follows:

- By representing the erosion-sediment process, the tank model developed for the prediction of sediment discharge in the Watershed is the Tank 1 Model with 3 (three) cascade series tanks and Model Tank 2 with 2 (two) cascade tanks.
- The optimum arrangement or configuration of the tank model gives a predictive result of sediment discharge in the watershed that is not good is the Model Tank 1 and the Tank Model 2. This is applied in the Kreo Sub-watershed, the Lesti River Sub-watershed in East Java. This can be seen from the accuracy of the model, namely the value of the volume error (VE), the relative error value (RE), the mean least squares error value (RMSE) and the correlation coefficient value (R), the Kreo Sub-watershed, Table I and Table II: Sub-watersheds Lesti River, Table III, that conditions in the field of sediment discharge process still occur with the value of sediment discharge is quite large, even though the rain stops or decreases, but the simulation results of the value of sediment discharge decreases. Optimal model parameters on Model Tank 1 and 2 tank models have RE values between 924.51% - 1049.26%, VE values between 50.81% - 121.42%, value R = 0.9 and values RMSE between 6703.59 - 17,297.85. The values of the parameters and the constant of the tank model in the River Flow area have different values according to the conditions of the Watershed reviewed.

It should be noted about the distribution of rain according to the time and distribution of rain (space function) actually, which is related to the regional average rainfall analysis as input data in the tank model modeling program for the prediction of sediment discharge in the watershed. Timeliness for measurement and data retrieval must be the same and the length of data taken, especially rainfall data, and data on suspension sediment discharge and discharge data at the watershed location reviewed. Further research is needed in the development of a tank model to predict sediment discharge, if there is a change in land use in the watershed. Further research is needed, the development of a tank model for predicting sediment discharge in the watershed monitoring system with a couple of information technology.

References

[1] Soenarno S 2001 Integrasi dan Keterpaduan Pengelolaan Sumberdaya Air yang Berkelanjutan dalam Era otonomi Darah *Prosiding Diskusi Panel Nasional Kerjasama Universitas*

Diponegoro dengan Departemen Energi dan Sumberdaya Mineral (Semarang) 21

[2] Tarigan S D 2009 Pengembangan Sistem Informasi Spasial Berbasis Web (Web GIS) untuk Sinergi Rehabilitasi DAS Kritis Nasional *Prosiding Seminar Nasional "Kebijakan dan Aplikasi Teknologi Informasi dan Komunikasi untuk Peningkatan Daya Saing Agribisnis Indonesia* (Bogor: Himpunan Informatika Pertanian – Institut Pertanian Bogor – Departemen Pertanian Republik Indonesia)

[3] Sustayo M O 2016 Pendekatan Spasial Ekologis dan Skala Multidimensi Dalam Pengelolaan DAS Berkelanjutan *Disertasi* (Yogyakarta: Fakultas Kehutanan, Universitas Gadjah Mada)

[4] Suripin S 2002 *Pelestarian Sumber Daya Tanah dan Air* (Yogyakarta: Penerbit Andi)

[5] Nearing M A, Foster G R, Lane L J and Finkner S C 1989 A process-based soil erosion model for USDA-Water Erosion Prediction Project technology *Transaction of the American Society of Agricultural Engineers* **32** 1587-1593

[6] Woolhiser D A, Smith R E and Goodrich D C 1990 *KINEROS, A Kinematic Runoff and Erosion Model:Documentation and User Manual* (ARS-77. Fort Collins, Colo.: USDA Agricultural Research Service)

[7] Beasley D B and Huggins L F 1991 *ANSWERS. User's Manual* (Agricultural Engineering Department, Purdue University, West Laffayete, Indiana)

[8] Young R A, Onstad C A, Bosch D D and Anderson W P 1987 *AGNPS, Agricultural nonpoint-source pollution model:A watershed analytical tool. Conservation Research Report No. 35* (Washington, D.C.: USDA)

[9] Young R A, Onstad C A, Bosch D D and Anderson W P 1989 AGNPS: A nonpoint-source pollution model for evaluating agricultural watersheds *J. Soil and Water Conservation* **44** 2 168-173

[10] Arnold J G, Srinivasan R, Muttiah R S and Williams J R 1998, Large-area hydrologic modeling and assessment: Part I. Model development *J. American Water Resources Assoc.* **34** 1 73-89

[11] Borah D K and Bera M 2003 Watershed-Scale Hydrologic and Nonpoint-Source Pollution Models : Review of Mathematical Bases *American Society of Agricultural Engineering* **46** 6 1553-1566

[12] Kothyari U C, Tiwari A K and Singh R 1996 Temporal Variation of Sediment Yield *Journal of Hydrologic Engineering* 169-176

[13] Yasuto T, Takahiro S and Kaoru T 2008 Lumping a Physically-based Distributed Sediment Runoff Model with Embedding River Channel Sediment Transport Mechanism *Annuals of Disas. Prev. Res. Inst.* **51 B** 103-117

[14] Takahiro S, Yasuto T and Kaoru T 2007 The Spatio-Temporal Predictions of Rainfall-Sediment-Runoff Based on Lumping of a Physically-based Distributed Model *Annuals of Disas. Prev. Res. Inst.* **50 B** 79 - 94

[15] Takahiro S, Yasuto T and Kaoru T 2008 Lumping of a Physically-based Distributed Model For Sediment Runoff Prediction in a Catchment Scale *Annuals Journal of Hydraulic Engineering, JSCE* **5** 43–48

[16] Hudson N 1971 *Soil Conservation* (London: BT Bats ford Limited)

[17] Soewarno S 1991, *Hidrologi Pengukuran dan Pengolahan Data Aliran Sungai (Hidrometri),* Nova, Bandung, hal. xx + 825.

[18] Rao S S 1977 *Optimization Theory and Application* (New Delhi: Wiley Eastern Limited)

[19] Cooper V A, Nguyen V T V and Nicell J A 1997 Evaluation of Global Optimization Methods for Conceptual Rainfall-Runoff Model Calibration *Water Sci. Technol.* **36** 5 53-60

[20] Chen R S, Pi L C and Hsieh C C 2005 A Study on Automatic Calibration of Parameters in Tank Model *Journal of the American Water Resources Association (JAWRA)*

[21] Setiawan B I, Fukuda T and Nakano Y 2003 Developing Procedures for Optimization of Tank Model's Parameter *CIGR Journal of Scientific Research and Development*

[22] Setiawan B I 2003 Optimasi Parameter Tank Model *Buletin Keteknikan Pertanian* **17** 1 8-16

[23] Kuok K K, Harun S and Shamsudin S M 2010 Global Optimization Methods for Calibration and

Optimization of the Hydrologic Tank Model's Parameter *Canadian Journal on Civil Engineering* **1** 1 1-14
[24] Kodoatie R J and Sjarief R 2010 *Tata Ruang Air* (Yogyakarta: Penerbit Andi)
[25] Yang C T 1996 *Sediment Transport Theory and Practice* (Singapore: The McGraw-Hill Companies, Inc. Printers Pte Ltd)
[26] Senturk F and Simons D B 1992 *Sediment Transport Technology - Water and Sediment Dynamics* (USA: Water Resources Publication, Littleton, Co)
[27] Soewarno S 1991 *Hidrologi Pengukuran dan Pengolahan Data Aliran Sungai (Hidrometri)* Bandung: Nova)
[28] Linsley J R, Kohler M and Paulhus J 1986 *Hidrologi untuk Insinyur* Edisi Ketiga (Jakarta: Erlangga)
[29] Solaimani K 2009 Rainfall-runoff Prediction Based on Artificial Neural Network (A Case Study: Jarahi Watershed) *American-Eurasian J. Agric. & Environ. Sci.* **5** 6 856-865
[30] Lee Y H and Singh V P 2005 Tank Model for Sediment Yield *Water Resources Management (2005)* **19** 349-362
[31] Chen R S and Pi L C 2004 Application of diffusive Tank Model in Drainage analysis of Paddy Fields *Journal of The American Water Resources Assosiation* **40** 1 33-41
[32] Takahiro S, Yasuto T and Kaoru T 2012 Spatial lumping of a distributed ranfall-sediment-runoff model and its effective lumping scale *Hydrological Processest.* **26** 855-871
[33] Yasuto T, Takahiro S and Kaoru T 2008 Lumping a Physically-based Distributed Sediment Runoff Model with Embedding River Channel Sediment Transport Mechanism *Annuals of Disas. Prev. Res. Inst.* **51 B** 103-117
[34] Takahiro S, Yasuto T and Kaoru T 2007 The Spatio-Temporal Predictions of Rainfall-Sediment-Runoff Based on Lumping of a Physically-based Distributed Model *Annuals of Disas. Prev. Res. Inst.* **50 B** 79 - 94
[35] Takahiro S, Yasuto T and Kaoru T 2008 Lumping of a Physically-based Distributed Model For Sediment Runoff Prediction in a Catchment Scale *Annuals Journal of Hydraulic Engineering, JSCE* **5** 43-48

IOP Conf. Series: Materials Science and Engineering 1108 (2021) 012011 doi:10.1088/1757-899X/1108/1/012011

Electronic system of rubber belt electro-mechanical continuously variable transmission for motorcycle applications

B Supriyo[1,*], S Ariyono[2] and S Sihono[3]

[1] Magister Terapan Teknik Telekomunikasi, Politeknik Negeri Semarang, Semarang, Indonesia
[2] Program Studi Teknik Mesin, Politeknik Negeri Semarang, Semarang, Indonesia
[3] Program Studi Teknik Elektronika, Politeknik Negeri Semarang, Semarang, Indonesia

*bsupriyo2015@polines.ac.id

Abstract. This paper deals with the design of electronic system for changing the transmission ratio of a rubber belt electro-mechanical continuously variable transmission (CVT) for motorcycle applications. In this CVT, the rubber belt is used as a medium to transfer both the speed and torque from the input pulley to the output pulley based on contact friction force existed between the belt and both pulley surfaces. Each pulley set consists of one fixed and one sliding pulley sheaves. The axial position of the input sliding pulley sheave is detected using a two channel incremental rotary encoder and controlled by an electro-mechanical shifter system utilizing DC motor, actuator gear set and cam mechanism. The sliding sheave of the output pulley is axially pushed by contra spring to provide sufficient belt compression for preventing belt slip. The CVT ratio is changed by varying the input and output belt-pulley contact radii. It can be done by controlling the axial position of sliding input pulley sheave, the angular speeds of input and output pulleys are sensed using one channel incremental rotary encoders. By knowing these two speed values, the CVT ratio can be computed. The microcontroller system of Arduino is programmed to detect and control the axial position of the input sliding pulley sheave to regulate the CVT ratio. The experimental results show that the electronic system can perform the CVT ratio changes adequately.

1. Introduction

The use of two-wheeled motorcycles as a means of personal transportation in Indonesia has increased significantly recently, especially for rubber belt Continuously Variable Transmission (CVT) type motorcycle which is generally called as automatic motorcycle. This kind of motorcycle becomes very popular because of good driveability, smooth transmission shifting and easy to drive. To drive this motorcycle, the driver only turns the gas handle appropriately to control the speed of the motorcycle without manually adjusting the gear shifting clutch, because the clutch engages automatically when the engine speed reaches its predetermined speed. However, the motorcycle equipped with CVT is still less efficient than that with manual transmission, due to some power losses inherent in the CVT itself [1-3]. To cope with this problems, some extensive studies on improving the transmission efficiency by reducing the losses have been done lately in [4-6].

The three main components of the basic CVT: input pulley, output pulley and rubber belt. The belt transfers speed and torque from the input pulley to the output pulley by means of the belt-pulleys' friction forces, as shown in figure 1. When the belt rotates without slipping on both surfaces of input and output pulleys, then the tangential velocities of these three components are the same. The CVT ratio (*rcvt*) can be defined as in (1), while the geometrical speed ratio (*rgs*) can be represented as in (2).

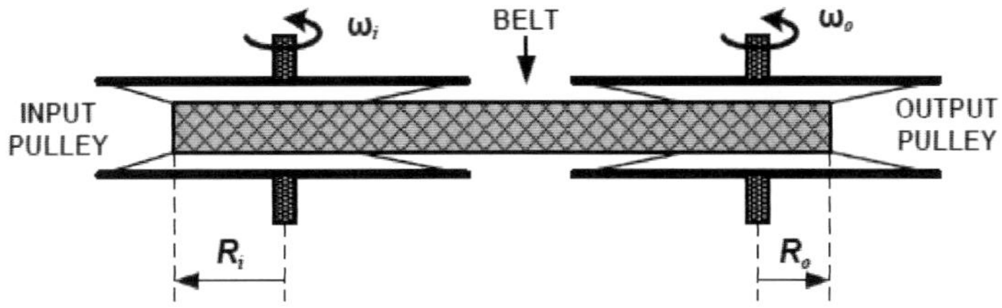

Figure 1. Basic CVT.

$$r_{cvt} = R_o / R_i \quad (1)$$
$$R_{gs} = \omega_i / \omega_o \quad (2)$$

where R_i, R_o, ω_i and ω_o are contact radius developed between input pulley and belt, contact radius developed between output pulley and belt, angular speed of input pulley and angular speed of output pulley, respectively. Each pulley set of this CVT consists of one fixed and one sliding pulley sheaves facing to each other forming the 'V' letter attached to its respective shaft, as shown in figure 2 and figure 3. In this case, both the left side of the input pulley sheave and the right side of the output pulley sheave are sliding sheaves, whereas the other two sheaves are fixed. Figure 2 shows the changing CVT ratio process from under drive position ($r_{cvt} < 1$) to over drive position ($r_{cvt} > 1$), while figure 3 shows the changing CVT ratio process from over drive position to under drive position.

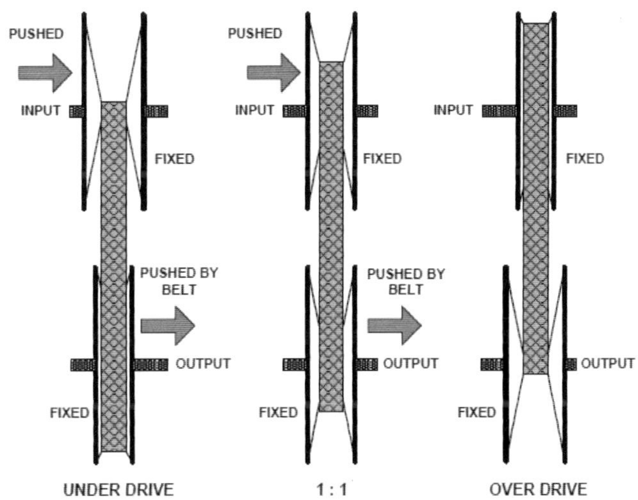

Figure 2. Changing CVT ratio from under drive to over drive.

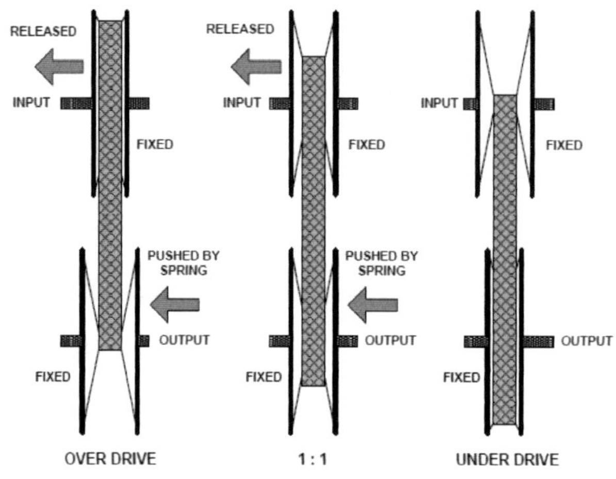

Figure 3. Changing CVT ratio from over drive to under drive.

Generally, in a traditional automatic CVT motorcycle as shown in figure 4, the crankshaft of the motorcycle engine drives the CVT input pulley shaft, while the CVT output pulley shaft is connected to the input part of the centrifugal clutch system. The output part of the clutch system is then connected to the finar gear of the rear wheel axle. The power source from the engine in the form of speed and torque is regulated by the automatic transmission (CVT) to drive the rear wheel. By regulating the speed and torque optimally, the efficiency of fuel use on motorcycles can be improved significantly [7].

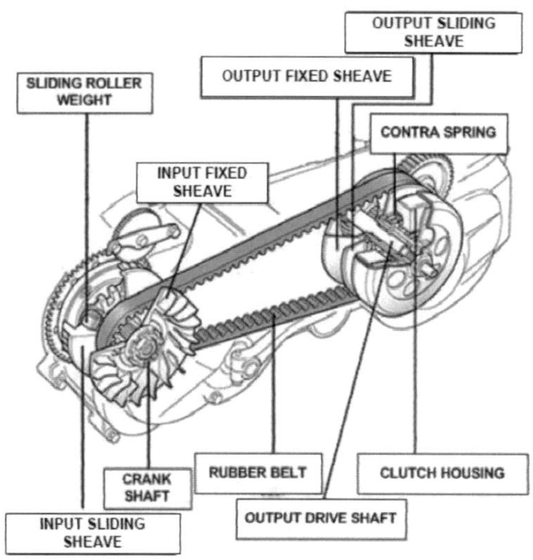

Figure 4. Traditional rubber belt CVT.

The shifting of the input sliding pulley sheave is regulated by the centrifugal force generated by the sliding roller weight placed on the grooves inside the input sliding pulley sheave during high engine speed. The shifting of the output sliding pulley sheave is regulated by the combination of axial forces resulted from the belt and contra spring placed on the back of this sliding pulley sheave. The spring provides sufficient belt compression to prevent belt slip.

When the input pulley rotates at high speed, the centrifugal force acting on the sliding roller weights produces the axial force which pushes the input sliding pulley sheave inward, clamps and shifts the belt upward accordingly on its respective pulley surfaces, hence increasing the input belt-pulley contact radius. On the output side, the belt slides down and push the output sliding pulley sheaves outward along its shaft, thus widening the pulley gap. It decreases its output belt-pulley contact radius, hence reducing its CVT ratio.

If input pulley rotates at lower speed than before, the smaller centrifugal force acting on the sliding roller weight is produced, resulting in a smaller axial force, releasing the input sliding pulley sheave outward and causing the belt to slide downward on the pulley surfaces, hence decreasing its input belt-pulley contact radius. On the output side, the belt loses and the spring pushes back the output sliding pulley sheave inward to push the belt upward tightly on the pulley surfaces, increasing its output belt-pulley contact radius, hence increasing its CVT ratio.

The development of electronic control for DC motor-driven CVT ratio shifter has been studied in several research [8-10]. In this study, the prototype of an electro-mechanical rubber belt CVT test rig for motorcycle application has been developed. In this prototype, the shifting position of the input sliding pulley sheave is sensed using an incremental rotary encoder and electromechanically controlled using the combination of a gear-train set and a DC motor as its actuator. By regulating the position of the input sliding pulley sheave, the CVT ratio is controlled to achieve its predetermined set point. The aim of this study is to design and develop an electronic system to perform the CVT ratio controller. The Arduino based microcontroller is programmed to detect and control the axial position of the input sliding pulley sheave for regulating the CVT ratio as well as to transmit all measurement data to computer.

2. Materials and methods

The concept of the electro-mechanical rubber belt CVT has been introduced in Cholis et al [11]. The electro-mechanical rubber belt CVT test rig is presented in figure 5. The mechanical set of the electromechanical CVT consists of input fixed pulley sheaves, input sliding pulley sheave, output fixed pulley sheaves, output sliding pulley sheave, rubber belt, input shaft, output shaft, output contra spring and actuator cam mechanism equipped with 4-gear train-set. The cam mechanism adjusts the axial position of the input sliding sheave to control the width of the input pulley gap. The output contra spring compresses the output sliding pulley sheave to prevent the belt from slipping.

In this test rig, the 3-phase AC motor is used to represent the motorcycle engine which drives the shaft of the input pulley set. The speed controller drive is used to control the speed of the AC motor. In this experiment, the speed of the primary pulley shaft is maintained to about 1000 rpm. The shaft of the AC motor is connected to a fixed ratio pulley-belt speed reducer system consisting of primary pulley (diameter of 60mm) as input, rubber belt (770mm in length) as connecting belt and secondary pulley (diameter of 127mm) as output. The secondary pulley of the speed reducer is directly attached to the shaft of the CVT input pulley.

The input speed sensor uses the 400-ppr incremental rotary encoder in which its shaft is connected to the shaft of the CVT input pulley, while the output speed sensor also uses the 400-ppr incremental rotary encoder, but its shaft is connected to the shaft of the CVT output pulley. Each encoder directly monitors the angular speed of its respected pulley, by first counting the number pulses produced by the encoder for 1-second gating time, then converting it to its respective value in rpm unit. The shifting actuation of the input sliding pulley sheave is performed by the cam mechanism which is driven by the DC motor via the actuator gear set having speed reducer of about 10.7. An H-bridge motor driver is used to drive the DC motor. Another 400-ppr incremental rotary encoder is used as both the speed sensor for the DC motor and the input sliding pulley position sensor. The shaft of this encoder is connected to the output shaft of the DC motor. This encoder has two output channels which produces two continuous pulses with 90 degrees out of phase to each other. Based on these two pulses, both the direction and the number of the DC motor shaft rotations can be determined, hence the axial position of the input sliding pulley can be indirectly calculated in terms number of pulses. In this study, the DC motor speed is regulated using PWM technique.

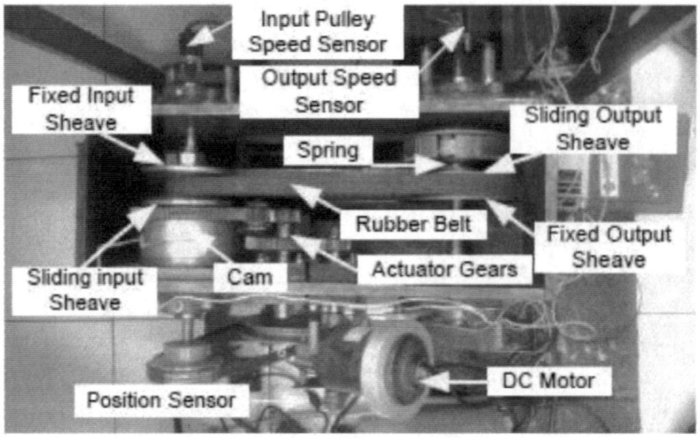

(a)

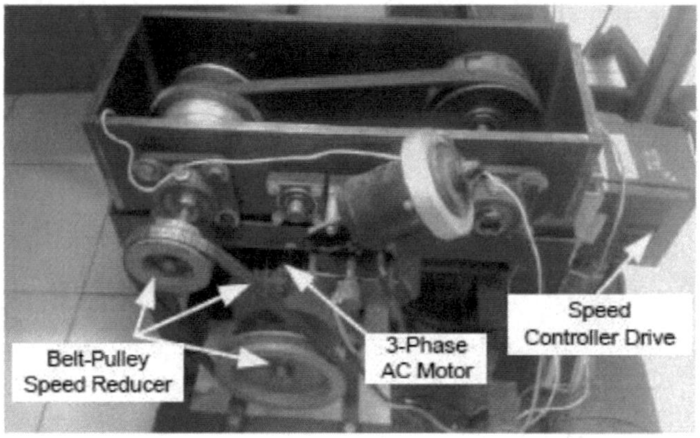

(b)

Figure 5. (a) Mechanical and electronic system (b) Power source system.

The electronic block diagram of the CVT test rig is presented in figure 6. This consists of speed and position sensors, microcontroller, DC motor driver, DC motor and computer. Each block of the electronic component can be described as follows:

2.1. Position sensor of input sliding pulley sheave

A two channel incremental rotary encoder of 400 pulses per revolution (ppr) is attached to the output shaft of the actuator DC motor to indirectly track the axial position of the input sliding pulley sheave and detect the effective input belt-pulley contact radius. These measurement data of the input sheave position are then correlated to the measurement data of the CVT ratio derived from the speed measurements of both input and output pulleys.

2.2. Speed sensors

The angular speeds of input and output pulley shafts are sensed using 400-ppr incremental rotary encoders, namely input pulley speed sensor and output pulley speed sensor. The encoder produces 400 pulses per one shaft rotation. By recording the total number of pulses for each speed sensor for every 1-second gating time, then the angular speed of each pulley shaft can be calculated rpm. By knowing both input and output pulley speeds experimentally, the geometrical speed ratio can be obtained by using formula in (2). If there is no slip between the belt and pulley surfaces, then experimentally this geometrical speed ratio can represent the real CVT value.

2.3. Arduino Uno

Arduino Uno is an 8-bit microcontroller board based on ATMega328. It has 14 digital input/output pins, in which six of them can be configured for pulse width modulation (PWM) outputs. In addition, it has 6 pins configurable for analog inputs. Arduino Uno is very popular and considerably cheap in price. It can be used to control less complicated application satisfactorily. In this study, Arduino Uno is used as a main controller system programmed using Arduino IDE. It communicates with computer via USB to record, save and display the measurement data of input pulley position, input and output shaft speeds and to control the DC motor for adjusting the input pulley position accordingly.

2.4. DC motor and driver

A DC motor is used to drive the input shaft of the 4-gear train-set to actuate the cam mechanism and finally to control the CVT ratio based on the ratio of the effective belt-pulley contacts of input and output pulleys. The speed of this DC motor is controlled using PWM signal from Arduino microcontroller via the H-bridge DC motor driver.

2.5. Computer

Computer communicates with Arduino Controller via USB port to record, save and display the measurement data consisting of input pulley position, input and output shaft speeds and the real CVT ratio.

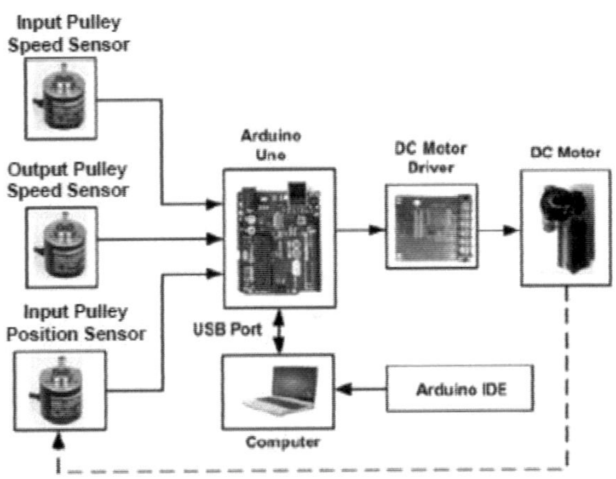

Figure 6. Block Diagram of the electronic system.

The research method is focusing on how to prepare the CVT test rig and to conduct experiment works for obtaining the measurement data consisting of axial position of input sliding pulley sheave, speed of the input pulley and the speed of output pulley. The method to conduct the experimental works can be presented as follows:

- The experiment starts by initially regulating the speed of the AC motor as the power source to the CVT test rig such that the speed of the input pulley shaft is about 1000 rpms
- Changing the CVT ratio by actuating the DC motor to shift the axial position of the input sliding pulley sheave. The axial position is sensed using the incremental rotary based position sensor and represented in the number of pulses. The CVT ratio is changed for every 100 pulse position increment starting from the under drive to over drive position.
- Reading the encoder pulses of input and output pulleys' speed sensors for 1-second gating time and converting these data to their respective units in rpms.
- Calculating the CVT ratio based on the input and output speeds obtained from step 2.
- Save the data of axial position of the input sliding pulley, angular speed of input pulley, angular speed of output pulley and calculated CVT ratio to the computer.

- Draw the graph of pulley speeds versus steply axial position of input pulley sheave.
- Draw the graph of CVT ratio versus steply axial position of input pulley sheave.

3. Results and discussion

The relationship between both angular speeds of input and output pulleys and steply axial position of input sliding pulley sheave which is represented by the number of encoder pulses is shown in figure 7. The speed of the input pulley is actually kept constant at about 1000 rpms. This speed is supplied by the AC motor representing the engine of the motorcycle. The speed is adjusted using a speed controller drive. However, the real speed supplied to the input pulley shaft is a bit lower due to the increasing clamping force developed by the contra spring which cause the increasing friction force developed between the belt and the pulley surfaces. The closer the speed of the output pulley approaching to the speed of the input pulley, the heavier belt-pulley friction load should be encountered by the AC motor causing the speed of the input pulley drops more significantly.

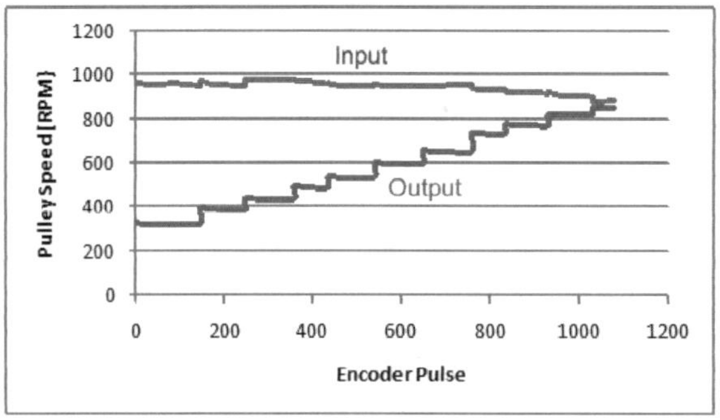

Figure 7. Pulley speeds versus steply axial position of input pulley sheave.

The relationship between CVT ratio and steply axial position of input sliding pulley sheave representing by the number of encoder pulses is shown in figure 8.

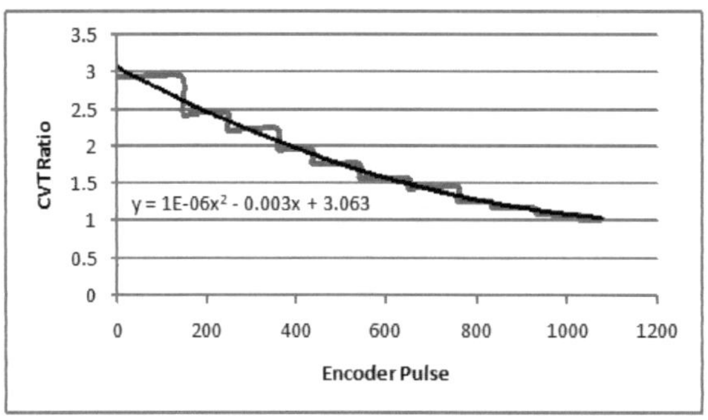

Figure 8. CVT ratio versus steply axial position of input pulley sheave.

4. Conclusion

The design and development of electronic system of rubber belt electro-mechanical Continuously Variable Transmission for motorcycle applications has been carried out successfully. The results show that there is a relationship between the CVT ratio and the axial position of input sliding input detected using a 400-ppr incremental rotary encoder position sensor, which follows a second order polynomial equation. Based on this equation, it is possible to approximate the CVT ratio by utilizing the number of pulses produced by the encoder based position sensor. The detected effective CVT ratio is between 3.0 (lowest speed) and 1.0 (highest speed). The input angular speed of the CVT system slightly decreases as the output pulley speed starts approaching the input speed due to the significant increase in belt clamping force exerted by the output contra spring.

References

[1] Zhu C, Liu H, Tian J, Xiao Q and Du X 2010 Experimental investigation on the efficiency of the pulley-drive CVT *International Journal of Automotive Technology* **11** 2 257–261

[2] Grzegożek W and Kot A 2016 The experimentalanalysis of the slip in the rubber belt CVT *IOP Conf. Series: Materials Science and Engineering* **148** 1 1-8

[3] Bertini L, Carmignani L and Frendo F 2014 Analytical model for the power losses in rubber V-belt continuously variable transmission (CVT) *Mechanism and Machine Theory* **78** 1 289–306

[4] Chen T F, Lee D W and Sung C K 1998 An experimental study on transmission efficiency of a rubber V-belt CVT *Mechanism and Machine Theory* **33** 4 351–363

[5] Chen T F and Sung C K 2000 Design considerations for improving transmission efficiency of the rubber V-belt CVT *International Journal of Vehicle Design* **24** 4 1-16

[6] Shirasuna T, Hatakeyama R and Sakai Y 2016 Establishment of Fuel Economy Estimation Method Focused on Transmission Efficiency of Rubber Belt Type CVT *SAE International Journal of Engines* **9** 4 2396-2415

[7] Tsuji T and M.Takeoka M 1996 Study of Fuel Consumption Improvement of the Car with the Dry Hybrid Belt CVT *Society of Automotive Engineers of Japan, JSAE Review* **17** 1 381-385

[8] Tsutsumikoshi S, Tomita K and Ohashi K 2002 Development of Motorcycle Using Electronic Controlled Continuously Variable Transmission *SAE Technical Paper 2002-32-1771*

[9] Asumi M, Ishikawa H, Mori K and Matsudaira N 2005 Development of Electronically Controlled Belt-type CVT for Motorcycles *SAE Technical Paper 2005-32-0024*

[10] Supriyo B, Ariyono S, Tjahjono B and Sumiyarso B 2019 Electro-mechanical Transmission Ratio Shifter of Rubber Belt Continuously Variable Transmission for motorcycle applications *Journal of Physics: Conference Series* **1273** 1 1-6

[11] Cholis N, Aryono S and Priyandoko G 2015 Design of single acting pulley actuator (SAPA) continuously variable transmission (CVT) *Energy Procedia* **68** 1 389–397

Design and development of student attendance information system using QR code in accounting department of Politeknik Negeri Semarang

S Sarana*, A Sadida, A W Mansyur and A Suwondo

Politeknik Negeri Semarang, Semarang, Indonesia

*sarpolines@gmail.com

Abstract. This study aims to streamline the lecture attendance system and overcome indications of cheating on student absences in the lecture process in class. Manual attendance list causes lost time of lectures and waste of paper, and therefore the need for an attendance system to reduce these problems. Higher education with all its strategic activities cannot be separated from the importance of using information technology. One of the important activities routinely carried out by a tertiary institution is to record Academic Administration activities (teaching and learning process), this routine activity involves many interests both from the lecturers, administration and from the learning facilities and infrastructure they have. This activity must be made an information system in order to facilitate all parties and produce information that is fast and valid. One such information is student information about absence which can be caused by permission, illness or even negligence. This activity is carried out every day so that if and error occurs or negligence will harm all parties involved in the teaching and learning process therein. This article discusses how to design and build an Academic recording information system in the form of student attendance that can produce accurate and timely information available to meet the needs of students, Academic Administration and department managers so that the information system can be aligned with academic regulations and can be accepted by system users. The object used is the Semarang State Polytechnic Accounting Department. The method used is to use the Unified Model Language approach which consists of Use Case Diagrams, Class Diagrams, and Data Dictionaries. While the development model using waterfall, the stages include analysis of the existing system, concept design, physical design, implementation and testing. The output of the research is an Android-based student attendance information system application using QR Code which generates information for students, Academic Administration and department managers.

1. Introduction

Politeknik Negeri Semarang (Polines) is a vocational-based educational institution, this learning system is more focused on work or practice skills compared to theory. The lecture system can work well if the recording of student arrivals is effective and efficient. Politeknik Negeri Semarang (Polines) is currently in the midst of students still using manual recording that is students who have permission to register every student who receives lectures and lecturers record the absence of students in class at lecturer teaching hours. The student participation recording system used today is ineffective and inefficient because the opportunity for students to manipulate the release participation data still occurs and sometimes the lecturer does not record the absence of the student. Teaching and Learning Management

(PBM) also corrects the difficulty in calculating the hours of absence of students at one time that will be sent to the Student Administration section to determine whether students can get an exam, whether students get a Warning Letter 1 (SP1), Warning Letter 2 (SP2) The Semarang State Polytechnic is in accordance with the Academic Regulations (Perak) which have often been a mismatch of data in the absence of student absence between PBM data and the results of calculations in the Student Administration Section (BAAK).

Technology that continues to develop nowadays can be used to improve the quality of the systems needed in recording student attendance at the Polytechnic. One technology that can be applied is Rapid Response technology or known as QR Code. Application of this technology as practiced by Arulogun. The application of QR Code for monitoring student attendance developed and used in this study is able to eliminate time wasted during manual attendance gathering and opportunities for education administrators to capture classroom face-to-face statistics for allocation of attendance scores that are appropriate and for further managerial decisions [1]. Also implemented in Malaysia, a systematic student participation management system based on the QR Code allows many easy methods to receive meetings [2-5].

This application is expected to reduce and even eliminate the manipulation of data received by students in the Politeknik Negeri Semarang, especially Computerized Accounting Study Program and facilitate the calculation of student absence hours that are expected to be the same data in PBM as the data in BAAK. Application software development in various fields such as academic, construction, industry, etc. uses the SDLC (System Development Life Cycle) method, whereas for system development using the Use Case model, and to get improved system performance developed using the UML model with case study on Amazon.com. The development of student attendance recording systems is also carried out with this SDLC method as designed by Hanisah [2] which states "in collaboration with developing a system development cycle (SDLC), this system has been fostered using a web-based application". Supporting technology in academic support systems as outlined above is an interesting challenge to apply to overcoming difficulties in the Politeknik Negeri Semarang.

Regarding the problem discussed in this study is How to discuss Academic recording information systems that include student attendance that can produce accurate information and be available in a timely manner to meet the needs of students, PBM and department managers make information systems in line with system regulations.

2. Literature review

2.1. *Waterfall system development model*

Software engineering method used by researchers is the waterfall method. According to Pressman et all waterfall is a classic model that is systematic, sequential in building software [3]. Here are descriptions of the waterfall model.

Phases in the waterfall model according to the Pressman reference:

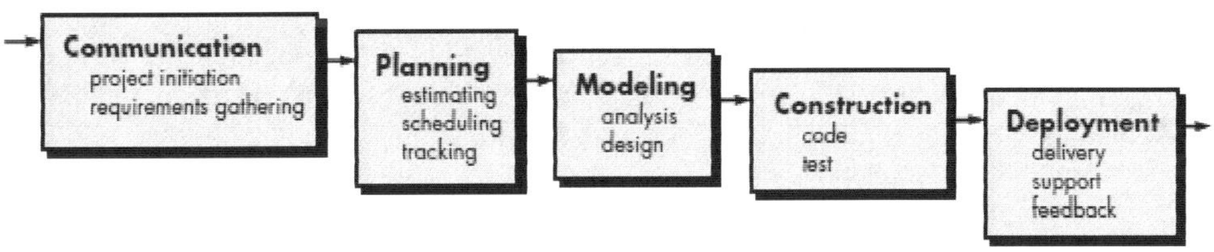

Figure 1. Waterfall Pressman.

2.1.1. Communication. This step is an analysis of software requirements, and the stage for conducting data collection by meeting with customers, as well as collecting additional data both in journals, articles, and from the internet.

IOP Conf. Series: Materials Science and Engineering 1108 (2021) 012012 doi:10.1088/1757-899X/1108/1/012012

2.1.2. Planning. The planning process is a continuation of the communication process (requirements analysis). This stage will produce user requirements documents or can be said as data related to the user's wishes in making software, including plans to be carried out.

2.1.3. Modeling. This modeling process will translate requirements into a software design that can be estimated before coding is made. This process focuses on the design of data structures, software architecture, interface representation, and procedural details (algorithms). This stage will produce a document called software requirements.

2.1.4. Construction. Construction is the process of making code. Coding is the design translation in a language that can be recognized by a computer. The programmer will translate the transaction requested by the user. This stage is the actual stage in working on a software, meaning that the use of computers will be maximized in this stage.

After the coding is complete it will be tested on the system that was made earlier. The purpose of testing is to find errors in the system and then it can be fixed.

2.1.5. Deployment. This stage can be said to be final in making a software or system. After conducting analysis, design and coding the finished system will be used by the user. Then the software that has been made must be done regularly maintenance.

2.2. Quick Response code (QR code)
QR Code, short for Quick Response Code, is a two-dimensional image that has the ability to store data. QR Code is commonly used to store data in the form of text, be it numeric, alphanumeric, or binary code. QR Code is widely used for commercial purposes, especially in Japan, usually containing a URL link to a particular address or just text containing advertisements, promotions and others. One that is not commonly used in QR Code is inserting images into the information that is stored.

Figure 2. QR code.

3. Research methodology

3.1. Research flowchart
The research flowchart is shown in Figure 3 below:

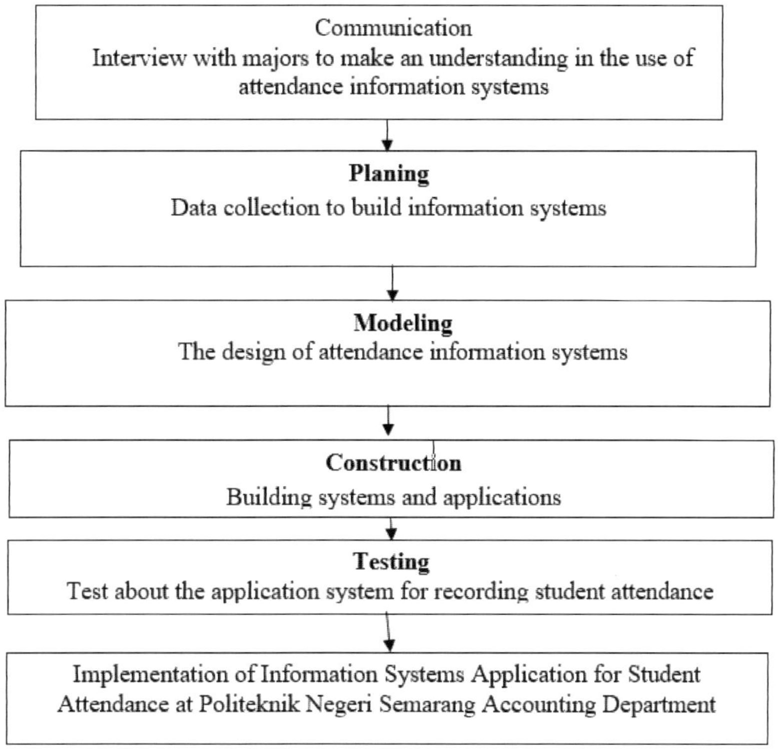

Figure 3. Research flowchart.

3.2. Research stages
The research was made based on the development of the waterfall software. The stages are requirements analysis, design, implementation and testing.

3.2.1. Communication. This stage approaches and talks with the Department of Accounting to get an understanding of the use of this system.

3.2.2. Planning. At this stage the necessary data is collected as input material to create PBM activity recording applications namely course data, lecturer data, class data, room data, student data and class schedule data. The process for processing input data is with the help of RFID devices. The output produced is in accordance with what is expected.

3.2.3. Modeling. Modeling or Design consists of 3 (three) stages namely process modeling with the aim of designing data flow diagrams (DAD), entity relationship diagrams (ERD) and database tables, and designing user interfaces aimed at designing interfaces / displays of input and output systems on screen using GUI (Graphical User Interface) principles that are easily understood by system users.

3.2.4. Construction. Construction in another sense is the implementation of system design into a program module (coding program). In this process, the design will be converted into coding operations by using certain programming languages which are based on the use of genetic algorithms for the process of preparing lecture and examination schedules.

3.2.5. Testing. The testing process is carried out on internal logic to ensure all statements have been tested. Functional external testing to find errors and ensure that the input will provide the actual results as needed. Testing in this study took a case study of PBM Recording in the academic year 2018/2019 at the Semarang State Polytechnic Accounting Department.

3.3. PBM activity recording system
The description of the working order of the system or the system algorithm can be described simply as follows in figure 4:

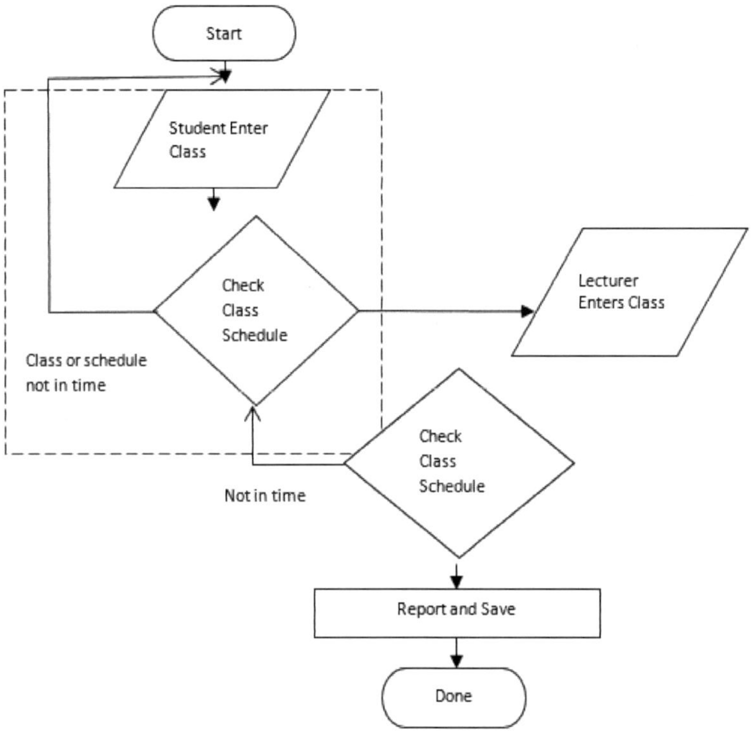

Figure 4. Algorithm for recording PBM activities.

4. Results

4.1. System design

4.1.1. Use case diagram. Use Case is used to describe the interaction between an actor and a system that is built. Usecase's description as follows:

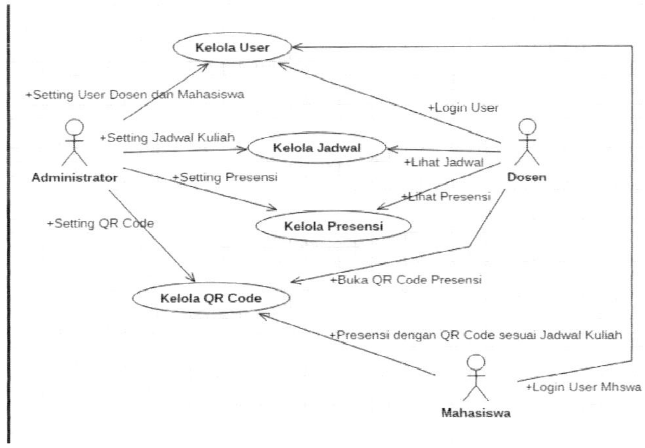

Figure 5. Use case diagram.

4.1.2. ER diagram.

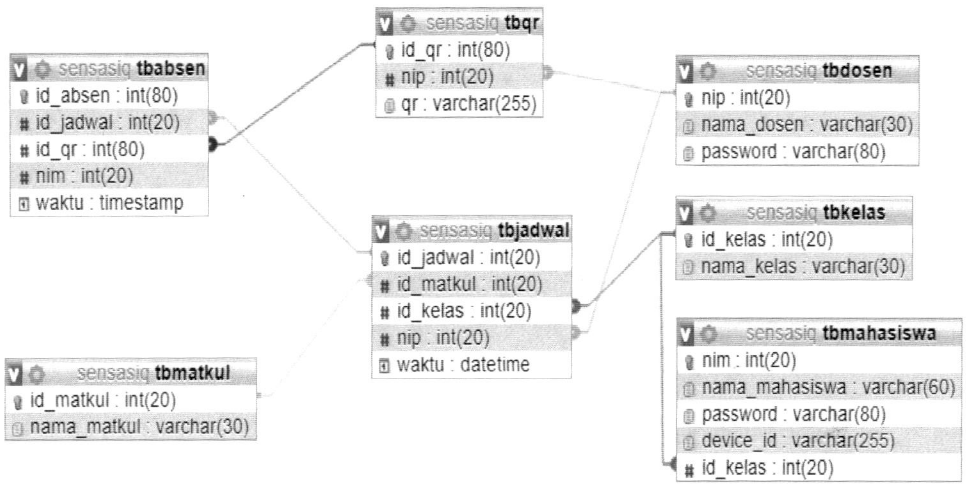

Figure 6. ER diagram.

Diagram Dictionary for ER Diagram design:

tbabsen

Kolom	Jenis	Tak Ternilai	Bawaan	Tautan ke	Komentar	MIME
id_absen *(Utama)*	int(80)	Tidak				
id_jadwal	int(20)	Tidak		tbjadwal -> id_jadwal		
id_qr	int(80)	Tidak		tbqr -> id_qr		
nim	int(20)	Tidak				
waktu	timestamp	Tidak	CURRENT_TIMESTAMP			

Indeks

Nama kunci	Jenis	Unik	Dipadatkan	Kolom	Kardinalitas	Penyortiran	Tak Ternilai	Komentar
PRIMARY	BTREE	Ya	Tidak	id_absen	18	A	Tidak	
id_jadwal	BTREE	Tidak	Tidak	id_jadwal	9	A	Tidak	
id_qr	BTREE	Tidak	Tidak	id_qr	18	A	Tidak	
nim	BTREE	Tidak	Tidak	nim	6	A	Tidak	

tbdosen

Kolom	Jenis	Tak Ternilai	Bawaan	Tautan ke	Komentar	MIME
nip *(Utama)*	int(20)	Tidak				
nama_dosen	varchar(30)	Tidak				
password	varchar(80)	Tidak				

Indeks

Nama kunci	Jenis	Unik	Dipadatkan	Kolom	Kardinalitas	Penyortiran	Tak Ternilai	Komentar
PRIMARY	BTREE	Ya	Tidak	nip	11	A	Tidak	

tbjadwal

Kolom	Jenis	Tak Ternilai	Bawaan	Tautan ke	Komentar	MIME
id_jadwal *(Utama)*	int(20)	Tidak				
id_matkul	int(20)	Tidak		tbmatkul -> id_matkul		
id_kelas	int(20)	Tidak		tbkelas -> id_kelas		
nip	int(20)	Tidak		tbdosen -> nip		
waktu	datetime	Tidak				

Indeks

Nama kunci	Jenis	Unik	Dipadatkan	Kolom	Kardinalitas	Penyortiran	Tak Ternilai	Komentar
PRIMARY	BTREE	Ya	Tidak	id_jadwal	18	A	Tidak	
id_matkul	BTREE	Tidak	Tidak	id_matkul	18	A	Tidak	
id_kelas	BTREE	Tidak	Tidak	id_kelas	18	A	Tidak	
nip	BTREE	Tidak	Tidak	nip	18	A	Tidak	

tbkelas

Kolom	Jenis	Tak Ternilai	Bawaan	Tautan ke	Komentar	MIME
id_kelas *(Utama)*	int(20)	Tidak				
nama_kelas	varchar(30)	Tidak				

Indeks

Nama kunci	Jenis	Unik	Dipadatkan	Kolom	Kardinalitas	Penyortiran	Tak Ternilai	Komentar
PRIMARY	BTREE	Ya	Tidak	id_kelas	7	A	Tidak	

tbmahasiswa

Kolom	Jenis	Tak Ternilai	Bawaan	Tautan ke	Komentar	MIME
nim *(Utama)*	int(20)	Tidak				
nama_mahasiswa	varchar(60)	Tidak				
password	varchar(80)	Tidak				
device_id	varchar(255)	Ya	*NULL*			
id_kelas	int(20)	Tidak		tbkelas -> id_kelas		

Indeks

Nama kunci	Jenis	Unik	Dipadatkan	Kolom	Kardinalitas	Penyortiran	Tak Ternilai	Komentar
PRIMARY	BTREE	Ya	Tidak	nim	16	A	Tidak	
tbmahasiswa_ibfk_1	BTREE	Tidak	Tidak	id_kelas	16	A	Tidak	

tbmatkul

Kolom	Jenis	Tak Ternilai	Bawaan	Tautan ke	Komentar	MIME
id_matkul *(Utama)*	int(20)	Tidak				
nama_matkul	varchar(30)	Tidak				

Indeks

Nama kunci	Jenis	Unik	Dipadatkan	Kolom	Kardinalitas	Penyortiran	Tak Ternilai	Komentar
PRIMARY	BTREE	Ya	Tidak	id_matkul	14	A	Tidak	

tbqr

Kolom	Jenis	Tak Ternilai	Bawaan	Tautan ke	Komentar	MIME
id_qr *(Utama)*	int(80)	Tidak				
nip	int(20)	Tidak		tbdosen -> nip		
qr	varchar(255)	Tidak				

Indeks

Nama kunci	Jenis	Unik	Dipadatkan	Kolom	Kardinalitas	Penyortiran	Tak Ternilai	Komentar
PRIMARY	BTREE	Ya	Tidak	id_qr	38	A	Tidak	
nip	BTREE	Tidak	Tidak	nip	38	A	Tidak	

4.2. Application

4.2.1. Admin web page. Polines Attendance Application is an application that aims to create a Smart Anti-Fraud based Solution of QR Code. Just as the name suggests, Attendance Polines function to be a substitute for a conventional attendance system that takes queuing time and costs for procuring sheets of attendance paper. After logging in via the form that appears, you will be directed automatically to the lecturer dashboard page. It is on this Lecturer page that data management is carried out.

Figure 7. Home page.

If there are no errors in the username / password that you fill in the login form, then you will automatically be directed to the Dashboard page

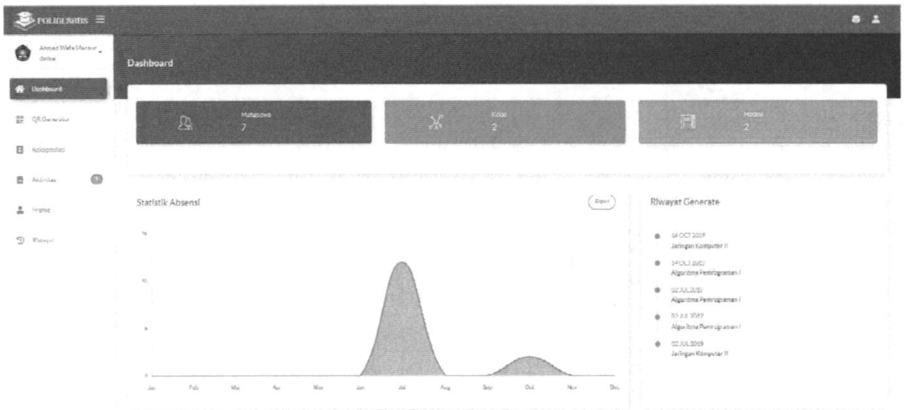

Figure 8. Lecturer page.

The QR Generator menu functions to create a QR code which later must be scanned by students.

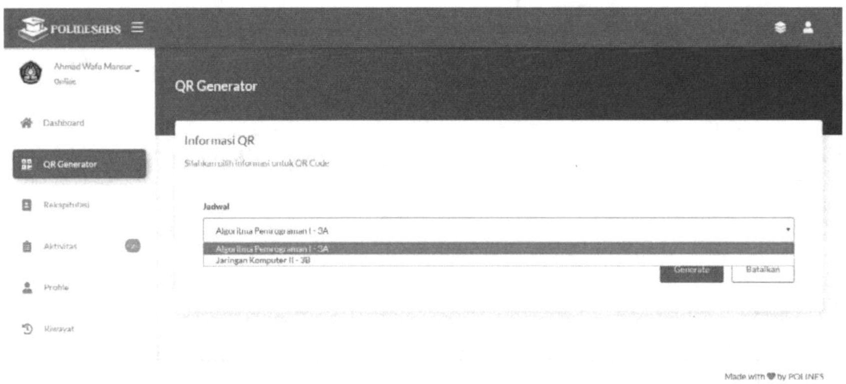

Figure 9. QR generator.

The QR Code results will be displayed to the screen.

Figure 10. QR code.

4.2.2. Student app. After successfully logging in the application, students are then required to attend the attendance process. Click the menu in the upper left corner and select "Scan Attendance QR".

IOP Conf. Series: Materials Science and Engineering 1108 (2021) 012012 doi:10.1088/1757-899X/1108/1/012012

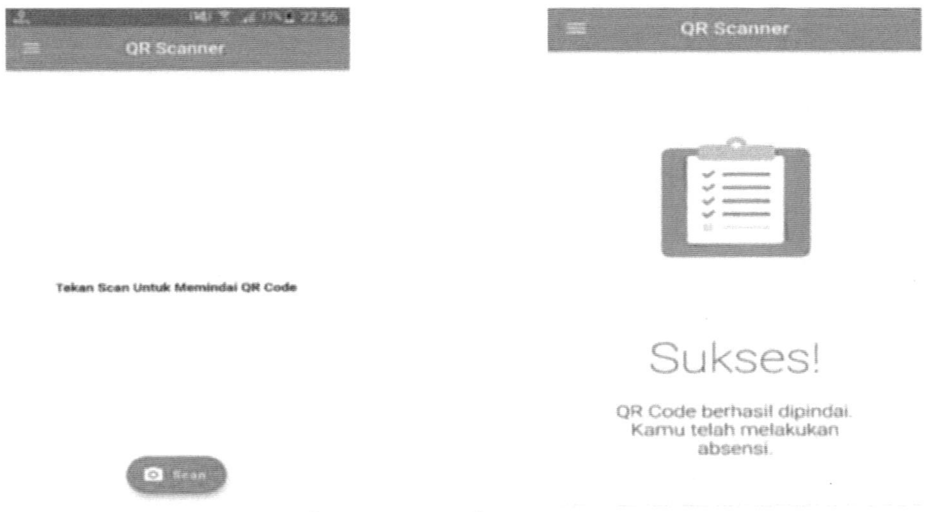

Figure 11. Student app.

If successful in the application a message will appear that the attendance that you have done has been successfully carried out. With the success of the QR code scan, student data will be saved to a database on the server.

4.3. Test system

The system that has been successfully created needs to be tested to see whether the system's performance meets the planned specifications and further known User acceptance of the system and the need for improvement to improve the system itself.

Table 1. Activities / activities tested for each role.

No	Role	Activities	Checklist
1	Admin	Check and manage to ensure all activities run smoothly	Checklist system → System components
2	Lecturer	Check the role and function of making QR Code and report / recap of attendance	QR code generator Attendance Recap
3	Student	Scan QR code Enter absent data recorded	Scan QR code Success Report

5. Conclusion

Making attendance systems with QR Code is done through a design that includes Usecase, ER Diagrams, Database Design and Data Dictionary. This design is based on the needs concluded from the FGD.

The results of the design are implemented into an Android-based application that is divided into 2 stakeholders, namely Lecturers and Students. Lecturers use the application to generate and display QR Codes on the screen in front of the class. Student uses an android application on their smartphones to scan the QR Code projected on the screen and send their data to the server as proof of his presence. The application has been carried out a System Test whose results indicate that in general the Application can be accepted and utilized for the implementation of the Teaching and Learning activities.

Suggested improvements are concluded from entering the implementation of the system test for future system improvements including integration with the existing Schedule Application, the provision of a Local Server that can be more stable and faster access and a more comprehensive attendance report for all lecturers as attendance reports and students as a recap of performance and compensation sanctions from absence.

References

[1] Arulogun O T, Olatunbosun A, Fakolujo O A and Olaniyi O M 2013 RFID-Based Students Attendance Management System *International Journal of Scientific & Engineering Research* **4** 2

[2] Hamid H B 2010 *Rfid Based Systematic Student's Attendance Management System* (Thesis Faculty of Electrical and Electronic Engineering, Universiti Malaysia Pahang)

[3] Pressman J G, Richardson S D, Speth T F, Miltner R J, Narotsky M G, Hunter and Krasner S W 2010 Concentration, chlorination, and chemical analysis of drinking water for disinfection byproduct mixtures health effects research: US EPA's four lab study *Environmental science & technology* **44** 19 7184-7192

[4] Aksamit D, Morris M and Leuenberger J 1987 Preparation of student services professionals and faculty for servicing learning-disabled college students *Journal of College Student Personnel* **28** 53-59

[5] Bowe F 2000 *Universal Design in education. Teaching nontraditional students* (Westport, CT: Bergin and Garvey)

Prediction effective capacity battery of cellular BTS for an effective maintenance

M B Harijadi*, T Andromeda and M Somantri

Electrical Engineering Department, Diponegoro University, Semarang, Indonesia

*burhanudin.harijadi@gmail.com

Abstract. A cellular base transceiver station (BTS) systems must be operated continuously in transmitting signals to the mobile stations. Reliability operation must be performed by this BTS to serve users without any interruptions. In several cases the BTS are found failed due to several disturbances that occurred during operation. One of the disturbances is caused by electrical energy supply interruptions. Although the cellular BTS power supply system is usually equipped by battery as a secondary power supply, the reliability can be influenced by the health of the battery. The health of battery is inclined by instability of main electricity supply, temperature fluctuation of the battery environment, increment of battery aging, the number of charging and discharging cycles (CDC), and the number depth of discharging conditions (DOD) that occurred. The impact of these factors will affect to the strength of battery. This paper proposes an investigation method to the battery health to ensure the performance of BTS. The method is applied by considering the charging and discharging characteristics of the battery. Since the characteristic signals are not linear, the calculation is performed using fuzzy method. Early results show that this prediction method can be applied to support maintenance work. The proposed method gives an overview of the battery health that can be followed up by technician to ensure the reliability of the BTS.

1. Introduction

Cellular telecommunications network has been grown rapidly and widely used by the public today. The communication signal service is provided by a BTS transmitter that directly serves to the user's handset involved in the telecommunications connection process. The operation of BTS uses the main energy supply from PLN electricity that is converted to dc voltage by a rectifier device. And to maintain the continuity of the electricity supply, BTS's are backed up by batteries as secondary energy [1,14,15]. Furthermore, the availability of the BTS transmitter is expected to reach high availability value, and only a few percent allowed to be failed. The connection between the BTS and the battery shown in figure 1.

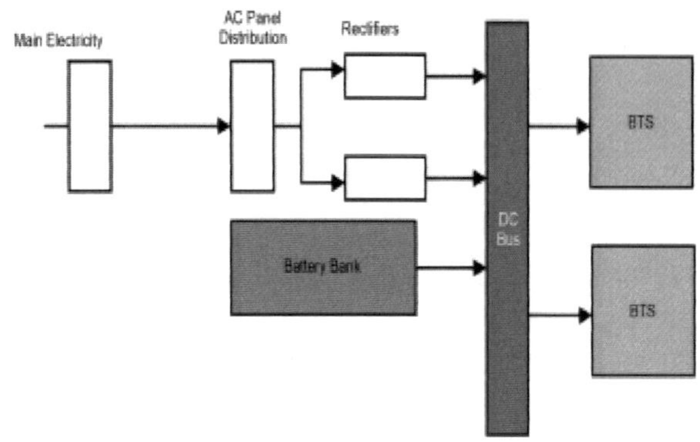

Figure 1. Diagram battery vs BTS [1].

Batteries as secondary energy is a device that converts stored chemical energy into electrical energy. Each cell has a positive pole (cathode) and a negative pole (anode). A pole with a positive sign indicates that it has higher potential energy than a pole with a negative sign. A negative sign is a source of electrons that when connected to an external circuit will flow and provide energy to external equipment. When a battery is connected to an external circuit, electrolytes can move as ions inside it, so that chemical reactions occur in both poles. The transfer of ions in the battery will provide an electric current as an output that produces electrical energy [2].

Battery Charge Discharge Cycle (CDC) is a process of discharge cycles from 100 percent of battery capacity to the lowest specified capacity limit. The discharge process is not recommended until the battery capacity is empty, so it is necessary to set the value of a certain amount, for example, 10 percent or 20 percent, etc. called the State of Charge (SOC). The existence of certain SOC values is done to maintain the health of each battery cell [11-13]. As the opposite of the SOC value, then the value of the amount of battery discharge is usually referred to as the Depth of Discharge (DOD) which is presented as a percentage of Discharge AH / Real Capacity [2].

Battery performance influenced by several external factors i.e. due to the instability of PLN electricity supply give impact to increasing temperature of the battery environment, the number of charging and discharging cycles, and the number DOD will impact to decreasing battery performance [2-7,14,16-19]. Another factor which influence battery performance as well as an internal factor i.e. self-discharging which will reduce the battery capacity in a certain period [15]. And the correlation between battery capacity with Charge Discharge Cycle shown in figure 2.

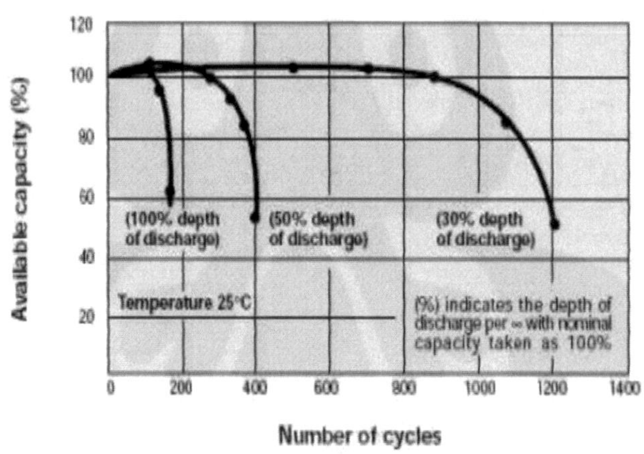

Figure 2. Correlation of battery capacity vs Charge Discharge Cycle [7].

IOP Conf. Series: Materials Science and Engineering 1108 (2021) 012013 doi:10.1088/1757-899X/1108/1/012013

Several researchers calculate an estimation of the battery capacity using artificial intelligence, including using the fuzzy logic method [13]. The fuzzy logic method has three main processes in implementing it on a device, i.e. fuzzification, rule, and inference system, and defuzzification [8-10]:

- Fuzzification is a process to change an input from a strict form (crisp) to fuzzy which is usually presented in the form of fuzzy sets with a function of each member.
- Inference System is as a reference to explain the relation- ship between input and output variables which are processed, and generated variables are fuzzy. To explain the relationship between input and output usually use" IF-THEN".
- Defuzzification is the process of converting these fuzzy variables into definite data (crisp) that can be sent to control equipment.

The fuzzy membership function is shown in figure 3. This paper is divided into five sections. The first section describes the background and purpose of the paper. The second section describes the research methodology used fuzzy logic in this paper. The next section describes the result and discussion comparison of the Mamdani defuzzification and Sugeno defuzzification. Finally, this paper is closed by a conclusion.

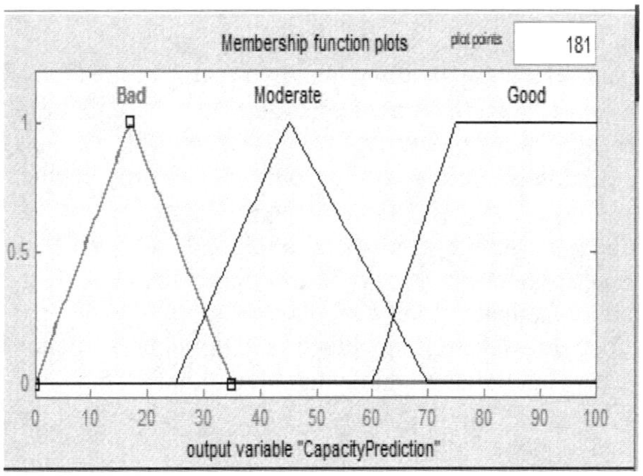

Figure 3. Fuzzy membership function [9].

2. Literature review

Recently many electronic devices such as cellular devices, electric vehicles, notebooks, cameras, music players, etc. have used lithium-ion batteries as secondary energy because it has many advantages compared to other types of batteries such as Nickel Cadmium [7,14,15,17]. Several kinds of research have been done to explore the performance of lithium-ion batteries. And in the energy sector, many studies have been carried out recently to observe lithium-ion batteries used in electric vehicles [6]. But just a few studies carried out to observe lithium ion batteries performance used in cellular system. In electric vehicle usage, the battery works after being fully charged and then used by the electric vehicle so that the SOC (State of Charge) voltage can be measured periodically. After completing the use of electric vehicles, the DOD (Depth of Discharge) value can be measured as well. The value of DOD is the opposite of SOC value. Another researcher evaluates the correlation of battery capacity with various temperatures, shown in figure 4.

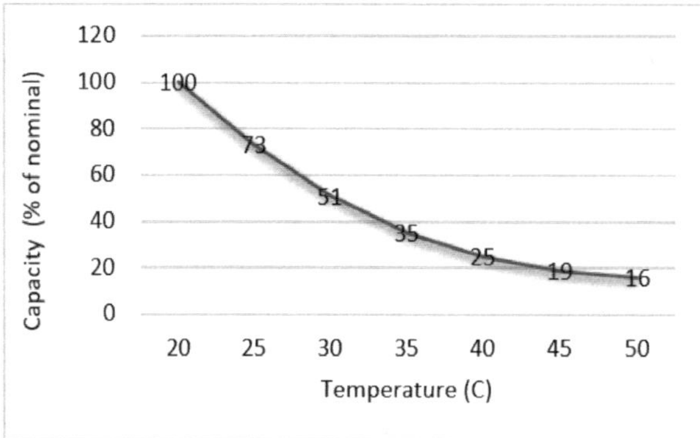

Figure 4. Battery capacity vs various of temperature [16].

Several researchers calculate the SOC value of batteries using artificial intelligence, including using the fuzzy logic method [6]. The use of the fuzzy method is done with consideration of intelligent fuzzy systems that can describe, knowing, and modeling human thought processes and design a system so that it can describe human behavior. By using a fuzzy process consisting of fuzzification, inference and defuzzification, the level of correctness of the determination of the output value obtained is more accurate. Another researcher performs monitoring performance of lithium-ion batteries used in electric vehicles [6]. The battery performance calculated by modeling and simulate the SOC value, cause of this manner considered more efficient than direct measurement such as open connection voltage.

By consideration until now most of the research is conducted on the performance of electric vehicles battery, and currently still not so much research on the performance of cellular BTS batteries, making it an interesting challenge to investigate it further. The cause of the battery of cellular BTS in the application is deployed widely in wide areas to cover all users in which every area has complex mains electricity conditions. Such as batteries in the urban area usually have more stable electricity than suburban areas. This means batteries in urban areas have less Depth of Discharge than suburban areas.

3. Methodology

This research was performed by several steps i.e.

- Designing diagram of the system
- Establishing the connection between sensors and the server
- Collecting data measurements
- Designing the Matlab simulink modeling
- Designing fuzzy logic functions
- Running fuzzy logic functions
- Representing the result

Here we measured several parameters of the battery, i.e. voltage value, current value, temperature value and the value of DOD. The parameter values are obtained from the sensors that have been installed which will send the measurement data periodically at a predetermined period that is every 5 minutes. Monitoring is carried out within 30 days of measurement. The data obtained then processed to become a simulation material using Simulink Matlab to determine the member function and designing crisp and rules used as input for the defuzzification process. With variations of these input parameter values and simulated with predetermined rules, a result of the fuzzy is obtained.

3.1. Parameters

Fuzzy logic member function represents facts and certain criteria according to the four measurement objects i.e. the voltage value, current value, temperature value, and DOD value. The measurement data we got then in the next step we use to create a member function of each input parameter with a certain value such as low, moderate and high, shown in table 1. Noted here we do not apply the internal discharge in the calculation cause of the value is very small just in micro ampere unit.

Table 1. Variable and category each parameter.

No	Parameter	Weight %	Range	Criteria	Score	Output
			33-60	Hot	1	Bad
1	Temperature	25	23-35	Moderate	2	Moderate
			10-25	Cold	3	Good
			40-45	Low	1	Bad
2	Voltage (V)	25	44-49	Moderate	2	Moderate
			48-55	High	3	Good
			18-40	High	1	Bad
3	DOD (%)	25	8-20	Moderate	2	Moderate
			0-10	Low	3	Good
			90-150	High	1	Bad
4	Current (A)	25	50-100	Moderate	2	Moderate
			30-60	Low	3	Good
			0-35	Low	1	Bad
5	Capacity prediction (%)	25	25-70	Moderate	2	Moderate
			60-100	High	3	Good

We define rule function with various of four parameter inputs which each parameter has three various memberships. All the various input parameters processed with IF-THEN operation to obtain the output. Totally here in this measurement we use 3x3x3x3 rules equal to 81 rules is shown in table 2 (part of total rules).

Table 2. Rule function (part of total rules).

Rule	Temp (C)	Voltage (V)	DOD (%)	Current (A)	Capacity Pred (%)
1	Cold	Low	Low	Low	Moderate
2	Cold	Low	Low	Moderate	Moderate
3	Cold	Low	Low	High	Moderate
4	Cold	Low	Moderate	Low	Moderate
5	Cold	Low	Moderate	Moderate	Moderate
6	Cold	Low	Moderate	High	Bad
7	Cold	Low	High	Low	Moderate
8	Cold	Low	High	Moderate	Moderate
9	Cold	Low	High	High	Bad
10	Cold	Moderate	Low	Low	Moderate
11	Cold	Moderate	Low	Moderate	Moderate
12	Cold	Moderate	Low	High	Moderate
13	Cold	Moderate	Moderate	Low	Moderate
14	Cold	Moderate	Moderate	Moderate	Moderate
15	Cold	Moderate	Moderate	High	Moderate
16	Cold	Moderate	High	Low	Moderate
17	Cold	Moderate	High	Moderate	Moderate
18	Cold	Moderate	High	High	Moderate
19	Cold	High	Low	Low	Good
20	Cold	High	Low	Moderate	Moderate
21	Cold	High	Low	High	Moderate
22	Cold	High	Moderate	Low	Moderate
23	Cold	High	Moderate	Moderate	Moderate
24	Cold	High	Moderate	High	Moderate
25	Cold	High	High	Low	Bad
26	Cold	High	High	Moderate	Moderate
27	Cold	High	High	High	Bad

The system design comprised of fuzzification process, rule function and defuzzification process is shown in figure 6.

IOP Conf. Series: Materials Science and Engineering 1108 (2021) 012013 doi:10.1088/1757-899X/1108/1/012013

3.2. Scenarios

In in this paper we collect 500 data inputs from each sensor. To process the data which has been measured and collected from sensors in each of the parameters we process using Matlab simulink. The first step here is process various data input using fuzzy toolbox. The input data measurements are shown in table 3 (part of total data measurement).

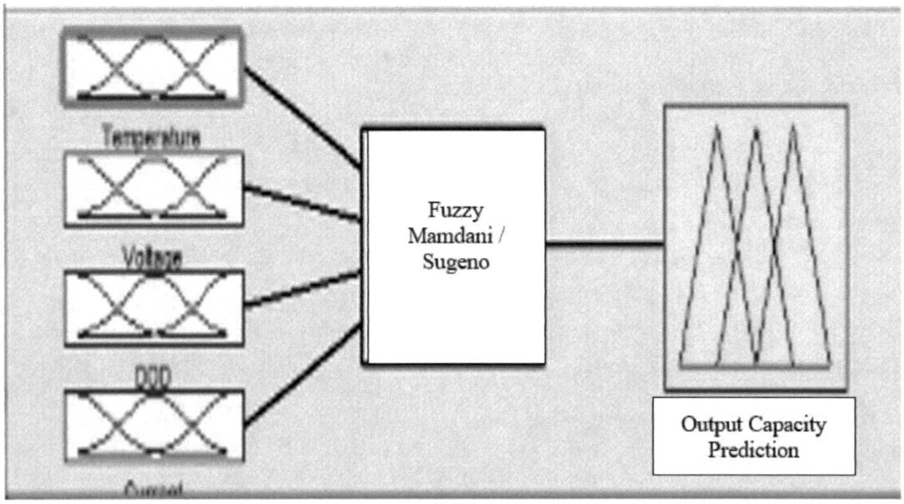

Figure 5. System design.

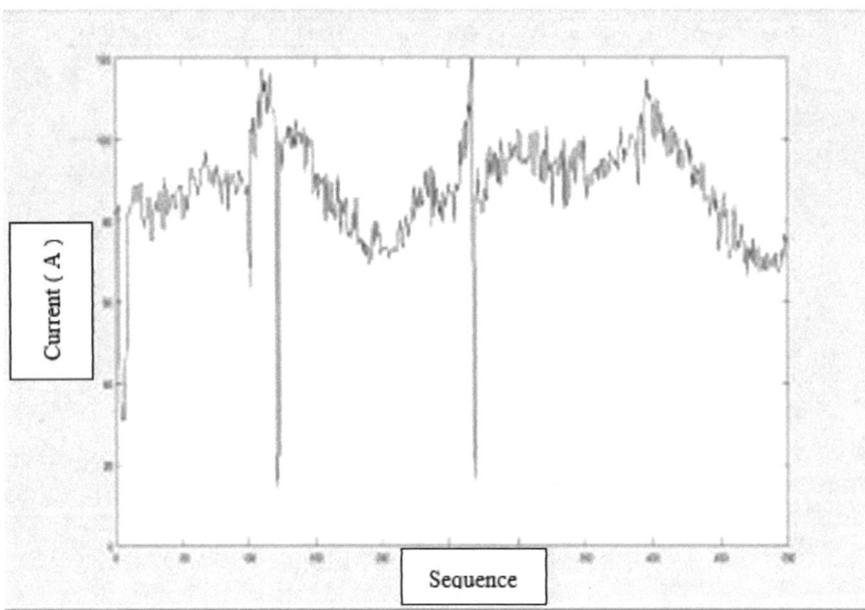

Figure 6. Current data input.

Table 3. Data measurement.

Data	Temp (C)	Voltage (V)	DOD (Times)	Current (A)
1	29.9	53.6	2	87.4
2	29.8	53.6	0	82
3	29.8	53.6	18	78.7
4	29.9	53.7	31	75.8
5	29.8	53.6	34	81.6
6	29.8	53.6	29	81.6
7	29.8	53.7	4	82.2
8	29.8	53.6	38	76.2
9	29.8	53.6	28	76.2
10	29.8	53.7	40	81.6
11	29.8	53.7	10	78.1
12	29.8	53.7	6	78.1
13	29.7	53.7	17	85.2
14	29.7	53.7	1	85.2
15	29.8	53.7	36	73.6
16	29.8	53.7	24	73.6
17	29.7	53.7	22	76.3
18	29.7	53.7	36	74.2
19	29.7	53.7	1	73.7
20	29.7	53.7	16	78.9
21	29.7	53.7	26	79.1

- Fig. 6 shown the various current data input of total 500 data.
- Fig. 7 shown the various temperature data input of total 500 data.
- Fig. 8 shown the various voltage data input of total 500 data.
- Fig. 9 shown the various Depth of Discharge data input of total 500 data.

In next step we perform defuzzification process by implementing Mamdani method and Takagi Sugeno Kang method. The consideration of implementation these 2 methods is to have comparison of the 2 outputs result. Further-more we could have recommendation which method more optimum to the battery prediction measurement.

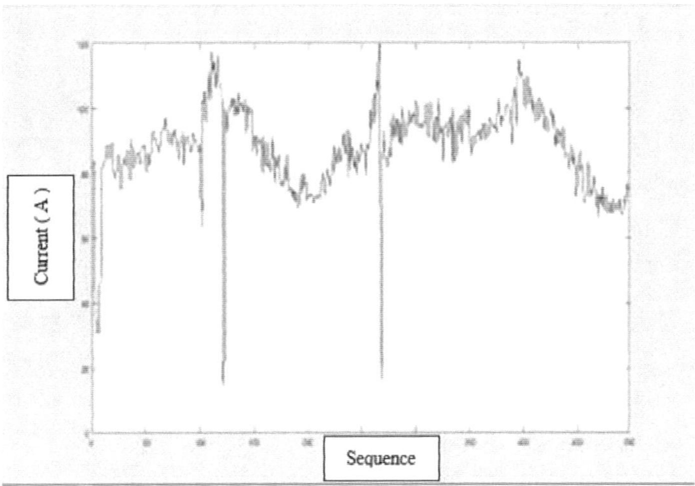

Figure 6. Current data input.

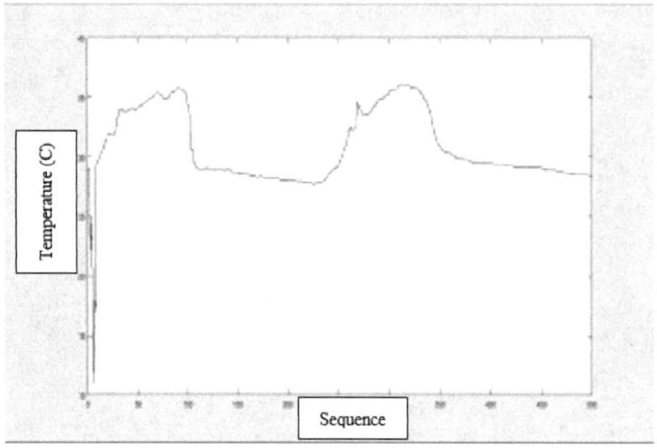

Figure 7. Temperature data input.

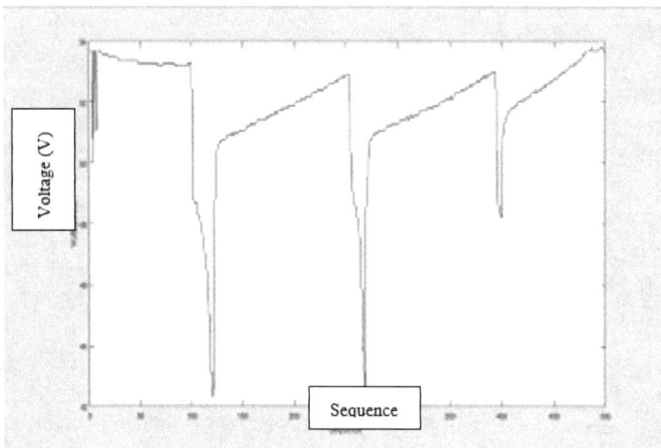

Figure 8. Voltage data input.

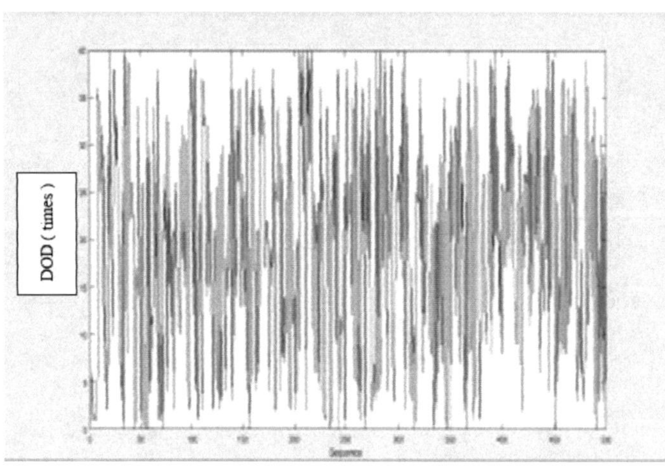

Figure 9. Depth of discharge data input.

4. Result and discussion

We calculate various of four inputs criteria in fuzzification process and rule inference process, shown in figure 10.

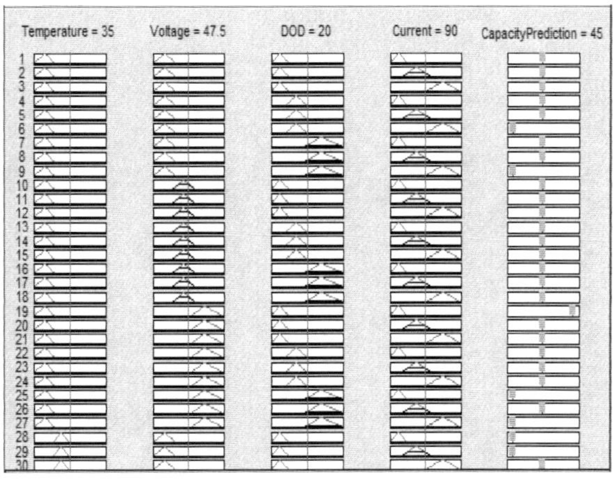

Figure 10. Rule and inference process.

We see in figure 11 the Mamdani output battery capacity prediction with various parameter input criteria i.e. various current criteria, various temperature criteria, various DOD criteria, and various voltage criteria. The temperature range criteria shown mostly have a not radical fluctuation pattern. In other side voltage and current range criteria sometimes have a radical fluctuation pattern because of discharge occurred in the battery. And the DOD range criteria are shown several fluctuation patterns. Finally, we get the output with various criteria. The DOD range fluctuation shown is an important factor impact on capacity reduction. The high DOD range shown give high impact on the battery capacity reducing, with the lowest capacity is 30 percent, we know is a bad condition. In order to have comparison of the defuzzification output result we implement another method i.e. Sugeno method, shown in fig. 12. We see the output pattern of Sugeno method is most similar with Mamdani method. But Sugeno method have a lower range of the output criteria than Mamdani method. This occurred because Sugeno method apply different weighting factor in defuzzification process.

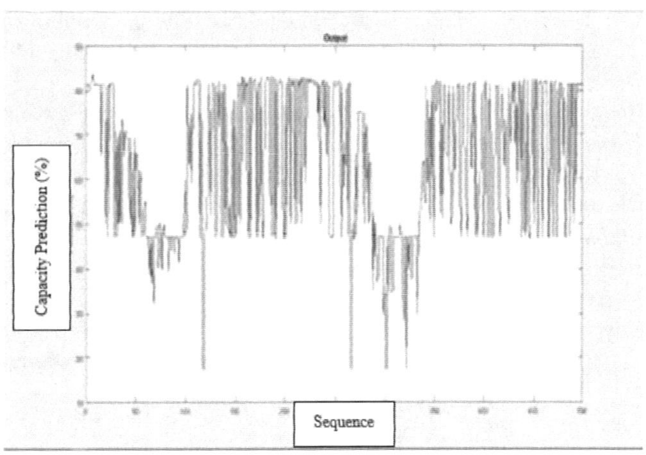

Figure 11. Mamdani output battery capacity prediction.

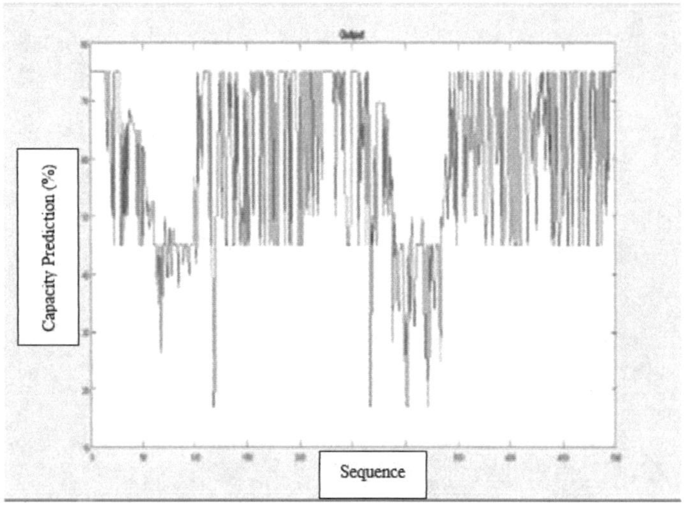

Figure 12. Sugeno output battery capacity prediction.

5. Conclusion

From the analysis it can be concluded that the prediction effective capacity of the BTS battery can be calculated by performing a fuzzy process of the input parameters in the form of current value criteria, temperature value criteria, voltage value criteria, and DOD value criteria.

Defuzzification with the Mamdani and Sugeno methods give almost the same output pattern, but the Mamdani method provides a larger output range. The fluctuation of the input range will give impact to the battery capacity, and from the graph shown the high Depth of Discharge gives a high impact on battery capacity reducing. The lowest prediction shown battery capacity prediction is 30 percent of nominal capacity, and this is a bad battery condition. Furthermore, after the cellular engineer knows the battery capacity condition, they should take proper action for example replace the bad battery with the new one by on-time action.

References

[1] Mohammed H A, Rosdiadee N and Mahamod I 2016 Green wireless network optimisation strategies within smart grid environments for Long Term Evolution (LTE) cellular networks in Malaysia *Renewable Energy Journal* **85** 157-170

[2] Han X, Lu L, Zheng Y, Feng X, Li Z, Li J and Ouyang M 2019 A review on the key issues of the lithium ion battery degradation among the whole life cycle *ETransportation* **1** 100005

[3] Yi L, Kailong L, Aoife M F, Alana Z, Maitane B, Elise N M, Joeri V M and Harry E H 2019 Data-driven health estimation and lifetime prediction of lithium-ion batteries: A review *Renewable and Sustainable Energy Reviews Journal* **113**

[4] Shen S, Sadoughi M, Chen X, Hong M and Hu C 2019 A deep learning method for online capacity estimation of lithium-ion batteries *Journal of Energy Storage* **25** 100817

[5] Yingzhi C, Chunyu D, Geping Y, Yunzhi G, Lingling Z, Ting G, Lijie Y and Fuping W 2015 Multi-stress factor model for cycle lifetime prediction of lithium ion batteries with shallow-depth discharge *Journal of Power Sources* **279** 123-132

[6] Seyed M R, Zongchang L, Yan C and Jay L 2014 Review and recent advances in battery health monitoring and prognostics technologies for electric vehicle (EV) safety and mobility *Journal of Power Sources* 110-124

[7] Seung-Wook E, Min-Kyu K, Ick-Jun K, Seong-In M, Yang-Kook S and Hyun-Soo K 2007 Life prediction and reliability assessment of lithium secondary batteries *Journal of Power Sources* **174** 954–958

[8] Hannan M A, Lipu M H, Hussain A and Mohamed A 2017 A review of lithium-ion battery state of charge estimation and management system in electric vehicle applications: Challenges and recommendations *Renewable and Sustainable Energy Reviews* **78** 834-854
[9] Pritpal S, Ramana V, Xiquan W and David R 2006 Design and implementation of a fuzzy logic-based state-of-charge meter for Li- ion batteries used in portable defibrillators *Journal of Power Sources* **162** 829–836
[10] Panchal S, Mathew M, Dincer I, Agelin-Chaab M, Fraser R and Fowler M 2018 Thermal and electrical performance assessments of lithium-ion battery modules for an electric vehicle under actual drive cycles *Electric Power Systems Research* **163** 18-27
[11] Chen Y, Huang D, Zhu Q, Liu W, Liu C and Xiong N 2017 A new state of charge estimation algorithm for lithium-ion batteries based on the fractional unscented kalman filter *Energies* **10** 9 1313
[12] Xia B, Lao Z, Zhang R, Tian Y, Chen G, Sun Z and Wang H 2018 Online parameter identification and state of charge estimation of lithium-ion batteries based on forgetting factor recursive least squares and nonlinear Kalman filter *Energies* **11** 1 3
[13] Ali M U, Zafar A, Nengroo S H, Hussain S, Junaid A M and Kim H J 2019 Towards a smarter battery management system for electric vehicle applications: A critical review of lithium-ion battery state of charge estimation *Energies* **12** 3 446
[14] Zubi G, Dufo-López R, Carvalho M and Pasaoglu G 2018 The lithium-ion battery: State of the art and future perspectives *Renewable and Sustainable Energy Reviews* **89** 292-308
[15] Prakash S, Mustain W E and Kohl P A 2009 Performance of Li-ion secondary batteries in low power, hybrid power supplies *Journal of Power Sources* **189** 2 1184-1189
[16] Raijmakers L H J, Danilov D L, Eichel R A and Notten P H L 2019 A review on various temperature-indication methods for Li-ion batteries *Applied energy* **240** 918-945
[17] Bubbico R, Greco V and Menale C 2018 Hazardous scenarios identification for Li-ion secondary batteries *Safety science* **108** 72-88
[18] Liu X, Ai W, Marlow M N, Patel Y and Wu B 2019 The effect of cell-to-cell variations and thermal gradients on the performance and degradation of lithium-ion battery packs *Applied Energy* **248** 489-499
[19] Raijmakers L H J, Danilov D L, Eichel R A and Notten P H L 2019 A review on various temperature-indication methods for Li-ion batteries *Applied energy* **240** 918-945

Sediments discharge analysis using tank model for disaster mitigation

T Mulyono*, J Jumi and A Azizah

Politeknik Negeri Semarang, Semarang, Indonesia

*teted64@gmail.com

Abstract. Environmental degradation as a result of deforestation carried out in the Catchment resulted in a decrease in its ability to store water. This has the effect of increasing the amount of sediment discharge. The process of estimating sediment discharges is very difficult because the data input variables are many and varied, usually, the data are very limited, because the erosion process occurs until the sediment discharge mechanism is quite complex. The process of sediment discharges in Catchment s is influenced by rain and surface runoff and is represented in the storage type. In this study, an approach using the Tank Model was conducted. The purpose of this study is to develop a tank model for sediment discharge analysis in disaster mitigation. The steps are setting the field experiment for collecting rain and discharge sediment data as the model input and setting the model analysis by making the structure and formulation of the tank model. There are 3 proposed tank models namely Tank Model 1 (three tanks, series, and cascade), Tank Model 2 (two cascade tanks), and Tank Model 3 (three cascade tanks). Model parameters are determined using the Genetic Algorithm (AG) method optimization approach. The analysis shows that Tank Model 3, composed of 3 (three) cascade tanks, represents a Catchment better than the other 2 tank models. This can be seen from the value of the accuracy of the model, namely the value of volume error (VE), the value of relative error (RE), the value of the mean least square error (RMSE), and the value of the correlation coefficient (R). But still has a range of differences for the value of sediment discharges, the cause may be a factor in the pattern of rain spread in the hydrological process, synchronization of the measurement process and data length, and the possible assumptions of the model parameters.

1. Introduction

Environmental degradation as a result of deforestation carried out in the Catchment resulted in a decrease in its ability to store water. Irresponsible deforestation has resulted in 58 critical Catchments in Indonesia [1], then increased to 62 critical Catchments [2], and currently, there are 108 Catchments in critical condition [3]. Critical Catchment indications increase with high sediment flow rates, resulting in floods, landslides, this causes damage to water structures and reservoir sedimentation [4].

The process of estimating sediment discharges is very difficult because the data input variables are many and varied, usually, the data are very limited, because the erosion process occurs until the sediment discharge mechanism is quite complex. Sediment discharge analysis is usually done by making a sedimentary curve (Sediment Rating Curve) which requires a fairly long measurement data that covers a variety of small to large discharge ranges and if carried out measurements require time, money and energy [5]. Sediment erosion-discharge models, such as the WEPP model [6], KINEROS model [7], ANSWERS model [8], AGNPS model [9,10] and SWAT model [11,12] include erosion and sediment discharge analysis in regional land areas river flow caused by rainfall and land surface flow [13]. Apply the Time-Area method for prediction of sediment yield with time variations [14,15] with a lumped sediment runoff model analyzing sediment flow covering areas on sloped land and in rivers in Catchments.

Application of the tank model for sediment yield, with analysis using 3 (three) cascade tank arrangements and the same tank model parameters [16], both for surface ruoff analysis in the form of discharge [17-20], and analysis of sediment results, namely discharge multiplied by sediment concentration. The application of the tank model to the sediment yield assumes that sediment concentrations undergo infiltration, percolation, and actual conditions are unlikely to occur in such a process, this is a weakness in the model. Application of the tank model uses Kalman filter for sediment yield, with analysis using the arrangement of 3 (three) cascade tanks and the same tank model parameters, both for rain-flow analysis in the form of discharge, and analysis of sediment yield i.e. discharge multiplied by sediment concentration [21]. The tank model parameters are determined by the vector condition in the model system on the Kalman filter using the optimal sequential estimation technique and the Kalman filter algorithm is formed by three components: (1) the system model; (2) measurement models; and (3) Kalman filter. Sediment flow analysis uses the lumped sediment runoff model which includes areas in sloped land and rivers in river basins and analyzes sediment discharges based on the rain-flow model, which is a tank model by incorporating sediment elements with 2 (two) tank arrangements) [22], but not yet implemented.

There is a significant difference from the results of previous studies with the research offered, that this study represents the process of erosion - sediment discharges in the Catchment in the form of reservoirs that is modifying the tank model developed for prediction of sediment discharge, namely Model Tank 1 with 2 (two) cascade tanks, Tank Model 2 with 3 (three) cascade tanks and Tank Model 3 with 1 (one) tank. Here this study aims to develop a tank model with predictive analysis of sediment discharges in Catchments for disaster mitigation by modifying tank structures based on erosion and sediment discharge processes in Catchments. Limitation of the study Sediment discharge referred to in this study is the result of erosion caused by rainfall and surface runoff in the Catchment (sloping land, grooves, and ditches). The implication of the model development is to provide supporting information, especially to the stakeholders of the Catchment management in carrying out sediment discharge monitoring to determine environmental degradation in the Catchment and the planning of water structures.

2. Methods

This research is located in the Kreo sub-catchment which has an area of 1692,812 hectares (Figure 1). The research location is the entry in the territory of Kendal district and Semarang city, with a description of the research areas, that the Kreo sub-catchment has a tropical climate that is influenced by monsoons. The rainy season occurs from November to May and the dry season from June to October.

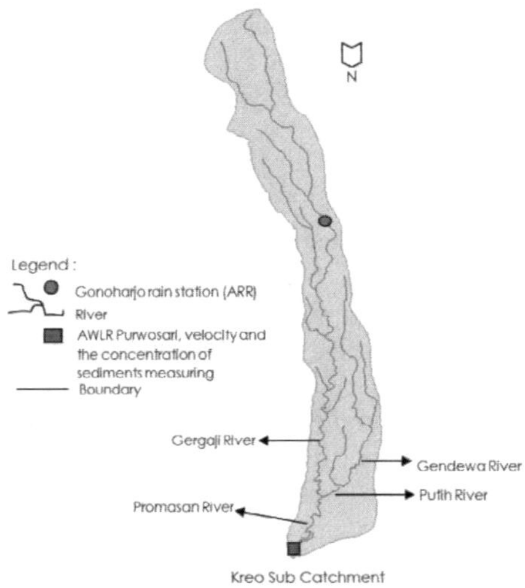

Figure 1. Location of research in Kreo Sub catchment.

The method in this study is a computational simulation to obtain sediment discharges with optimum tank model parameters, through calibration with measured sediment discharge data on the Catchment, to produce the best value of sediment discharge from the model output. Computational simulations are carried out on tank models using the Genetic Algorithm method. Computational simulations for tank models are carried out to analyze the prediction of sediment discharges in the Catchment in the Kreo sub-Catchment with model input data in the form of measured rain and measured sediment discharges. The steps are setting the field experiment for data retrieval as the model input data, and setting the model analysis by making the structure and formulation of the tank model. Research steps as shown in Figure 2.

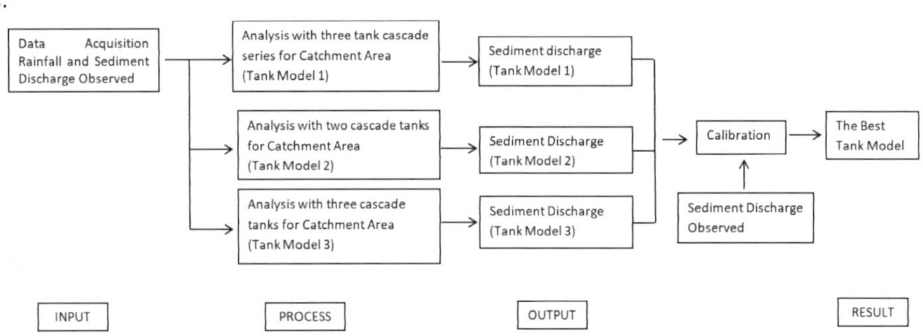

Figure 2. Research steps.

Data collection includes rain data Automatic rainfall recorder (ARR) at Sta. Gonoharjo and measured sediment discharge data at an estimated post or river flow observation station (SPAS) in Purwosari Village, Gunungpati District, Semarang City. The examples of rainfall data and sediment discharge data that are used as model input data are shown in Table 1.

Table 1. Rain data and sediment discharge data.

Time (minutes)	0	10	20	30	40	50	80	110	150	180
Rainfall (mm)	0,5	0,5	0,5	1	5	11,5	8	9	2,5	1
Sediment Discharge(ton/day)	45,30	67,55	422,13	1063,24	1176,26	1243,66	748,824	679,67	173,12	90,20

There are 3 tank models with the proposed parameters, namely Tank Model 1 (three cascade series tanks), Tank Model 2 (two cascade tanks), and Tank Model 3 (three cascade tanks) Figure 3.

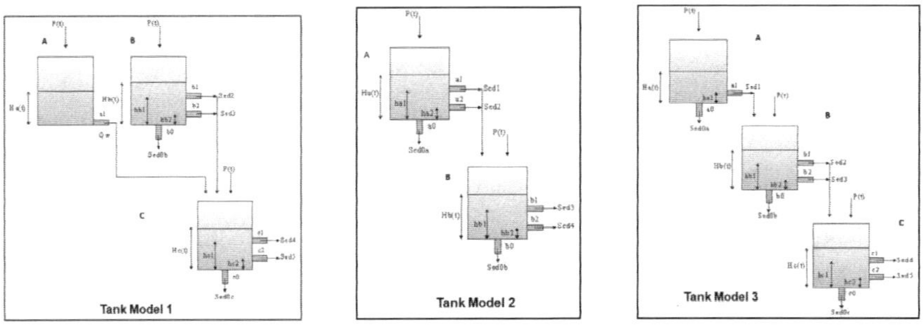

Figure 3. Tank model structure with parameters for prediction sediment discharge in Kreo sub catchment.

where: P(t) is rainfall (mm); Sed_1 (t), Sed_2 (t), Sed_3 (t), Sed_4 (t), Sed_5 (t) is the Sediment Discharge (ton/day); Sed_{0a}, Sed_{0b}, Sed_{0c} is Sediment Discharge deposition (ton/day); Q_w is the subsurface flow (mm); Ha(t), Hb (t), Hc(t), is storage height the Sediment Discharge (mm); a_1, a_2, b_1, b_2, c_1, c_2 are parameters model sediment discharge; a_0, b_0, c_0 is the parameters model sediment deposition; ha_1, hb_1, hc_1 is high of upper hole the sediment discharge on a model (mm); ha_2, hb_2, hc_2 is high of bottom hole the sediment discharge on a model (mm).

The process of obtaining sediment discharge is stimulated by the release of soil by rain and surface runoff. The basic assumption of this model is that sediment is transported and generated when surface runoff occurs. Thus, the carrying capacity of the surface runoff is estimated to simulate the sediment discharge process. Soil release, sedimentation, and sediment discharge are handled with the equation of sediment flow continuity as in Equation 1 [23-25].

$$\frac{\partial(h_a C)}{\partial t} + \frac{\partial(q_a C)}{\partial x} = e(x,t) \qquad (1)$$

where: C = sediment concentration in the flow (kg/m^3)
h_a = depth of surface flow (m)
q_a = discharge per unit width (m^2/second)
e (x,t) = erosion by rainfall and surface runoff (kg/m^2/hour)

To calculate the sediment discharge from the three tank models using equations 2 through equation 25 which is the development of equation 1

2.1. Tank model 1

$Q_w(t) = [(Ha(t)+P(t)] \times a_1$ (2)

$Sed_2(t)=[((Hb(t)+P(t)+ Q_w(t)) \times C_h) - hb_1] \times b_1$ (3)

$Sed_3(t)=[(Hb(t)+P(t)+ Q_w (t)) \times C_h) - hb_2] \times b_2$ (4)

$Sed_4(t)=[(Sed_2(t)+Sed_3(t))+((Hc(t)+P(t)) \times C_N))\times \quad hc_1] \times c_1$ (5)

$Sed_5(t)= [(Sed_2(t)+Sed_3(t))+((Hc(t)+P(t)) \times C_N)) \times hc_2] \times c_2$ (6)

$Sed_{0b}(t)=[(Hb(t)+P(t)+ Q_w (t)) \times C_h] \times b_0$ (7)

$Sed_{0c}(t)=[(Hc(t)+P(t)) \times C_N] \times c_0$ (8)

Total sediment discharge on the right side of the tank = Sed_{total} = $Sed_4(t)+Sed_5(t)$ (9)

2.2. Tank model 2

$Sed_1(t) = [((Ha(t)+P(t)) \times C_h) - ha_1] \times a_1$ (10)

$Sed_2(t) = [((Ha(t)+P(t)) \times C_h) - ha_2] \times a_2$ (11)

$Sed_3(t) = [(Sed_1(t)+Sed_2(t)+((Hb(t)+P(t)) \times C_N)) \times hb_1] \times b_1$ (12)

$Sed_4(t) = [(Sed_1(t)+Sed_2(t)+((Hb(t)+P(t)) \times C_N)) - hb_2] \times b_2$ (13)

$Sed_{0a}(t)= [(Ha(t)+P(t)) \times C_h] \times a_0$ (14)

$Sed_{0b}(t)= [(Hb(t)+P(t)) \times C_N] \times b_0$ (15)

Total sediment discharge on the right side of the tank = $Sed_{total} = Sed_3(t)+Sed_4(t)$ (16)

2.3. Tank model 3

$Sed_1(t) = [((Ha(t)+P(t)) \times C_h) - ha_1] \times a_1$ (17)

$Sed_2(t) = [(Sed_1(t)+ ((Hb(t)+P(t)) \times C_h)) - hb_1] \times b_1$ (18)

$Sed_3(t) = [(Sed_1(t))+ ((Hb(t)\ P(t)) \times C_h)) - hb_2] \times b_2$ (19)

$Sed_4(t) = [(Sed_2(t)+Sed_3(t)+((Hc(t)+P(t)) x C_N)) - hc_1] \times c_1$ (20)

$Sed_5(t) = [(Sed_2(t)+Sed_3(t)+((Hc(t)+P(t)) x C_N)) - hc_2] \times c_2$ (21)

$Sed_{0a}(t) = [(Ha(t)+P(t)) \times C_h] \times a_0$ (22)

$Sed_{0b}(t) = [(Hb(t)+P(t)) \times C_h] \times b_0$ (23)

$Sed_{0c}(t) = [(Hc(t)+P(t)) \times C_N] \times c_0$ (24)

Total sediment discharge on the right side of the tank = $Sed_{total} = Sed_4(t)+Sed_5(t)$ (25)

All of parameters non negative : $0 < a_0+a_1+a_2 \leq 1$; $0 < b_0+b_1+b_2 \leq 1$; $0 < c_0+c_1+c_2 \leq 1$; $a_1 \geq a_2$; $b_1 \geq b_2$; $c_1 \geq c_2$; $ha_1 \geq ha_2$; $hb_1 \geq hb_2$; $hc_1 \geq hc_2$

To get a tank model for prediction of sediment discharges that represent actual Catchment conditions using the research criteria volume error (VE), relative error (RE), correlation coefficient (R) and root mean squares error (RMSE) shown by equations 26 through equation 29, with a limit value of research criteria for VE < 5%, RE -10% to 10%, R > 0.7 and RMSE close to 0. Its formula is as follows:

$$VE = \frac{\sum_{i=1}^{N} Sed_{obs} - \sum_{i=1}^{N} Sed_{sim}}{\sum_{i=1}^{N} Sed_{obs}} x\,100 \quad (26)$$

$$RE = \frac{1}{N}\sum_{i=1}^{N} \frac{\left|Sed^i_{sim} - Sed^i_{obs}\right|}{Sed^i_{obs}} x\,100 \quad (27)$$

$$R = \frac{\sum\left(Sed^i_{sim} - \overline{Sed^i_{sim}}\right)\left(Sed^i_{obs} - \overline{Sed^i_{obs}}\right)}{\sqrt{\sum\left(Sed^i_{sim} - \overline{Sed^i_{sim}}\right)^2 \sum\left(Sed^i_{obs} - \overline{Sed^i_{obs}}\right)^2}} \quad (28)$$

$$RMSE = \sqrt{\frac{\left(\sum Sed^i_{sim} - Sed^i_{obs}\right)^2}{N}} \quad (29)$$

where: VE is volume error [%]; RE is relative error [%]; R is correlation coefficient; RMSE is the root mean squares error; Sed_{sim} is sediment discharge simulation period-i [t/d]; Sed_{obs} is sediment discharge measurable period-i [t/d]; N is the number of data; $\overline{Sed_{sim}}$ is average of sediment discharge simulation period-i [t/d]; $\overline{Sed^i_{obs}}$ is the average of sediment discharge measurable period-i [t/d].

3. Results and discussion

The results of the simulation of 3 tank models (Figure 3) for prediction of sediment discharge in the Kreo sub-Catchment based on rain input data and sediment discharge data (Table 1), on January 6, 2014, and January 26, 2014, is shown in Figure 4:

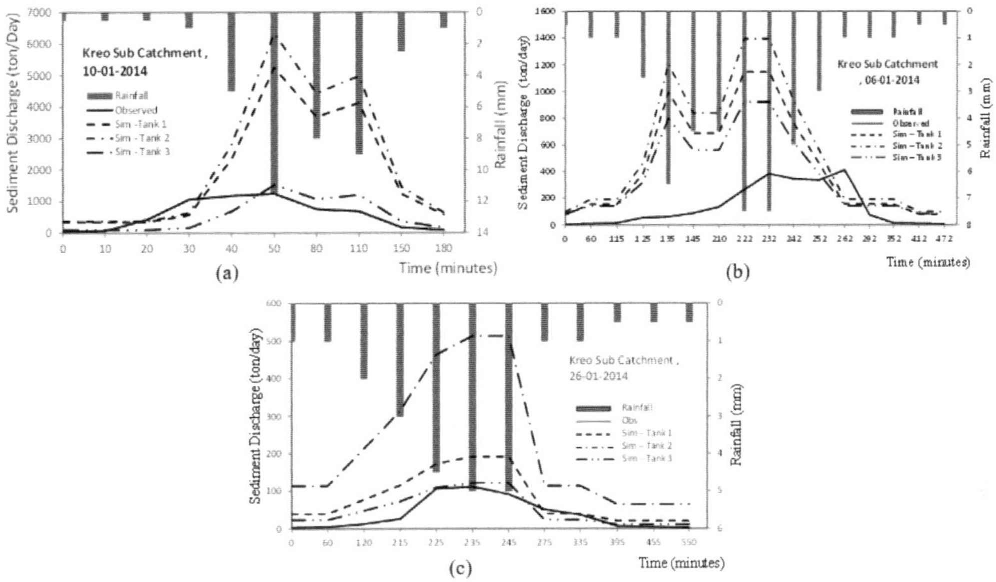

Figure 4. Calibration Results of Observation and Simulation Sediment Discharge using 3 Tank Models in Kreo Sub-Catchment on (a) 10 January 2014, (b) 6 January 2014 and (c) 26 January 2014.

Calibration results using 3 tank models are shown in Table 2.

Table 2. Parameter 3 tank models.

rain events / parameters	January 6, 2020 Tank Model 1	Tank Model 2	Tank Model 3	January 10, 2020 Tank Model 1	Tank Model 2	Tank Model 3	January 26, 2020 Tank Model 1	Tank Model 2	Tank Model 3
a0	-	0.094	0.012	-	0.192	0.459	-	0.061	0.001
a1	0.761	0.319	0.028	0.289	0.188	0.466	0.168	0.263	0.527
a2	-	0.160	-	-	0.015	-	-	0.015	-
b0	0.264	0.235	0.047	0.098	0.065	0.068	0.164	0.022	0.011
b1	0.423	0.372	0.400	0.122	0.130	0.427	0.113	0.342	0.551
b2	0.217	0.152	0.145	0.080	0.130	0.296	0.075	0.003	0.037
c0	0.122	-	0.256	0.039	-	0.000	0.165	-	0.013
c1	0.152	-	0.516	0.578	-	0.532	0.507	-	0.259
c2	0.081	-	0.228	0.058	-	0.375	0.026	-	0.188
ha1	-	57.490	0.920	-	58.952	0.718	-	47.285	0.274
ha2	-	0.091	-	-	0.164	-	-	0.303	-
hb1	58.130	56.055	49.146	59.833	51.271	58.496	59.879	41.489	38.456
hb2	0.650	0.104	0.441	0.201	0.091	0.002	0.200	0.269	0.799
hc1	56.611	-	33.002	37.476	-	40.406	32.399	-	47.278
hc2	0.969	-	0.233	0.016	-	0.338	0.006	-	0.442

The value of the research criteria for the calibration results of the three tank models to show an accurate tank model is shown in Table 3.

Table 3. Value criteria accuracy of the three tank models.

Rain fall	Relative Error(RE)			Volume Error(VE)			Correlation Coefficient (R)			RMSE		
	Tank Model 1	Tank Model 2	Tank Model 3	Tank Model 1	Tank Model 2	Tank Model 3	Tank Model 1	Tank Model 2	Tank Model 3	Tank Model 1	Tank Model 2	Tank Model 3
06/01 /2014	413,63	277,67	254,13	119,73	57,08	50,75	0,52	0.52	0.52	275,4	183,11	176,74
10/01 /2014	154,41	300,64	135,43	62,63	203,70	58,36	0,65	0,65	0,65	742,6	1851,7	744,14
26/01 /2014	433,74	375,67	104,35	165,54	129,69	16,23	0,89	0,89	0,89	81,71	63,64	20,95

The results showed that the tank model 3 is better than the other 2 tank models. This is evidenced by the results of the calibration using the parameter values in Table 2, with the sediment discharge values and the accuracy criteria values are shown in Figure 4 (a), (b), (c) and Table 3. The calibration results on the three tank models show that tank model 2 is less representative of the actual Catchment conditions.

The results of the analysis of the tank model for prediction of sediment discharge, the rising limb and peak value of the simulation sediment discharge is greater than the measured sediment discharge value, this rainfall tends to be high, of course, the amount of surface flow increases so that the output value of sediment discharge in the tank model becomes greater (Figure 4 (b), (c)) and recession limb simulation sediment discharge value decreases, this is because rainfall tends to decrease and surface flow also decreases so that the output value of sediment discharge in the tank model becomes small. Except for Figure 4 (a) there may be a disturbance upstream of the tentative measurement, the measured sediment discharge value is greater than the simulated sediment discharge value. Another cause is the pattern of rainfall distribution in the hydrological process, synchronization of the measurement process, and the length of the data and the possible assumptions of the model parameters.

4. Conclusion

By representing the process of erosion - sediment discharges in the River Basin, the three tank models can be used for prediction of sediment discharges: Tank Model 1 with 3 (three) series cascade tanks, Tank Model 2 with 2 (two) cascade tanks, Tank Model 3 with 3 (three) cascade tanks. The arrangement or configuration of the tank model that provides predictive results of sediment discharges close to the measured sediment discharge is Tank Model 3, composed of 3 (three) cascade tanks. the mean least square error (RMSE) and the correlation coefficient (R). But it still has a different range for sediment discharge values. This is possible because the factors are the pattern of rainfall distribution in the hydrological process, synchronizing the measurement process and the length of the data as well as the possible assumptions of the model parameters.

Acknowledgment

Thank you to DRPM, Ministry of Research and Technology for funding this research through the PTUPT scheme. This 2019's competitive applied research team would like to express its gratitude and appreciation to:

- DRPM, Ministry of Research and Technology which has funded this activity.
- Politeknik Negeri Semarang which has funded this activity.
- The director of Politeknik Negeri Semarang which has facilitated this activity
- The leader of research and community service (P3M) of Politeknik Negeri Semarang as the facilitator of this activity.

References

[1] Soenarno S 2001 Integrasi dan Keterpaduan Pengelolaan Sumberdaya Air yang Berkelanjutan dalam Era otonomi Darah *Prosiding Diskusi Panel Nasional Kerjasama Universitas Diponegoro dengan Departemen Energi dan Sumberdaya Mineral* (Semarang)

[2] Tarigan S D 2009 Pengembangan Sistem Informasi Spasial Berbasis Web (Web GIS) untuk Sinergi Rehabilitasi DAS Kritis Nasional *Prosiding Seminar Nasional "Kebijakan dan Aplikasi Teknologi Informasi dan Komunikasi untuk Peningkatan Daya Saing Agribisnis Indonesia" Himpunan Informatika Pertanian – Institut Pertanian Bogor – Departemen Pertanian Republik Indonesia* (Bogor)
[3] Sustayo M O 2016 *Pendekatan Spasial Ekologis dan Skala Multidimensi Dalam Pengelolaan DAS Berkelanjutan* (Yogyakarta: Fakultas Kehutanan, Universitas Gadjah Mada)
[4] Suripin S 2002 *Pelestarian Sumber Daya Tanah dan Air* (Yogyakarta: Penerbit Andi)
[5] Suripin S 2002 *Kelemahan Lengkung Sedimen dalam Memprediksi Sedimen layang* 237-248
[6] Nearing M A, Foster G R, Lane L J and Finkner S C 1989 A process-based soil erosion model for USDA-Water Erosion Prediction Project technology *Transaction of the American Society of Agricultural Engineers* **32** 1587-1593
[7] Woolhiser D A, Smith R E and Goodrich D C 1990 *KINEROS, A Kinematic Runoff and Erosion Model:Documentation and User Manual* (ARS-77 Fort Collins, Colo.: USDA Agricultural Research Service)
[8] Beasley D B and Huggins L F 1991 *ANSWERS. User's Manual,* Agricultural Engineering Department, Purdue University (West Laffayete, Indiana)
[9] Young R A, Onstad C A, Bosch D D and Anderson W P 1987 *AGNPS, Agricultural nonpoint-source pollution model:A Catchment analytical tool. Conservation Research Report No. 35* (Washington, D.C.: USDA)
[10] Young R A, Onstad C A, Bosch D D and Anderson W P 1989 AGNPS: A nonpoint-source pollution model for evaluating agricultural Catchment s *J. Soil and Water Conservation* **44** 2 168-173
[11] Arnold J G, Srinivasan R, Muttiah R S and Williams J R 1998, Large-area hydrologic modeling and assessment: Part I. Model development *J. American Water Resources Assoc.* **34** 1 73-89
[12] Borah D K and Bera M 2003 Catchment -Scale Hydrologic and Nonpoint-Source Pollution Models: Review of Mathematical Bases *American Society of Agricultural Engineering* **46** 6 1553-1566
[13] Kothyari U C, Tiwari A K and Singh R 1996 Temporal Variation of Sediment Yield *Journal of Hydrologic Engineering* 169-176
[14] Takahiro S, Yasuto T and Kaoru T 2007 The Spatio-Temporal Predictions of Rainfall-Sediment-Runoff Based on Lumping of a Physically-based Distributed Model *Annuals of Disas. Prev. Res. Inst.* **50 B** 79-94
[15] Yasuto T, Takahiro S and Kaoru T 2008 Lumping a Physically-based Distributed Sediment Runoff Model with Embedding River Channel Sediment Transport Mechanism *Annuals of Disas. Prev. Res. Inst.* **51 B** 103-117
[16] Lee Y H and Singh V P 2005 Tank Model for Sediment Yield *Water Resources Management* **19** 349-362
[17] Sugawara M, Watanabe W, Ozaki E and Katsuyama Y 1984 *Tank Model with Snow Component* (Japan: National Research Center for Disaster Prevention)
[18] Sugawara M and Ozaki E 1991 Runoff analysis of the Chang Jiang (the Yangtze River) *Hydrological Sciences Journal* **36** 2 135-152
[19] Sugawara M 1979 Automatic calibration of the tank model *Hydrology. Sci. Bull.* **24** 375–388
[20] Phien H N and Pradhan P S S 1983 The tank model in rainfall-runoff modelling *Water S.A.* **9** 3 93-102
[21] Lee Y H 2007 Tank Model Using Kalman Filter for Sediment Yield *Journal of the Environmental Sciences* 1319-1324
[22] Takahiro S, Yasuto T and Kaoru T 2007 The Spatio-Temporal Predictions of Rainfall-Sediment-Runoff Based on Lumping of a Physically-based Distributed Model *Annuals of Disas. Prev. Res. Inst.* **50 B** 79-94
[23] Lopes V L and Lane L J 1998 Modelling sedimentation processes in small Catchment s

Sediment Budgets (Proceedings of the Porto Alegre Symposium) IAHS Publ. **174**
[24] Jayawardena A W and Bhuiyan R R 1999 Evaluation of an interrill soil erosion model using laboratory catchment data *Hydrol. Process.* **13** 89-100
[25] Takahiro S, Yasuto T and Kaoru T 2008 Lumping of a Physically-based Distributed Model For Sediment Runoff Prediction in a Catchment Scale *Annuals Journal of Hydraulic Engineering, JSCE* **5** 43-48

Speed control system in mobile robot based on Bezier curve trajectory

M F Alfatih*, M A Riyadi and I Setiawan

Department of Electrical Engineering, Faculty of Engineering, Universitas Diponegoro, Semarang, Indonesia

*muhammadfaizalfatih@gmail.com

Abstract. The development of robot technology is currently growing rapidly. Mobile robot is one robot with many applications in everyday life. One of the important things in the field of mobile robots is the problem of trajectories or trajectories, in this case how robots can track approved paths. In this study, the mobile robot control system has been tested to track tracks. The path used in this study is the third-order Bezier Order, with control points and end points of the predetermined pathway. Based on the results of simulation tests on the development of mobile robot trajectories, based on the results of simulation tests on mobile robots, it is found that the performance of the track tracking control system is very dependent on the placement of the Bezier curve control points or in other words the actual path formed by the robot is greatly influenced by the shape of the Bezier trajectory that was designed.

1. Introduction

Robot is a mechanical device that can perform physical tasks [1], either using human supervision and control, using pre-defined programs or utilizing the principles of artificial intelligence [2]. Research on mobile robots has been intensively discussed [3–5]. Furthermore, many mobile robots have been used for various purposes such as industry [6], defense, places that have not been touched by humans (e.g. the Mars Rover Robot is used for planetary exploration, Mars) [7], or even for educational purposes, such as robots educative and also likes things like robot ball competition, grass cutting robot competition, fire source search robot competition [8].

One of the main tasks that must be completed by each mobile robot is to get to the expected locations and find out their position and orientation at all times (navigation system) [9]. A simple technique that can be used is by moving the robot to follow a straight path to the destination to rotate in place according to the expected orientation [10]. Because of the simplicity of control, in general this method is widely used in practical robots, such as those used in nursing robots (nursing robots) designed by Johann Borentein [11]. One of the real disadvantages of using this technique is that robots cannot move smoothly for points that must be passed. So as to avoid this, the robot trajectory designed must be continuous and smooth.

Bezier curves are commonly used in the field of computer graphics and in some literature have also been successfully implemented as robot trajectory [12,13]. Bezier curves can be used to optimize robot paths that are calculated with respect to length and curvature [14]. For robots to follow lane, it is important for the speed of the left and right wheel drive angle to adjust gently. The Bezier curve method can be applied to create curved lines to implementation continuous angular velocity [15].

IOP Conf. Series: Materials Science and Engineering 1108 (2021) 012015 doi:10.1088/1757-899X/1108/1/012015

The drive system basically functions to move to drive the robot from one location to another. In this case there are two types of driving systems on popular mobile robots: Ackerman driving which is commonly used in automobile vehicles and differential driving (often found in wheelchairs with automatic drive).

Based on previous research on mobile robots, there are still shortcomings in the form of reference tracks that are still inaccurate in providing reference links to mobile robots. This paper will present a construction of trajectories for mobile robots based on the Bezier curve method, with differential robot type mobile robots which generally have a degree of flexibility in maneuvering or rotating in their movements.

2. Methods

2.1. Mobile robot kinematics model

One type of mobile robot that is commonly used, especially for indoor operation is a mobile robot with a steering or differential drive system (differential drive) [16]. The main reason is because it is relatively more flexible in maneuvering and ease of control.

This type of robot basically has two main wheels, each of which is driven by a separate drive (generally in the form of a permanent magnet DC motor with gear-reducing which serves to strengthen the motor torque), as a counterweight generally this robot is also equipped with one or two castor wheels which is placed at the back of the robot. Figure 1 shows the robot architecture seen from the top: If the two cogs are rotating at the same speed, the robot will move in a straight direction, whereas if the speed of one of the wheels is slower than the robot will move to form a curve with the direction of the path to one the slower-moving wheels.

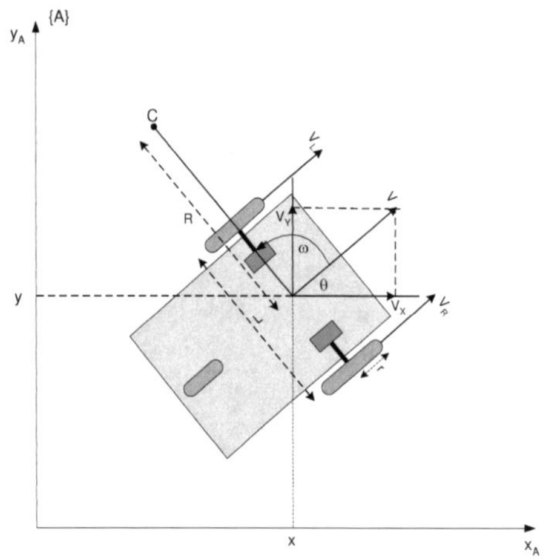

Figure 1. Mobile robot position and orientation in Cartesian coordinate systems [16].

By its nature, this type of mobile robot is included in the category of non-holonomic robot: The position and direction (orientation) are in a coupled state, in this case the robot cannot practically translate to the left (or right) without first changing the robot's orientation.

Based on Figure 1 for the length of the wheel radius r, as well as the rotation speed of each right wheel ωR and left wheel ωL respectively, then the linear speed of the right wheel $v_R(t)$and the left wheel $v_L(t)$can be found using equation (1) and (2) following [17]:

$$v_R(t) = r\omega_R(t) \quad (1)$$

$$v_L(t) = r\omega_L(t) \quad (2)$$

When the robot rotates for a moment with the length of the radius R measured from the center of rotation and the center of the two wheels (Figure 3 can be used as a reference), the rotation speed at each point of the robot will always be the same (the robot is a rigid mechanical system), so the following equations (3) and/or (4) apply to calculate the rotational speed of the robot.

$$\omega(t) = \frac{v_R}{R + L/2} \tag{3}$$

$$\omega(t) = \frac{v_L}{R - L/2} \tag{4}$$

Based on equations (3) and (4) the rotation speed of the robot can be calculated only based on information from the two linear wheels' speeds of the robot:

$$\omega(t) = \frac{v_R(t) - v_L(t)}{L} \tag{5}$$

Whereas the radius of the path can be searched by substituting equation (5) into equation (6), and solving it for R:

$$R = \frac{L\,(v_R + v_L)}{2\,(v_R - v_L)} \tag{6}$$

Seen from equation (6), the radius of the momentary circle trajectory is inversely proportional to the difference in the speed of the two robot wheels, the smaller the difference between the two wheel speeds, the radius of the momentary circle formed by the robot trajectory is getting longer, and vice versa, whereas if $v_R = v_L$, then $R = \infty$R, or practically the robot will move to form a straight path. For the robot to rotate at the center of its axis ($R = 0$), based on equation (6), the speed of the two wheels must be the same in the opposite direction or $v_R = -v_L$

Based on equations (5) and (6), the linear velocity of the robot can be calculated using the following equation (7):

$$v(t) = \frac{v_R(t) + v_L(t)}{2} \tag{7}$$

For simplicity, equations (6) and (7) can be collected in the form of matrix-vector equations as shown in equation (8) below:

$$\begin{bmatrix} v(t) \\ \omega(t) \end{bmatrix} = \begin{bmatrix} 1/2 & 1/2 \\ 1/L & -1/L \end{bmatrix} \begin{bmatrix} v_R(t) \\ v_L(t) \end{bmatrix} \tag{8}$$

Equation (8) above shows the direct kinematics relation between the linear velocity of the robot wheels and the linear and angular speed of the robot, while equation (9) below shows the opposite relation:

$$\begin{bmatrix} v_R(t) \\ v_L(t) \end{bmatrix} = \begin{bmatrix} 1 & L/2 \\ 1 & -L/2 \end{bmatrix} \begin{bmatrix} v(t) \\ \omega(t) \end{bmatrix} \tag{9}$$

By knowing the linear and angular velocity of this robot at any time, then the speed at each Cartesian axis can be found by projecting the robot velocity vectors to these axes, this is shown in equation (10) below:

$$\begin{bmatrix} \dot{x}(t) \\ \dot{y}(t) \\ \dot{\theta}(t) \end{bmatrix} = \begin{bmatrix} \cos(\theta(t)) & 0 \\ \sin(\theta(t)) & 0 \\ 0 & 1 \end{bmatrix} \begin{bmatrix} v(t) \\ \omega(t) \end{bmatrix} \tag{10}$$

From equation (10) it can be seen that for each moment the position and orientation of the robot in the Cartesian coordinate system can be calculated by integrating each equation term:

$$x(t) = \int_0^t v(\tau)\cos(\theta(\tau))d\tau \quad (11)$$

$$y(t) = \int_0^t v(\tau)\sin(\theta(\tau))d\tau \quad (12)$$

$$\theta(t) = \int_0^t \omega(\tau)d\tau \quad (13)$$

The results of solving the equations above are given below:

$$\theta(t) = \frac{(v_R(t) - v_L(t))t}{L} + \theta_0 \quad (14)$$

$$x(t) = x_0 + \frac{v(t)}{\omega(t)}[\sin(\theta(t)) + \theta_0) - \sin(\theta_0)] \quad (15)$$

$$y(t) = y_0 - \frac{v(t)}{\omega(t)}[\cos(\theta(t)) + \theta_0) - \cos(\theta_0)] \quad (16)$$

2.2. Motion control

As mentioned earlier, the main purpose of a robot control system is to move the robot from its initial state (P0) to its final location (P3) by following the expected trajectory. This is illustrated in Figure 2:

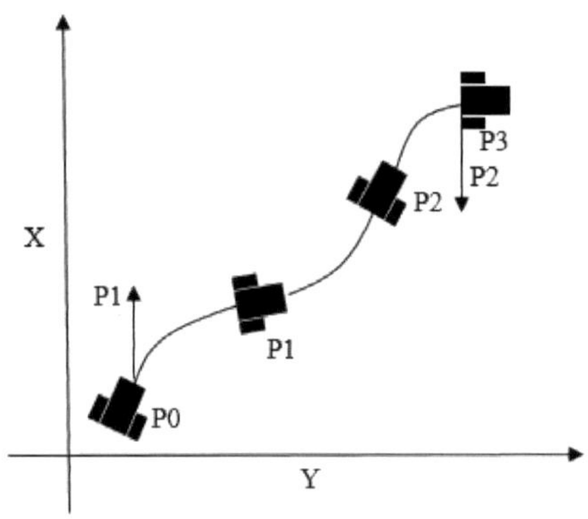

Figure 2. Illustration of robot trajectory.

To achieve these objectives a motion control system is needed. At the lowest level, this system basically regulates the rotational speed of each robot's wheels in accordance with the Reference as shown in Figure 3:

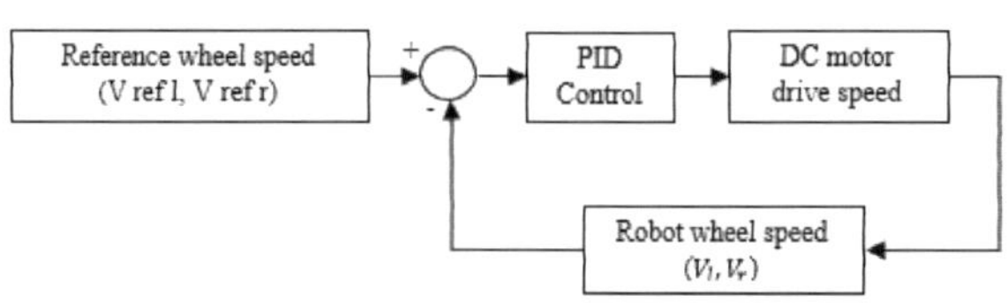

Figure 3. Block diagram control of DC motor in robot mobile.

As discussed earlier, the feedback control system that will be used to control the DC motor is the PID control, with the equation of the transfer function.

$$CO(t) = K_P e(t) + K_I \int_0^t e(\tau)d\tau + K_D \frac{de(t)}{dt} \tag{17}$$

With $CO(t)$= Output controller
K_p = Proportional constant
T_i = time integral
T_D = time derivative
K_I = gain integral (K_P/T_I)
K_D = gain derivative $(K_P T_D)$

While the DC motor itself basically has the following order 1 transfer function:

$$H(s) = \frac{b}{Ts+1} \tag{18}$$

with
b = gain static DC motor

T = time constant DC motor

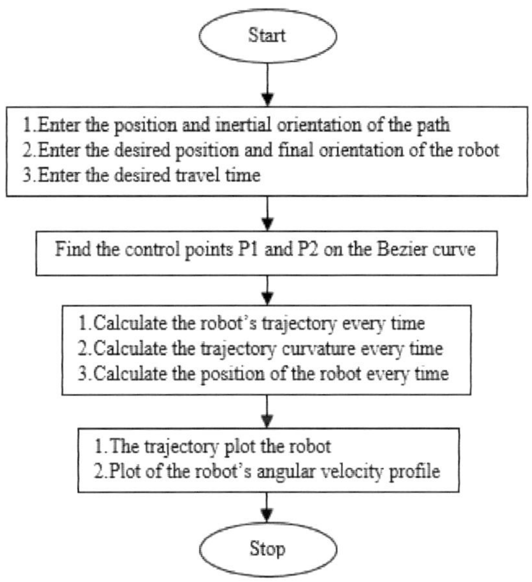

Figure 4. Flowchart simulation trajectory.

When the system starts or starts, the operator must input the values of X and Y starting position, end position and control points and robot travel time. The system will calculate the robotic path, curvature and rotational speed of the robot from the start point to the end point. After that the system will run the Bezier curve trajectory generator algorithm to graph the robot's trajectory based on the input value from the operator.

2.3. Simulation of the overall mobile robot system control block

In general, the control system model that was simulated is shown in Figure 5.

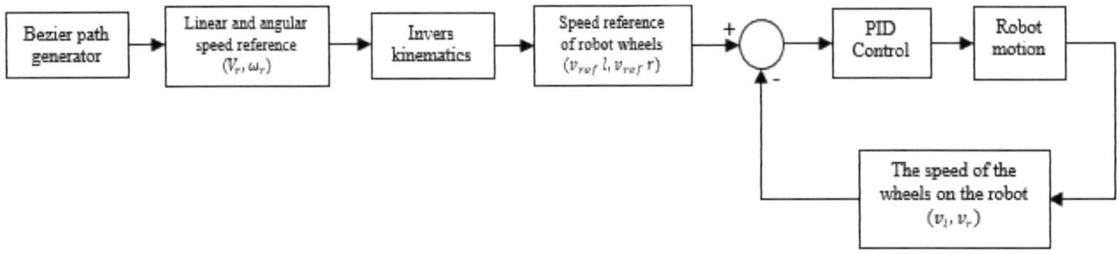

Figure 5. Robot control system block.

In this simulation the quantities of the PID control amplifier was searched empirically for the values of the given DC motor parameters. The ultimate goal of this simulation is basically to look at the dynamics of the robot's trajectory to reference the expected trajectory.

Bezier curves are parametric curves that are often used in graphical applications and related fields [4,14,18–20]. This curve is determined by several control points and always the initial and final control points are determined. The shape can be changed by moving the control point. The two-dimensional Bezier curve of order n is represented in the following equation:

$$P(t) = \sum_{i=0}^{n} \left(\frac{n!}{i!(n-i)!}\right) t^i (1-t)^{n-i} P_i, \qquad t\epsilon[0,1] \tag{19}$$

Where $P(t) = [x(t), y(t)]^T$ and P_i are two-dimensional Bézier curves and control points. Bezier curves have geometrical properties that are useful for path planning:

- Each Bezier curve starts at $P_0$and ends atP_n.
- The curve is tangent to the line connected by $P_0$and P_1 and the line connected by P_n and P_{n-1}respectively at P_0andP_n:
  $T(0) = n(P_1 - P_0)$ (20)
  $T(0) = n(P_n - P_{n-1})$ (21)
- The second derivative in P_0must be determined by P_0, P_1andP_2,and the second derivative in P_n must be determined by $P_{n-2}$and P_{n-1}.

$$S(0) = n(n-1)(P_2 - 2P_1 + P_0) \tag{22}$$

$$S(1) = n(n-1)(P_5 - 2P_4 + P_3) \tag{23}$$

According to the above properties, the control points of the Bezier curve are calculated below. The curvature at any point on the Bezier curve is determined by the first and second derivatives with respect to the parameter u as [21]:

$$P(q) = P_0(1-q)^3 + P_1 q(1-q)^2 + 3P_2 q^2(1-q) + P_3 q, q \in [0.1] \tag{24}$$

To achieve the path of continuous curvature, we mainly focus on ensuring continuity of curvature at the waypoint connecting two curves.

3. Results and discussion

The application of the Bezier curve method for the trajectory of the mobile robot trajectory was tested with different starting points (P0), control points 1 and 2 (P1, P2) and end points (P3). The observed results are the accuracy of the track shape with the starting point (P0), control points 1 and 2 (P1, P2) and end points (P3) entered, the robot's rotational speed profile, robot wheel speed and robot travel time from starting point (P0) to end point (P3).

3.1. Simulation with starting point ($P_1 = 0.0$) and end point ($P_3 = 400.400$) with some varied control points (P_1, P_2), and speed 10 cm/s

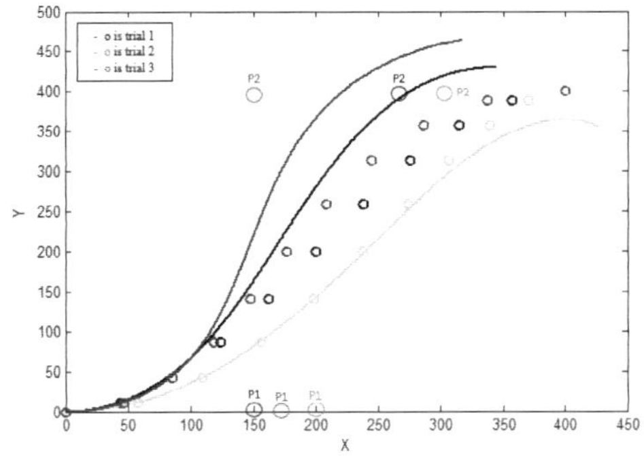

Figure 6. Bezier curve paths with P0 (0,0) and P3 (400,400) and velocity of 10 cm/s.

With reference to equation (24):

$$P(q) = P_0(1-q)^3 + P_1q(1-q)^2 + 3P_2q^2(1-q) + P_3q, q \in [0.1]$$

Figure 6 shows that the Bezier curve is formed with P0 (0,0), P3 (400,400) and P1 and P2 which are varied and the robot speed is 10 cm / s. From this simulation the system that was built successfully created a robot car reference path based on input parameters x and y. The results of tracing the actual trace of the curve attempted experienced significant differences. As shown in the figure, the final coordinates of each experiment differ. for the first experiment with P1 (150.0), p2 (250,400) detected endpoints (350,410). for the second experiment with P1 (200.0), p2 (300,400) the endpoints detected were (420,320), then for the third experiment with P1 (170.0), p2 (150,400) detected endpoints (300,470).

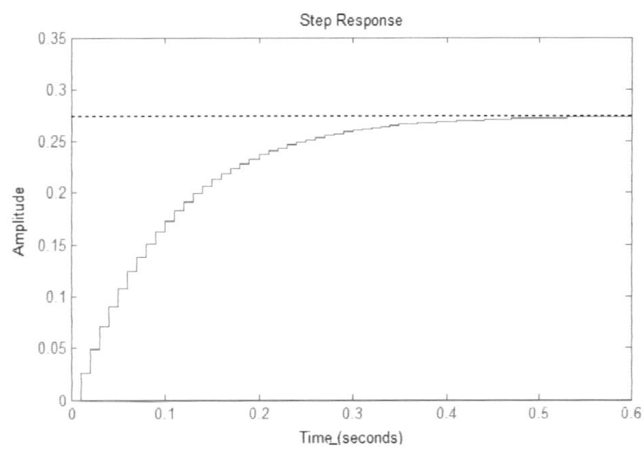

Figure 7. Step response in the PID control scheme with the track reference in Figure 6.

In Figure 7, referring to the PID control equation in equation 17:

$$CO(t) = K_P e(t) + K_I \int_0^t e(\tau)d\tau + K_D \frac{de(t)}{dt}$$

And with equation 18 for a first order DC motor equation:

$$H(s) = \frac{0.0274}{0.1 + 1}$$

The response control system that is applied to the dynamics of the trajectory that is designed has a good and fast level of lift.

3.2. Simulations on variations in parameters P3 with fixed speed and control points (P1, P2)

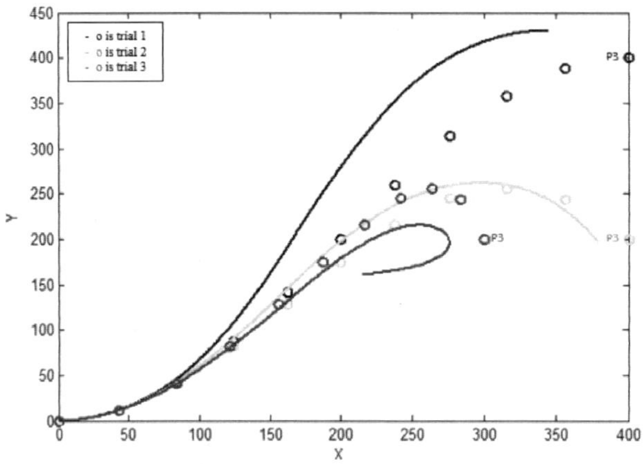

Figure 8. Bezier curve paths with variation parameters*P3* with fixed speed and control points (P1, P2), with Speeds of 10 cm/s.

Figure 8 shows the Bezier curve with the endpoint P3 that must be traveled by a robot that is located differently. with the same control points, namely p1 (150.0), P2 (250,400) and P0 (0.0), and also the robot's fixed speed is 10 cm / s. the graph shows that the reference path can be made accurately, but the robot is not able to get to the final point correctly. for the first experiment with P3 (400,400) the robot goes to P3 (348,448), for the second experiment with P3 (400,200) the robot goes to P3 (389,210) and for the third experiment with P3 (300,200) the robot goes to P3 (210,151).

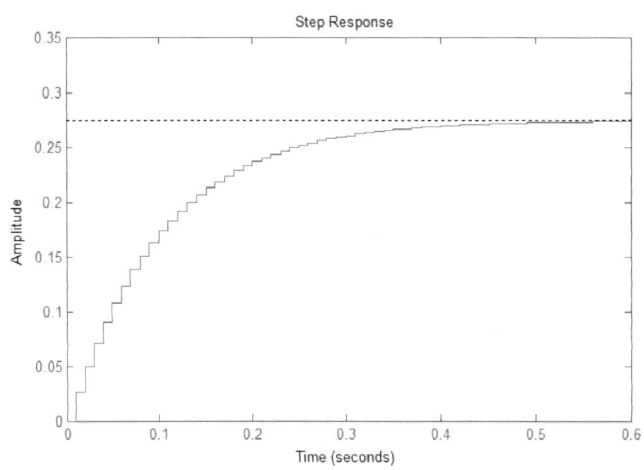

Figure 9. Step response in the PID control scheme with the track reference in Figure 8.

Figure 9 the response of the control system applied to the dynamics of the designed path has a stable lifting rate, and the results are the same as from the first experiment.

3.3. Simulations with variations in speed, but with the same control points (P1 and P2) and end points (P3)

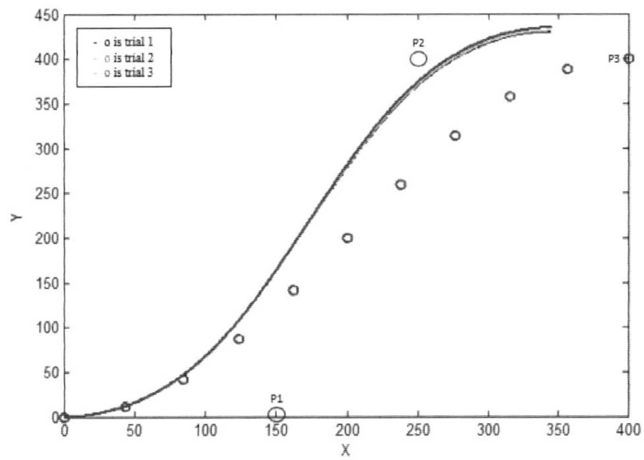

Figure 10. Bezier curve paths with parameters P0 (0.0), P1 (150.0), P2 (250.400), P3 (400.400) and with variation of speeds.

Figure 6 shows that the Bezier Bezier curves are formed with P0 (0,0), P1 (150,0), P2 (250,400), P3 (400,400) and different robot speeds. In this experiment the system that was built succeeded in making the reference robot car line in accordance with the input parameters x and y. The results turned out to have significant similarities, so the difference in speed of the robot did not affect the shape of the bezier curve.

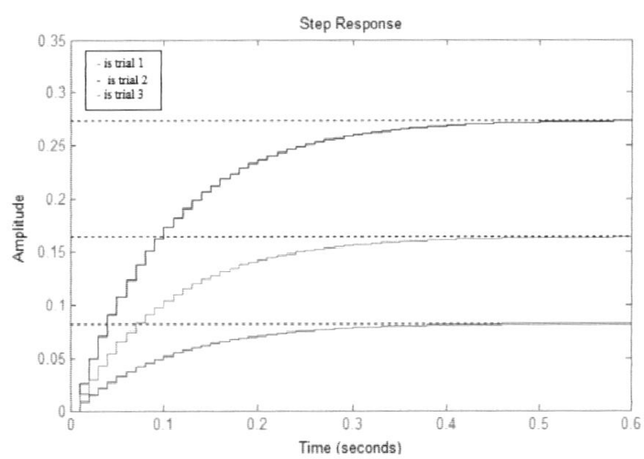

Figure 11. Step response in the PID control scheme with the track reference in Figure 10.

In Figure 11, the control system response that is applied to the dynamics of the designed path has a smooth lift rate and different response times in the stability of the control system. For a speed of 10 cm / s with an amplitude of 0.08 at 0.4 seconds, then for a speed of 6 cm / s the output amplitude is 0.20 at 0.5 seconds, and for a speed of 3 cm / s the output amplitude is 0.50 at 0.6 seconds.

4. Conclusions

Based on testing and analysis conducted in this study, the performance of a track tracking mobile robot control system is strongly influenced by the shape of the Bezier trajectory or trajectory given as a reference. For the coordinates of the endpoint to which X_{final}, Yfinal, are, the relatively consistent control points that give the final result with minimal errors are $P_1=X_{final/2}$,0 and $P_2=X_{final/2}$, Yfinal. For differential type Mobile Robot, there is a restraint of the Y axis's final coordinate destination which is the maximum $X_{final/2}$.

References

[1] Adriansyah A 2014 Perancangan Localization Menggunakan Metode Dead Reckoning *Sinergi* **18** 25–30

[2] Bounini F, Gingras D, Pollart H and Gruyer D 2017 Modified artificial potential field method for online path planning applications *IEEE Intelligent Vehicles Symposium (IV)* 180-185

[3] Yu W, Shuo W, Rui W and Min T 2017 Generation of temporal – spatial Bezier curve for simultaneous arrival of multiple unmanned vehicles *Inf. Sci. (Ny)* 418–419 34–45

[4] Simba K R, Uchiyama N and Sano S 2013 *Real-Time Trajectory Generation for Mobile Robots in a Corridor-Like Space Using B ´ ezier Curves* 37–41

[5] Dinçer Ü and Çevik M 2019 *Improved trajectory planning of an industrial parallel mechanism by a composite polynomial consisting of Bézier curves and cubic polynomials* **132** 248–263

[6] Li H and Savkin A V 2018 An algorithm for safe navigation of mobile robots by a sensor network in dynamic cluttered industrial environments *Robot. Comput. Integr. Manuf.* **54** 65–82

[7] Yang G J and Choi B W 2013 Implementation of joint space trajectory planning using RTO swith considering velocity constraints for mobile robots *IEEE ISR* 1-3

[8] Xianglin Y, Zhiwen Z, Junhao X and Zhiqiang Z 2015 *Trajectory Planning for RoboCup MSL Mobile Robots Based on Bézier Curve and Voronoi Diagram* 2552–2557

[9] Dixon J and Henlich O 1997 *Mobile Robots Navigation*

[10] Illah R 2004 *Autonomous Mobile Robots* (London, England: MIT Press)

[11] Borenstein J and Koren Y 1985 A Mobile Platform for Nursing Robots *IEEE Trans. Ind. Electron.* **IE-32** 2 158–165

[12] Huang J, Farritor S M, Qadi A and Goddard S 2006 *Localization and Follow-the-Leader Control of a Heterogeneous Group of Mobile Robots* **11** 2 205–215

[13] Hwang J H, Arkin R C and Kwon D S 2003 Mobile robots at your fingertip: Bezier curve on-line trajectory generation for supervisory control *IEEE Int. Conf. Intell. Robot. Syst.* **2** 1444–1449

[14] Costanzi R, Fanelli F, Meli E, Ridolfi A and Allotta B 2016 Mobile Robots Based on Bezier Curves *IFAC-PapersOnLine* **49** 15 145–150

[15] Ishikawa H, Noguchi K, Maki R and Naitoh H 2009 *Path generation with clothoid curve using image processing for two-wheel-drive autonomous mobile robots* **42** 16

[16] Koren Y and Borenstein J 1991 Potential field methods and their inherent limitations for mobile robot navigation *Proc. - IEEE Int. Conf. Robot. Autom.* **2** 1398–1404

[17] Jusuf Dwi Kurnianto N H, Ali H A and Fernando A 2012 *Navigasi Mobile Robot Berbasis Trajektori dan Odometry dengan Pemulihan Jalur Secara Otomatis* **66** 37–39

[18] Xu L, Wang D, Song B and Cao M 2017 Global smooth path planning for mobile robots based on continuous Bezier curve *Proc. - 2017 Chinese Autom. Congr. CAC* 2081–2085

[19] Zhou F, Song B and Tian G 2017 *Bezier Curve Based Smooth Path Planning for Mobile Robot*

[20] Hassani V and Lande S V 2018 Path Planning for Marine Vehicles using Bézier Curves *IFAC-PapersOnLine* **51** 29 305–310

[21] Zhang L, Sun L, Member I, Zhang S, Liu J and Member I 2015 *Trajectory Planning for an Indoor Mobile Robot Using Quintic Bezier Curves* 757–762

IOP Conf. Series: Materials Science and Engineering 1108 (2021) 012016 doi:10.1088/1757-899X/1108/1/012016

VeRO: Smart home assistant for blind with voice recognition

L Triyono*, T R Yudantoro, S Sukamto and I Hestinigsih

Informatics Engineering Study Program, Politeknik Negeri Semarang, Semarang, Indonesia

*liliek.triyono@polines.ac.id

Abstract. In daily life, people with visual impairment have difficulty in their activities, especially at home activities. People with visual impairment as of 2017 reached 1.5 percent or around 3.75 million people from the population of Indonesia and more than 253 million people worldwide. The number of blind people who are not small need help to get a good quality of life and independent. Friendly home design for the blind is not yet available specifically. It is a particular difficulty for blind people to access the latest electronic goods or information (news). Then this becomes the basis of this research on how to provide easy access for blind people. This research provides a solution that is designing a prototype of a house that is visually impaired in addition to the use of electronic equipment. This system starts from the microphone as a sensor in charge of picking up sound and then the results are processed by Android and then connected to NodeMCU via Wi-Fi. The microphone used is the default microphone of the Android smart phone. By using the speech recognition method, the command is enough to say the keywords, Android will manage the sound into a code that will be sent to the microcontroller to perform the task. There are several categories of commands that can be handled, including (a) Turning on / off electronic devices, (b) Providing news information from popular news web portals, (c) Entertainment by playing streaming radio and (d) General information on hours and dates. All planned functions are running properly, there are some features that might be developed further. Adding sensitivity to starting voice commands and adding the ability to read long stories can be further research.

1. Introduction

The sense of sight is one source of information that is vital for humans, because most of the information obtained by humans from the five senses, while the rest comes from other senses. So, with a problem in vision will make a decrease in the ability of its activities, different from normal conditions. People with visual impairments or people with visual impairment can be divided into two types of groups: Total Blindness (Blind) and Low Vision. Blind Total (Blind) is a person who is blind who really cannot see clearly what is around or can see light. Low Vision is a type of blind person who can still see even though the condition is still not alert or with certain distance restrictions.

This problem makes the decline in activities for the blind, especially activities at home. Based on the results of interviews with Pertuni (Indonesian Association of the Blind) Central Java Regional Office in November 2018, it was stated that the number of blind people in Central Java reached 600 people consisting of 25 branches spread in Central Java. While the number of blind people in 2017 reached 1.5 percent or around 3.75 million people of the Indonesian population. Based on WHO data, more than 253 million people worldwide suffer from visual impairment or visual impairment [1]. However, with such a number of people with visual impairment, homes for friendly blind access are not yet available.

IOP Conf. Series: Materials Science and Engineering 1108 (2021) 012016 doi:10.1088/1757-899X/1108/1/012016

Then this makes the conversation for easy access for blind people. With the strengthening of Law No. 8 Article 16 of 2016 concerning the equalization of persons with disabilities, the true housing for persons with disabilities should be made with disability-friendly aspects.

Alexa and Google Home are one of the technologies that have implemented Voice Recognition, with voice recognition expected to make it easier to implement in homes that are disabled by people, especially those who are blind. With a cylindrical shape, Alexa and Google Home have good assistant features, and despite the fact that both of these technologies cannot yet be implemented to support Indonesian, and have not yet been implemented into a house with disabilities friendly.

Reporting from www.forbes.com in the first quarter of 2017, Amazon echo controls 80 percent of the market in the smart speaker market, while Google 19.3 percent. With these two smart speaker giants, Google grew to 483 percent, compared to Amazon's 8 percent [2]. Users in America alone have reached 90 million who already take advantage of the benefits provided by smart speakers and will increase in the next quarter [3]. This fact confirms the acceptance of Voice Recognition technology is highly accepted by millennial users as it is today. For alexa prices range from $ 45.00 - $ 51.00, while google home ranges between $ 11.00 - $ 11.50 survey from *www.alibaba.com*.

From these data and facts, the longer the number of users is increasingly accepted by the market, it is necessary to develop a tool that can be used to facilitate or help people with disabilities in this case the blind need to develop tools that are simple, cheaper and easier. Ease of access control for persons with disabilities to electronic goods, making the house occupied more comfortable and easier in terms of accessibility of the use of electronic devices. Starting with this problem, we designed a prototype of a house that is not yet blind to the use of electronic equipment, namely "VeRO: Assistant Smart House for the Blind with Voice Recognition".

2. Research methods

2.1. Problem analysis

To find out the problems that occur in everyday life it is necessary to observe by conducting surveys in the environment, taking and analyzing data, which are then compared according to the level of urgency that exists. As a result of this step the researchers found problems arising in people with visual impairment such as (a) Turning on / off electronic devices, (b) Providing news information from popular news web portals, (c) Entertainment by playing streaming radio and (d) General information time and date. Thereby encouraging researchers to determine more appropriate solutions to be implemented, namely by using speech recognition to turn on lights and other home devices. Then it is processed with the required logic outputs.

From the plan, then the group determines the hardware requirements needed such as an Android operating system smart phone as data input; NodeMCU as hardware control; 8 channel relay to turn on or turn off the appliance according to the switch function; Wi-Fi for data transfer intermediaries and some additional equipment such as LED lights for indicators, etc.

2.2. Design

At the design stage, the most optimal preparation of processes, data, process flow and data relationships to meet the needs in accordance with the results of the needs analysis. The interface design will be designed as attractive as possible so that users are more interested in using it. The design to be used is an environmentally friendly design that is easily accessible and understood by users. The prototype design plan for the application can be seen in Figure 2.

IOP Conf. Series: Materials Science and Engineering 1108 (2021) 012016 doi:10.1088/1757-899X/1108/1/012016

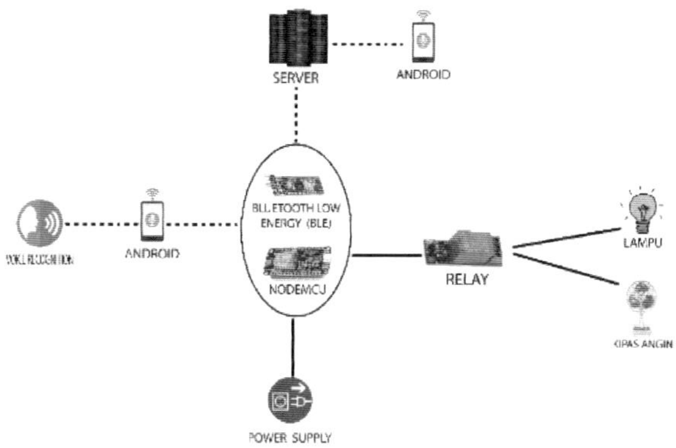

Figure 1. Schematic of the Vero system.

How to use the blind Android application:

- The application will be ready for use the first time the phone is turned on.
- Users just shake their cellphones to activate the application.
- Users mention keywords (using voice commands) that have been determined to turn on / turn off electronic equipment in the house or select a menu that has been prepared.
- The application will return to standby after receiving an order and the order is processed.

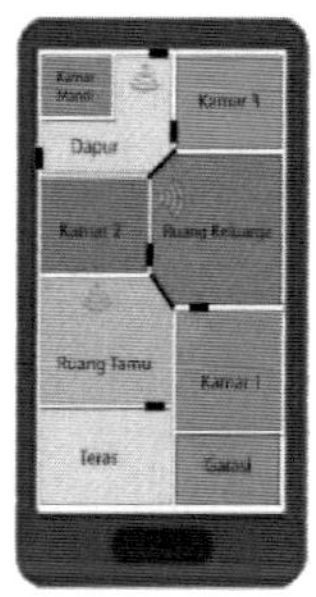

Figure 2. Home maid application design visually impaired friendly.

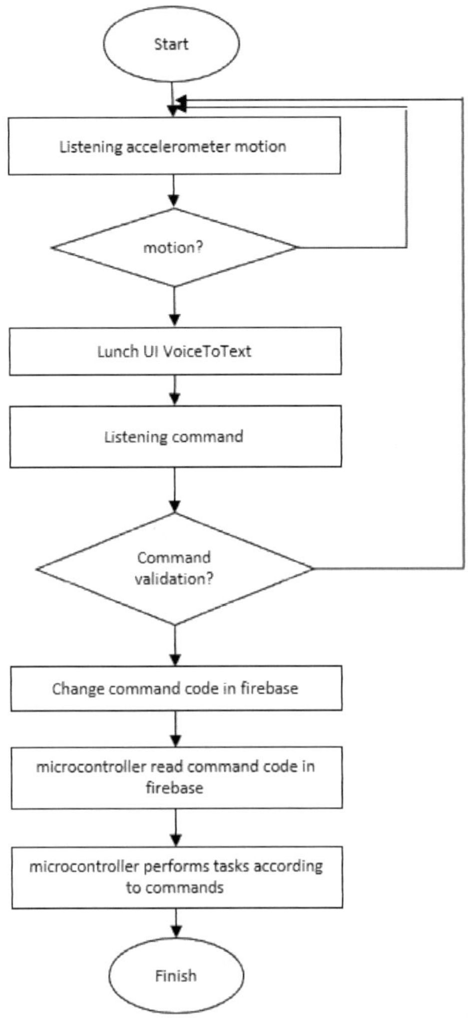

Figure 3. Vero system flowchart.

2.3. Implementation

In this implementation phase, the design will continue with writing the program code, the interface design that has been designed will be translated into a language that can be recognized by the computer so that it can work in accordance with the functions that have been designed.

2.4. Testing method

After the code writing phase is complete the testing of the system / application will be made. Blackbox Testing Method is used in this study, this test is based on functionality and output tests on software design in accordance with standards and reactions that can run well. In addition, this stage also serves to find errors in the system and can then be corrected.

(a) Android with Vero Application (b) Application Running

Figure 4. Vero application.

2.5. Maintenance

The last stage is maintenance, the software to be designed is as good as possible so that it has fulfilled the functions and uses needed by the user, and if it needs to be upgraded, the system will be done in the long term.

Figure 5. Implementation of electronic circuits.

3. Results and discussion

To run this application, make sure the application on the mobile used to be connected to the internet, without an internet connection this application will not work. The way this application works is to listen to instructions (which have been implemented into this application) and then execute them. To start giving instructions to this application, follow these steps:

- Move the cell phone right and left until your cellphone vibrates and a dialog box appears as shown below.
- After the dialog box appears, the application will start listening to the instructions (in the form of sound) that we provide.

IOP Conf. Series: Materials Science and Engineering 1108 (2021) 012016 doi:10.1088/1757-899X/1108/1/012016

Figure 5. Giving voice commands "*nyalakan lampu*" (turn on the lamp).

- After we give instructions, the dialog box will disappear and the application will execute the instructions.

Figure 6. The light is on the mockup.

Some instructions that can be used in this application are as follows:

1. *Info aplikasi* (Application info)
2. *Jam* (clock)
3. *Tanggal* (date)
4. *Nyalakan radio* (turn on radio)
5. *Baca berita popular* (read popular news)
6. *Baca berita pertama* (read first news)
7. *Baca berita kedua* (read second news)
8. *Baca berita ketiga* (read third news)
9. *Baca berita keempat* (read forth news)
10. *Baca berita kelima* (read fifth news)
11. *Nyalakan lampu* (turn on lamp)
12. *Matikan lampu* (turn off lamp)
13. *Nyalakan kipas* (turn on fan)
14. *Matikan kipas* (turn off fan)
15. *Nyalakan air* (turn on water)
16. *Matikan air* (turn off water)
17. *Nyalakan semua alat* (turn on all device)
18. *Matikan semua alat* (turn off all device)

The system can provide output according to what was planned beforehand, there are some notes in testing.

- Voice to Text release application using the Accelerometer takes a long time and requires swing / wobble accuracy when you want to enter voice commands,
- When the Application reads the news with text that has more than 4999 characters, it will be cut at 5000 characters because the API cannot support more than the maximum number.
- It is still difficult to use Mobile as a blind tool in terms of relatively large device dimensions.

4. Conclusion

In designing the system and tools in this study, it can be concluded that the development of both the application and the embedded system that has been prepared has worked well. Performance that has been successfully carried out by the application include: (1) The system can apply voice recognition in the android application to the commands that will be performed. (2) The system has been able to carry out orders to turn on and turn off electronic devices using internet media. (3) The system has been able to play radio streaming in accordance with the programmed radio channels. (4) The system has worked to help blind people follow news from popular news portals.

From the testing that has been done there are several developments that can be developed in the future implementation. The application can be applied in real terms to the blind to find out problems that arise after use. Applications can be developed for wearable devices for easy access.

References

[1] WHO 2018 *Blindness and vision impairment* [Online] Retrieved from: https://www.who.int/en/news-room/fact-sheets/detail/blindness-and-visual-impairment
[2] Koetsier J 2018 *Massive Reversal: Google Home Sales Explode 483% To Beat Amazon Echo For Smart Speaker Crown* [Online] Retrieved from: https://www.forbes.com/sites/johnkoetsier/2018/05/25/massive-reversal-google-home-sales-explode-483-to-beat-amazon-for-smart-speaker-crown/#6c5c20c7508c
[3] Koetsier J 2018 *Smart Speaker Users Growing 48% Annually, To Hit 90M In USA This Year* [Online] Retrieved from: https://www.forbes.com/sites/johnkoetsier/2018/05/29/smart-speaker-users-growing-48-annually-will-outnumber-wearable-tech-users-this-year/#5e087f235dde
[4] Barata M, Galih Salman A, Faahakhododo I and Kanigoro B 2018 Android based voice assistant for blind people *Libr. Hi Tech News* **35** 6 9–11
[5] Bansode M, Jadhav S and Kashyap A 2015,Voice Recognition and Voice Navigation for Blind using GPS," *Ijireeice* **3** 4 91–94
[6] Nilesh J, Alai P, Sapnil C and Bendre M 2014 Voice Based System in Desktop and Mobile Devices for Blind People *International Journal of Emerging Technology and Advanced Engineering (IJETAE)* **4** 2 404-407
[7] Ran L, Helal S and Moore S 2004 Drishti: an integrated indoor/outdoor blind navigation system and service. In Second IEEE Annual Conference on Pervasive Computing and Communications *Proceedings of the* 23-30
[8] Barata M, Oktavia O, Sugianto S and Salman A G 2014 *Aplikasi Voice Assistant Untuk Penyandang Tuna Netra Berbasis Android*

Bandwidth constrained distributed estimation for wireless sensor networks

E D Wardihani*, A Suharjono and I Sayekti

Politeknik Negeri Semarang, Semarang, Indonesia

*edwardihani@polines.ac.id

Abstract. The focus of this paper is power constrained in wireless sensor networks. We purpose an adaptive transmit power levels based on sensors noise variance and channel conditions. We also investigate its impact on energy saving. First, the measurement results of the sensors are quantized into discrete messages. Second, the quantized data are transmitted to the fusion center where a final estimate is generated. The optimal transmit power levels for each sensor is determined by the sensor noise levels and channels conditions from sensor to the fusion center. The goal is minimized the total transmitting power, while ensuring a given Mean Squared Error (MSE) performance. The sensor will be active when the measurement results of the sensors have low noises variances and the condition of the channel between the sensor and the FC is good and if the conditions are otherwise the sensor is not active with the aim of saving power. For the remaining active sensors, their optimal transmit power levels are determined jointly by individual channels gain, local observation noise variance sensor and the targeted MSE performance. Numerical examples show that an adaptive power levels achieves significantly smaller MSE than uniform power levels for the same average power consumption.

1. Introduction

A wireless sensor network is a network that consists of a number of sensors deployed in the certain areas, which are called the field of sensors. The main task of WSN is to detect natural phenomena such as temperature, humidity, pressure, movement of vehicles, and others. The sensor consists of three components, sensing, processing data and communication devices. The development of digital and wireless communication technologies makes it possible to produce many sensors, which are small sized, multi-function, low power and be able to communicate over short distances [1].

Recently the wireless sensor network (WSN) technology has developed rapidly because these is the one of the solutions to challenging problem in various fields such as environmental, military, health, home, and other commercial applications. However, the main challenge in the implementation of WSN is the limitation of energy and bandwidth of communication because the sensor is powered by battery, which has a limited lifetime [2-5].

A common Wireless Sensor Networks (WSN) architecture consists of a fusion center and a number of geographically distributed sensors. In this paper consider decentralized estimation of unknown parameter by a set of distributed sensor nodes and fusion center. The problem of decentralized estimation has been extensively studied, first in context of distributed control, later in tracking and data fusion [6]. Most of this works assume that the joint distribution of sensor observations is known and the real messages can be sent from sensor to the fusion center without distortion. These assumptions are unrealistic for practical sensor networks since the wireless links between the sensors and the fusion center invariably suffer from adverse channels effects such as attenuation and fading.

This paper considers the model of the wireless link between sensor and fusion center as additive white Gaussian noise (AWGN) channels under suitable channels path loss. To minimize the total energy consumption, each sensor optimally chooses the number of quantization levels and transmitting power levels by taking into account both their local SNRs and individual channel path losses.

This paper is organized as follow: section II methods. The design of optimal quantization scheme for each sensor based on their SNRs and channels path loss described in Section III. The numerical analysis is presented in section IV and finally concluding remark are given in Section V.

2. Methods

The system in this study is a WSN with a star topology, one hop, where a set of sensors takes measurements and sends quantified measurement results to a data processing center or fusion center (FC). However, the results of this study can be used for other topologies and multi hops. The system image used is shown in Figure 1. Consider a set of K distributed sensors, each making an observation on a random signal estimated θ. The observations are corrupted by additive noises and are described by Luo [7]:

$$x_k = \theta + n_k\,,\ k = 1,\ldots,K \tag{1}$$

The noise n_k : k = 1,2,..., K are zero mean spatially uncorrelated with variance σ_k, but otherwise unknown. Each sensor performs a local quantization of x_k and generates a quantization message $m_k(x_k, L_k)$ of L_k bits. Each message is then transmitted to the FC through a separate AWGN channels with a known path loss coefficient, and the FC generates the final estimate of θ based on the received signals. In this paper use the Mean Squared Error (MSE) criterion to measure the quality of an estimator for θ.

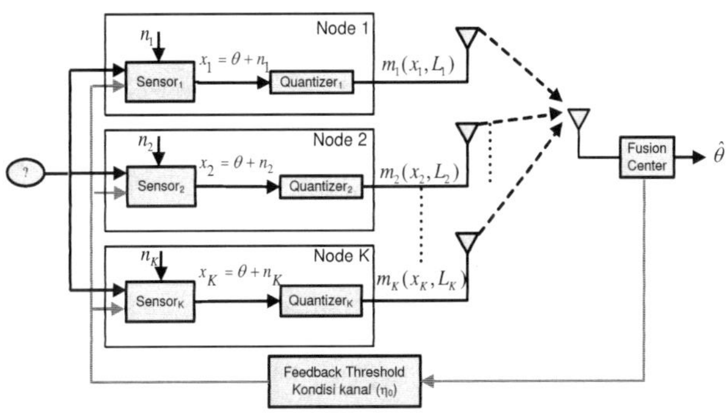

Figure 1. Decentralized estimation with adaptive quantization.

2.1. Power allocation

Each sensor locally quantizes the analog observations x_k into discrete message $m_k(x_k)$, of length L_k bits. These local estimates are transmitted to FC. The channel between sensor and FC is corrupted with additive white Gaussian noise (AWGN). The channels are assumed to be loss and the path loss is given by $a_k = d_k^{a}$ *where* d_k is the transmission distance and α is the path loss exponent. At the FC final estimation is performed by combining these messages.

The goal is optimally choosing quantization and transmits power levels at each sensor (P), while ensuring a targeted MSE performance (D_0). The formula of power optimization is:

$$min\|P\|_2 \tag{2}$$
$$s.t.\, D' \leq D_0 \quad L_k \geq 0$$

Where *D'* are the achieved MSE performance. L_k is an integer signifying the number of quantized bits per sample at the sensor *k*.

The implement the described adaptive quantization scheme sees Fig.2. The FC broadcast the threshold η_o whose value is based on the collected network information. Each sensor decided the quantization message length L_k according to its own local information (σ_k, a_k) and η_o.. The sensor with $\sigma_k \leq W$ and $a_k \leq \eta_o$ will transmit a message with length $L_k \geq 1$, L_k is determined from simulation. Otherwise sensor with $\sigma_k \geq W$ and $a_k \geq \eta_o$ should be inactive to save energy consumption.

3. Results and discussion

The main task of the sensor network is to monitor the physical condition of an environment and communicate the results to other sensors or to a data processing center which is often called the Fusion Center (FC) [8-9]. Distributed estimation systems that do not require knowledge of the distribution of sensor measurements that are famous for the Universal Decentralized Estimation (UDES) system have been proven to be able to overcome the challenges of resource limitations in JSN [7]. In addition, the centralization system requires a large communication bandwidth and allows the accumulation of data in FC [10]. These results indicate that this system has very good prospects for further development. This chapter describes the initial stage of the research, namely the further evaluation of the problem of resource limitations in the distributed estimation system. The aim is to obtain estimation parameters that can be improved or extended that contribute to improving resource efficiency and improving estimation performance. The main purpose of this section is to determine the number of quantization bits and the optimal amount of transmit power if a specific MSE target is desired. Implementation of the proposed algorithm.

Figure 2 shows FC sends the channel condition threshold η_0 to all sensors. The value of this threshold is based on information collected from the network. Each sensor determines the length of the quantization bit L_k based on the information it has, namely the measurement noise variant σ_k, path loss a_k, and the threshold condition of the channel η_0 sent by FC. Because there is no knowledge about pdf sensor noise then probabilistic quantization is used which can work for all types of pdf sensor noise and produces unbiased messages.

To strengthen the analytical result, it was simulated using following parameters. The total number of sensors $K = 100$, the range of sensor measurement W = 1, where $W=1$ and sensor observation noise variances to be Chi-squared distribution with degree 1. Also, the channel path loss coefficients and the distance between sensor and FC are generated by a uniform distribution [0,10]. Each simulation is carried out 10^6 iterations and the results displayed are the average values of the simulation results. With the power scheduling method, a smaller number of sensors is needed to achieve the desired MSE target. The proposed power efficiency method is the optimal power distribution method where the bit length quantization and transmission power allocation are determined based on variants of measurement losses and the channel condition threshold between the sensor and FC which can achieve a better MSE compared to the uniform method where the network power is shared equally to all sensors, shown in Figure 3.

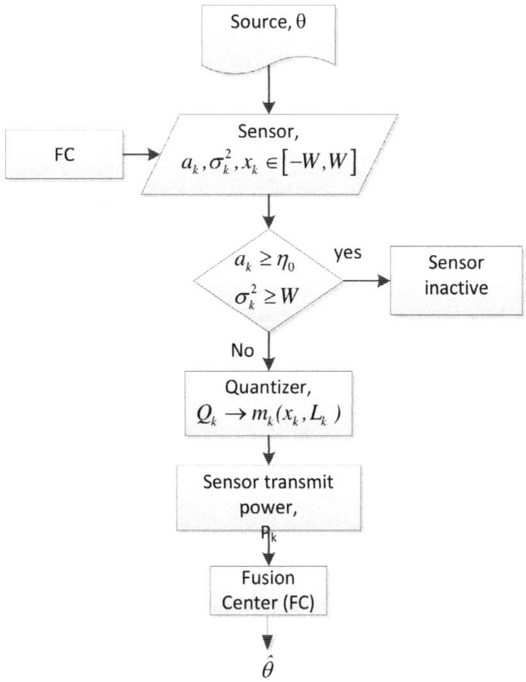

Figure 2. The algorithm of adaptive quantization.

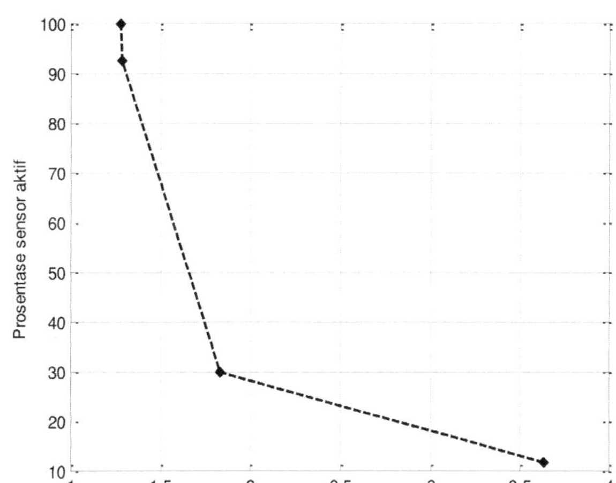

Figure 3. Active sensor percentage if the MSE target P_T=500 Watt, K=100, $d_k \in [1,10]$, $\sigma_k^2 \in [0,1]$.

The simulation result show that when target D_0 increases, more sensors will become inactive. Such inactive sensors neither perform quantization nor transmit any message to the FC in order to conserve energy. Figure 3 plots the percentage of active sensors versus some target MSE $D_0 \geq D$BLUE, where DBLUE is MSE of the centralized BLUE defined in Wardihani et al [2]. This paper compared the adaptive quantization with the uniform quantization, where each sensor quantizes their measurement into the same number of bits to achieve the same target MSE. Figure 5 shows that adaptive quantization has smaller MSE than uniform quantization.

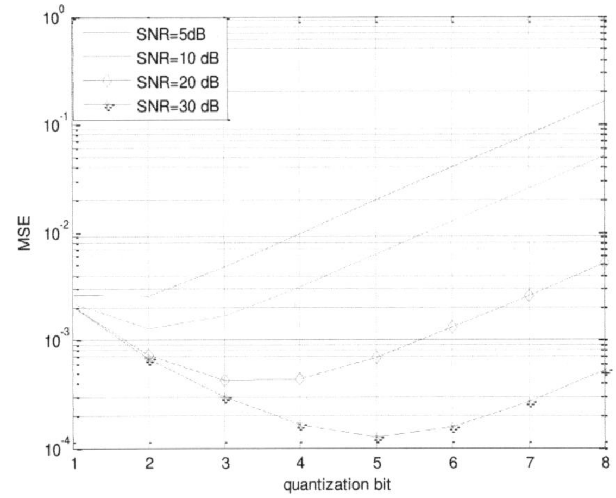

Figure 4. Number of quantization bit at the different SNR of local sensor with total energy constraint = 500 W.

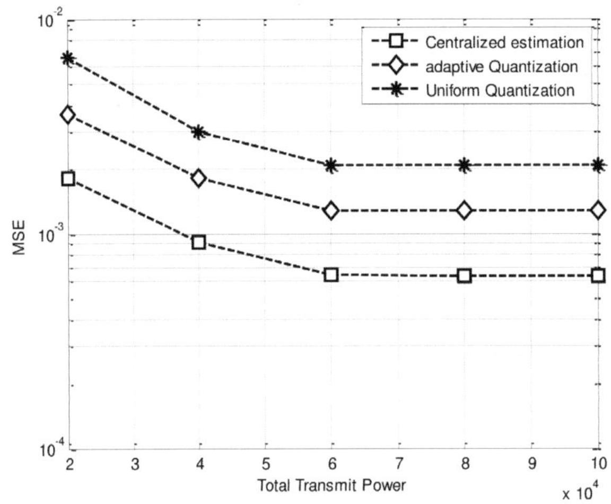

Figure 5. MSE achieved by estimator if given the same total transmit power with different quantization method.

4. Conclusion

This paper derived the optimal transmit power levels in sensor networks to minimize energy consumption while satisfying a certain distortion constraint. The optimal transmission power levels for each sensor can be determined jointly in term channels path loss and the observation noise sensor. Numerical examples show that optimal power levels achieve significantly smaller MSE than uniform transmit power levels for the same average power consumption.

References

[1] Akyildiz I F, Su W, Sankarasubramaniam Y and Cayirci E 2002 A survey on sensor networks *IEEE Commun. Mag.* **40** 8 102–114

[2] Wardihani E D, Wirawan W and Hendrantoro G 2015 Optimal Energy Allocation Scheme in Distributed Estimation for Wireless Sensor Networks over Rayleigh Fading Channels, Optimal Energy Allocation Scheme in Distributed Estimation for Wireless Sensor Networks

over Rayleigh Fading Channels *Int. J. Distrib. Sens. Netw. Int. J. Distrib. Sens. Netw.*
[3] Wardihani E D, Wirawan W and Hendrantoro G 2010 Energy Constrained Distributed Estimation In Wireless Sensor Networks Over AWGN Channels *International Student Conf. on Adv. Sci. and Tech. (ICAST)* (Kumamoto Univ, Japan)
[4] Wardihani E D, Wirawan W and Hendrantoro G 2014 Impact of Propagation Parameters on energy efficiency in virtual MIMO-Based Wireless Sensor Network *Jurnal Kursor* **7** 4
[5] Wardihani E D, Sindung H W S and Subuh P 2016 Power Constrained Distributed Estimation For Wireless Sensor Networks *International Conference on Science, Technology and Interdisciplinary Research (IC-STAR 2016)*
[6] Sani A and Vosoughi A 2016 Distributed vector estimation for power-and bandwidth-constrained wireless sensor networks *IEEE Transactions on Signal Processing* **64** 15 3879-3894
[7] Luo Z -Q Universal decentralized estimation in a bandwidth constrained sensor network *IEEE Transactions on Information Theory* **51** 6 2210–2219
[8] Li J and AlRegib G 2007 Rate-Constrained Distributed Estimation in Wireless Sensor Networks *IEEE Transactions on Signal Processing* **55** 5 1634–1643
[9] Li J and AlRegib G 2009 Distributed estimation in energy-constrained wireless sensor networks. *IEEE Transactions on Signal Processing* **57** 10 3746-3758
[10] Xiao J -J, Cui S, Luo Z -Q and Goldsmith A J 2008 Linear Coherent Decentralized Estimation *IEEE Transactions on Signal Processing* **56** 2 757–770

Online monitoring system for emergency diabetes mellitus patients

R Kurniawati[1,*], S Suryono[2] and S Ismail[3]

[1] Master Student Adult Nursing, Diponegoro University, Semarang, Indonesia
[2] Department of Physic, Diponegoro University, Semarang, Indonesia
[3] Emergency and Critical Nursing, Department of Nursing Science, Adult Nursing Division, Faculty of Medicine, Diponegoro University, Semarang, Indonesia

*ratnaummudzaky@gmail.com

Abstract. In this research, an online monitoring system was designed to detect emergency due to uncontrolled blood glucose fluctuations in patients with diabetes mellitus. Hyperglycemia in the morning causes dawn phenomenon and Somogyi effect. This incident is very fast, and its uncontrollable time of occurrence can inflict a fatal condition causing death. Unfortunately, there are no systems able to detect such events, currently. This study proposes an online monitoring system with the input of the patient's blood glucose measurement into the smartphone. Blood glucose values are sent and stored in the webserver, and the results are used to determine the incidence of dawn phenomenon and the Somogyi effect. The study was able to produce an online monitoring system which can be monitored by nurses, doctors, and administrators via the dashboard.

1. Introduction

Diabetes mellitus is a serious health problem [1]. An estimated 30.3 million or 9.4% of people of all ages of the population in the United States was suffering from diabetes in 2015 [2]. Around the world, every seven seconds, a person dies due to the diabetes, approximately four million deaths every year [3,4]. The hyperglycemic crisis is one of the most serious complications in patients with diabetes mellitus [5,6]. The risk of death in diabetes mellitus varies greatly with age, glycemic control, and renal complications [7]. Uncontrolled blood glucose jeopardizes the body organs especially the nervous and cardiovascular systems [4]. The primary key preventing chronic complications is suppressing the rise in blood glucose, blood pressure, and lipids [8].

Hyperglycemia in the morning causes dawn phenomenon [9] when it is preceded by a hypoglycemia condition, it is called Somogyi effect [10]. Both conditions can result the patient's blood sugar instability if they are not monitored [11]. Blood glucose management is the best way to prevent disease progression. However, blood glucose monitoring as a prevention or cure for diabetes has not been widely developed [12,13].

Monitoring blood glucose levels was proved to prolong life expectancy by allowing diabetics controlling hypoglycemic episodes or hyperglycemia, thus it provides better control of their condition, and the diabetics can manage the condition of their hypoglycemia or hyperglycemia thereby it provides more control to prevent complications. Currently blood glucose monitoring involves the invasive method [10,14].

Management of diabetes mellitus uses a comprehensive approaches including medical care, psychosocial, lifestyle changes, education, continuous monitoring, and self-management [15,16]. Those targets cannot be achieved without the involvement of patients through diabetes self-care management for the diabetes mellitus treatment [17].

Diabetes self-care management is an important part of diabetes care. Some research showed diabetes self-care management improves life quality and prevents the severity of long-term diabetes complications [12]. However, self-management is a complex process. The patient is responsible for treating her illness, and requiring commitment, special skills, knowledge, and confidence. Diabetes self-care management is long life and complicated treatment, where patients need to adapt to new lifestyles, make decisions, and perform certain tasks [17]. The stability of blood glucose levels within normal limits is an important goal for the patient [18].

All this time, the monitoring of the dawn phenomenon and the Somogyi effect is very rare to be conducted. The invasive glucometer method makes it difficult for patients to measure blood glucose regularly. Routine blood glucose measurement using the invasive method is not recommended for all people with diabetes mellitus [19]. Unfortunately, blood glucose measurement of invasive methods cannot provide real-time information with data accessed by health personnel on a daily basis. The problem can be solved by regular, fast, and accurate monitoring for patients.

This paper proposes wireless area health technology from electronic medicine that is under the domain of telemedicine. This method is effective for improving communication between health care providers and patients with diabetes mellitus. Wireless technology can be used worldwide, and transmit the information about health [4].

System realizations

The built system consists of devices those are remote areas adjacent to the patient, and *data station* located in the cloud web service. The online system of emergency monitoring in patients with diabetes mellitus proposed in this paper is shown in Figure 1.

2. System realizations

The built system consists of devices those are remote areas adjacent to the patient, and *data station* located in the cloud web service. The online system of emergency monitoring in patients with diabetes mellitus proposed in this paper is shown in Figure 1.

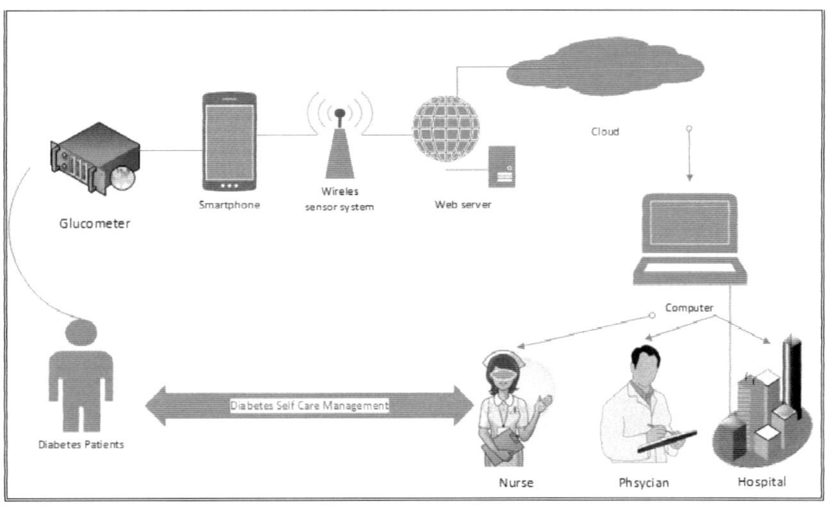

Figure 1. Diagram integration of dawn effect emergency online monitoring system.

Blood glucose levels of patients with diabetes mellitus was measured using noninvasive method. Blood glucose value data was sent via smartphone to web service. There are three categories of data in each patient, first, the demographic data include the patient's identity, age, body mass index, occupation, education. Second datum is blood glucose values before and after meals, accompanied by measurement time and normal range of blood glucose. This measurement is performed periodically at the time of a diabetic patient's crisis. Third datum is self-management education feedbacks matched suitable to the patient's condition. Web services receive data and save to web data base. The measurement results are shown to the dashboard. Access rights holders to the website are patients, nurses, doctors and administrators. The nurse collaborates directly with the doctor and provides feedback on the received blood glucose result.

3. Results

Total of 83 diabetic patients were regularly monitored for three days using an online monitoring system. There were 67 patients that had dawn phenomenon, 6 patients had Somogyi effect, and 10 patients were died. The average speed of blood glucose data transfer from measurement to the graph in the smartphone is 15 seconds. The process of measuring blood glucose using smartphones and web services employed "my sugar" applications designed by researchers. Patient data input includes identity, body mass index, gender, education, employment and self-care management education was conducted to prevent the occurrence of diabetes mellitus emergency. In the measurements using non-invasive glucometer, the value of blood glucose is automatically appeared on the smartphone screen, and stored in the web database www.mysugar.cf in the form of numbers and graphs. This application was equipped with measurement time, normal range of blood glucose, and self-management education that can be given directly to the patient. The measurement system of dawn phenomenon and Somogyi effect using my sugar application can be seen in Figure 2.

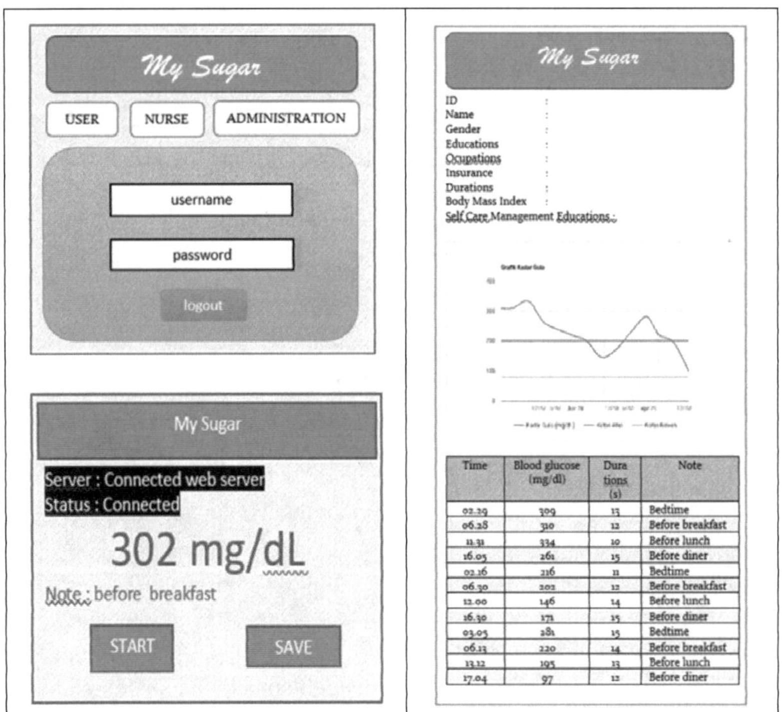

Time	Blood glucose (mg/dl)	Durations (s)	Note
02.29	309	13	Bedtime
06.28	310	12	Before breakfast
11.31	334	10	Before lunch
16.05	261	15	Before diner
02.16	216	11	Bedtime
06.30	202	12	Before breakfast
12.00	146	14	Before lunch
16.30	171	15	Before diner
03.05	281	15	Bedtime
06.13	220	14	Before breakfast
13.12	195	13	Before lunch
17.04	97	12	Before diner

Figure 2. Web-based graphic displays online monitoring system for diabetes mellitus patients.

Dawn phenomenon begins with an increase in blood glucose above the normal limit of 126 mg/dL [20], this condition is called hyperglycemia in the morning. Patient 1's blood glucose value at midnight measurements reached 348 mg/dL, and patient 2's blood glucose value reached 280 mg/dL, then in the morning before meal, the blood glucose value for both patients were decreased but still above the normal value of 220 mg/dL and 171 mg/dL, while at noon and the afternoon, the values began to decline. This case occurred on the second day with blood glucose values of 241 mg/dl and 251 mg/dL higher than normal, day and afternoon began to decline near normal. The third day of measurement was also found 239 mg/dL and 279 mg/dL, so that f both respondents experienced an increase in blood glucose or three days' measurements.

Online monitoring system using noninvasive method to make fluctuations in blood glucose value can be monitored automatically whereas this phenomenon is not monitored properly, previously. Detection results of dawn phenomenon are shown in Figure 3.

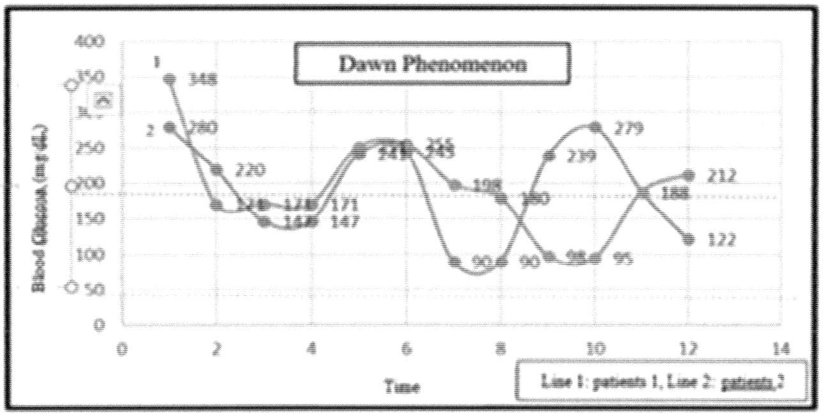

Figure 3. Dawn phenomenon warning detections.

The second emergency condition is the Somogyi effect. The occurrence of Somogyi effect is different from the dawn phenomenon. It begins with a decrease in blood glucose levels below normal (70 mg/dL) [20]. Blood glucose value of patient 9 was at 40 mg/dL, increased to 190 mg/dL decreased to 70 mg/dL and increased again to 122 mg/dL, diminished to 74 mg/dL on the third day, and rose again to 204 mg/dL. This condition was observed for three days. It can be seen initial blood glucose value of Patient 6 was at 59 mg/dL then that is increased to 98 - 139 mg/dL, decreased again to 44, 49, 65 mg/dL on the second day, and fixed blood glucose value was fixed below the normal range then increased, and became on its normal limits on the third day. The use of an online monitoring emergency system is supporting the control of the patient's blood glucose fluctuations. The result of detection of Somogyi effect can be seen in figure 4.

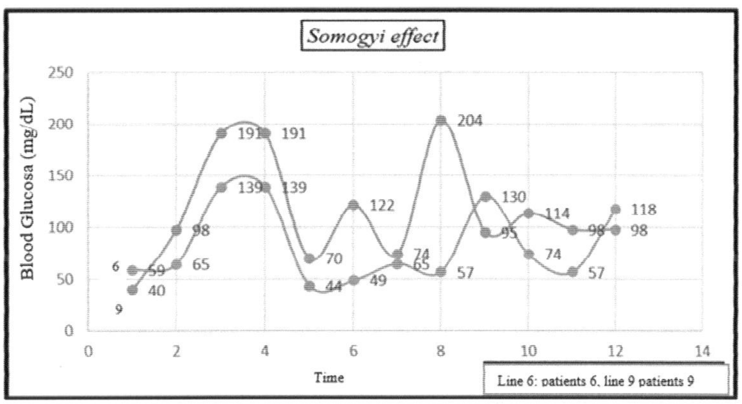

Figure 4. Somogyi effect warning detections.

4. Discussion

Telemedicine is telecommunication to support health care. Telemedicine timely transmission and remote interpretation of the patient provide follow-up and preventive interventions. By using various forms of information technology such as internet, cell phone calls, short messaging services and television programs, the provision of care is more effective. The media used to distribute health care are mobile phones, digital diaries, computers, and laptops. Most of the electronic healthcare and telemedicine services provided in developed countries are based on the diagnosis and clinical management of chronic diseases [4].

In addition to medical feedback, patient satisfaction associated with tissue plays an important role in the management of blood glucose. Blood glucose monitoring based on smartphone applications becomes more effective, and it is important to serve patients with high-quality and well-functioning devices that can increase patient satisfaction and willingness to use them [21].

The use of online monitoring helps each patient to routinely measure their blood glucose four times a day without hurting the patient's fingers. This is in line with *NICE Guideline* [22]. Blood glucose monitoring results good glycemic improvement in patients with diabetes mellitus without increasing the risk of hypoglycemia, prevents cardiovascular risk, and improves life quality [23]. The ignorance to general self-care among patients with type 2 diabetes results severe complications [24].

Advantage of Online monitoring system in diabetes mellitus emergency is *self-care* based *web service* program that can be done easily, not injuring, and designed in such a way that diabetes patients can be trained independently using glucometer, with time automatically performed periodically in one day that is very comfortable. Diabetes mellitus patients is receiving intensive care with insulin performed *self-monitoring* appropriately that is significantly reduced the patient's blood glucose level [19,25]. Remote health monitoring for home-stay patients can reduce health care costs [26].

Advances in *smartphone* technology combined with data connectivity and *web services* in *mobile* applications support self-management of diabetes. The mobile application offers several advantages for patient upgrading, such as converting record book to an electronic system. Data stored on mobile devices can be sent by *online* to health care workers or diabetes educators [27].

The web-based blood glucose monitoring system facilitates the feedback of health care providers. Patients regularly monitor their glucose level, and upload the result data to the *service website*. The data is reviewed by a professional health team, which provides feedback on outcomes, offers encouragement, and recommends changes to patient monitoring practices, insulin administration, or diet [23,27].

Mobile devices can not only be employed to monitor health conditions, but also as patient interactive interfaces to communicate with nurses on how to deal with certain health or emergency situations. The provision of *real-time* data is of significantly benefit, especially to patients who have chronic conditions requiring attention, remote locations that are not accessible or disaster situations [26].

Mobile devices provide low cost, and efficient remote health monitoring through service provisioning *mobile web* approach. This method is more agile, flexible, and economical [26].

Web-based health monitoring allows faster data access. The privacy of patient data is maintained because each patient can only see the results of his own measurement. *Web* application is an independent *platform* compared to original mobile apps (like *Android* or iPhone application). Patients who have chronic conditions and require 24-hour monitoring [26].

Web-based applications support diabetes self-management and decisions (*decisions support system)*, and functions as an electronic notebook, automated analysis, and device to predict glucose level based on the obtained data. This application is used to process large, complex data and provide feedback related to insulin dose and healthy behavior.

Web-based self-care management of diabetes patients research resulted that the *smartphone* applications provide moderate benefits for the management of type 2 diabetes mellitus. However, further studies with valid study designs and longer follow-up are required to evaluate the impact of effective mHealth applications for self-care management for diabetes mellitus patients. Web-based diabetes treatment strongly supports the self-management of patients with diabetes mellitus [28].

Smartphone interventions vary in approach to strengthen lifestyle changes, healthy diet, and physical exercise. These results indicate that mobile utilization leads to improved A1C and self-management in diabetes care [29]. Diabetes mellitus patients are advised to check glucose levels at least four to five times a day to avoid hyperglycemic/hypoglycemic. Glucose monitoring often plays an important role to improve the life quality [30]. Internet-based monitoring can provide long-term glycemic benefits. After 5 years of *follow-up,* patients in the Internet intervention group showed a mean decrease in HbA1c levels from 7.43% to 7.09% compared to 7.45% to 7.38% in the conventional care group [23].

Patients can read the results through a device connected to a telephone line transmitting all data to a special officer which analyzes in graphical, numerical and static formats with specific programs. Doctors can determine new therapies, previous therapy changes, or just check the same information. The usefulness of this system in the treatment of diabetes mellitus is to increase the information available for therapeutic adjustment, improve communication between patient and physician, improve the education and empower the patients, positive trend toward increased metabolic control, and blood glucose stability, especially in patients with diabetes mellitus [31].

Remote monitoring improves clinical outcomes, care coordination, and patient satisfaction, enabling patients to inform health care providers about glycemic control, and diabetes educators or other providers to monitor glucose test results, and to provide real patient feedback. This is the time to improve knowledge, *self-efficacy*, and behavioral change. The repetition of this physiological data collection and interpretation feedback combined with feedback to the patient enables a care plan modified by health care providers, and improved outcomes [32].

The effectiveness of web-based care programs in controlling diabetes were compared to regular care. There was a much greater decrease in HbA1c levels in persistent users compared to intermittent users among web-based treatment patients [33]. Other clinical outcomes also depend on patient compliance with *telemonitoring* [34].

The results show that conventional blood glucose measurement is less able to describe plasma glucose changes from high to low values characteristic of type 1 diabetes in children. Hyperglycemia *Postprandial* plays very important role in the development of vascular complications in diabetes. In contrast to methods *non-invasive*, continuous glucose monitoring for 3 consecutive nights provides a much more complete description of the blood glucose [35]. Periodic blood glucose measurement is the most useful method for evaluating glycemic variability in detail. Predicting glycemic variability in the morning before and after meals based on patient characteristics without frequent finger injury, this information can provide appropriate guidance in drug selection. [36] CGMS, a *non-invasive method,* is a useful tool for diagnosing asymptomatic nocturnal hypoglycemia [37].

Clinical studies show that hypoglycemia is strongly associated with death in diabetes mellitus patients. Nocturnal hypoglycemia is not only related to hyperglycemia during the day, but also increasing glucose levels in the morning. The relationship between glucose levels before and after breakfast and nocturnal hypoglycemia in patients with type 2 diabetes were studied. Reduction glucose levels before and after meals increased glycemic variability for 24 hours was performed by adjusting patient care to avoid nocturnal hypoglycemia. Knowledge of morning glycemic variability is important to increase glycemic variability for 24 hours. Measurement of blood glucose levels before and after breakfast is complicated for the patient. The *non-invasive* method predicts morning glycemic variability more easily. This method employs univariate and gradual (increasing and decreasing variable method) multivariate regression analysis. The p value (<0.05) was considered significant [36]. Nocturnal hypoglycemia often occurs in prolonged duration, related to glucose values before sleep (≤ 100 to 150 mg/dL), especially occurring in the early part of the night [37]. Research on the difference in intravenous blood glucose measurement is better than conventional in patients with critical conditions [38].

5. Conclusion

Blood glucose monitoring is an important component in the diabetes mellitus patient's treatment. Online monitoring system for blood glucose measurement can detect the occurrence of blood glucose fluctuations in patients with diabetes mellitus. Online monitoring system monitoring model contributes the early detection of emergency conditions due to dawn phenomenon and Somogyi effect in diabetes mellitus patients. The advantages of online monitoring system are a modality to support the patient's decision in real-time as an alarm for hypoglycemia or hyperglycemia, and an integrated device connected the patient monitoring system and health services, easy tools to use, wireless connectivity, and online recorded data.

References

[1] Factor S I 2016 *Somogyi Effect In Type 2 Diabetes Mellitus : A Case Report* **2** 2 32–33

[2] Centers for Disease Control and Prevention 2016 *National Diabetes Statistics Report: Estimates of Diabetes and Its Burden in the United States* (Atlanta, GA: Centers for Disease Control and Prevention, US Department of Health and Human Services)

[3] Kayar Y, Ilhan A, Kayar N B, Unver N, Coban G, Ekinci I and Eroglu H 2017 *Relationship between the poor glycemic control and risk factors, life style and complications* **28** 4 1581–1586

[4] Bano G, Ali S A, Mawani M and Ali S A 2016 Burden of Diabetes Mellitus and Role of Telemedicine in its Management : Narrative Review *Ann. Clin. Lab. Res.* **4** 3 113 1–8

[5] Weng S and Tsai K 2015 *Long-term Mortality Risk After Hyperglycemic Crisis Episodes in Geriatric Patients With Diabetes : A National Population- Based Cohort Study* 1–6

[6] Zannad F 2015 Heart failure and mortality outcomes in patients with type 2 diabetes taking alogliptin versus placebo in EXAMINE : a multicentre , randomised , double-blind trial *Lancet* **6736** 14 1–10

[7] Wedel H 2015 *Excess Mortality among Persons with Type 2 Diabetes*

[8] Raballo M 2012 A study of patients' perceptions of diabetes care delivery and diabetes: Propositional analysis in people with type 1 and 2 diabetes managed by group or usual care *Diabetes Care* **35** 2 242–247

[9] Monier L 2012 *Frequency and Severity of the Dawn* **35** 3–5

[10] Minicucci W J, Maia F F, Neto A M and Zantut-wittmann D E 2015 Somogyi effect as the most common cause of fasting hyperglycemia in T1D patients *Diabetol. Metab. Syndr.* **7** 1 A62

[11] O'Neal L E T B 2017 Dawn Phenomenon *Pubmed*

[12] Doñate-Martínez A, Ródenas F and Garcés J 2016 Impact of a primary-based telemonitoring programme in HRQOL, satisfaction and usefulness in a sample of older adults with chronic diseases in Valencia (Spain) *Arch. Gerontol. Geriatr.* **62** 169–175

[13] Wu W, Huang Y, Qiu J, Sun J and Wang H 2017 Self-Monitoring of Blood Glucose to Assess Dawn Phenomenon in Chinese People with Type 2 Diabetes Mellitus *Int. J. Endocrinol.*

[14] Narkhede P, Dhalwar S and Karthikeyan B 2016 *NIR Based Non-Invasive Blood Glucose Measurement NIR Based Non-Invasive Blood Glucose Measurement*

[15] NICE 2015 Managing type 1 diabetes in adults *NICE Pathways* 1–21

[16] Hou Y, Li W, Qiu J and Wang X 2014 ScienceDirect Efficacy of blood glucose self-monitoring on glycemic control in patients with non-insulin-treated type 2 diabetes *Int. J. Nurs. Sci.* **1** 2 191–195

[17] Alrahbi H 2014 Diabetes self-management (DSM) in Omani with type-2 diabetes *Int. J. Nurs. Sci.* **1** 4 352–359

[18] Wolf A, Caithaml V and Nestával L 2011 *Telemonitoring of Blood Glucose and Energy Expenditure of Patients with Diabetes Mellitus to Support Optimization of Insulin Dosing* (Czech Technical University in Prague, Faculty of Biomedical Engineering) 395–397

[19] Geng Z, Tang F, Ding Y, Li S and Wang X 2017 Noninvasive Continuous Glucose Monitoring Using a Multisensor-Based Glucometer and Time Series Analysis *Sci. Rep.* **7** 1 1–10

[20] American Diabetes Association 2017 Standards of Medical care in diabetes *J. Clin. Appl. Res. Educ.* **40** 1–142

[21] Kim H S 2014 Efficacy of the smartphone-based glucose management application stratified by user satisfaction *Diabetes Metab. J.* **38** 3 204–210

[22] Wróbel M S 2016 Non-invasive blood glucose monitoring with Raman spectroscopy: Prospects for device miniaturization *IOP Conf. Ser. Mater. Sci. Eng.* **104** 1
[23] Tildesley H D, Po M D and Ross S A 2015 Internet Blood Glucose Monitoring Systems Provide Lasting Glycemic Benefit in Type 1 and 2 Diabetes: A Systematic Review *Med. Clin. North Am.* **99** 1 17–33
[24] Nelson L A, Wallston K A, Kripalani S, Greevy Jr R A, Elasy T A, Bergner E M and Mayberry L S 2018 Mobile phone support for diabetes self-care among diverse adults: Protocol for a three-arm randomized controlled trial *JMIR research protocols* **7** 4 e92
[25] Veras V S, Regina C, Teixeira D S, Antônio M and Teresa M 2014 Glycemic Profile Of Persons With Diabetes Mellitus In A Home Blood Glucose Self-Monitoring Program *The Self-Monitoring of Blood Glucose* **23** 3 609–616
[26] Elgazzar K, Aboelfotoh M, Martin P and Hassanein H S 2012 Ubiquitous health monitoring using mobile web services *Procedia Comput. Sci.* **10** 332–339
[27] El-Gayar O, Timsina P, Nawar N, Eid W and Elizabeth S 2103 Mobile applications for diabetes self-management: status and potential *J Diabetes Sci Technol* **7** 1 247–262
[28] Cui M, Wu W, Mao J, Wang X and Nie M 2016 T2DM self-management via smartphone applications: A systematic review and meta-analysis *PLoS One* **11** 11 1–15
[29] Tran J, Tran R and White J R 2012 Smartphone-based glucose monitors and applications in the management of diabetes: An overview of 10 salient 'apps' and a novel smartphone-connected blood glucose monitor *Clin. Diabetes* **30** 4 173–178
[30] Yadav J, Rani A, Singh V and Murari B M 2015 Prospects and limitations of non-invasive blood glucose monitoring using near-infrared spectroscopy *Biomed. Signal Process. Control* **18** 214–227
[31] Montagnana M, Caputo M, Giavarina D and Lippi G 2009 Overview on self-monitoring of blood glucose *Clin. Chim. Acta* **402** 1–2 7–13
[32] Coughlin S S 2017 Mobile technology for self-monitoring of blood glucose among patients with type 2 diabetes mellitus *mHealth* **3** 47–47
[33] Franc S 2014 Telemedicine and type 1 diabetes: Is technology per se sufficient to improve glycaemic control? *Diabetes Metab.* **40** 1 61–66
[34] Kim K M 2016 Efficacy of a New Medical Information system, Ubiquitous Healthcare Service with Voice Inception Technique in Elderly Diabetic Patients *Sci. Rep.* **5** 1 18214
[35] Elizabeth B, Teresa M and Maria D 2001 Limitations of conventional methods of self-monitoring of blood glucose *Diabetes Care* **24** 11
[36] Takeishi S, Mori A, Kawai M, Yoshida Y, Hachiya H, Yumura T and Kawai H 2017 Investigating the relationship between morning glycemic variability and patient characteristics using continuous glucose monitoring data in patients with Type 2 Diabetes *Internal Medicine* **56** 12 1467-1473
[37] Kaufman F R 2022 Nocturnal hypoglycemia detected with the continuous glucose monitoring system in pediatric patients with type 1 diabetes *J. Pediatr.* **141** 5 625–630
[38] Combs-cantrell D T 2009 New England Journal *Media* **362** 1 1597–1605

Survey of smart parking application deployment

D Nugraha*, F Y H Ahmed, M I Abdullah and M G M Johar

Management and Science University, Shah Alam, Malaysia

*dian.nugrahatca@gmail.com

Abstract. Number of vehicles continues to increase in the urban place parallel with the infrastructure advancement. The increase of vehicles on road causes various problems and obstacles to find an empty parking slot. This problem often causes congestion, waste more effort and time-consuming process especially during rush hours when many people are driving for various reasons. In this paper, we explore the concept of smart parking based on the application of their categories and explain the classification from various parking support technologies. Present developed parking system supported by various technologies and the category. Classified the application function of this survey. The application of this smart parking application might assist in reducing congestion and other issue by manage the parking process. Achieve the key objective of this research as stated by various scientist and researcher to deliver an advanced parking system. Conducted by developing an application interface base on a mobile, desktop and web. It is an excellent determination for the parking management to deliver a real-time information update on the parking area, which is the point of interest as an improvement to the existing system and research.

1. Introduction

Population growth in urban areas that affect land reduction. At the same time, it increases the use of private vehicles for various purposes such as going to work [1], shopping centers [2] and others. This affects the increase in parking space to overcome parking problems with the number of vehicles on the streets. By increasing the number of parking lots, whether only to help the driver or vice versa will limit the availability of public land for residents in the city [3]. Finding an empty parking spot in a metropolitan region turn into a troublesome task for drivers particularly in rush hours [4]. During rush hours in most big cities, cars that are searching for parking spaces often takes 40% of the total traffic This causes waste in effort and fuel, increases congestion [5], and results the discharge of CO_2, and influences the economical-impact [6].

As shown in Figure 1 below, the process for users to find available parking space on average takes 3.5 to 14 minutes [7–9]. This situation cause more than 30-40% of traffic on the road. Thus, the parking management plays a vital role in reducing the bottleneck traffic [3]. Moreover, implementing smart parking in the urban city it could bring such a significant number of advantages for parking management along with the city arrangement due to meet the purpose of the research development. One of the methods that possible to implement is a reserve for a parking space and show the direction as a guide to the selected parking spot by systematically. In a purpose to facilitate the drivers in finding the expected parking lot.

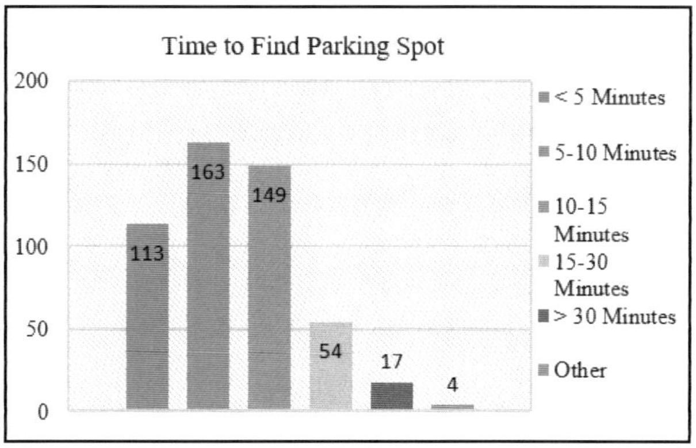

Figure 1. Searching time for a parking spot [7].

Further, introducing smart parking in various public facilities like airport [10], railway stations, malls, hotels, hospitals and so on is an important asset to note. Smart parking technology will help their visitor and customer, to get an effective solution in finding the parking space base on parking applications which provide reliable information on the parking area in these public facilities [11]. At the same time, it will improve their service and user satisfaction [12]. Based on the results of the evaluation and research the current advancement of smart parking applications by its compatibility to run across the operation of systems and devices.

Smart parking is implemented with a myriad of technologies facilitate the drivers in finding available parking lots within minimum waiting period. This paper aims to define advancements of technology based on the smart parking framework that discussed related work around the deployment of smart parking application, which categorized based on implemented software framework, service, application deployment and others that show in table 2. Utilized a correlation of advances as a part of each framework and employs the use of the correlation in various technologies of smart parking. Demonstrated a hypothetical correlation as part of the discussion.

2. Smart parking

The headline of the smart parking system is a reservation system that provides information on the destination parking lot and provides direction for the driver to find a parking space. Intended the users to get desired parking lot. Components of a smart parking system are. The user, parking management and parking lot and its facilities as illustrated in figure 2.

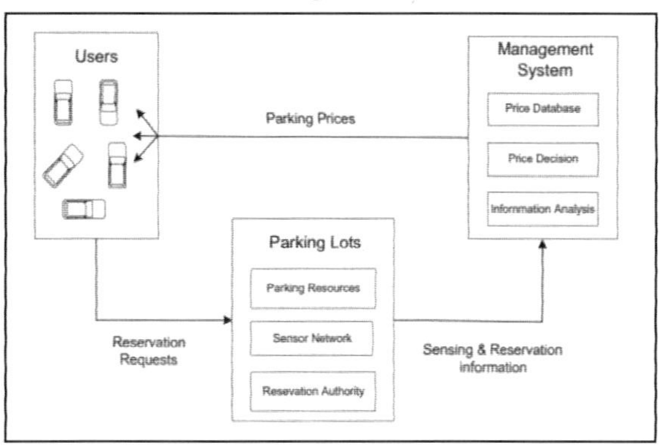

Figure 2. Integrated smart parking management system [13].

As explained in the introduction above, the growing number of vehicles is increasing as a result the need of people for vehicles to transport them from one location to another. Parking lots are the first place they are looking for when they arrive at their destination. of the accumulated data, the cause of congestion in the parking area as a result of a parking process [14]. Against this background, it spurred researchers and scientists in various fields to solve this problem and develop smart parking systems. Proposed scheme from surveyed paper mostly figure as figure 3 below.

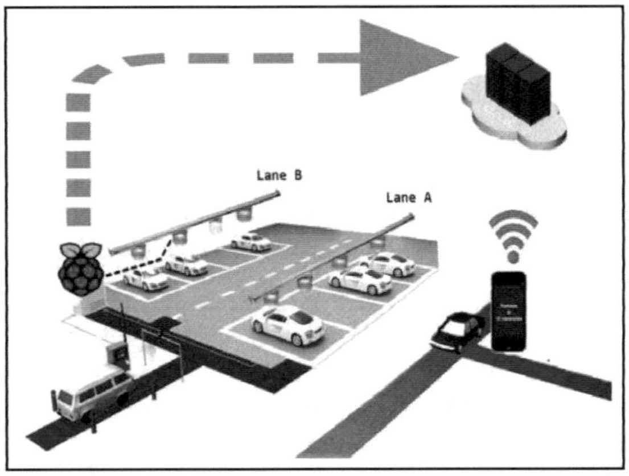

Figure 3. Integrated smart parking management system [13].

Smart parking is claimed by the expertise as a solution and clever approach to tackle the parking issue [15]. Smart parking technology aims to bring the ease for any driver in order to face the parking problem as efficiently as possible. Goals of this system can be achieved by utilizing an integrated information from parking framework advancement system that has been developed in various innovation [16]. Some aspect to consider in developing a smart parking system is should to reduce the time for seeking empty parking space [17] and shorten the tasks of parking frameworks for the driver [13] which is an extraordinary improvement for urban city development.

Evaluation approach of smart parking system

In preparation to this smart parking survey, it is consisting of two case situations such as parking determination based on reference case and proposed case. The reference case is a situation where the parking location can be determined without an application assistance. In opposite way, the purposed case the determine the parking location based on the application [18]. Both are the result of parking decisions based on the decisions and actions of the users. The difference between these two situations by the influence of the smart parking system on the decisions and actions taken by the driver. The proposed case model and the intelligent parking effect are two subjects that need to be evaluated in a smart parking application especially in this survey paper. In short, the statement above can be figured by the component flow and the evaluated approach as shown in Figure 4.

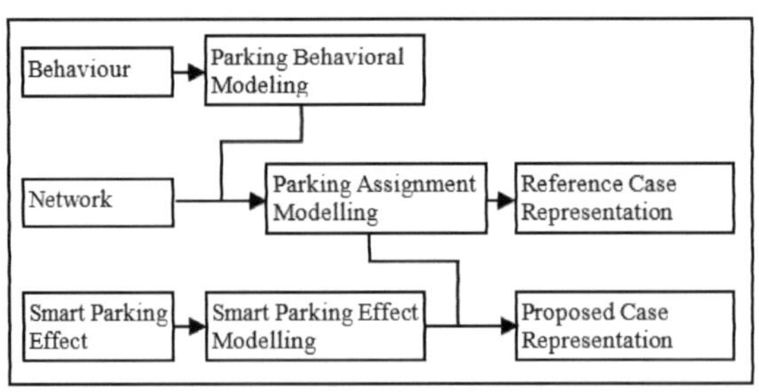

Figure 4. Component and approach evaluation [18].

Determination of the parking decision process taken from the results of several investigated papers. This investigation based on user behavior when they will park their vehicles in urban areas. An essential smart parking includes information flow and traffic flow as shown in figure 5 below.

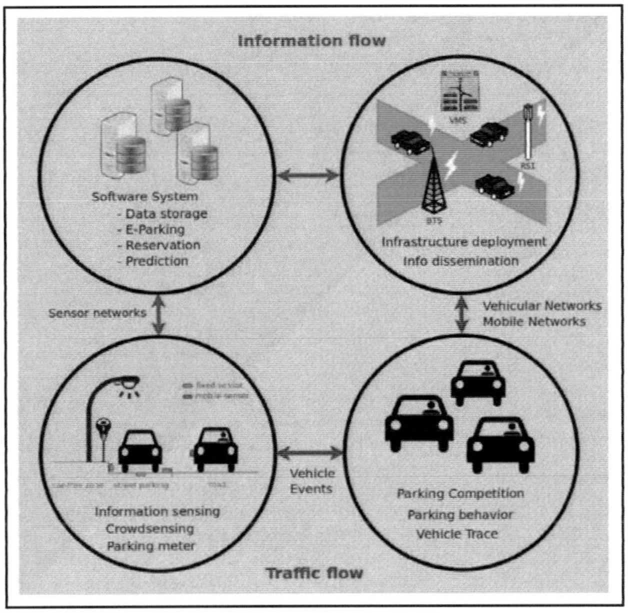

Figure 5. Smart parking overview [16].

Flow of information and traffic is an important data in smart parking systems that connect to each other, to produce effective solutions to find an available parking space. Decisions to the parking area could be taken before the vehicle arrives at destination lot or pre-trip process decision.

Parking behaviors are variety dependent on the driver's information as well as the distance and time to parking destiny [19]. Integrated parking network is involved to steering the rider to the parking space in an on-trip process decision. When the vehicle leaves parking space, information on parking space will be updated and space will provide to another vehicle.

Next section will explain numerous innovation of empirical study improvement from smart parking system by various science field. Here authors purposed to do a survey on the related study due to clarify all implemented systems on surveyed paper. Overview of the surveyed paper discusses smart parking area will be describing in section 3 below.

3. Literature survey of smart parking

This section is shown selected papers to be surveyed from smart parking base on their application. As a reference to this survey author has collecting paper from Science Direct, IEEE, and google scholar. Here the author takes 17 papers from 70 paper that collected from those 3 databases. Selected papers on this survey, contains papers merely discussing of smart parking application. To show the variety of paper in this area here author has summarized 9 paper as a sample. Rest of 17 surveyed paper will present on discussion section that will describe and compare it in detail each of their project.

Outline of gathered information and examine performance on pertinent information is taken on this section. In introducing the outcomes of each papers, with a specific goal to assess and decipher their result. Table 1 below, shown as an empirical study of this research area.

Table 1. Overview surveyed smart parking application.

Author and Year	Purpose	Summary
[20] 2017	Design a multiplatform mobile application concept.	This work is presenting a mobile application to solve the parking problem in the Virginia Tech area. As the result participant is not certain application effectiveness to find parking spaces. UI of ParkinVT was still in progress, relative lower satisfaction for understandable and need a corrected in the final design.
[4] 2017	A smart parking multiagent approach to providing decision a real-timely.	Optimizing the compromise from the set of efficient solutions, which is determined by means of a multi-criteria ranking method ELECTRE III. Simulation results is widely applicable to the approach in real cases of parking solution.
[21] 2016	An Intelligent parking assistance as future smart cities component.	A prototype parking system based on MEAN. API getRegionData is fetching parking information from google places near-by-search AJAX and the API (createMarker) as a marker on maps. Information flow acquired from a pattern recognition subsystem and stored parking information in MongoDB.
[22] 2016	An adaptive mechanism for smart parking by using smart-phones and the Internet of Things.	A centralized system to forecast available indoor parking spaces. To produce a low-cost mobile application to obtained data from parking occupancy. Conducting computer simulations and real experimentation with a preliminary implementation have reduced the time and energy costs to find available parking spaces.
[23] 2016	A smart parking system that provides occupancy estimation, payment method, permit management, and parking analytics.	Validating limited parking hypotheses in Helsinki area and recruitment of participants to the pilot of the project. A smart parking system based on LoRa (Long-Range communication technologies) to support the IoT systems. The challenges have been resolved by validating an associate parking spot in a city, allow drivers to check for available parking spaces, able to pay for the parking fee, get electronic parking permits, and the city authority gets parking analytics for the city planning.
[24] 2015	Introducing a mobile-based parking reservation system.	Here author develop a mobile smartphone reservation framework, by utilizing SDK Xcode combine with SQLite and debugging real-timely with Firefox Web program. An application model has presented by considering nearby parking service providers, to satisfy customers requirements and will reserve the best parking for the use.
[25] 2014	A computer program for processing a parking lot image and produce a vacant parking spaces information.	In this works, using a camera as a sensor to take photos and use the information to steer the driver into a vacant car park. The developed system in both software Matlab and camera as a software and hardware platform. An automatic parking system make the whole process of parking cars more efficient and less complex for both drivers in parking his car.
[10] 2013	Provide an online car parking reservation desktop application service using XAMPP.	This Application provides a vacant parking lot, the payment method, the booking period, expire date, and other alternatives before their arrival at the airport. Connecting the user to the webpage through an interface generated by PHP and JavaScript. MySQL and DBMS applied to store related data of the car park. To connect the DBMS and PHP, APACHE is used as a bridging linker that supported by XAMPP.
[26] 2011	A Mobile Applications system to help the drivers find available parking spots quickly.	The vehicle mostly trips and parked in car parks. Then this author develops a smartphone application with PhoneGap and web technologies to achieve scalability on the back end of Google App Engine. The result of deploying this application shows that the system works and can indeed provide real-time information about the availability of the parking area.

Summarization of above surveyed studies, implemented system from each study base on smart parking application is different to each other in some aspect. In every single surveyed paper has had a different purpose, method, platform, framework and strategy in order to against the parking problem. Reach the goal to save time to get the expected parking lot and make it easier for the consumer without wasting any valuable resources.

3.1. Advantages and disadvantages

In every survey on the smart parking application, not always be perfect mostly paper it will always leave a gap for future improvement by them self or to another researcher. From surveyed discussion and project, it has advantages and disadvantages of their projects to make their method better than another.

Here I have figured nine sample of papers from 17 papers that have surveyed. In this section, we draw conclusions from the advantages and disadvantages of the parking system in the study sample that we survey. Table 2 below will show in detail of each surveyed paper of smart parking advantages and disadvantages.

Table 2. Advantages and disadvantages from surveyed papers.

Author	Advantages	Disadvantages
[11]	• Predicting accurately and sense spot/vehicle occupancy real-time. • Decreasing number of on-street parking. • Suitable for the money during utilization.	• Device and Environment should integrate which support the system. • The parking should be cleared vision.
[20]	• Real-time sensor detection. • Already implemented and tested in the real world. • A user interface that is the focus to be a user-friendly application, improve the parking experiences and efficiency.	• While in progress project and upgrading UI process. • Demands to installing an additional equipment on campus parking lots. • Consuming time in the process of installing this app.
[21]	• Routing information is shown using GPS to save travel time. • Implement MEAN stack, which faster and real-time compare to traditional web. • Figured details of available parking.	• Inconvenience application, because it is just a simulation instead of data from the real world. • Not embedding intelligent component. • Priority has not provided to an emergency.
[27]	• Lifting car park management into a computerial service. • Provide bidirectional communication service to get accurate data simultanly. • Economical resources for parking space owner.	• Component to build this application is has not selected. Software and hardware to construct this project are not decided • A general discussion that makes it blur to be realism in the real world.
[22]	• QR codes and NFC tags are cheap technologies to implement. • QR codes, NFC tags, and BLE Physical Web Beacons to determine actual occupancy of parking space accurately. • A lightweight client program on a smartphone allows accessing the service through the Internet.	• The sensor at a key location divides a car park into smaller zones. • Requires a smartphone that supports the technology is applied. • Still, a concept has not come out with the entire application to show the entire system.
[23]	• REST approach through HTTP level appliances it provide security, caching, and proxy capabilities. • Convenience in finding a parking space and paying the parking fees	• WinNode needs to full charge when spend less time in the car. • Require higher transmitting point to improve communication range and reduces the number of gateways nSeeded.
[24]	• License plate recognition via the smartphone camera and register a new car automatically. • The system could feature a long-term parking such as weekly or monthly reservations.	• Need intuitive and flawless designs in this system. • Requires certain type of smartphone to use this app (IPhone).
[25]	• The camera is moveable to detect every space in parking area. • Using data from the image to guide a driver to an available car park. • Using camera is efficient and low-cost technology.	• The weather conditions affect the visibility of the park lots. • The camera should be placed in a perfect place where can clearly see all the car parks.
[28]	• Using android application allow a user to scan a QR code and pay for their parking space	• Only one type of user abstraction in the system. • Integration Google maps API or GPS is unavailable to visualizing parking spaces.
[10]	• Minimum cost is involved because the internet is used. • The online desktop application is easy to deploy in across device and operating system.	• Physical hardware or parking map to figure vacant parking space. • Secure socket layer was not set up to encrypt data on credit card during for payment process

Above article exposures, has to compare and analyze every pros and con in every work. Even though they have put everything into the research they are working on, applying the latest methods and the best technology to solve the existing parking problems. However, from the surveyed paper, leaves gaps for further development that has not addressed by the researchers. In addition, those researches remain gaps to improve the works that they have not achieved in some sectors. This provides an opportunity for other researchers to conduct research in the same field to fill the gaps that still exist in this research and make advancements on this research area.

4. Discussion

4.1. Smart parking classification

Smart parking also called intelligent parking it consists of three components: an intelligent user/device, an intelligent environment, and intelligent vehicle & environment. An intelligent device is the application base on a device to get information about the car parks. An intelligent environment is an application in the parking environment; a user gets to get the information about the vacant park when they arrive at the parking area. Last is an intelligent device and environment is a combination of both component before, a user can get information about vacant park accurately base on communication machine to machine or M2M [29]. Table 3 below is the classification category of surveyed smart parking application paper by its integration. Artificial neural networks are a reliable and sustainable method, as in a number of emerging field technology fields such as the Deep learning machine, cloud computing and information security [30].

Table 3. Classification of smart parking papers.

ref	Deployment		Service		
	NetU	*NetS*	*ID*	*IE*	*IDE*
[20]	-	-	√	-	-
[11]	-	WiFi, NFC	√	√	√
[21]	-	-	√	-	-
[27]	WiFi, CN	WSN	√	√	√
[22]	WiFi	NFC, BT	√	√	√
[23]	LoRaWAN Network, WinNode-WinSpot		√	√	√
[9]	WiFi, CN		√	√	√
[7]	WSN	CN	-	√	-
[31]	WiFi, through MQTT protocol		√	√	√
[24]	Wire & WiFi	CN	√	-	-
[25]	WiFi	-	-	√	-
[32]	WiFi, NC		√	√	√
[28]	-	-	√	-	-
[10]	-	-	√	-	-
[33]	WiFi, CN		√	√	√
[5]	WiFi, CN		√	-	-
[26]	WiFi, CN		√	-	-

Deployment: NetS (Network of Sensor), NetU (Network of user).
CN (Cellular Network) BT (Bluetooth)and WSN (Wirless Sensor Network)
Service: ID (Intelligent Device), IE (Intelligent Environment), and
IDN (Intelligent Device and Environment)

From the collection of 17 papers that have been surveyed, have been getting 3 clusters of smart parking system. A total of 8 papers apply Intelligent Device, 2 papers using intelligent environment, and the remaining 7 papers combine intelligent device and environment. From the sample data that has been obtained, then some of the paper will be described in the following section.

4.1.1. Intelligent vehicles/device are the vehicle. The application detects parking information automatically, without any integration between the environment and the vehicles/user. Here some sample paper which adopts a system intelligent vehicle/device as shown in table 3.

Intelligent device application has implemented in "ParkinVT" by Liang [20]. Purpose of this paper is to improve the parking experiences and efficiency in Virginia Tech campus. The most important feature of "ParkinVT" is the notification function. Once the users set their favorite on-campus parking lots, and the remaining parking spaces in these parking lots reach a critical amount. Such as less than five spaces remain in a parking area, "ParkinVT" will automatically send a reminder to the users based on the traffic flow and distance between users' departures and destinations. Due to user satisfaction as a key objective of this research, Liang has divided the difficulty levels to find the parking lot into easy, medium and hard. For a guest and staff will be given the difficulty in level easy, for the faculty will be given the difficulty in level medium and to the student will be given the difficulty in level hard. The level is determined by the user interests on the Virginia Tech campus. As a future improvement Liang has conduct a survey to evaluate the inefficient component on her application.

A mobile application which develops by Bernspång [26] which is an intelligent device application that been implemented in his system. Deployment of this work is to make an interconnection between JSON (JavaScript Object Notation)-encoded that it is an application Engine to the Smartphone application which means this work is based on the intelligent device of the smart parking system but in this case, for the device is using a smartphone. Connect sensors and user connections through application engine of JSON is by transmitting the sensor that collect the data. Then, sending an actual URL to represent the data that in this case represents from readings of the simulated magnetometer sensor. This application starts suggesting possible match from the parking space by using Google Geocoding API.

4.1.2. Intelligent environment. Application of this intelligent is the environment/building have been conducting and implement the smart system can detect the available parking and show it to the user. The intelligent environment detects sensing result from the sensor in a parking area and then stores in on their database system to be accessed by the driver in the same area with this intelligent environment/ building.

Application base on the intelligent environment has implemented on a paper from Al-Kharusy [22]. This intelligent parking, develop a figuring of the occupied parking space using wireless signal transmission. An obtained data of parking information is a result from image processing using a camera that has spread across several points in the parking lot at Massey University. The capture a predicted image and classify empty parking spots. Then transfer captured information wirelessly using transmitter from a radio transmitter, send it to the receiver in control room to pre-processing segment, and filter the image result using a computer with FPGA (Field-Programmable-Gate-Array) program to filter the noise.

4.1.3. Intelligent device and environment. Application base on this category is device and the environment already engaging to each other, just said that the vehicle it is been one of a member of it intelligent environment. In another word, the vehicles it's already integrated with the environment using a device to communicate with the intelligent environment. Here are samples from surveyed papers, which is an application base on an intelligent device according to data from table 3. Atif [27] and Charles [23] elaborate a dimension that in their system that includes machine learning, applies the algorithm to achieve adaptive systems that can respond to environmental changes that can have an impact on the smart parking system they develop.

Application base on an intelligent device and environment has been deployed in a paper from Caballero-Gil [22]. Purpose of this project is to provide a communication between smartphone and Server on parking space and then guide the driver to the recommendation parking space. Communicate the parking areas of the smartphone through embedded technology in the smartphone that is BLE, NFC, QR, and Bluetooth. Data readied by the system on the server and from the user application. Directly the data sent from application to the server as well as vice versa, so that data from the location of the vehicle can be stored on integrated systems.

In addition, Ahad [11] has implemented an intelligent device and environment in his project "Park Easy". This project proposed to provide the shortest way and provide information from predicted availability parking obtained from accurate sensor detection updated. Park Easy is using wireless NFC (Near Field Communication) to detect the availability of parking space by capturing the image by integrated cameras in the parking area. Assist a camera sensor with ultrasonic sensors for detection from the specified parking space. Then, transmits the data to the control unit into the next is instantly sent to the application of users who are approximately the parking area via user mobile devices.

4.2. Deployment of smart parking base on device platform

The current smart parking technology has made it possible to apply in various media. Establishment of smart parking framework in actualized on registering gadgets which cover smartphones, integrated sensors, desktop computer and smart display increasingly turned out to be littler, less expensive and more powerful [13]. The following application of smart parking application in various media that have been collected from surveyed papers as shown in table 4 below.

Table 4. Smart parking application base on device platform and feature.

ref	Author	Device			Feature		
		Mobile App	*Desktop App*	*Web App*	*QR*	*Dir*	*Map*
[20]	Liang (2017)	√	-	-	-	√	√
[11]	Ahad (2016)	√	-	-	-	-	-
[21]	Narasimha-Mohanasamy (2016)	-	-	√	-	-	√
[27]	Atif (2016)	√	-	-	-	-	-
[22]	Caballero-Gil (2016)	√	-	-	√	-	-
[23]	Charles (2016)	-	-	√	-	√	√
[9]	El-Seoud (2016)	√	-	√	-	-	-
[7]	Alkheder (2016)	√	-	-	-	-	√
[31]	Khanna (2016)	√	-	-	-	-	-
[24]	Baitalmal (2015)	√	-	-	-	-	-
[25]	Al-Kharusi (2014)	-	√	-	-	-	-
[32]	Livio (2014)	√	-	-	√	-	√
[28]	Cosgrove (2013)	√	-	√	√	-	-
[10]	Hussien (2013)	-	√	-	-	-	-
[33]	Geng (2013)	√	-	√	-	-	√
[5]	Trusiewicz (2013)	√	-	-	-	-	√
[26]	Bernspång (2011)	√	-	-	-	-	√

Feature: QR (QRcode), Dir (Direction)

Here we will explain in detail group of smart parking application that classified base on a web, mobile, and desktop. In addition, here are samples of surveyed papers is described referring to the table above.

4.2.1. Web application. Computer and online oriented application is the most appreciated applications nowadays [34]. TheWeb applications are discussed and argued by any development community as a cross-platform application. That's means, no specific device and the specific browser is required in deployment. In addition, this application can be easily run on various operating like IOS, Android, Windows, Linux and other [35]. Web applications are applications that are accessed using the web browser via internet or intranet network. The web application is also a computer software encoded in a programming language that supports web-based software such as HTML, JavaScript, CSS, Ruby, Python, Php, Java and other programming languages.

Smart parking base on the web application as done by Narasimha-Mohanasamy and Jenq [21] in a project of "ParkIt", here author deploys an application base on a web application to deliver a ParkIt web-based application. This app can be easily redirected the driver in order to show indoor parking garage which associated with a system of ParkIt. Parking information can be acquired from a pattern recognition subsystem.

Web application from smart parking also has implemented in Hussien project [10]. This work is an online car booking system, this app instantly adds value to client website with a graphical user interface and web users access to book a parking space, pay online, update the booked space before the exit date, finding the way to available vacant space and the status of pre-booking before them heading to the airport.

4.2.2. Mobile application. Users of smart phones and mobile applications have increased over the past decade around the world and have become a leading technology segment. Which consists of software to do certain tasks for mobile users [36]. Multi-platform mobile application development advancement system is picking up prevalence due to their trademark to order the application source code for multi-OS (operating system) [37]. The relationship between mobile application web applications is so close to each other, with parking coordination on this makes it applicable to a smart device due to its mobility that is easy to deploy [38]. To determine the absence of functionality equipment, at the same points to even now fulfill the craving to convey normal Web innovation, cross approaches developed as a mix of Web technology and native function [39]. The most noticeable example of this approach is PhoneGap.

Smart parking application which stands on a mobile application has developed by Khanna [31] in 2016 by installing the application before driver uses it, then select the destination area. The mobile application taking apart to interfacing end-users to interact with the parking system, select amount of time, make a payment using e-wallet or credit card. The application also acts as an alarm and a reminder when the driver failed to park the occupancy then the alarm will ring assume that driver/user has mistaken the parking place. If the reservation has reach, time limit has then the notification sent to user/driver, giving an option whether they want to extend their parking time or leaving the parking area.

Another paper that has implemented the mobile application on smart parking system is Bernspång [12]. By selecting a parking space base on a nearby the location and the address field. When destination address has typed, then Google Geocoding API starts suggesting possible matches parking area with contains geographic coordinates. The application will try to get the current location of the user/driver and show theirs specify coordinate. Then plotted parking space result on a map if user/driver using the near-by feature from this mobile application. This paper gives a driver that spotlights on real-time data, for example, the accessibility of free space in the parking house through which it goes through the applications introduced on the client's smartphone.

4.2.3. Desktop application. Apart from the above platform, there is also another media interface deployment that uses a desktop application to process and interfacing the data. In this case, desktop application is implemented to processes the data of available parking space. Desktop Application is an application that can be operated independently without using a browser or internet connection on an autonomous computer.

A desktop application it is has done by Al-Kharusi [25] on his work as shown in table 4. This desktop application of smart parking system is purposed to find the availability of a car park at Massey University. The project is using an FPGA program base on MatLab programming. FPGA analyzed an aerial view method, test a car park by different occupancy level. This application shows an availability car park, giving a coordinate and steer the driver to find the closest space to the parking entrance. Empty space is detected using the camera as a sensor to get image detection from a vacant car park.

4.3. Smart parking application base on framework platform

A design for the reconciliation of a various advancement is generally displayed. Determining the interface design and information content of a website is important to arouse consumers' initial interest in exploring the site further [40]. Future programming stacks have proposed a reservation, installment, expectation, keen parking application and even the framework readiness. The product frameworks are displaced by various business arrangements, such as the UI and the best approach to display parking data is very comparable. A gathering of parking places is amassed and just gives a likelihood of accessibility to drivers. The following data is collected from the framework is used by each paper to build the system they have proposed is shown in table 5 below.

Table 5. Comparison smart parking base on framework platform.

ref	Used Technology	Service			
		EP	*RS*	*PS*	*NT*
[20]	JavaScript, Ionic, Google Maps API	√	-	-	√
[11]	-	√	-	-	-
[21]	MEAN Stack, Mootools, Google Maps API	√	-	-	-
[27]	-	√	-	-	-
[22]	MEAN Stack, Bootstraps	√	-	-	-
[23]	RESTful, Spring, HTML5, CSS3, Django	√	-	√	-
[9]	-	√	√	-	-
[7]	Google App Inventor	-	√	-	√
[31]	Apache Cordova, Angular Js, Javascript, IBM MQTT server	√	√	-	-
[24]	Apache Server, PHPMyAdmin, MySQL	-	√	-	-
[25]	MathLab, FPGA platform	√	-	-	-
[32]	MySQL, JSON RPC, Apache Cordova, jQuery and jQuery mobile, Apache Tomcat, HTML 5	√	√	√	√

Table 5. Cont.

ref	Used Technology	Service			
		EP	*RS*	*PS*	*NT*
[28]	MEAN Stack, Bootstraps, Zxing	-	√	√	-
[10]	XAMPP, PHP MyAdmin, MySQL, HTML5	-	√	√	-
[33]	-	√	√	-	-
[5]	C#, ASP.NET MVC 3, AJAX, JSON, JavaScript	-	√	-	-
[26]	Python, Django-nonrel, HTML5, CSS 3	√	√	-	-

Service: EP (E-Parking), RS (Reservation), PS (Payment System), and NT (Notification Time Parking).

While improving a web application, an essential part of the framework development is to pick a right innovation stack that permits fast prototyping, consistent emphasis, code reuse, most extreme proficiency, and vigor. An essential part of developing an application is selecting a right framework innovation like a LAMP or a MEAN stack framework. LAMP is most popular, free of charge, open source, web development stack is now facing a new trending technology stack, the MEAN stack [41]. It is an important aspect but difficult to learn and comprehend by the designers dealing with the front-end and the back-end. Based on follows issue then, the idea of Full Stack JavaScript was created as a solution [42]. Here the collected data, divided as shown in table 5 the data have been the framework into PHP & MySQL Base Application and Node.JS & MongoDB Base application. Descriptions from both frameworks is explain in following section.

4.3.1. PHP and MySQL base application. LAMP stack framework is a combination of Linux, Apache, MySQL, PHP/Perl and (Java EE, Spring) which a comprises various programming language. LAMP stack framework of web development has implemented by web programmers across the community even around the world. The light stack tentatively changed to a WAMP (Windows, Apache, MySQL, PHP/Perl) stack by supplanting Linux with Windows in considering the programmer experience, the PHP can be substituted with Python [43]. However, javascript comprehended the multi-language paradigm by presenting a MEAN stack development system, in a collaboration of MongoDB, Express, AngularJS, and Node.js. Which this framework is stands on a single programing language of a JavaScript. In additon, both framework LAMP/WAMP and MEAN have the same structure for building a great web application [44].

An application that developed by Hussien [10] based on the LAMP stack framework. In this thesis, authors implement two component of an architectural system, which is the architecture of DBMS (database management system) and the application for parking management based on the user interface. PHP and JavaScript generate an interface as a connection from the webpage. MySQL database store the data of the car park and implement the DBMS to manage them. Connecting DBMS to PHP bridge linker is used which it's a part of APACHE that supported by XAMPP package. The entire desktop application construction is J2EE, PHP and DBMS based.

4.3.2. NodeJS and MongoDB base application. Currently, most developers are switching to javascript based application programming. Examples of framework applications by utilizing the javascript language is a web programming based on MEAN stack. MEAN are a framework which consists of function i.e., node.js [45] to provides the platform for the framework, MongoDB [46] to provides the backend data store, Express [47] to provides the web server, and AngularJS [48] to provides the client-side framework for modern web applications [49]. To support a Real-time data updates, almost all of developer and researcher put AJAX (Asynchronous JavaScript and XML) which a system for sending and receiving data from a server without page refresh [50]. The "Asynchronous" section refers to the fact that when JavaScript delivers an AJAX call to the web server, it keeps running until it gets a response, it does not block and quit when data is being processed by the server.

This framework has been implemented by Jenq [21] that is application based on Node.js and MongoDB. In this work, the authors build client-side applications that build web browsers using JQuery and Mootools to provide PakitPage, ParkitRegion and as bookmarks. In server-side programming, nodeJS has chosen as follows with express, MongoDB, and Mootools. Using GoogleAPI Maps to indicate where to park. Saving APIs retrieves parking information using Google's near-by-search AJAX spots and APIs (createMarker) to bookmark maps. Using MongoDB database to store parking space availability.

5. Conclusion

The number of vehicles is increasing along with population growth in urban areas. During the process of parking their vehicles, it causes various negative impacts such as congestion in the effect of inefficient from parking management. This impact has contributed to 30-40% of the congestion on the street. A solution to overcome this problem is to apply the advancement system of smart parking. The scientists in have advanced a lot of innovative methodologies smart parking any interdisciplinary of science. With the aim to address this parking problem, to support the increasing mobility in big cities.

This survey paper has been discussed the deployment of smart parking systems based on the application. By combining and summarizing the fundamental philosophies used as part of the present work. Reveal a comparative representation of combined ideas for utilizing methods in the current parking application system. For examples are LoRaWAN network communication, multi-criteria ranking method ELECTRE III, expert system, GPS-based, vision-based correspondence which captured by an integrated camera, wireless communication base on QR codes, NFC tags, and BLE technology. The implemented framework be able to encourage the public and even the parking sector in terms of financial, social, and welfare. Implementing this system is supposed to be an eco-friendly system, saving fuel, time and effort. Improving future development of smart parking application system in accordance with the latest situation. Utilizing technological advancements that vary and integrate various parking systems in innovatively different technologies from mechanisms in smart parking areas. This paper intends to determine the most efficient framework technology as a solution covering several aspects such as smoothness, security, reliability in deployment to overcome parking problems.

Acknowledgment
Authors are grateful to School of Graduate Studies (SGS), Research Management Centre (RMC), and Faculty of Information Science Engineering (FISE) of Management and Science University (MSU) Malaysia for their support.

References

[1] Shao C, Yang H, Zhang Y and Ke J 2016 A simple reservation and allocation model of shared parking lots *Transp. Res. Part C Emerg. Technol.* **71** 303–312
[2] Butowsky A, Gai K, Coakley M, Qiu M and Tappert C C 2016 City of White Plains Parking App: Case Study of a Smart City Web Application *Proc. - 2nd IEEE Int. Conf. Cyber Secur. Cloud Comput. CSCloud 2015 - IEEE Int. Symp. Smart Cloud, IEEE SSC 2015* 278–282
[3] Robert P 2015 *Good Reason for Parking Management* 20
[4] Boudali I and Ben Ouada M 2017 Smart Parking Reservation System Based on Distributed Multicriteria Approach *Appl. Artif. Intell.* **31** 5–6 518–537
[5] Trusiewicz P and Legierski J 2013 Parking Reservation-application dedicated for car users based on telecommunications APIs *Comput. Sci. Inf.* 865–869
[6] Lin T S 2015 *Smart Parking: Network, Infrastructure and Urban Service*
[7] Alkheder S A, Al Rajab M M and Alzoubi K 2016 Parking problems in Abu Dhabi, UAE toward an intelligent parking management system 'ADIP: Abu Dhabi Intelligent Parking *Alexandria Eng. J.* **55** 3 2679–2687
[8] Liu R, Yang Y, Kwak D, Zhang D, Iftode L and Nath B 2017 Your Search Path Tells Others Where to Park: Towards Fine-Grained Parking Availability Crowdsourcing Using Parking Decision Models *Proc. ACM Interact. Mob. Wearable Ubiquitous Technol* **1** 3 1–27
[9] El-Seoud S A, El-Sofany H and Taj-Eddine I 2016 Towards the development of smart parking system using mobile and web technologies *2016 Int. Conf. Interact. Mob. Commun. Technol. Learn.* **978** 10–16

[10] Hussien H W 2013 *Online Car Parking Reservation System*
[11] Ahad A, Khan Z R and Ahmad S A 2016 *Intelligent Parking System*
[12] Salehi M, Mirzaei H, Aghaei M and Abyari M 2012 Dissimilarity of E-marketing VS traditional marketing *Int. J. Bus. Soc. Sci.* **2** 1 510–515
[13] Wang H and He W 2011 A reservation-based smart parking system," *Comput. Commun. Work. (INFOCOM WKSHPS) 2011 IEEE Conf.* 690–695
[14] Dokur O 2015 *Embedded System Design of a Real-time Parking Guidance System*
[15] Faheem F, Mahmud S A, Khan G M, Rahman M and Zafar H 2013 A survey of intelligent car parking system *J. Appl. Res. Technol.* **11** 5 714–726
[16] Lin T S, Shuenying T, Smart L and Lin T S 2016 *Smart parking : Network , infrastructure and urban service*
[17] Chouhan S and Sandhya P 2017 *Internet Of Thing Based Car Parking System* 6–9
[18] Chaniotakis E 2014 *Parking Behavioural and Assignment Modelling*
[19] Aalsalem M Y and Khan W Z 2017 CampusSense - A smart vehicle parking monitoring and management system using ANPR cameras and android phones *Int. Conf. Adv. Commun. Technol. ICACT* **5** 2 809–815
[20] Liang J 2017 *'ParkinVT', A Concept Mobile Application for Improved Parking at Virginia Tech*
[21] Narasimha-Mohanasamy N and Jenq J 2016 An intelligent parking assistant *20th World Multi-Conference Syst. Cybern. Informatics, WMSCI 2016* **1** 174–177
[22] Caballero-Gil C, Caballero-Gil P and Molina-Gil J 2016 Cellular automata-based application for driver assistance in indoor parking areas *Sensors (Switzerland)* **16** 11
[23] Charles S W 2016 *The design and implementation of a smart-parking system for Helsinki Area*
[24] Baitalmal A M 2015 *Mobile Application Based Parking Reservation System* 80
[25] Al-Kharusi H and Al-Bahadly I 2014 Intelligent Parking Management System Based on Image Processing *World J. Eng. Technol.* **2** 2 55–67
[26] Bernspång F 2011 *Smart Parking using Magnetometers and Mobile Applications*
[27] Atif Y, Ding J and Jeusfeld M A 2016 Internet of Things Approach to Cloud-based Smart Car Parking *Procedia Comput. Sci.* **58** 193–198
[28] Cosgrove D 2013 *SmartPark*
[29] Högskola M, Grubisic M and Marsic T 2015 *System collaboration and information sharing through Internet of Things* **96**
[30] Zajmi L, Ahmed F Y H and Jaharadak A A 2018 *Concepts , Methods , and Performances of Particle Swarm Optimization , Backpropagation , and Neural Networks*
[31] Khanna A and Anand R 2016 IoT based Smart Parking System *2016 Int. Conf. Internet Things Appl.* 266–270
[32] Livio H, Fabian H and Samuel L 2014 *Parking Monitoring and Management Solution (PMMS)*
[33] Geng Y and Cassandaras C G 2013 New 'Smart Parking' System Based on Resource Allocation and Reservations *Intell. Transp. Syst. IEEE Trans.* **14** 3 1129–1139
[34] Khanim S G and Ahmed F Y H 2018 *Utilization of Short Message Services (SMS) for Library Notification System* **13** 9 6503–6513
[35] Fransson R and Driaguine A 2017 *Comparing Progressive Web Applications with Native Android Applications : An evaluation of performance when it comes to response time* 5
[36] Fong S L, Elfaki A O, Johar M G M and Kevin L T A 2012 Investigative Study towards the Development of Mobile Language Translator *Int. J. Digit. Content Technol. its Appl.* **6** 22 11–20
[37] Palmieri M, Singh I and Cicchetti A 2012 Comparison of cross-platform mobile development tools *2012 16th Int. Conf. Intell. Next Gener. Networks, ICIN 2012* 179–186
[38] Hasan M M and Haque M A 2016 *Mobile Application Development Approaches: Recommendation for E-commerce Enterprises*
[39] Heitkötter H, Hanschke S and Majchrzak T A 2012 Comparing Cross-Platform Development Approaches for Mobile Applications *Proc. 8th Int. Conf. Web Inf. Syst. Technol.* 299–311

[40] Johar M G M and Awalluddin J A A 2011 *The Role of Technology Acceptance Model in Explaining Effect on E - Commerce* **3** 3
[41] Lennon J 2014 *Build a real-time polls application with Node . js , Express , AngularJS , and MongoDB* 1–16
[42] Nghi A and Thanh L 2016 *Mean Stack Web Development*
[43] Agrawal S and Gupta R D 2014 Development and Comparison of Open Source based Web GIS Frameworks on WAMP and Apache Tomcat Web Servers *ISPRS - Int. Arch. Photogramm. Remote Sens. Spat. Inf. Sci.* **XL-4** 1–5
[44] Adhikari A 2016 Full Stack JavaScript Development With MEAN *ACM SIGACT News* **40** 2 33–35
[45] Pasquali S 2015 *Deploying Node.js.*
[46] Ghasemi P 2013 *Migration of Legacy Web Application Using NoSQL Databases* 53
[47] Arbenz M 2011 *Express In Action : Writing, Building, and Testing Node.JS Application*
[48] Dao V 2016 *Development of a front-end application using AngularJS : 1UP Media company case* 1–47
[49] Dayley B 2014 *Node. js, MongoDB, and AngularJS Web Development*
[50] Molin E 2016 *Comparison of Single-Page Application Frameworks* 53

Web-based rule-based expert system to screen postpartum depression

D Mufidati[1,*], M N Widyawati[1] and S Suryono[2]

[1] Postgraduate Program in Applied Midwifery Science, Poltekkes Kemenkes, Semarang, Indonesia
[2] Department of Physics, Faculty of Science and Mathematics, Diponegoro University, Semarang, Indonesia

*andrimufidati@gmail.com

Abstract. The high prevalence of postpartum depression significantly increases maternal morbidity. Lack of diagnosis for postpartum depression results in mothers missing early treatment necessary. The traditional screening method using questionnaire presents its own disadvantage in diagnosis. Therefore, development of postpartum screening method that focuses on ease in implementation is very much needed to improve postpartum maternal health. This paper proposes a screening information system for postpartum depression to help its proper treatment. This research employed the rule-based method to diagnose postpartum depression and the expert system to input symptoms data that will result in diagnoses. Respondents input symptoms data to be diagnosed by the system. The system then automatically generate output of suggested treatment based on expert input associated with the diagnoses the system is working on. This research employed the Randomized Control Trial method with Control Group Pre and Post Tests Design. Samples of this research were 52 postpartum mothers consisting of 26 respondents in the system group, and the other 26 in the manual group. Results show that 42.31% (n=11) of mothers suffer from postpartum depression. There is a significant difference in EPDS score prior to and after the use of the information system (P=0.000). The system also successfully generated very high TAM score, meaning that it is capable of postpartum screening along with ease of use and some other advantages. Therefore, screening for postpartum depression using information system is very effective and ensures proper and timely treatment.

1. Introduction

Screening is the best way to figure out postpartum depression to speed up treatment process [1]. Late diagnosis and treatment may result in lower life quality and they also affect overall physical and mental health. Therefore, it is important to further study and improve screening methods for postpartum depression. This process will lead to a more effective and easy method of screening to be implemented by health professionals, as well as patients. One attempt to improve health care for postpartum mothers is the use of information system. The advent of new technology and the availability of information means that patients are tend to be more proactive in dealing with their health issues. Patients are directly involved in their personal treatment to better maintain their condition and to have better prognosis [2].

Depression is one of the public health issues that cause morbidity among women with twice prevalence to that of men [3]. Postpartum depression results in psychological disorder either in short or long term and not only for the mother, but also for the family members. Children of depressed mother tend to suffer from retardation in psychological, cognitive, neurological, and motor developments. This retardation experienced by children and their depressed mother may last until they reach the age of 4 to 8 years. There are reports suggesting that depressed mothers are 3 times more likely to have emotional problems with their children and are 10 times more likely to have bad relationships with their children [4]. Considering the serious consequences of postpartum depression, it is such a shame that the level of proper care and treatment is still below expectation [5]. This discouraging fact closely relates to difficulties in diagnosis as screening is still carried out manually, and difficulties in administering treatment by health professionals [6,7].

Determination of postpartum depression status still relies on paper-based questionnaires with which health professionals estimate the status of postpartum depression by manual calculation of scores from the questionnaire. This is further hampered by the many types of questionnaires available that makes it hard for them to decide which one to use. Moreover, health professionals do not have ample time to conduct screening, and at times they do not even know how to deal with positive screening results [1,8]. Lack of knowledge from the mothers themselves in understanding what they are going through tend to make them passive as they regard postpartum depression as a normal phase after giving birth. This may result in depressed mothers not getting proper treatment as their depression status is not known [6,7]. Information system has great potential in supporting treatment of postpartum depression in mothers. A web-based system can be used effectively in improving diagnosis potential for postpartum depression. Web-based screening is effective as it comes with ease of use and it provides suggestions that match the problems [2].

2. Methods

With the information system, screening for postpartum depression can be done anytime and anywhere. Patients can readily know their status from diagnosis results displayed. Diagnosis in this information system is based on scores from compulsory questions that patients must answer. Meanwhile, treatment suggestions are matched to patient`s diagnosis results. A diagram of the information system is shown in Figure 1.

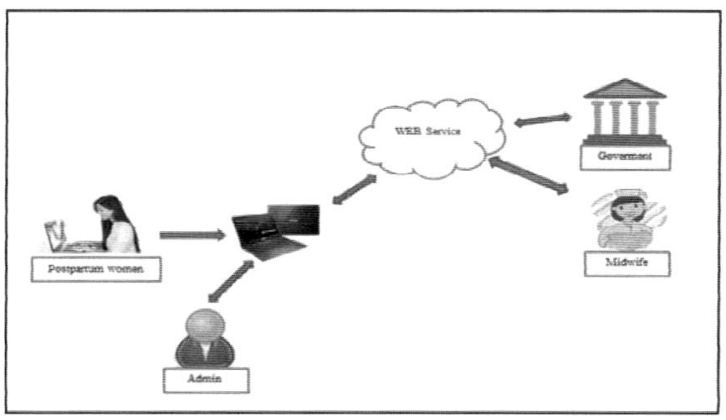

Figure 1. Information system scheme.

Users access website from any browser and come to the home page. An administrator can use access rights by logging in with user name and password. The administrator is in charge of analyzing data and providing suggestion for treatment from experts. Meanwhile, users can access the system by first registering on the registration menu and filling out characteristics biodata.

Afterwards, users can log in with their user name and password on the main menu. Users can see their postpartum depression status by answering 10 questions available in the system. Unregistered users cannot gain access from the main menu and hence, cannot perform screening for postpartum depression. This research employed the Randomized Control Trial method with Control Group Pre and Post Tests Design. Samples were 52 respondents, 26 of which are in the intervention group and the remaining 26 are in the control group in Public Health Center.

The intervention group underwent diagnosis using the information system, while the control group went through the manual method. Intervention for both the intervention and manual group was conducted for 2 weeks. Pretest was performed for both groups on the first visit, whereas posttest was carried out on the second visit, which is two weeks after the first one. Comparison was then conducted for pretest and posttest results, both within each group and between the two groups.

The resulting system was tested against real data of the object using the EPDS questionnaire. Testing was performed in line with the rule of the system and the results from information system and manual methods were then compared. System testing was performed using the 10 EPDS questionnaire data available from diagnosis by a nurse. Calculation for system accuracy value is gained from division of the number of matching data to that of the total data, and then times 100%.

3. Results and discussion

3.1. Information system to identity status of postpartum depression

Results of data testing show that system output is in accordance to that of EPDS questionnaire administered by nurses, which means that the system is valid. Diagnostic test for both system and manual diagnoses is given in Table 1.

Table 1. System and manual diagnostic test.

EPDS Questionnaire Score	Post-Partum Period/days	System Diagnosis	Manual Diagnosis	Explanation
11	3	Postpartum blues	Postpartum blues	Valid
10	3	Postpartum blues	Postpartum blues	Valid
4	21	Normal	Normal	Valid
13	32	Suspected postpartum depression	Suspected postpartum depression	Valid
6	13	Normal	Normal	Valid
8	14	Normal	Normal	Valid
0	2	Normal	Normal	Valid
12	17	Suspected minor postpartum depression	Suspected minor postpartum depression	Valid
3	10	Normal	Normal	Valid
9	5	Normal	Normal	Valid

In terms of information system, diagnosis for postpartum depression is obtained after input of symptom data is done. Samples for normal diagnosis are shown in Figure 2.

Identification for postpartum depression status in this research is validated by EPDS score and postpartum period. A description of EPDS questionnaire scores and postpartum period for the group screened with information system method is given in Table 2.

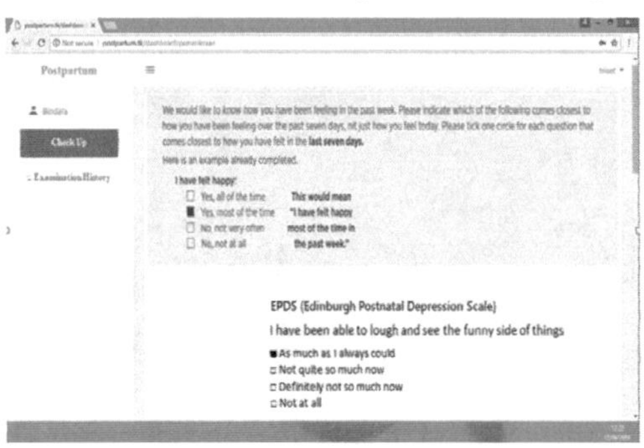

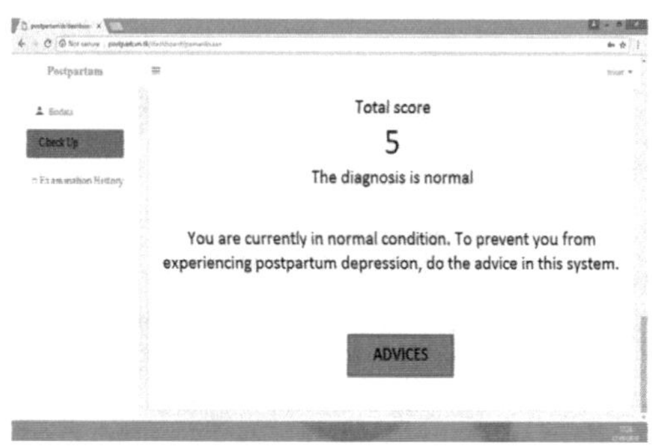

Figure 2. Postpartum depression diagnosis using information system.

Table 2. EPDS score for the group screened with information system method.

Variable	EPDS Questionnaire Score Before using the information system method (n=26)	EPDS Questionnaire Score Before using the manual method (n=26)
EPDS Questionnaire Score		
Mean ±SD	9.04±3.33	8.23±3.23
Minimum	4	3
Maximum	17	15
Postpartum Period/days		
Mean ±SD	17.77±9.56	18.35±8.74
Minimum	4	3
Maximum	37	32
Postpartum depression category		
Normal	15 (57.69%)	16 (61.54%)
Postpartum blues	7 (26.93%)	6 (23.08)
Suspected minor postpartum depression	4 (15.38%)	4 (15.38%)
Suspected postpartum depression	0 (0%)	0 (0%)
Suspected postpartum psychosis	0 (0%)	0 (0%)

It can be seen in Table 2 that the mean score for EPDS from respondents belonging to the information system group is 9.04, while it is 8.23 from the manual group. There is no difference in EPDS scores from both groups prior to intervention. This is evident from the same mean EPDS score. In terms of percentage, respondents that do not suffer from postpartum depression are 15 mothers (57.69%) for the information system group and 16 mothers (61.54%) for the manual group.

Meanwhile, respondents diagnosed with postpartum blues are 7 mothers (26.93%), in the information system group, and 4 mothers (15.38%) are suspected of minor postpartum depression. On the other hand, respondents diagnosed with postpartum blues are 6 mothers (23.38%), in the manual group, and 4 mothers (15.38%) are suspected of minor postpartum depression. More respondents are diagnosed with postpartum depression in the information system group (42.31%), compared to the manual group (38.46%).

Varied status of postpartum depression is understood to be due to different postpartum periods among respondents in both the information system and the manual groups. Postpartum blues tend to be experienced by mothers on the third day to the first two weeks after giving birth [9,10].

Meanwhile, postpartum depression has its onset from the first two weeks to the first year after giving birth [11]. This vast time range of recruitment for respondents results in varied diagnoses.

Identification for postpartum depression with both information system and manual methods show similar results. This is due to the same aspect and characteristics of questions in the questionnaires [12]. Results of the use of information system in this research agree with those of other researches. An earlier research comparing screening for depression and anxiety using online based system and paper system on grown-ups with cystic fibrosis show similar internal consistency between online-based and paper-based measurements [13].

Some other research suggest that online questionnaires provide valid results as are paper-based questionnaires [14,15]. Researchers over the past decade have revealed that online questionnaires are capable of generating more accurate information compared to its paper-based counterpart [7]. Moreover, with respondents independently filling out the questionnaires (instead of undergoing interviews) both for online and paper-based format, yield comparative results [16].

Development in information system provides easy access to all kinds of health information, including postpartum depression. The advent of a postpartum depression screening instrument has paved the way for helping mothers assessing their postpartum depression status. Screening using information system can be widely used by the community and has been proven to be valid and reliable [12]. This system is interesting for mothers having difficulties in understanding how they feel and dealing with the anxiety of being labeled having mental disorder. This means that mothers are now more proactive in making use of postpartum depression screening instrument that will provide them with privacy and comfort.

Screening online comes with more advantages than manual screening. Ease in filling out forms, increased data completeness, minimized entry error, are among the advantages of screening with information system. Other than those, filling out questionnaires using information system is faster than its paper-based counterpart [16]. This fact is supported by an earlier research stating that online questionnaire completion takes approximately 2 minutes, while manual questionnaire completion needs around 5 minutes [7,17].

The use of information system to identify postpartum depression will also be beneficial for women with high risks as they do not have to visit health professionals to do so. They can easily access information of postpartum depression status from the comfort of their homes [18]. Earlier screening using information system also results in better prognosis. Moreover, this method eases the burden of health professionals with its easy and effective diagnosis. Results show that screening using information system can be used to recruit and filter mothers with symptoms of postpartum depression. Therefore, the use of information system to help find out depression status of postpartum mothers, is an innovative solution [7].

3.2. Information system for suggested treatment for postpartum depression

In terms of information technology, suggested treatment for postpartum depression is based on system diagnoses. Samples of suggested treatment for normal diagnosis are shown in Figure 3.

Figure 3. Suggested treatment from the information system.

Suggested treatment for postpartum depression is depicted in different pretest EPDS scores resulting from screening in the information system groups is given in Table 3.

Table 3. EPDS score for the group screened with manual method.

Variable	Group	
	Information System	Manual
Pre-EPDS Questionnaire Score		
Mean ±SD	9.04±3.33	8.23±3.23
Minimum	4	3
Maximum	17	15
Post-EPDS Questionnaire Score		
Mean ±SD	3.35±2.50	4.85±3.21
Minimum	0	0
Maximum	10	12
P value	0.000	0.000

Ability of the information system in giving suggested treatment for postpartum depression is shown in differences in scores prior to and after its use. It can be seen in Table 3 that pre and post intervention test show p-value of 0.000, with $p < 0.05$ which can be concluded that both groups have difference in EPDS score changes. This shows the effect of intervention, either with the information system or with the manual method.

Psychological therapy for postpartum depression is usually administered face to face. However, this method of counseling and using leaflets is prone to patients forgetting and losing the suggestions. Other than that, health personnel also have limitations in providing treatment for postpartum depression. In line with technological development, change in lifestyle, and real life needs, psychological therapy based on information technology has become a viable alternative for postpartum depression treatment in many countries. Overcoming access difficulties for postpartum depression treatment is another advantage of online therapy. As long as internet connection is available, therapy can be accessed anytime, anywhere [19].

The information system group has more mothers diagnosed with postpartum depression compared to its manual counterpart. This is perhaps because mothers in the information system group is more active in screening for postpartum depression. It is suggested in one earlier research that individuals with chronic condition tend to seek for more health information from the web. Mothers who have just given birth and are suffering from symptoms of postpartum depression tend to be more active in using the internet to help them access health services not available traditionally, as they want speedier treatment for postpartum depression [20,21].

This is in line with Lazakidou who said that one advantage of web-based information system is its easy access from just about anywhere. The web site developed here is very practical and can be readily accessed with any device, including smart phones [22]. Treatment for postpartum depression given in this research is designed by an expert (psychologist) to help midwives providing care for postpartum depression. Treatment for postpartum depression using information system is seen as an effective approach in terms of cost and labor, compared to treatment with manual method. In agreement with an earlier research suggesting that online psychological therapy has proven to be effective in handling and reducing the risk of postpartum recurrence. Significant results in this research proves the system's capability in providing effective are to help mothers coming to terms with their postpartum depression.

3.3. Effectiveness of information system in using information system

Testing for the system built employed TAM questionnaire. Results of TAM questionnaire analysis based on two perceptions of Perceived Ease of Use and Perceived Usefulness of the system developed are shown in Figure 4.

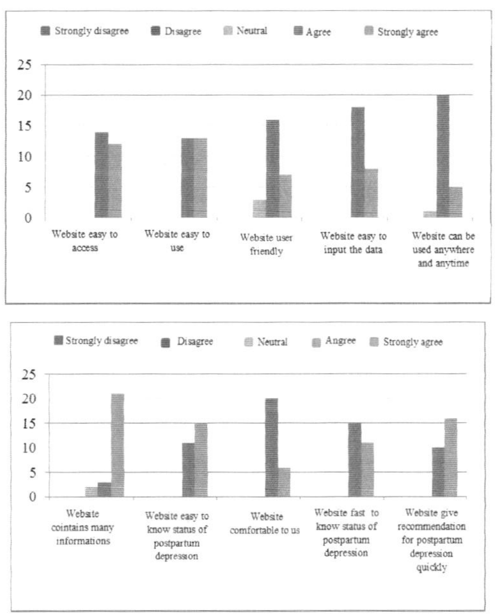

Figure 4. Diagram of TAM questionnaire answers of perceived ease of use and perceived usefulness variable.

Assessment for effectiveness of screening system of postpartum diagnosis is based on acceptance of the system built. In its detail, acceptance of information system is seen from two aspects of perceived usefulness and perceived ease of use. It can be seen from Figure 4 that variable Perceived Ease of Use gets a maximum answer of 5 and the minimum answer 3. Meanwhile the most common answer is 4, which means agree. This means that respondents agree that the system for postpartum depression screening is really easy to use.

The variable of Perceived Ease of Use gets a maximum answer of 5 and the minimum answer 3. Meanwhile the most common answer is 4, which means agree. This means that respondents agree that they gain advantage from the system of postpartum depression screening. Hence, both results indicate that respondents find easy and useful intervention.

Results in this research are in line with those of Cujipers et al on global scale online depression prevention measures The system developed here is very effective in helping mothers understand their postpartum depression status and know how to deal with it [23].

4. Conclusion

The screening system developed here has been validated and is proven to be valid in identifying status of postpartum depression. Identification relies on the 10 EPDS questions automatically set in the system. And, the screening system developed here has been validated and is proven to be valid in providing suggested treatment for postpartum depression, based on knowledge input from experts in related diagnoses generated by the system. Therefore, screening for postpartum depression using information system is very effective and ensures proper and timely treatment.

References

[1] Venkatesh K K, Zlotnick C, Triche E W, Ware C and Phipps M G 2014 Accuracy of brief screening tools for identifying postpartum depression among adolescent mothers *Pediatrics* **133** e45-e53

[2] Pandya B D, Mital D P, Srinivasan S and Haque S 2014 A web-based system for early detection of symptom of depression *International Journal of Medical Engineering and Informatics* **6** 65-73

[3] Norhayati M, Hazlina N N, Asrenee A and Emilin W W 2015 Magnitude and risk factors for postpartum symptoms: a literature review *Journal of affective Disorders* **175** 34-52

[4] Gjerdingen D K and Yawn B P 2007 Postpartum depression screening: importance, methods, barriers, and recommendations for practice *The Journal of the American Board of Family Medicine* **20** 280-288

[5] Hewitt C, Gilbody S, Brealey S, Paulden M, Palmer S and Mann R 2009 Methods to identify postnatal depression in primary care: an integrated evidence synthesis and value of information analysis *Health Technology Assessment Journal* **13** 1-230

[6] Stewart D, Robertson E, Dennis C, Grace S and Wallington T 2003 Detection, prevention and treatment of postpartum depression *Postpartum depression: Literature review of risk factors and interventions* 71

[7] Drake E, Howard E and Kinsey E 2014 Online screening and referral for postpartum depression: An exploratory study *Community mental health journal* **50** 305-311

[8] Kingston D, Janes-Kelley S, Tyrrell J, Clark L, Hamza D and Holmes P 2015 An integrated web-based mental health intervention of assessment-referral-care to reduce stress, anxiety, and depression in hospitalized pregnant women with medically high-risk pregnancies: a feasibility study protocol of hospital-based implementation *JMIR research protocols* **4**

[9] Wahyuntari E *Postpartum Affective Disorders*

[10] Mansur H 2009 *Psikologi Ibu dan Anak untuk Kebidanan* (Jakarta: Salemba Medika)

[11] Sundaram S, Harman J S and Cook R L 2014 Maternal morbidities and postpartum depression: an analysis using the 2007 and 2008 Pregnancy Risk Assessment Monitoring System *Women's health issues* **24** e381-e388

[12] Le H -N, Perry D F and Sheng X 2009 Using the internet to screen for postpartum depression *Maternal and Child Health Journal* **13** 213-221

[13] Cronly J, Duff A J, Riekert K A, Perry I J, Fitzgerald A P and Horgan A 2018 Online versus

paper-based screening for depression and anxiety in adults with cystic fibrosis in Ireland: a cross-sectional exploratory study *BMJ open* **8** e019305

[14] Wagner B, Horn A B and Maercker A 2014 Internet-based versus face-to-face cognitive-behavioral intervention for depression: a randomized controlled non-inferiority trial *Journal of affective disorders* **152** 113-121

[15] Holländare F, Andersson G and Engström I 2010 A comparison of psychometric properties between internet and paper versions of two depression instruments (BDI-II and MADRS-S) administered to clinic patients *Journal of Medical Internet Research* **12**

[16] Wu R C, Thorpe K, Ross H, Micevski V, Marquez C and Straus S E 2009 Comparing administration of questionnaires via the internet to pen-and-paper in patients with heart failure: randomized controlled trial *Journal of medical Internet research* **11**

[17] Castle J 2008 Early detection of postpartum depression: Screening in the first two to three days *J Lancaster Gen Hosp. Winter* 4

[18] Ryan A and Wilson S 2008 Internet healthcare: do self-diagnosis sites do more harm than good? *Expert Opinion on Drug Safety* **7** 227-229

[19] Andersen C E 2015 *A Web-Based Intervention for Postpartum Depression. An Assessment of User Acceptability and Information System Evaluation*

[20] Wisner K L, Logsdon M C and Shanahan B R 2008 Web-based education for postpartum depression: conceptual development and impact *Archives of women's mental health* **11** 377

[21] Danaher B G, Milgrom J, Seeley J R, Stuart S, Schembri C and Tyler M S 2012 Web-based intervention for postpartum depression: Formative research and design of the MomMoodBooster program *JMIR research protocols* **1**

[22] Lazakidou A 2010 *Web-Based Applications in Healthcare," in Web-Based Applications in Healthcare and Biomedicine* (Springer) 143-155

[23] Cuijpers P, Beekman A T and Reynolds C F 2012 Preventing depression: a global priority *Jama* **307** 1033-1034

Web-based rule-based system for early detection of anemia among pregnant mothers

S Y Veronica[1,*], M N Widyawati[2] and S Suryono[3]

[1] Midwifery Program, STIKes Aisyah Pringsewu, Lampung, Indonesia
[2] Applied Midwifery Science, Poltekkes Kemenkes, Semarang, Indonesia
[3] Department of Physics, Diponegoro University, Semarang, Indonesia

*verosakha@yahoo.com

Abstract. Anemia among pregnant mothers is a serious issue that may cause pregnancy complications, premature birth that can lead to mortality, and low birth weight (BBLR). Monitoring for anemia among pregnant mothers is quite a phenomenon for midwives and health professionals due to time constraint as it takes time to establish a proper diagnosis. Late diagnosis establishment results in late treatment. Therefore, this paper proposes a web-based rule-based information system using the rule-based algorithm to ease monitoring and detection of anemia among high risk pregnant mothers. The method employed is online rule-based information system to diagnose anemia among pregnant mothers. Respondents input data of symptoms that will then be processed by the information system to generate diagnoses. Resulting diagnoses are displayed as they are. The rule-based method in the information system validates data of symptoms and diagnoses with experts in the field to serve as reference in providing health care information. This is a pre-experimental research with one posttest group. The intervention group (N=10) was given treatment using the information system, while the control group (N=10) was given treatment manually. Interventions were conducted for 20 days, and measurements were carried out on the 21st day using the Technology Acceptance Model (TAM). The anemia detection system for pregnant mothers developed here is proven to be effective in helping anemia detection and has been validated by experts (midwives) and hence, been declared valid. With effectiveness score of 90 %, this web-based rule-based information system is effective in detecting, monitoring, and reporting anemia among pregnant mothers.

1. Introduction

Anemia among pregnant mothers is a serious issue that may cause pregnancy complications, premature birth that can lead to mortality, and low birth weight (BBLR) [1]. Establishing anemia diagnosis by health professionals requires considerable amount of time [2]. Late diagnosis establishment triggers worsening condition in patients as it will lead to late treatment [3,4]. Clinical diagnosis of all types of anemia is not always easy to perform due to some basic reasons. Once anemia starts, the cause has to be determined to ensure specific and proper therapy. Detection, classification, and treatment of anemia are the priorities of the health system [5].

Health Information System has been used to support everyday health care services administered in health facilities such as Public Health Center and hospitals to provide treatment and perform intervention on health issues [6,7]. However, reporting of health care services from administrative areas to data centers has not been able to produce accurate and timely data/information [8].

On the other hand, anemia is often taken for granted as it is regarded as a common condition. Common people should have proper knowledge on its types and symptoms and how to prevent or treat anemia to minimize its risks. This can be achieved by making use of information technology for health care [9]. Especially in terms of its function as media to disseminate health information and diagnose illness. This will certainly be helpful for everyone in preventing and treating health problems. Those are some of the reasons why the researchers are interested in conducting this research of early detection of anemia among pregnant mothers using rule-based information system.

2. Methods

This research employed web-based rule-based information system to improve effectiveness of anemia detection and monitoring for pregnant mothers. Diagram of the information system is shown in Figure 1. The web-based rule-based information system is an innovation that helps ease of detection for risky pregnancy. This system can be accessed with a computer or smart phone and it features automated detection of risky pregnancy, once all input data are obtained, simply by visiting its web address.

Pregnant mothers have to fill out forms of biodata and medical, then the information system displays some questions related to symptoms and they must choose some questions concerning what they experienced by clicking the proper questions and then click on the Expert Diagnosis menu to see the results.

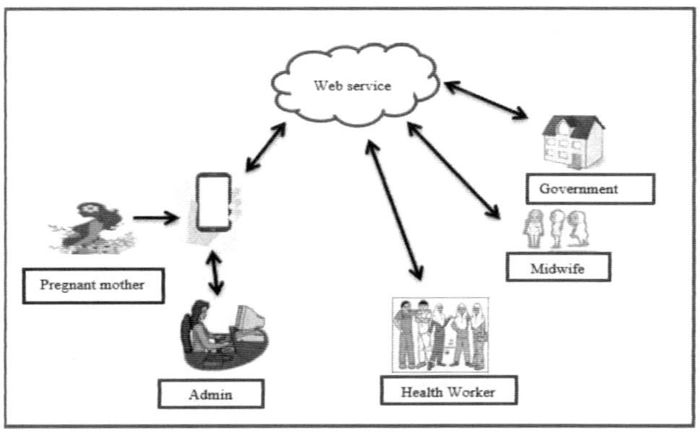

Figure 1. Information system diagram.

This system can also be accessed by health professionals, midwives, and government officials. Registration only requires logging in as an administrator by choosing a user name and a password before clicking user registration to be enrolled by the administrator of this information system. Once registered as an administrator, a user can log in and access recap of pregnant mother data and graphs and also classifications of anemia among pregnant mothers provided on a monthly basis. These data and graphs are updated and edited by the administrator.

This research involved two groups intervention, which is treated with the rule-based information system, and control, which is treated with the manual method. The dependent variables are anemia detection and effectiveness of information system, while the independent variable is web-based rule-based information system. This is a pre-experimental research using only one posttest group design. Samples of this research were midwives fulfilling inclusion criteria of 20-35 years old age range, able to operate computers or use smart phones, and willing to participate. The number of midwives as samples were 20.

The research instrument was web-based rule-based information system. Meanwhile, the materials used was TAM checklists to measure effectiveness of the information system. Detection results using this information system has been validated by experts (midwives). Diagnoses of anemia symptoms from the information system match those established by experts (midwives) and help to provide health care for pregnant mothers suffering from anemia.

3. Results and discussion

3.1. Displayed results from information system
Detection of anemia among pregnant mothers using web-based rule based information system is an innovation from the currently available system. This innovation to detect anemia among pregnant mothers helps with easy detection of risky pregnancy. It can be accessed via computers or hand phones with the advantage of automatic detection of risky pregnancy, once all input data are in. Recap menu on the information system is shown in Figure 2.

3.2. Graph menu of anemia among pregnant mothers on the information system
The graph shows automatic numbers of pregnant mothers registered in the system. The graph will also show interpretation and classification of anemia, which means whether a mother is suffering from iron deficiency, folic acid deficiency, B12 anemia, hemolytic anemia, aplastic anemia, or even just having a normal pregnancy. This menu helps with easy and speedy reporting of data of pregnant mothers with anemia by midwives. In terms of information system, diagnosis for anemia is obtained after input if symptoms data is done. Symptoms of anemia is shown in Figure 3. Graph menu of anemia among pregnant mothers on the information system is shown in Figure 4.

3.3. Comparison of service speed between the web-based and the manual methods
Shows speedy detection of pregnant mothers with anemia using the web-based method is 23 minutes faster. It is 23 minutes faster for pregnant mothers with anemia to get treatment. This reduced time needed for diagnosis is the result of automated time from input data to the time they are converted into status of anemia among pregnant mothers. Longer detection affects further intervention and treatment. comparison of service speed between the web-based and the manual methods on the information system is shown in Table 1.

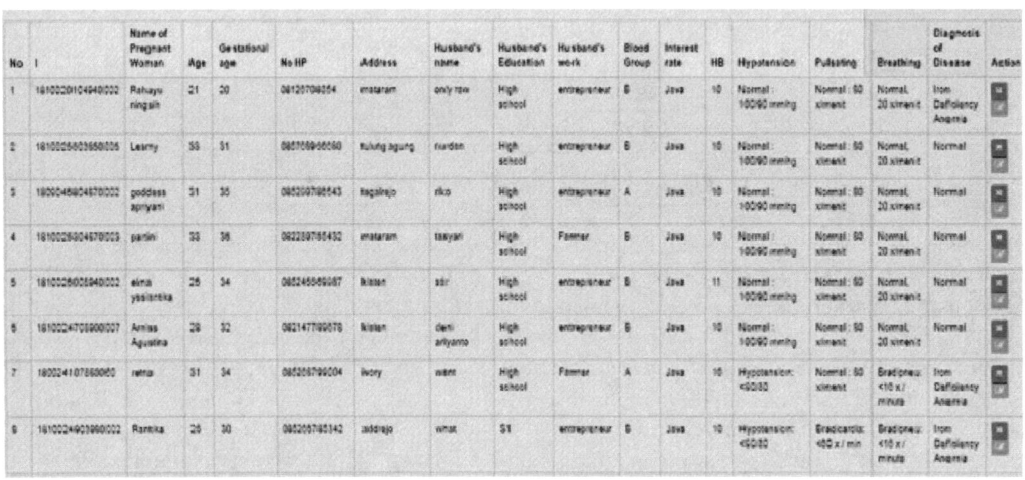

No	I	Name of Pregnant Woman	Age	Gestational age	No HP	Address	Husband's name	Husband's Education	Husband's work	Blood Group	Interest rate	HB	Hypotension	Pulsating	Breathing	Diagnosis of Disease	Action
1	[illegible]	Rahayu ningsih	21	30	[illegible]	mataram	only raw	High school	entrepreneur	B	Java	10	Normal : 100/90 mmHg	Normal : 90 x/menit	Normal, 20 x/menit	Iron Deficiency Anemia	
2	[illegible]	Leamy	33	31	[illegible]	kulung agung	nurdan	High school	entrepreneur	B	Java	10	Normal : 100/90 mmHg	Normal : 90 x/menit	Normal, 20 x/menit	Normal	
3	[illegible]	goddess apriyani	31	35	[illegible]	tegalrejo	riko	High school	entrepreneur	A	Java	10	Normal : 100/90 mmHg	Normal : 90 x/menit	Normal, 20 x/menit	Normal	
4	[illegible]	partini	33	36	[illegible]	mataram	tasiyan	High school	Farmer	B	Java	10	Normal : 100/90 mmHg	Normal : 90 x/menit	Normal, 20 x/menit	Normal	
5	[illegible]	elma yasilantika	25	34	[illegible]	tiktan	adir	High school	entrepreneur	B	Java	11	Normal : 100/90 mmHg	Normal : 90 x/menit	Normal, 20 x/menit	Normal	
6	[illegible]	Anniss Agustina	28	32	[illegible]	tiktan	deni ariyanto	High school	entrepreneur	B	Java	10	Normal : 100/90 mmHg	Normal : 90 x/menit	Normal, 20 x/menit	Normal	
7	[illegible]	retna	31	34	[illegible]	ivory	want	High school	Farmer	A	Java	10	Hypotension: <90/60	Normal : 90 x/menit	Bradipnea: <16 x / minute	Iron Deficiency Anemia	
8	[illegible]	Ramika	26	30	[illegible]	saddrejo	wmat	S1	entrepreneur	B	Java	10	Hypotension: <90/60	Bradicardia: <60 x / min	Bradipnea: <16 x / minute	Iron Deficiency Anemia	

Figure 2. Recap menu of anemia among pregnant mothers.

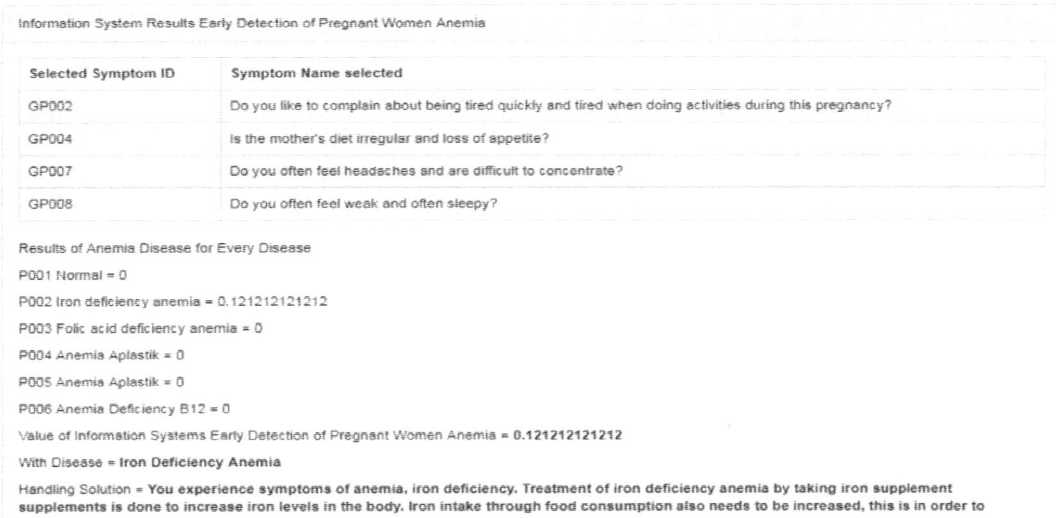

Figure 3. Symptoms of anemia.

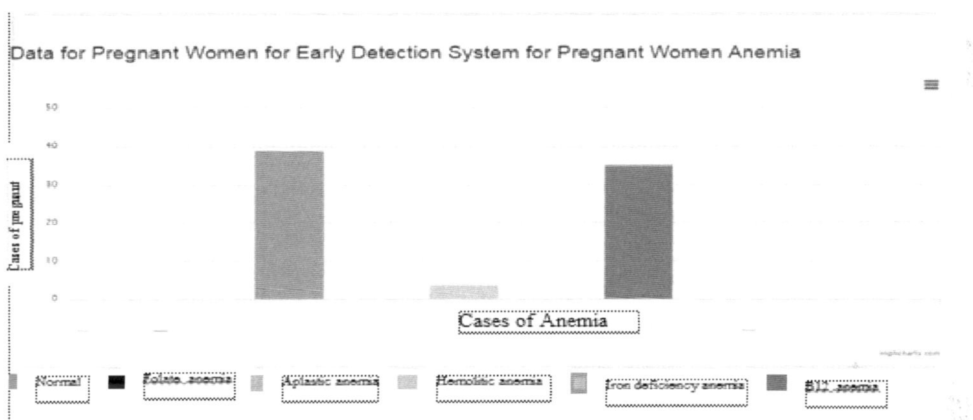

Figure 4. Graph menu of anemia among pregnant mothers.

Table 1. Comparison of service speed between the web-based and the manual methods.

Variable	Method	Mean Time
Speed	Manual Method	38 minutes
(minute)	Web-based Method	15 minutes

However, the speed of service from this web-based rule-based information system also depends on Internet connection. The better the Internet connection, the faster health care services can be provided for pregnant mothers. On the contrary, the worse the Internet connection, the slower health care services can be provided for pregnant mothers. Nonetheless, information on this web-based rule-based information system can still be accessed when the computer is offline or not connected to any Internet network. Hence, there is no need for real concern.

Information technology has transformed into many forms, including information system. Information system as part of information technology is required to improve organizational performance. Information that is assembled in orderly, clear, precise, and speedy way and is presented in a decent report certainly facilitate smooth operation of organizations and decision-making processes. The aforementioned analyses lead to a conclusion that early detection of anemia among pregnant mothers using information system built here can also serve as a means of recording and reporting [10].

Health Information System (SIK) has been used to support everyday health care services administered in health facilities such as Public Health Center and hospitals to provide treatment and perform intervention on health issues. However, reporting of health care services from administrative areas to data centers has not been able to produce accurate and timely data/information.

3.4. Accurate diagnosis of anemia detection among pregnant mothers using web-based rule-based information system. Accurate diagnosis from web-based rule-based information system as validated by experts (midwives)

Monitoring for pregnancy anemia is not only meant for mothers who are not well. It is aimed at knowing and dealing with health problems among pregnant mothers with the help of health professionals to maintain their healthy diet and lifestyle. Earlier accurate diagnosis is advantageous as it helps with efforts to prevent, stimulate proper treatment and referral to obstetricians for even better care. An earlier study comparing clinical diagnosis written manually in letters accepted by hospital consultants and those being given as inputs in a computer revealed that the latter is more satisfactory in many clinical settings. Diagnoses established by general practitioners are often not independently verified. They require further research to determine their accuracy [11,12].

Shows results from the web-based system and from expert (midwives) validation of 5 anemia classifications revealed an accuracy of 100 % valid. The system's diagnosis accuracy reveals good sensitivity scores of 83% and 80% specification. The information system developed is capable of accurate detection of anemia among pregnant mothers in terms of classifying types of anemia such as iron anemia, folic acid anemia, hemolytic anemia, aplastic anemia, and B12 anemia. Therefore, it can be said that the matches between diagnoses from the information system and those from the experts (midwives) are accurate. diagnosis accuracy of the web-based system and the experts (midwives) validation on the information system is shown in Table 2.

Table 2. Diagnosis accuracy of the web-based system and the experts (midwives) validation.

Anemia Score	Hb	System Diagnosis	Expert Diagnosis	Explanation
B (4)	10 g/dl	Iron anemia	Iron anemia	Valid
A (1)	11g/dl	Normal	Normal	Valid
B (5)	10g/dl	Iron anemia	Iron anemia	Valid
A (1)	11g/dl	Normal	Normal	Valid
A (1)	11g/dl	Normal	Normal	Valid
C (3)	9 g/dl	Aplastic anemia	Aplastic anemia	Valid
D (4)	9 g/dl	Hemolytic anemia	Hemolytic anemia	Valid
E (3)	10 g/dl	B12 anemia	B12 anemia	Valid
F (3)	9 g/dl	Folic acid anemia	Folic acid anemia	Valid
A (1)	12 g/dl	Normal	Normal	Valid
Total		10	10	
Diagnosis validation		100 %		Valid

3.5. Validation for effectiveness of the web-based rule-based information system is carried out using the TAM checklist that cover all related aspects

Shows validated effectiveness of the web-based rule-based information system from 10 experts (midwives) using the web-based information system who gave TAM score based on aspects of completeness, ease, match, accuracy, and precision, results in a total score of 90 %, which means that the web-based rule-based information system is proven to be effective. This proves that the web-based rule-based information system developed here is worth using to help with early detection of anemia among pregnant mothers. validation of web-based rule-based information system effectiveness on the information system is shown in Table 3.

The various facilities provided by the rule-based web-based information system for the detection of anemia in pregnant women by the user have a positive impact. Users no longer just judge, but users can make this rule-based web-based information system like a need to monitor, detect anemia of pregnant women in health services [13]. If someone feels the internet is easy to use, it will be more often or more willing to use it [14].

Table 3. Validation of web-based rule-based information system effectiveness.

Variable Effectiveness	Score										Total	
	SB		B		C		KB		BR			
	N	BB	N	BB	N	BB	N	BB	N	BB	BB	%
Completeness	5	5	5	4	-	3	-	2	-	1	4.5	90
Ease	5	5	5	4	-	3	-	2	-	1	4.5	90
Matches	4	5	6	4	-	3	-	2	-	1	4.4	86
Accuracy	5	5	5	4	-	3	-	2	-	1	4.5	90
Precision	6	5	4	4	-	3	-	2	-	1	4.6	92
Speed	5	5	5	4	-	3	-	2	-	1	4.5	90
Total effectiveness score											4.5	90

4. Conclusion

Results show that the web-based rule-based information system developed here is an innovation that helps the process of detection among pregnant mothers with 23 faster diagnoses to suggest proper treatments. The web-based rule-based information system has also been successfully validated by experts (midwives) and has been granted validity for its capability in automatically providing accurate diagnosis of pregnancy anemia categorization The web-based rule-based information system is also effective to detect, monitoring, and report cases of anemia among pregnant mothers for proper health care services to be administered.

References

[1] Fauziah S F, Widyawati M N and Amartha T A S 2017 Methods Of Postpartum Blood Loss Measurement In Indonesia Should Be Modified For Better Accuracy: A Literature Review *Proceedings of the International Conference on Applied Science and Health* 2 54-62

[2] Lopez A, Cacoub P, Macdougall I and Peyrin-Biroulet L 2016 Iron Deficiency Anaemia *The Lancet.* **387** 10021 907-916

[3] Cheng S, Chen W, Yang Y, Chu P, Liu X and Zhao M 2013 Effect Of Diagnostic And Treatment Delay On The Risk Of Tuberculosis Transmission In Shenzhen China: An Observational Cohort Study, 1993–2010 *Plos One* **8** 6 E67516

[4] Trayner I and Taylor L 2017 Nhs Western Isles-Involving Our Patients: Faster Access To Treatment For Hypertension In Primary Care *International Journal Of Integrated Care* **17** 5

[5] Kundrapu S and Noguez J 2018 *Laboratory Assessment Of Anemia. Advances In Clinical Chemistry* **83** 197-225

[6] Cahyanti A N and Purnama B E 2017 Pembangunan Sistem Informasi Manajemen Puskesmas Pakis Baru Nawangan *Speed-Sentra Penelitian Engineering dan Edukasi* **4** 4

[7] Han Nd, Han L, Tuan Dm, In Hp and Jo M 2014 A Scheme For Data Confidentiality In Cloud-Assisted Wireless Body Area Networks *Information Sciences* **284** 157-166

[8] Jones S S, Rudin R S, Perry T and Shekelle P G 2014 Health information technology: an updated systematic review with a focus on meaningful use *Annals of internal medicine* **160** 1 48-54

[9] Sumarno S and Roni P 2014 Sistem Pakar Untuk Mendiagnosa Penyakit Anemia *Prosiding Seminar Nasional* (Universitas Kanjuruhan Malang)

[10] Departemen Kesehatan Republik Indonesia 2012 *Pedoman Sistem Informasi Manajemen Puskesmas* (Jakarta: Depkes RI)

[11] Senewe F P and Wiryawan Y 2011 Pencatatan Dan Pelaporan Sistem Pemantauan Wilayah Setempat-Kesehatan Ibu Dan Anak Oleh Bidan Di Desa Di Puskesmas Sepatan Kabupaten Tangerang 2008 *Jurnal Ekologi Kesehatan* **10**

[12] Anderson A M, Matsumoto M, Saul M I, Secrest A M and Ferris L K 2018 Accuracy Of Skin

IOP Conf. Series: Materials Science and Engineering 1108 (2021) 012021 doi:10.1088/1757-899X/1108/1/012021

Cancer Diagnosis By Physician Assistants Compared With Dermatologists In A Large Health Care System *Jama Dermatology* **154** 5 569-573

[13] Fischer A J, Dart E H, Leblanc H, Hartman K I, Steeves R O and Gresham F M 2016 An Investigation Of The Acceptability Of Videoconferencing Within A School-Based Behavioral Consultation Framework *Psychology In The Schools* **53** 3 240-252

[14] Abdullah F, Ward R and Ahmed E 2016 Investigating The Influence Of The Most Commonly Used External Variables Of Tam On Students' Perceived Ease Of Use (Peou) And Perceived Usefulness (Pu) Of E-Portfolios *Computers In Human Behavior* **63** 75-90

IOP Conf. Series: Materials Science and Engineering 1108 (2021) 012022 doi:10.1088/1757-899X/1108/1/012022

Euclidean distance digital image processing for jaundice detect

E P Rahayu[1,*], M N Widyawati[1] and S Suryono[2]

[1] Postgraduate Program in Applied Midwifery Science, Poltekkes Kemenkes, Semarang, Indonesia
[2] Departement of Physics, Faculty of Science and Mathematics, Diponegoro University, Semarang, Indonesia

*etikaputrirahayu@gmail.com

Abstract. Jaundice is a serious health issue. Late treatment of jaundice cases in babies may result in neurodevelopmental disorder and irreversible brain damage. Diagnosis inaccuracy is usually caused by the fact that health professionals and health service providers often rely on visual observation instead of laboratory examination. Lack of expertise in detecting jaundice is a serious matter. This research proposes a method of web-based digital image processing as an alternative for early detection of jaundice based on babies' complexion. Images of babies' complexion and color calibration cards are taken to obtain images for online analysis. Determination of bilirubin levels is carried out using the method of Euclidean approximate distance of RGB values from babies' complexion and those of color calibration cards. Results show correlation of Euclidean distance to bilirubin level of babies of 0.93596 and web-based digital image processing accuracy of 90%. These mean that the information system developed here is capable of detecting jaundice cases. This research was performed observationally in high-risk prenatal ward involving 30 infants as samples.

1. Introduction

Lack of accuracy in detecting newborn jaundice is one of the reasons why newly born babies need special care [1,2]. Digital image processing requires the use of some image elements. Image elements of pel and pixel are some parts used to show the elements of digital images [3]. Stages of digital image processing includes color-based balancing, feature extraction, regression machine learning, and bilirubin level estimation [4].

Colored images are known as RGB image with red, green, and blue components for each pixel [5]. Digital image processing is carried out on photo therapy results including non-invasive system treatment performed by health professionals or family members used to periodically find out bilirubin levels before a patient needs doctor's attention [2,4]. Increased bilirubin level is experienced by babies due to heme degradation resulting from damages in erythrocyte that disrupts albumin transport to the liver [6,7].

Babies suffering from hyperbilirubin or jaundice are prone to hemolytic diseases, metabolic and endocrine disorder, liver anatomic disorder, infection, and asphyxia.

Digital image processing helps to monitor jaundice in babies. Earlier researches suggests that a mobile phone feature called BiliCam can be used to monitor jaundice in babies. However, BiliCam is only used for offline application [2]. Therefore, there is the need to develop an online application system that can be easily used anytime and anywhere [8]. Development of health monitoring system should result in an application that is easy to access, light, cheap, and comfortable to use. This is meant to allow monitoring from home [9]. Web-based digital image processing is capable of detecting jaundice cases.

2. Methods

Colored images are known as RGB image with red, this research employed observational method by detecting jaundice from babies' complexion. It was carried out in high-risk prenatal ward involving 30 babies that meet certain qualifications. There are four stages of jaundice detection: (a) image collection and complexion detection, (b) RGB digital image processing and Euclidean distance measurement, (c) web-based information system utilization, and (d) system accuracy validation. Euclidean distance is used to determine similarities in color between babies' complexion and color calibration cards. A diagram of the information system realized is shown in Figure 1.

2.1. Babies' image collection

Babies' images are taken using a 16 mega pixel camera with flash. Babies' are lied down with a calibration card put on their body. Complexion detection is performed using visual observation that later serves as comparison to that gained from web-based complexion detection using digital image processing.

2.2. Digital image processing

Euclidean distance is calculation of the distance between 2 points in the Euclidean space [4,10-12]. Value of Euclidean distance is gained from calculation between RGB (Red, Green, Blue) value of babies' complexion and color calibration cards. The color calibration cards have eight colors of black, grey, white, cyan, magenta, yellow, and two skin colors [2]. An example of a baby's image is shown in Figure 2.

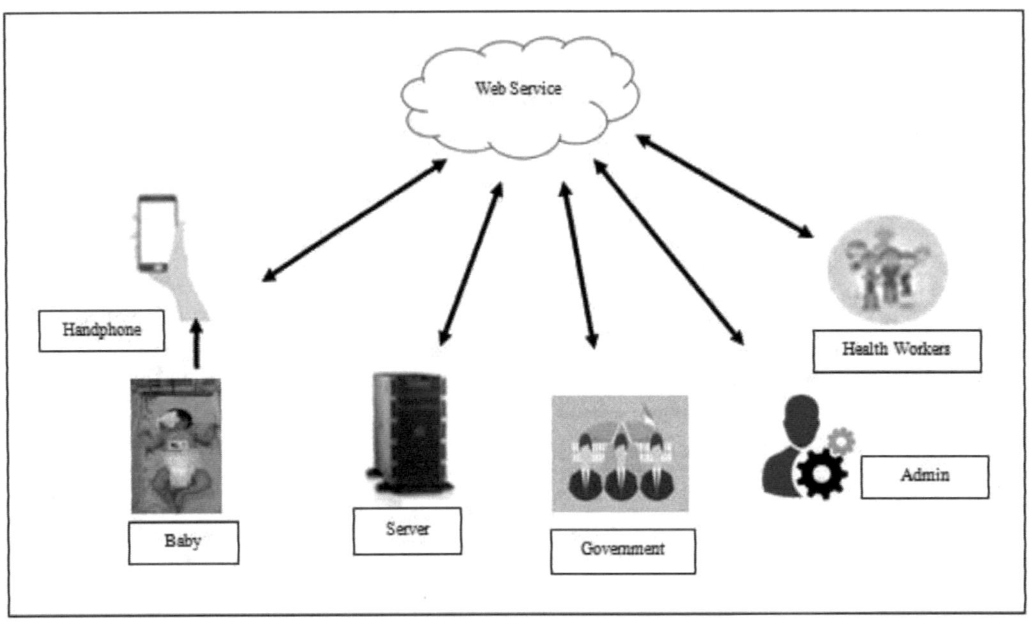

Figure 1. Diagram of realized information system.

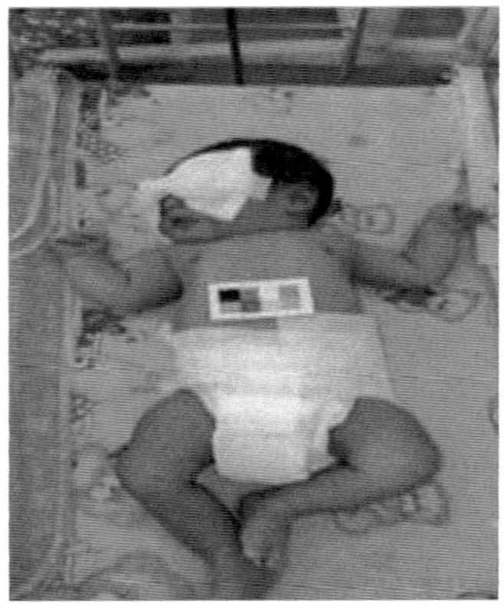

Figure 2. Example of a baby's image with a color calibration card.

2.3. Information system for jaundice case monitoring

The web-based information system is used to help detect jaundice cases effectively and as early as possible to prevent them become pathological issues. A scheme of the information system is given in Table 1.

3. Results and discussion

3.1. Correlation of baby's complexion and bilirubin level

Theoretically, skin complexion has Euclidean distance value that approximates the calibration card. Therefore, this research presents results of correlation between Euclidean distance of baby' skin complexion and calibration card, with that of bilirubin level from laboratory examination for cases of newborn jaundice, as given in Table 2.

Table 1. Information system scheme.

Input	Process	Output
a. Mother and baby identities b. Baby's clinical data c. Image of baby's body	Digital image analysis ↓ *Data matching* ↓ Notification	a. Jaundice diagnosis test result b. Table of baby's clinical data c. Geographical information of jaundice case

Table 2. Correlation of baby's complexion and bilirubin level.

	U1	U2	U3	U4	U5	R
Euclidean Distance	12.764	14.602	15.652	19.755	20.745	
Bilirubin levels (mg/dL)	9.31	9.74	10	11.45	13.42	0.93596

It can be seen in Table 2 that baby's complexion is significantly related to bilirubin level with R = 0.93596, which results from calculation of correlation between Euclidean distance and bilirubin level (from laboratory examination). The Euclidean value itself is taken from calculation of RGB value of baby's complexion and that of color calibration card attached to the baby's abdomen. Relationship equation of Euclidean distance and bilirubin level is shown in Figure 3.

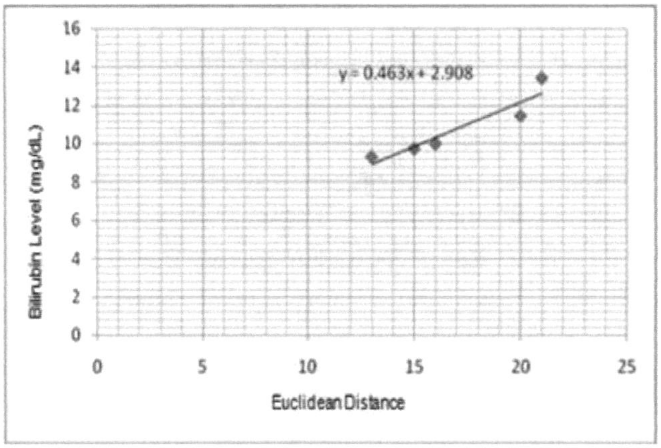

Figure 3. Relationship equation of Euclidean distance and bilirubin level.

It can be seen from Figure 3 that color matches indicated by Euclidean distance relates to bilirubin level gained from clinical observation. Correlation of Euclidean distance and bilirubin level is represented with y=0.463, which means that each 1 increase in distance equals an increase of 0.463 mg/dL in bilirubin level.

Digital image processing of baby's image results in RGB values that is converted to the Euclidean distance, which in turn is matched with the color calibration card attached to the baby's abdomen with the consent of a specialist in charge of the baby. Euclidean distance is the parameter of K Nearest Neighbor (KNN) classification that determines the match between objects being observed, hence, allowing the object to be used to measure its nearest neighbors [13]. Monitoring for jaundice cases can therefore use digital image processing by detecting baby's complexion [14,15].

3.2. Web-based jaundice case monitoring

Details of observed jaundice cases using the web-based information system is shown in Figure 4, which shows data of hyperbilirubin / jaundice cases based on baby's complexion. Diagnostic test for jaundice case uses the following categories: a) hyperbilirubin at bilirubin level of <10 mg/dL, b) hyperbilirubin at bilirubin level of 10-13 mg/dL, c) hyperbilirubin at bilirubin level of >13 mg/dL. The web-based information system with digital image processing is proven to be capable of effectively detect jaundice cases.

Figure 4. Web-based information system.

3.3. *System accuracy validation*

There were 20 cases belonging to the accurate category of the 30 samples in this study. Meanwhile the other 10 cases have approximate match (the least Euclidean distance value) and hence, have to be eliminated. So, there are 20 successful system users. Validation of system accuracy shows 90% case diagnosis accuracy. This means that the system developed here is capable of monitoring hyperbilirubin / jaundice cases. System accuracy validation is given in Table 3.

Table 3. System accuracy validation.

Name	Age	Address	Diagnosis	
			Visual	Information System
Patient 1	6 days	Brangsong	Hyperbilirubin	Hyperbilirubin
Patient 2	3 days	Kendal	Hyperbilirubin	Hyperbilirubin
Patient 3	3 days	Boja	Hyperbilirubin	Hyperbilirubin
Patient 4	5 days	Pageruyung	Hyperbilirubin	Hyperbilirubin
Patient 5	4 days	Rowosari	Hyperbilirubin	Hyperbilirubin
Patient 6	3 days	Cepiring	Hyperbilirubin	Hyperbilirubin
Patient 7	4 days	Kendal	Hyperbilirubin	Hyperbilirubin
Patient 8	4 days	Pegandon	Hyperbilirubin	Hyperbilirubin
Patient 9	7 days	Cepiring	Hyperbilirubin	Hyperbilirubin
Patient 10	4 days	Gemuh	Hyperbilirubin	Hyperbilirubin
Patient 11	3 days	Kangkung	Hyperbilirubin	Hyperbilirubin
Patient 12	7 days	Kangkung	Hyperbilirubin	Hyperbilirubin
Patient 13	4 days	Patebon	Hyperbilirubin	Hyperbilirubin
Patient 14	3 days	Patebon	Hyperbilirubin	Non -hyperbilirubin
Patient 15	4 days	Pageruyung	Hyperbilirubin	Hyperbilirubin
Patient 16	14 days	Singorejo	Hyperbilirubin	Hyperbilirubin
Patient 17	11 days	Patebon	Hyperbilirubin	Hyperbilirubin
Patient 18	4 days	Weleri	Hyperbilirubin	Non -hyperbilirubin
Patient 19	3 days	Kangkung	Hyperbilirubin	Hyperbilirubin
Patient 20	4 days	Weleri	Hyperbilirubin	Hyperbilirubin
Total			20	18
Accuracy validation				90%

4. Conclusion

Web-based digital image processing with matches measurement of Euclidean distance results in approximate values that can be used to detect jaundice cases based on bilirubin levels in babies. Digital image processing using Euclidean distance on bilirubin level results in correlation value of 0.93596, and correlation value of bilirubin level with Euclidean distance of y=0.463, which means that each 1 increase in distance relates to 0.463 mg/dL increase in bilirubin level. The web-based digital image processing method employed in this research has a validated accuracy of 90% in detecting jaundice cases.

References

[1] Amin S B and Lamola A A 2011 Newborn jaundice technologies: unbound bilirubin and bilirubin binding capacity in neonates *In Seminars in perinatology* **35**(3) 134-140

[2] De Greef L, Goel M, Seo M J, Larson E C, Stout J W, Taylor J A and Patel S N 2014 Bilicam: using mobile phones to monitor newborn jaundice *In Proceedings of the 2014 ACM International Joint Conference on Pervasive and Ubiquitous Computing* 331-342

[3] Gonzalez R C, Woods R E and Eddins S L 2004 Digital image processing using MATLAB *Pearson Education India*

[4] Aydın M, Hardalac F, Ural B and Karap S 2016 Neonatal Jaundice Detection System *J Med Syst* **40** 166

[5] MAthWorks 2020 *Image processing Toolbox : for use with MATLAB* [Online] retrieved from https://www.mathworks.com/help/images/getting-started-with-image-processing-toolbox.html

[6] Knutzen S E and Impey L 2015 The significance of base deficit in addemic term neonates *Am J Obstetri Gynecology* **213** 1-7

[7] Young P C, Korgenski K and Buchi K F 2013 Early readmission of newborns in a large health care system *Pediatrics* **131** e1538-e1544

[8] Yang L, Jiang L, Yue K and Pang H 2010 Design and implementation of the lab remote monitoring system based on embedded web technology *in Information Technology and Applications (IFITA)* 172-175

[9] Chen C M 2011 Web-based remote human pulse monitoring system with intelligent data analysis for home health care *Expert Systems with Applications* **38**

[10] Han J, Pei J and Kamber M 2011 *Data mining: concepts and techniques* (Elsevier)

[11] Nahari R V, Rachmad A, Haryanto H, Alfita R and Pramudia M 2016 Image segmentation based on quantization vectors *Nusantara of Engineering* **3**

[12] Masmawati D, Wirayuda T A B and Sthevanie F 2015 Copy-move detection on image forgery using Local Binary Pattern and Svd- matching *eProceedings of Engineering* **2**

[13] Azhari P F, Hidayat B and Rizal A 2015 Calculation of the degree of curvature of the spine in humans using the contourlet transform method and K-nearest Neighbor *Momentum* **11**

[14] Mansor M N, Hariharan M, Basah S N and Yaacob S 2013 New newborn jaundice monitoring scheme based on combination of pre- processing and color detection method *Neurocomputing* **120** 258-261

[15] Mansor M N, Yaacob S, Hariharan M, Basah S N, Ahmad Jamil S H F S, Mohd Khidir M L, Rejab M N, Ku Ibrahim K M Y, Ahmad Jamil A H F S, Junoh A K and Saad S A 2012 Jaundice in Newborn Monitoring using Color Detection Method *Procedia Engineering* **29** 1631-1635

GPS-based fire detection system (Global Positioning System) and SMS Gateway

A Aryanti[1,*], I Mekongga[2] and R S Dewi[1]

[1] Department of Electrical Engineering, Politeknik Negeri Sriwijaya, Palembang, Indonesia
[2] Computer Engineering, Politeknik Negeri Sriwijaya, Palembang, Indonesia

*Aryanti@polsri.ac.id

Abstract. This research aims to produce a GPS-based fire detection system (Global Positioning System) and SMS Gateway. The benefits of this detection system can detect early fire occurrence based on the detection of temperature conditions by accommodating the nature of the fire and able to detect any rise in temperature caused by the existence of the fire. This detection system must also be able to read any smoke produced by a fire. To realize the system, required sensors capable of reading the temperature and smoke. The Arduino Uno microcontroller is the brain control system of the system. At a temperature of> 35 C, the system will activate the DHT 11 and MQ 2 sensors that detect smoke> 50 ppm from fire. The system will activate Buzzer as a warning in the form of the next alarm sound Global Positioning System (GPS) will provide information in the form of coordinates of the location of the point of fire through GSM SIM900 Module Short Message Service (SMS) to the user. The results obtained mq2 = 128 ppm and temperature value = 38 ° C and GPS data with latitude of -3.04798388 and a longitude of 104.78263092. From the data it is seen that the mq2 value reaches> 50ppm and the temperature value reaches> 35 ° C, and the detector outputs buzzer sound and warning notification of coordinate point in the form of SMS containing the message "FIRE available" with the coordinates of the location of the fire detected by GPS.

1. Introduction

Fire is a disaster that can be caused by human factors, technical factors and natural factors that cannot be estimated when the occurrence. Fires that occur almost always caused by human activities, such as clearing land by burning the forest. Forest fires are a frequent fire in Indonesia [1].

The delay in knowing a fire is very fatal to the safety of the human soul and property. Previously, fire detectors were created with smoke detectors [2]. The sample used is transparency in lieu of smoke. The main component of this tool is the sensing that will convert the physical beam of light beam to an analog voltage signal, a transistor as a switch, a thread as a memory (latch), a multivibrator that will generate pulses and loudspeakers that produce sound as output. How the fire detector works with this smoke detector depends on the amount of light intensity received by the sensor. The sensor is the LDR (Light Dependent Resistor) which is inserted in the smoke cell [3]. The smokes cell is a box that smoke easily penetrates. In addition to LDR, on smeared cells are also installed light-emitting diodes. Sensory testing is done by inserting transparency in place of smoke in smoke cells that contain LED and LDR. The test results show that the sensor works in accordance with its function that is as a forming pulse are low and high logic [4].

The fire monitoring system should be able to accommodate the nature of the fire. By utilizing LM35 sensor that serves to observe the temperature rise [5] and AF30 sensor to observe the existence of smoke around it can be made by detecting temperature and smoke by using LM35 sensor and AF30 sensor based on microcontroller AT89S51 and equipped facility of sending SMS if dangerous situation or existence of indication will happen fire. In the test, by heating on the temperature sensor, side up to 55 ° Celsius, by providing thick smoke (cigarette smoke and combustion smoke gathered paper) [6].

In this research will be in Design of Fire Detection System based on GPS (Global Positioning System) [7,8] and SMS Gateway [9]. This fire detection system to be designed can detect early fire occurrence based on the detection of temperature conditions by accommodating the nature of the fire and able to detect the temperature rise caused by the existence of the fire. In addition, the system must also be able to read any smoke generated by the fire. To realize the system, required sensors capable of reading the temperature and smoke. This system is also equipped with a GPS system as a teller point to the coordinates of fire [10] to the user sent via short message (SMS) and buzzer as an alarm notification to the people around the place of fire.

2. Research methodology

The research method is illustrated in the form of block diagram as in the following figure.

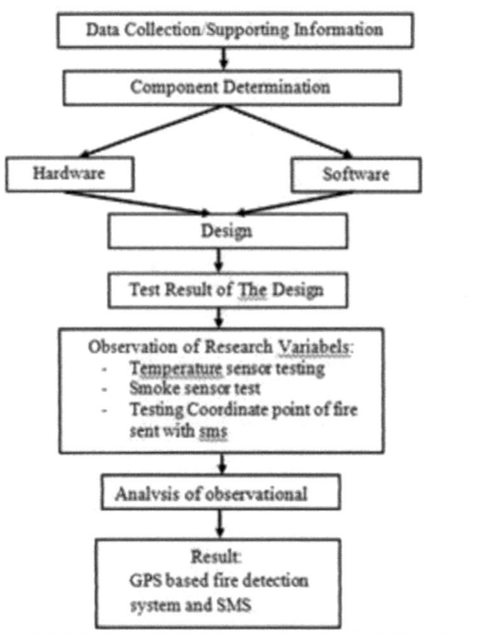

Figure 1. Research methodology.

Observation, field study, and literature study through books and articles from the internet, through the literature conducted on data collection. In the design of fire prediction system is required hardware and software components. The main hardware is Arduino ATmega328 [11], DHT11 temperature sensor [12,13], MQ2 smoke sensor [14] and supporting hardware such as GPS NE06MV4, GSM Module SIM 900 [15], buzzer, power bank solar cell. While the software used is programming algorithm using C language programming. The design results are tested and followed by data retrieval / research variables that include testing DHT11 temperature sensor, MQ2 smoke sensor testing, and fire point coordinate testing are sent by SMS.

The following is the block diagram of the circuit:

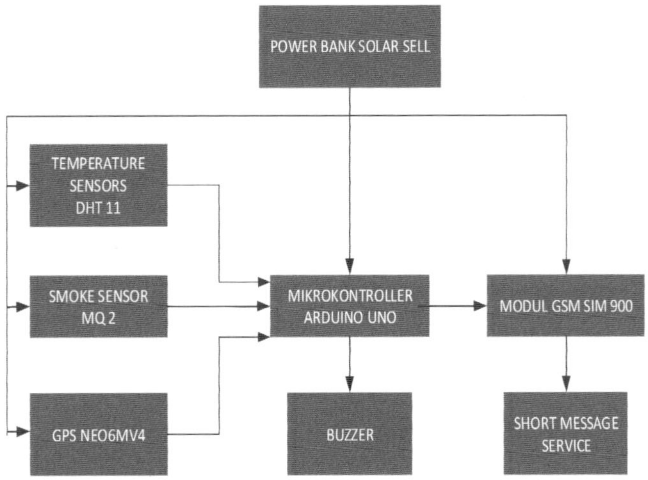

Figure 2. Block diagram.

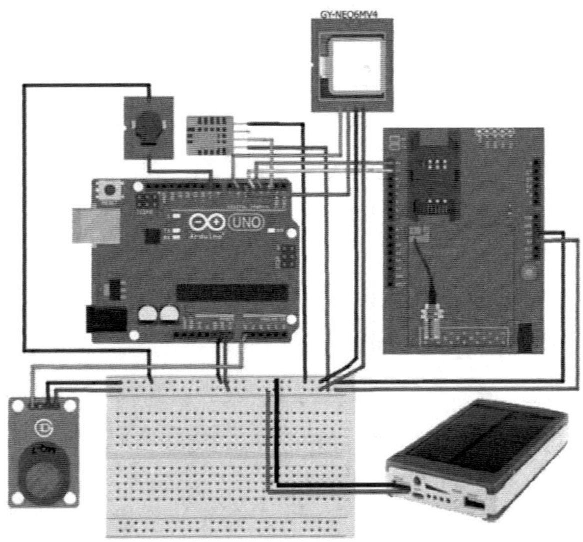

Figure 3. Overall circuit.

2.1. The working principle of the tool

Arduino Uno microcontroller is a brain control system work, where the microcontroller program application fire detection based on the detection of temperature conditions that must be able to accommodate the nature of the fire. Arduino Uno microcontroller must be able to detect any temperature rise caused by the existence of the fire. In addition, the system must also be able to read any smoke generated by the fire. At a temperature of> 35 C, the system will activate the DHT 11 and MQ 2 sensors that detect smoke> 50 ppm as a result of a fire fire, as input of the program input. The system will activate Buzzer as a warning in the form of alarm sound and then Global Positioning System (GPS) will provide information in the form of coordinate point of location of fire through SIMSI GSM Module in the form of Short Message Service (SMS) to the user.

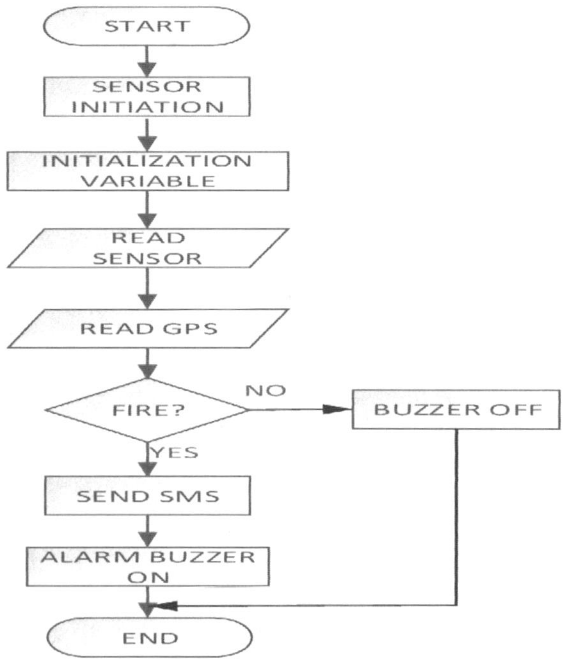

Figure 4. Flowchart series.

3. Results and discussion

Figure 5. Real mechanical design.

Figure 5 Shows the results of electronic design, software and mechanics that is the overall assembly of GPS-based Global Positioning System (GPS) and SMS Gateway. Electronic design includes all stages that are directly related to a system designed among them is to determine the properties and specifications of the tool, component selection, component installation and soldering. In designing software, input obtained from sensor then processed Arduino and output in the form of SMS which adjusted to program Arduino. In the mechanical design includes the selection of boxes to be occupied by a component, drilling, and the making of the box so that the desired tool.

3.1. Coordinate point notification via SMS when detected fire and smoke

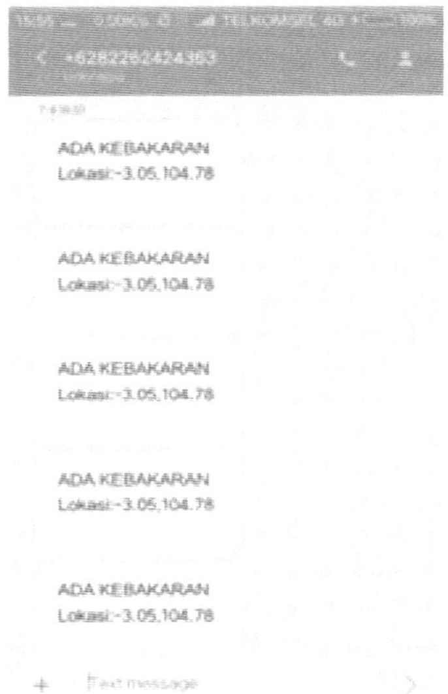

Figure 6. Fill message notification via SMS.

In figure 6 above is the message content of the coordinate point notification when a fire occurs. Coordinate point on the above notification is -3.05,104.78. Where the coordinate point consists of latitude, i.e. -3.05 and longitude, i.e. 104.78.

The first number is specified in the latitude coordinates between -90 and 90. While the first number is in longitude coordinates between -180 and 180.

To see the coordinates of the location is done by copying the coordinates in the notification, then open the Google Maps app and paste the coordinates in the search box at the top. Fig 7 is a coordinate location view on Google Maps.

Figure 7. GPS coordinate point display in Google Maps.

3.2. Display data on the Arduino serial monitor screen

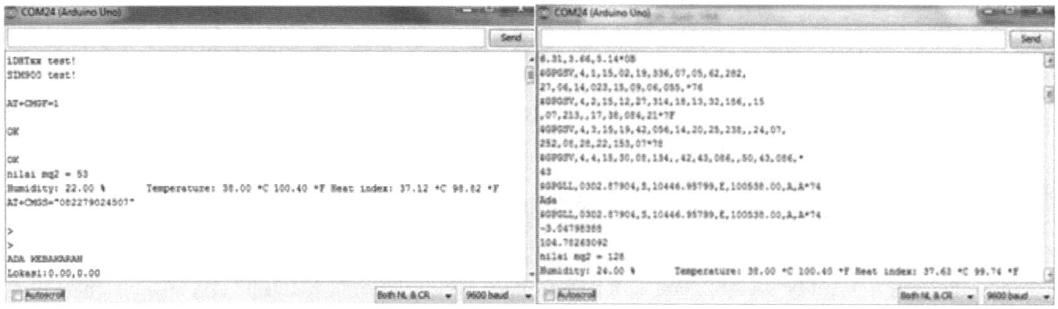

Figure 8. Display data on Arduino serial monitor screen.

Figure 8 on the monitor serial display above obtained results when tested on fire simulation using wax and anti-mosquito burn, that the value mq2 = 128 ppm and temperature value = 38 ° C and GPS data with latitude -3.04798388 and longitude coordinates 104.78263092. From the data it is seen that the mq2 value reaches> 50ppm and the temperature value reaches> 35 ° C, so the detector outputs buzzer sound and warning notification of coordinate point in the form of SMS containing the message "FIRE ADA" and with the coordinates of the location of the fire detected by GPS.

Table 1. Successful SMS submission.

Condition	SMS Timeframe	Successful
No Fire and Smoke Detected	No SMS	Successful
Detected Fire and Smoke	9.9 detik	Successful

4. Conclusion

This automatic fire detector works when the DHT11 sensor detects a temperature rise> 35 ° C and the MQ2 sensor detects smoke with a value> 50 ppm generated by the presence of fire. When the device detects a fire the buzzer will issue an alarm sound as a warning and the GSM SIM900A module will send SMS as a notification containing the coordinate points detected by the GPS. Preferably the use of this tool is placed open space so that GPS can accurately detect the point coordinates. For further development on the tool can be added application to monitor the circumstances around the tool with real time.

References

[1] One Data Indonesia 2014 *Data of Forest Fire Disaster 2011 – 2014* [online] retrieved from https://data.go.id/dataset/data-disaster-fire-forest-2/resource. Accessible date 23 March 2018

[2] Apryandi S 2013 *Design Build Fire Detector System via Mobile Based Microcontroller* (Pontianak: Tanjungpura University Final Report)

[3] Anonim 2014 *LDR (Light Dependent Resistor) Sensor* [online] retrieved from http:// basic electronics.web.id/ component /tranducer-sensor/-ldr-lightdependent-resistor/

[4] Widodo C E, Winarto and Sumariyah 2003 Making Fire Detector Equipment With Smoke Detector *Periodical physics* **6**(3) 51-54

[5] Anunggo R S 2012 *Temperature Monitoring System Based Sensor Temperature LM35 and GPS* (Yogyakarta: Vocational School UGM)

[6] Usuman I and Ardhi H 2010 Temperature And Smoke Detector System In Closed Space Utilizing Lm35 Sensor And AF30 Sensor *Periodical physics* **13**(2) B1-B6

[7] Andi 2009 *Global Positioning System* (Yogyakarta: Andi Publisher)

[8] Abidin H Z 2007 *"GPS Positioning"* (Bandung: Satellite Survey Lecture Module)

[9] Katankar K 2010 Short Message Service Using SMS Gateway *International Journal on Computer Science and Engineering* **2**(4)
[10] Habibi W 2011 Development of Tracking Systems And Tracking Device Software Based Global Positioning System (GPS) On Google's Software Platform
[11] Arduino 2016 *Arduino Uno Board* [online] retrieved from https://www.arduino.cc/en/Main/ArduinoBoardUno
[12] Prasetyo 2012 *DHT11 Temperature and Humidity Sensor* [online] retrieved from http://darysiupats.lt/dht11-temperature-and-humidity-sensor.html
[13] Adi 2012 *The temperature and humidity monitoring system is displayed on the LCD screen, utilizing Arduino as the brain of the system created and the DHT11 sensor as a temperature and moisture reader sensor* (Yogyakarta: UGM)
[14] SeedsStudio 2011 *Gas Sensor Datasheet And Smoke MQ-2* [online] retrieved from http://www.seeedstudio.com/depot/data sheet/MQ-2.pdf
[15] Boxall J 2014 *Arduino and SIM900 GSM Modules Tutorial* [online] retrieved from http://tronixstuff.com/2014/01/08/tutorial-arduino-and-sim900-gsm-modules/

Concrete mix design modelling based on variation of hidden layer and neuron of ANN for virtual learning development

S Santosa[1,*], S Suroso[1], M B Utomo[1], M Martono[1], M Mawardi[1] and Y P Santosa[2]

[1] Civil Engineering, Politeknik Negeri Semarang, Semarang, Indonesia
[2] Institute of Informatik Universität Bonn, Bonn, Germany

*stefanus.santosa@polines.ac.id

Abstract. Artificial Neural Network (ANN) is a Machine Learning (ML) algorithm which learn by itself and organize its thinking to solve problems. Although the learning process involves many hidden layers (Deep Learning) this algorithm still has weaknesses when faced with a high noise data. Concrete mixture design data has a high enough noise caused by many unidentified / unmeasurable aspects. Information needs about the compressive strength of early age concrete (under 28 days) are often needed while the construction process is still ongoing. H2O's Deep Learning type of ANN has been tried to predict the compressive strength of early age concrete, but the results are less than optimal. This study aims to improve the H2O's-ANN prediction model using Bagging to reduce the influence of noise and overfitting. The lowest RMSE that are able to be achieved in this research with Bagging is 6.385 while it is 6,674 without Bagging. This result proves that Bagging is significant to reduce the Deep Learning error rate for predicting the compressive strength of early age concrete. Future work this model, as a new innovation in machine learning in civil engineering vocational education subject, can be used to various concrete mixture design and learning about concrete. Because this concrete mix design model is digital, it can be used for virtual learning. Currently the author is conducting research to create a virtual concrete mix design learning model.

1. Introduction

Concrete is a mixture of cement, water and aggregates (coarse and fine), which mixed by some formula [1,2]. This mixture can be mixed with various chemical components and mineral admixtures to improve the quality of the concrete, the self-compacting [2-4], the ductility [5], and the durability.

Concrete compressive strength is measured on 28 days, but to control the quality of the concrete, the early measurements is needed [6]. Prediction of the compressive strength of concrete at the age of under 28 days is required when concrete is still under construction. Construction failures often occur because of overloading. This is because often the construction manager sets the loading without proper calculation due to the lack of information of the strength of the early age concrete. If the information about the strength of concrete at the age of under 28 days is easily obtainable, the construction manager can calculate the safety barrier of the load.

Variety of the mixture and the characteristic of the concrete produce many effects [7], for instance the same ingredients can produce different strength when the mixture design is different [3,8], including different self-compacting, ductility, durability [9], delivery, and curing. This shows that concrete design problem is very complex, thus many complex mixture designs were proposed. As a result, concrete compressive strength prediction becomes inaccurate.

IOP Conf. Series: Materials Science and Engineering 1108 (2021) 012024 doi:10.1088/1757-899X/1108/1/012024

In the recent development, several researchers used Computational Engineering to design concrete mixture, either for slump or compressive strength, by using linear regression, multiple linear regression, and non-linear regression [10-12]. Despite that, the results often not satisfying, conventional statistical approaches are often not a choice anymore.

Evolutionary model and deep learning also be used to improve the results. Nikoo created The Model of Prediction of Concrete Compressive Strength by Evolutionary Artificial Neural Networks which used Genetic Algorithm (GA) to find the optimal weight parameters for ANN [13]. Deng et al. used deep learning Convolution Neural Network (CNN) to recycled concrete compressive strength prediction. The results show that CNN has lower error than backpropagation neural network (BPNN) and support vector machine (SVM) [14]. Even though this is the case, CNN is used only when the positional information of a certain feature in the data is important such as in image processing or computer vision case rather than concrete mixture which the data is transactional.

Several other types of DNNs are popular as well, such as Multilayer Perceptron (MLP) and Recurrent Neural Networks (RNNs). However, for transactional data (tabular data), MLP works better than CNN or RNN. While CNN relies on parameter sharing for calculating grid or continuous signal data and RNN are strong in sequential data. For the purpose of this research, MLP implementation of H2O's deep learning will be used. The computational model obtained will be developed for virtual learning specifically about concrete mix designs

2. Research methods

This research will use H2O Deep Learning library to create a model. ReLU is chosen to be the activation function for the model and will be tested later in regards to the performance. The research method is described in Figure 1.

2.1. Data collection

A secondary data was collected from UCI Data Repository created by I-Cheng Yeh, Department of Information Management, Chung-Hua University (Republic of China). The data consists of concrete mix design with its concrete compressive strength on different age. These concrete mix designs were tested with destructive test to record the concrete compressive strength values. The composition of the mix, which is cement, blast furnace slag, fly ash, water, superplasticizer, coarse aggregate, fine aggregate, and age (3, 7, 14, 28 days), were varied to obtain various results.

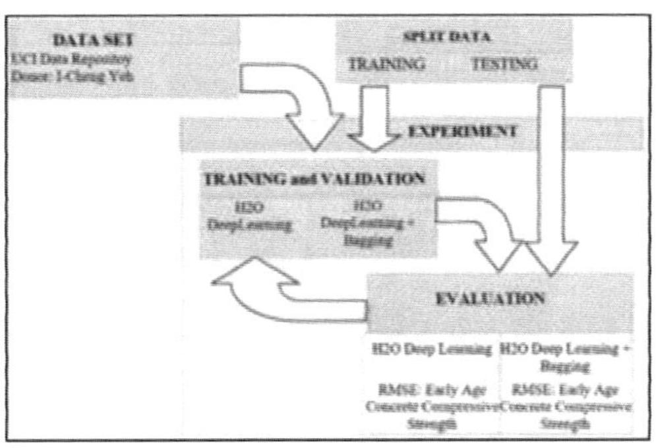

Figure 1. Research method.

2.2. Experiment

Before the experiment process were done, the data was split into training and test data. This training data then later on will be split again into train set and validation set by using cross validation by splitting the data into k subset of data. k-1 subset was used for training and the remaining subset was used for validation. For numerical stability, the data then normalized based on the training data. The data was split into 0.6 and 0.4 ratio for training and test data respectively.

Firstly, several parameters were defined while weights and biases are randomly generated. The parameters are number of hidden layers, number of hidden neurons, epoch, L1, L2, epsilon, and rho. Secondly, the model was trained with the train set to calculate the prediction. Then, this prediction will be used to calculate the loss by using root mean-squared error (RMSE). Afterward, the loss is back propagated to optimize the bias and weight. To check whether overfitting is occurred or not, the validation set is used to test the current parameters.

To improve the model over noisy data, Bagging was also used. The model was replicated b times and each model was trained using a randomly selected data with replacement. This stage is also called bootstrapping. Afterwards, the result of each model was aggregated.

2.3. Evaluation

After the training process was done, the test data was used to measure the accuracy of the model. In this step, the effect of Bagging was tested by training two model with and without Bagging then the test accuracies were compared.

3. Results and discussion

Before trying to find the best parameters for the prediction model, the data is split into 60/40 for training and testing respectively. Afterwards, the parameters are set to be the default value from H2O library. The summary of the parameters is described in Table 1.

Table 1. Parameters.

Parameter	Value
Epoch	20
L1	$1.0\varepsilon^{-5}$
L2	0
epsilon (ε)	$1.0\varepsilon^{-8}$
rho (ρ)	0.99

Firstly, the effect of the ReLU activation function over several architecture settings based on H2O's deep learning ANN architecture and H2O's deep learning ANN with Bagging architecture will be tested. The name of this architecture will be written as ANN and the others ANN+Bagging. The number of hidden layers' parameters are 2, 4, 6, 9, 12, and 15 with number of neurons for each layer 50. In order to pick the number of hidden layers, the model was tested with 50 neurons and then the hidden layer numbers were tested individually then compared. The results are shown in Figure 2 and Figure 3 below.

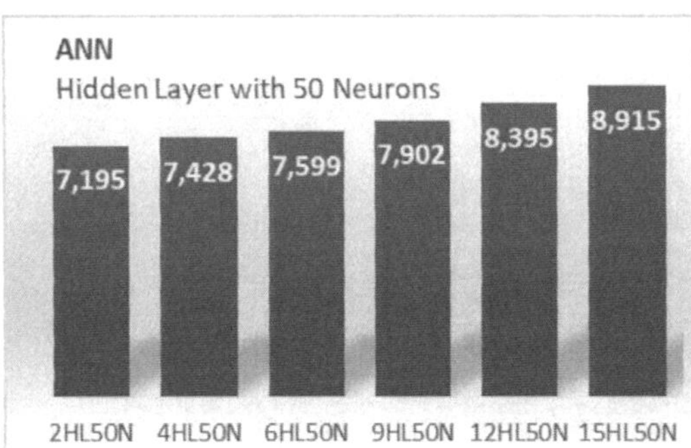

Figure 2. ANN hidden layer test results (lower is better).

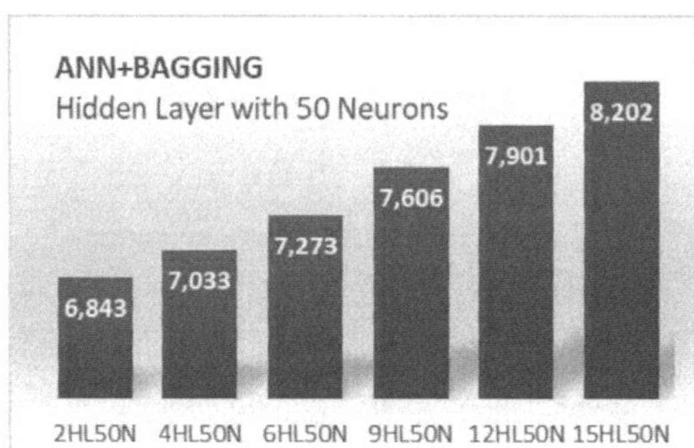

Figure 3. ANN+Bagging hidden layer test results (lower is better).

From the results, the lowest RMSE achieved by 2 hidden layers on ANN architecture with or without Bagging. The testing RMSE are 7,195 using ANN and 6.843 using ANN architecture with Bagging. The worst performance for both architectures is when the hidden layer is 15. This shows that deep architecture does not always guarantee lower error.

The second experiment is to test the optimal number of neurons in the hidden layer. Following the previous results, the number of hidden layers for ReLU activation function is 2. Then, the hidden neurons are 10, 30, 50, 80, and 100. The results are shown in Figure 4 for ANN architecture without Bagging and Figure 5 ANN architecture with Bagging.

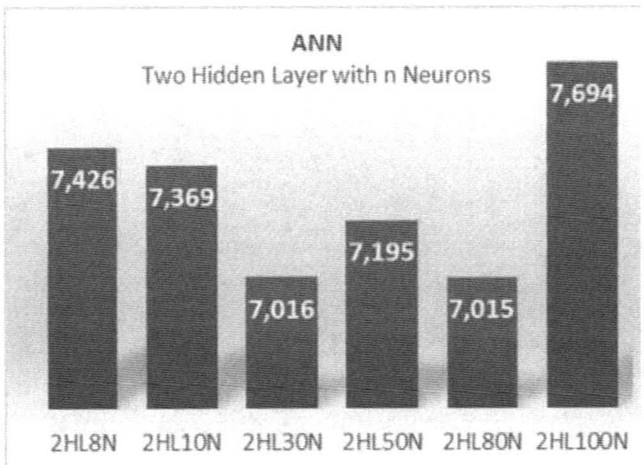

Figure 4. ANN Hidden neuron test results (lower is better).

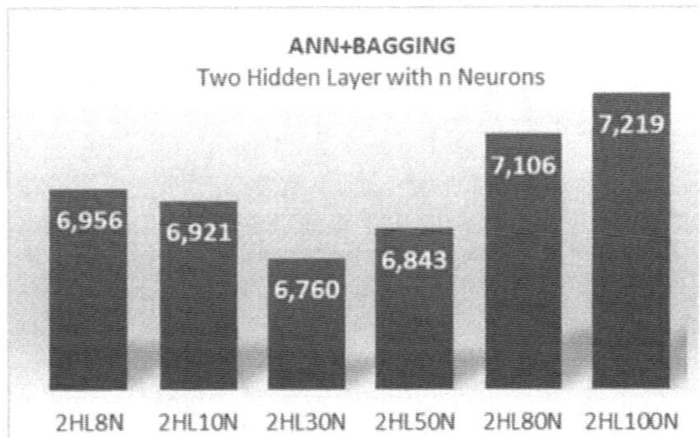

Figure 5. ANN+Bagging Hidden neuron test results (lower is better).

Architectural design experiments based on different numbers of neurons have different error patterns than previous experiments. Architecture with too little or too many neurons shows a high error rate. The lowest RMSE by ANN architecture without Bagging is 7,015 for the number of neurons 30 and 80 (Figure 4). The architecture of ANN with Bagging has the lowest error level for the number of neurons = 30, with RMSE 6,760 (Figure 5). Therefore, the experiment is continued by designing an architecture based on 30 neurons with the number of hidden layers be varied. The experiment was carried out using 2, 4, 6, 9, 12, and 15 hidden layers.

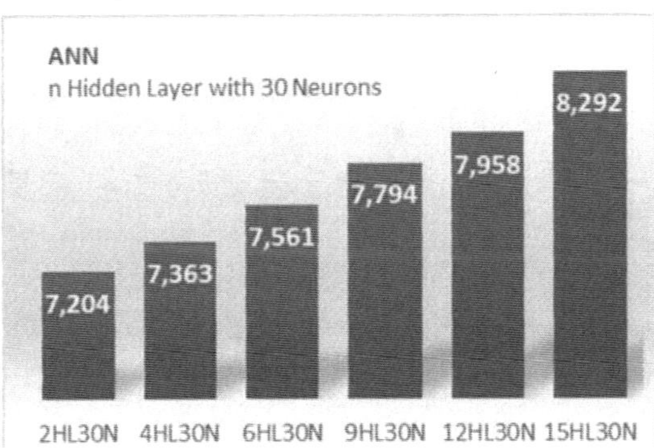

Figure 6. ANN hidden layer with 30 neurons test results (lower is better).

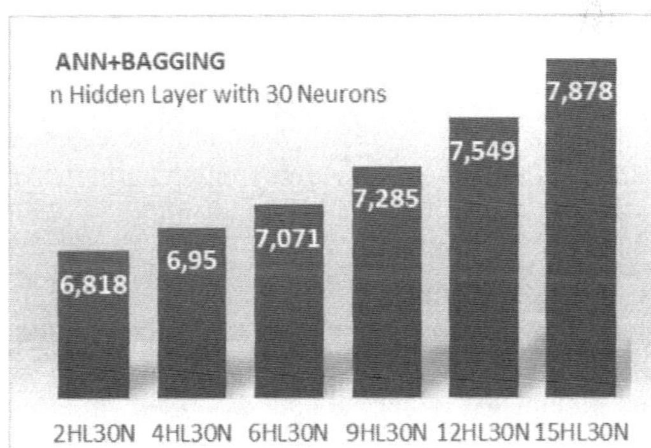

Figure 7. ANN+Bagging hidden layer with 30 neurons test results (lower is better).

The third experiment shows the results of RMSE has regular patterns. The lowest RMSE by ANN architecture without Bagging is 7,204 (Figure 6) and ANN with Bagging has the lowest error level RMSE 6,818 (Figure 7). Both experiments show that the best architecture is indicated by 2 hidden layers and 30 neurons. These results are achieved after both architectures have been trained and validated.

The next experiment is to test both architectures with data that has not been used for training and validation before. Testing uses 40% of the overall data. The test results are shown in Figure 8 and Figure 9.

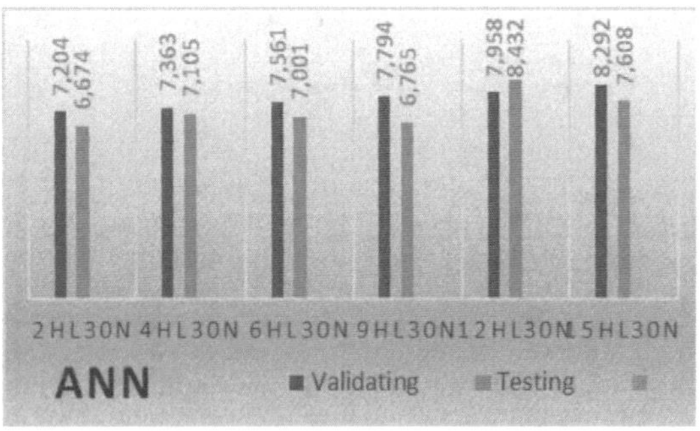

Figure 8. ANN Test Result (lower is better).

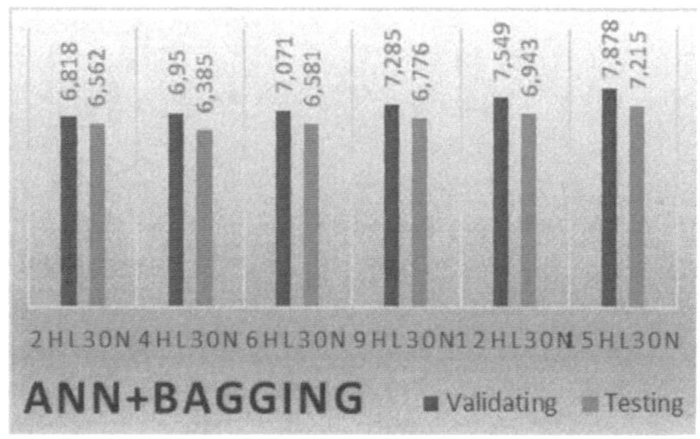

Figure 9. ANN+Bagging Test Result (lower is better).

The testing results of ANN architecture without Bagging and ANN architecture with Bagging show that both architectures have a good performance based on validation test result. The level of error shows that the test results are lower than the validation test. Figure 8 shows that ANN has lower test results than validation tests except for architectures with 12 hidden layers 30 neurons. The best architecture is achieved by 2 hidden layers 30 neurons arrangement with RMSE 6,674.

As shown in Figure 9 none of the test results ANN with Bagging has same or higher RMSE than the validation test. The highest error occurs in 12 hidden layers and 30 neurons architecture with RMSE 7,215. This proves that the depth of learning in the deep learning architecture as indicated by the number of hidden layers does not always guarantee better results. The lowest RMSE of ANN with Bagging architecture is 6,385 with 4 hidden layers 30 neurons arrangement.

As part of research on intelligent concrete mix design models, this paper aims to prove the role of Bagging for improving ANN performance, especially H2O's deep learning. Several architectural design experiments have been done by varying the number of hidden layers 2, 4, 6, 9, 12, 15 and the number of neurons 8, 10, 30, 50, 80, 100 with the ReLU activation function for data processing. The data was split into 0.6 and 0.4 ratio for training and testing data respectively. This training data was split again into train set and validation set using Cross Validation by splitting the data into k subset of data. k-1 subset was used for training and the remaining subset was used for validation. For numerical stability, the data then normalized based on the training data. From various experiments that have been done shows that an error rate test on architecture settings based on H2O's deep learning ANN and H2O's deep learning ANN with Bagging are as shown in Figure 10 below.

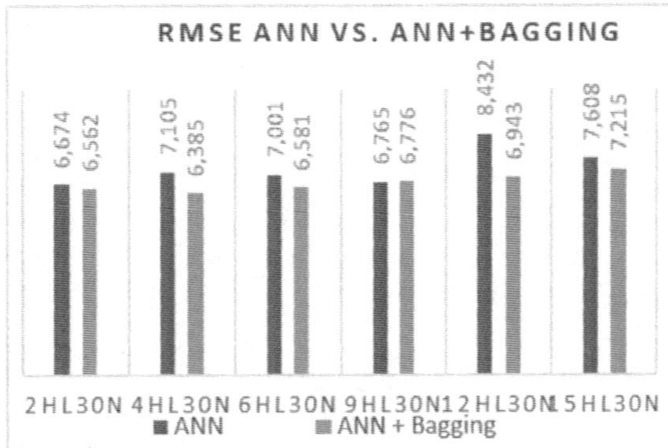

Figure 10. RMSE H2O's deep learning ANN vs. H2O's deep learning ANN with Bagging.

The experimental results show that 6 architectures based on H2O's deep learning ANN have higher errors than the 6 architectures based on H2O's deep learning ANN with Bagging. The best architecture is achieved by 4 hidden layers 30 neurons ANN with Bagging with RMSE value of 6,385 versus 6,674 by 2 hidden layers 30 neurons without Bagging. These results have proven the hypothesis of this study that Bagging has been able to reduce the error rate of the H2O's deep learning ANN algorithm to make predictions on a concrete mix design model.

Bagging is used due to the possibility of the noisy data. In this study Bagging has the ability to solve problems caused by concrete mix design data noise. Bagging is short for bootstrap aggregating which its purpose is to increase the stability and avoid overfitting by reducing variance. Bagging has the ability to solve problems caused by data noise. The role of Bagging has been proven in all architectures with lower RMSE values than without Bagging.

4. Conclusion

Based on various experiments that have been carried out, it can be concluded that:

- Artificial Neural Network (ANN), especially H2O's deep learning for predicting early age concrete mix design accuracy can be improved by using Bagging. As shown in the results above that the RMSE is lower when Bagging is used in the training process.
- The lowest RMSE that are able to be achieved in this research with Bagging is 6,385 achieved by 4 hidden layers 30 neurons architecture while it is 6,674 achieved by 2 hidden layers 30 neurons architecture without Bagging.
- The deepness of an Artificial Neural Network architecture not always guarantee optimum results. The model with 4 hidden layers is able to perform better than the model with higher number of hidden layers.

5. Future works

This model, as a new innovation in machine learning in civil engineering vocational education subject, can be used to various concrete mixture design and learning about concrete. The various composition of mixture design can be handled with care because this model learns from unseen condition.

Because this concrete mix design model is digital, it can be used for virtual learning. Virtual learning is very important to provide opportunities for students to explore without involving high costs due to material needs. Currently the author is conducting research to create a virtual concrete mix design learning model.

Acknowledgment

The authors would like to thank the Politeknik Negeri Semarang (Grant No.: SP DIPA-042.0.1.2.400997 / 2019) for the financial support.

References

[1] Chopra P, Sharma R and Kumar M 2016 Prediction of Compressive Strength of Concrete Using Artificial Neural Network and Genetic Programming *Advances in Materials Science and Engineering* 1–10

[2] Tangchirapat W, Rattanashotinunt C, Buranasing R and Jaturapitakkul C 2013 Influence of Fly Ash on Slump Loss and Strength of Concrete Fully Incorporating Recycled Concrete Aggregates *Journal of Materials in Civil Engineering* **25**(2) 243–251

[3] Najimi M, Sobhani J and Pourkhorshidi A R 2012 A comprehensive study on no-slump concrete: From laboratory towards manufactory *Construction and Building Materials* **30** 529–536

[4] Hüsken G and Brouwers H J H 2012 On the early-age behavior of zero-slump concrete *Cement and Concrete Research* **42**(3) 501–510

[5] Wendne R, Vorel J, Smith J, Hoover C G, Bažant Z P and Cusatis G 2015 Characterization of concrete failure behavior: a comprehensive experimental database for the calibration and validation of concrete models *Mater Struct* **48**(11) 3603–3626

[6] Ray N, Dedi P and Rizky D 2016 Correlation Coefficient Study of High Quality Concrete Compressive Strength Based on Age & Shape of SNI 03- 2847-2002 Standard Test Objects (Studi Angka Koefisien Korelasi Kuat Tekan Beton Mutu Tinggi Berdasarkan Umur & Bentuk Benda Uji Standar SNI 03-2847-2002) **1**(1)

[7] Kurt M, Kotan T, Gül M S, Gül R and Aydin A C 2016 The Effect of Blast Furnace Slag on the Self-compactability of Pumice Aggregate Lightweight Concrete *Sadhana* **41**(2) 253–264

[8] Riahi S and Nazari A 2011 Compressive strength and abrasion resistance of concrete containing SiO2 and CuO nanoparticles in different curing media *Sci. China Technol. Sci.* **54**(9) 2349–2357

[9] Hui C and Zhenyu L 2013 Research on the Experiment of Self-Leveling Concrete with Large Content of Fly Ash in *2013 Fifth International Conference on Measuring Technology and Mechatronics Automation* 934–936

[10] Jain A, Jha S and Misra S 2008 Modeling and Analysis of Concrete Slump Using Artificial Neural Networks *Journal of Materials in Civil Engineering - J MATER CIVIL ENG* **20**

[11] Yeh I C 2007 Modeling slump flow of concrete using second-order regressions and artificial neural networks *Cement and Concrete Composites* **29**(6) 474–480

[12] Yeh I C, Lien C, Peng C H and Lien L C 2010 Modeling concrete strength using genetic operation trees *2010 International Conference on Machine Learning and Cybernetics* **3** 1572–1576

[13] Nikoo M, Sadowski L and Torabian F 2015 Prediction of Concrete Compressive Strength by Evolutionary Artificial Neural Networks *Advances in Materials Science and Engineering*

[14] Deng F, He Y, Zhou S, Yu Y, Cheng H and Wu X 2018 Compressive strength prediction of recycled concrete based on deep learning *Construction and Building Materials* **175** 562–569

QoS analysis of WSN (Wireless Sensor Network) using node MCU and accelerometer sensors on bridge monitoring systems

D Junesco[1,*], E Supriyanto[2], A Hasan[2] and M Mukhlisin[3]

[1] Faculty of School of Graduate Studies, Management and Science University, Shah Alam, Malaysia
[2] Department of Electrical Engineering, Politeknik Negeri Semarang, Semarang, Indonesia
[3] Department of Civil Engineering, Politeknik Negeri Semarang, Semarang, Indonesia

*devanjunesco@gmail.com

Abstract. Based on the planning guideline of Health Monitoring System of Bridge Structure of Kementrian Pekerjaan Umum dan Perumahan Rakyat, one of the main components needed in bridge health monitoring system is a communication system. The application of the WSN (Wireless Sensor Network) function to perform monitoring can be used to create a more practical system. Based on the system, required hardware that can be integrated with the software and can connect on wireless. Proposed methods using WSN (Wireless Sensor Network) on devices, then do testing on connectivity to obtain network quality and data sensor. Data sensor will be uploaded on the cloud can be accessed from various places that connected to the Internet network thus forming an IoT (Internet of Things) based system with using Message Queue Telemetry Transport (MQTT) protocol. Experiments carried out in two ways of connectivity, point-to-point and star topology to determine the network quality of each device, which is Node MCU and Mi-Fi modems. Network quality parameters can be known by looking at the value of packet loss, delay, jitter, and throughput. In the results, the MQTT Protocol applied to build IoT system. In addition, proposed methods that used in this research has a great quality network with packet loss <25 ms. However, there's a big gap between serial data and online data that makes MQTT protocol is not suitable for monitoring systems that require fast data transfer.

1. Introduction

Bridges play a vital part in the economy of the country, as they facilitate the movement of goods and people around. These structures are subject to deterioration over time due to the continuous use and exposure to deicing salts, vibration, humidity, temperature variations, etc. The closure of a bridge due to extensive damage or failure may affect millions of people monetarily, or even cause loss of life in some cases. Hence it is important to evaluate the condition of the structure in order to avoid any failure and plan maintenance actions without disrupting its operations [1]. Indonesia as an archipelago has a unique topography in the form of lands separated by waters and rivers. This condition makes Indonesia need bridges as one of the important infrastructures in supporting the economy in Indonesia [2]. Maintaining the safety and reliable service of a large bridge over its relatively long life requires obtaining continuous and reliable data regarding its structure, including the damage caused by the temperature

IOP Conf. Series: Materials Science and Engineering 1108 (2021) 012025 doi:10.1088/1757-899X/1108/1/012025

gradient, cracking, fatigue, corrosion, of structures and the decrease of load capacity of the bridge, etc. which all should be carefully evaluated [3]. Technology development is very fast and has been widely applied in various fields. In the field of telecommunications, for example, the speed of data access and information is now a mandatory requirement for every society of the country. Development of infrastructure development and public facilities have also started to utilize telecommunication technology, one of them in the monitoring system of infrastructure development. As mentioned in [4], modern health monitoring systems usually consist of six general components:

- Sensor systems and data acquisition;
- Communication system;
- Data processing and analysis system;
- Data storage;
- Diagnostics (e.g. damage detection and algorithm models);
- Retrieval of required information.

To overcome this, the Internet of Things (IoT) can be used, which would provide flexibility to monitor structures (building, bridge) from anywhere [5]. The IoT brings new opportunities for our society. Yet, most of the existing Structural Health Monitoring (SHM) systems are not connected to the IoT. SHM is a system to monitor the structural health of buildings and bridges [6]. With the advancement of modern technology, structures like buildings, bridges, etc. are getting structurally complicated, and safety has become an issue. Structural Health Monitoring (SHM) with the Internet of Things (IoT) can help improving security and safety [7]. In Bridge-based Wireless Monitoring Systems usually use WSN (Wireless Sensor Network) where WSN is a network consisting of node sensors equipped with equipment such as memory, processor, ADC, and batteries. The small form of WSN has the disadvantage of limited energy and bandwidth so that the right topology is needed [8].

Based on the planning guideline of Health Monitoring System of Bridge Structure of Kementrian Pekerjaan Umum dan Perumahan Rakyat Republik Indonesia, this research focus on one of the main components needed for health monitoring system of the bridge structure, which is communication system. It is important to build a communication system that connected between sensors nodes are to be able to detect damage at each point of laying node sensor. Proposed method and devices are used to find the Quality of Service of the device itself to build WSN. Network analysis using QoS (Quality of Service) is defined as a mechanism or way that allows services to operate according to their own characteristics in the IP (Internet Protocol) network. Packet loss, delay, jitter, and throughput will be tested on this research as a measurement parameter of QoS. Data analysis that performed, will be compared with IEEE802.11b/g/n which is a standard provision of wireless networks. The network quality parameters generated from the node sensor network under test will be analyzed using QoS standards based on TIPHON.

2. Related works

In Satria et al. [9], experiments have been conducted using the MQTT protocol on Smart building. In the experiment parameter that used were delay, overhead, and packet loss. The experiment was conducted to find out Quality of Service (QoS) from MQTT. The virtual sensor was used to build a WSN based system. The experiment was done with change in the number of sensor used from 10, 100, and 1000. All of the data produce on the experiment was totally good with 0 packet loss, average overhead under 1 byte, and average delay under 1 second. Furthermore, on Wulandari [10] Quality of service analysis on internet network has been carried out at the UPT Loka Uji Teknik Pertambangan Jampang Kulon on the indoor area and provide a good result without any packet loss.

The great potential of Wireless Sensor Network is being seen in industrial, consumer and commercial application. The wireless technology is becoming one of the most prominent areas of research [11]. MQTT (Message Queue Telemetry Transport), an open protocol designed by IBM, was originally intended for unreliable networks with restricted resources such as low bandwidth and high-latency. MQTT consists of one broker server and two kinds of clients that are called Publisher (Publish client) and Subscriber (Subscribe client). Broker server acts as an intermediary for messages sent between Publish client and Subscribe client for the interesting topic. When the Publish client issues a topic and sends a message to the Broker server, the Subscribe client selects the topics which it finds interesting [12].

3. Proposed methods

Based on the related works that have been described in this research, microcontroller that integrated with sensor was made, a system with accelerometer sensor and node MCU ESP8266 or called as node sensor. Node sensors are set to connect wirelessly and form a wireless sensor network. Furthermore, node sensors are set to connect with internet to upload the data on cloud which is form an Internet of Things (IoT) system using MQTT protocol. In this research, the experiment carried out on bridge to test network quality that produce from node sensor. The draft and general overview of the bridge monitoring system is shown on Figure 1.

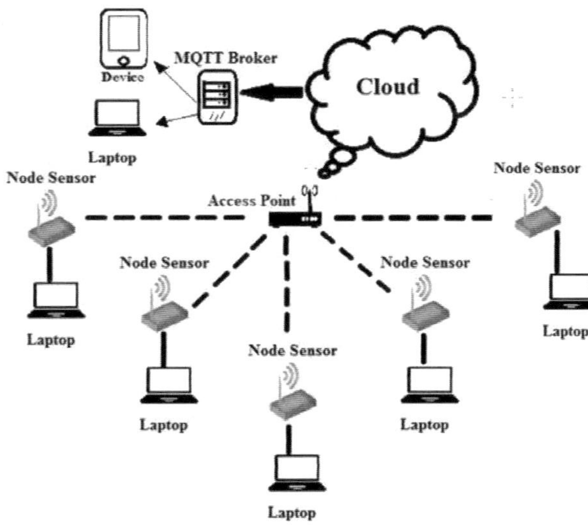

Figure 1. Design of SHM system.

3.1. Hardware

This system works on several nodes connected to the sensor access point. At each node sensor, it consists of Node MCU that connected to the accelerometer sensor and forms the M2M (Machine to Machine) component. The accelerometer sensor is used as sensor data generator which can indicates that on the network that have been made using Node MCU and access point there is transmitted data. The data sensor that generated by the accelerometer sensor will be processed and transmitted via Node MCU which connected to the access point which will connect directly to the internet network. The data transmitted will directly uploaded to the cloud internet using MQTT protocol. Each node sensor placed at different places to generate different quality of network. The detail of node sensor can be seen on Figure 2.

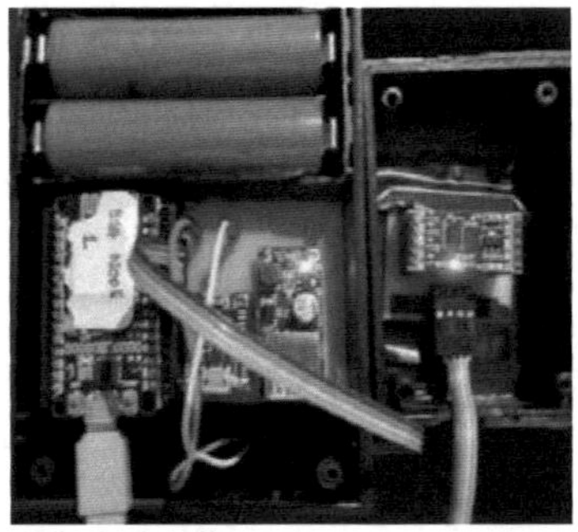

Figure 2. Hardware of node sensor.

The node sensors that are formed have several units: the microcontroller unit, the transceiver unit, the battery unit, and the sensor unit. This research uses Node MCU ESP8266 as microcontroller and also receiver device on node sensor. Node MCU ESP8266 already has a wireless module with Wi-Fi 802.11b/g/n, so it can directly transmit and receive data wirelessly. Node MCU ESP 8266 works at 32-bit resolution. In this paper, structural vibrations are sampled by accelerometer sensors. Accelerometer sensor used in this research is ADXL 345 which is has affordable price. In addition to low-cost but also has the low-power output of 3 axes X, Y, and Z with measuring range ± 16 g. The Node MCU and accelerometer work on a 3.3 volt. The sensor output of ADXL 345 is digital data. The minimum resolution that can be read by the accelerometer sensor is 4mg / LSB (Least Significant Bit).

3.2. Software

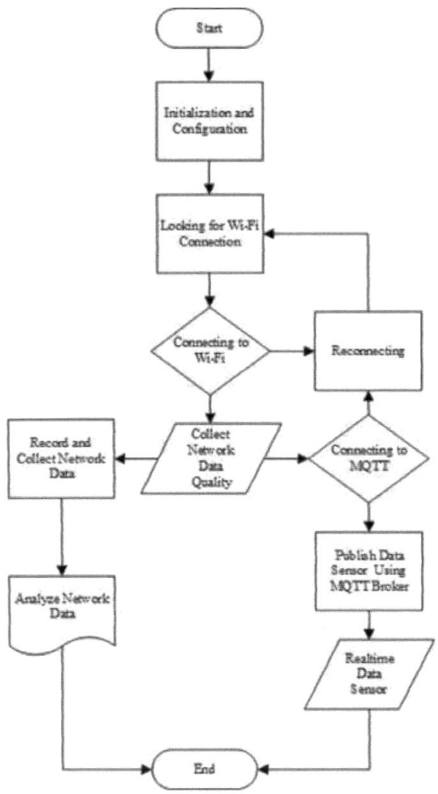

Figure 3. Flow diagram of the proposed system.

3.3. QoS (Quality of Service)
Quality of Service referred to in this research is network quality parameter that becomes indicator which shows that node sensor has good or bad performance in producing quality network with a certain distance in the bridge monitoring system applications.

3.3.1. Packet loss

$$PL = \frac{(packet\ data\ sent - packet\ data\ received) x\ 100\%}{Packet\ data\ sent} \tag{1}$$

Packet Loss is a parameter that describes a condition that shows the total number of packets lost. It can occur due to collision and congestion on the network and this affects all applications because retransmission will reduce the overall network efficiency even though the amount of bandwidth is sufficient for the application.

3.3.2. Delay/ Latency

$$\text{Average Latency} = \frac{total\ time\ to\ send\ packet}{total\ packet\ received} \tag{2}$$

Delay/latency is defined as the length of time required by the data packet to arrive at the destination. Latency is the time it takes data to travel from distance to destination. Latency can be affected by distance, physical media or long processing time.

3.3.3. Jitter

$$Jitter = \frac{Total\ Varian\ Delay}{total\ packet\ received} \tag{3}$$

Jitter is a variation of Delay, where there is a difference of delay in the packets sent in the same data flow stream. Jitter is commonly called delay variation, closely related to latency, indicating the amount of delay variation in data transmission on the network. Delay queue on the access point can cause jitter.

3.3.4. Throughput

$$Throughput = \frac{Packet\ data\ receivet}{Total\ time\ to\ send\ pocket} \tag{4}$$

Throughput is the number of successful packets sent at certain time intervals. Measurement at throughput is by dividing the speed (rate) of effective data transfer with a certain time interval (s) and measured in units of bps (bits per second) and mbps (megabits per second).

This research purpose to determine the performance of Node MCU as microcontroller and transceiver at node sensor and also to know accelerometer acceleration sensor data can detect parameter change on bridge by using serial monitor and by using MQTT protocol. Testing is divided into two methods with point-to-point connectivity and with star topology with Line of Sight (LoS) condition. With applying star topology, expected to show network quality data generated when there are collisions and congestion of data that indicate by looking packet loss, delay, and jitter quality.

4. Results and data analysis

4.1. Using point-to-point connectivity
In the experiment, point-to-point network quality obtained by adjusting the distance at the node sensor and access point as far as 10 m, 20 m, 30 m, up to 100 meters on the overpass bridge which is a hollow concrete bridge type. In point-to-point testing, data retrieval is done with 2 different devices, using Node MCU ESP8266 as transmitter and Mi-Fi modem. The testing repeated 5 times to get valid result. On point-to-point connections require only one node sensor and access point only. Overpass bridge can be seen on Figure 4.

IOP Conf. Series: Materials Science and Engineering 1108 (2021) 012025 doi:10.1088/1757-899X/1108/1/012025

Figure 4. Overpass bridge.

Here The result of the test that have been done by applying point-to-point connectivity, the data obtained can be seen on Table 1. Which is the result from using Node MCU ESP8266 device as access point, and Table 2. which is the result from using Mi-Fi device as access point. As mentioned before, the testing is done with Line of Sight.

Based on point-to-point connection using Node MCU ESP8266 as access point, at the farthest distance of testing the result of network quality that is not too bad. Based on TIPHON index, packet loss with percentage less than 25% included on average quality of network. Furthermore, the result of delay, jitter, and throughput is very good at all. But, there's anomalies when tested at a distance of 60 meters to 70 meters. When testing at a distance of 60 meters and 70 meters, sometimes surge of packet loss occurs. The surge sometimes effecting the result of throughput and sometimes it works normally. The index for all network quality test system category are given as in appendix.

Overall, based on point-to-point connection using Mi-Fi as access point, the quality of network produced is better than using Node MCU ESP8266. The anomalies that occur on Node MCU test, also happened in testing using Mi-Fi but need to be considered on packet loss testing. When on 80 meters the quality of packet loss was increased, but the numbers of delay, jitter, and throughput constantly decreasing. From the result, it can be analyzed that Mi-Fi have the ability to reconnect quickly and the quality of the throughput will continue to constantly decreased. That affected by the distance that causes the delivery and receipt time to take longer time.

Table 1. The result of point-to-point connection (Node MCU ESP8266).

Length (meters)	Packet Loss (%)	Delay (ms)	Jitter (ms)	Throughput (mbps)
10	0	1.188	0,008	4.843
20	0	1.92	0,015	5.632
30	0	2.006	0,019	4.005
40	0.4	2.704	0,200	4.603
50	2.8	3.454	0,577	1.97
60	6.8	4.893	0,757	1.8539
70	1.6	3.396	0,411	1.3801
80	2.4	3.363	0,304	1.5
90	12.4	4.033	0,713	1.5064
100	22	7.640	2,321	0.3986

Table 2. The result of point-to-point connection (Mi-Fi).

Length (meters)	Packet Loss (%)	Delay (ms)	Jitter (ms)	Throughput (mbps)
10	0	3.078	0.0054	10.381
20	0	3.742	0.0121	8.606
30	0	4.03	0.0621	6.694
40	0	4.384	0.0270	6.0278
50	2	4.289	0.0212	5.691
60	3.8	4.925	0.0474	2.737
70	4	4.879	0.0445	1.826
80	3	5.109	0.0905	1.2612
90	4	7.180	0.5036	0.6541
100	17.8	9.788	0.9577	0.4753

Based on the test data of connections that have been done, on each QoS parameters generated from both devices can be seen in Figure 5 until Figure 8.

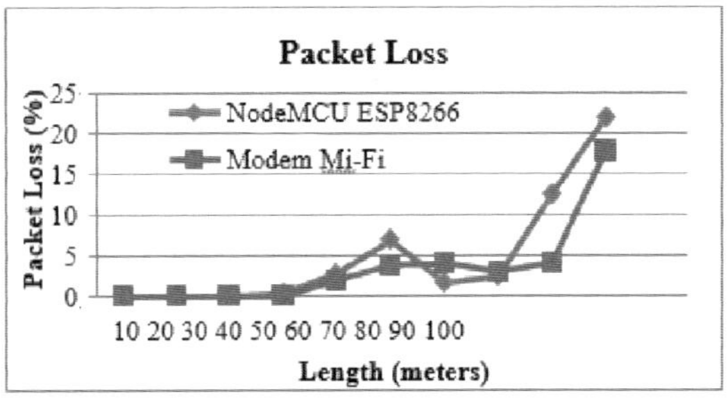

Figure 5. Packet loss chart.

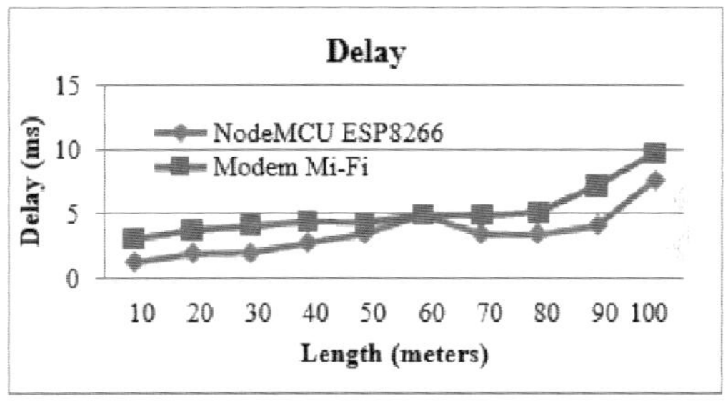

Figure 6. Delay chart.

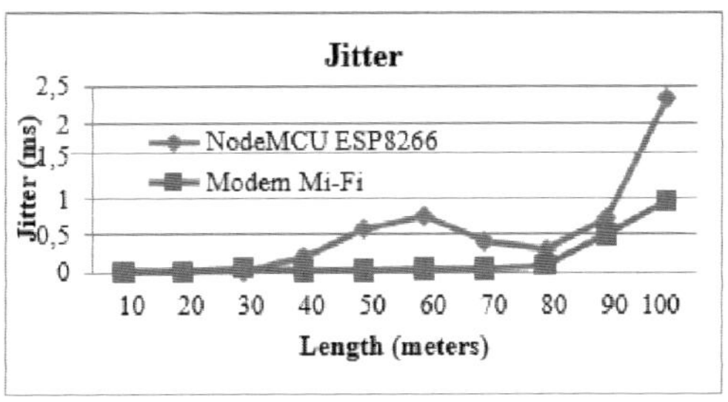

Figure 7. Jitter chart.

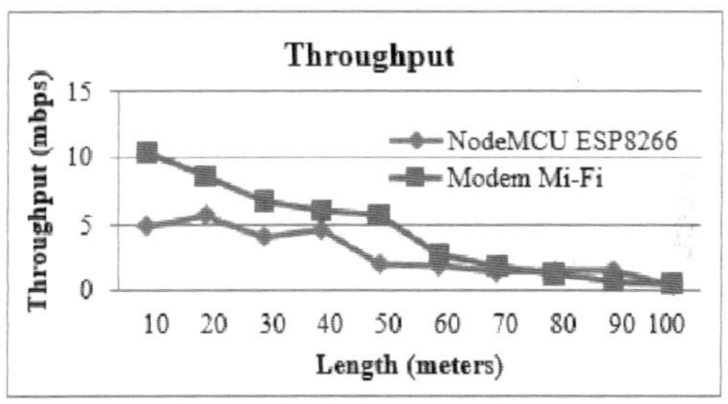

Figure 8. Throughput chart.

The bandwidth value of Node MCU ESP8266 conforms to the IEEE802.11b standard which has a maximum bandwidth of 11mbps and 6 mbps same as the program settings and at the time of practice. However, the anomaly occurs when the node sensor is 60 meters away, bridge connection and a bridge buffer. In this assumption, there's absorption on the signal that caused anomalies on overall data generated, based on the TIPHON standard, the resulting network quality is quite good because the overall packet loss result from both devices value is less than 25%.

4.2. Using star topology

In the test using a star topology, all of five node sensors are connected with one access point. Sensor nodes are placed in each different place, N1 at a distance 20 meters, N2 at a distance 40 meters, N3 at a distance 60 meters, N4 at a distance 80 meters, and N5 at a distance 100 meters. When applying star topology Mi-Fi is only the device that used for access point. That caused by Node MCU ESP8266 can't handle and serve more than 4 node sensors at the same time. Distribution of IP Address from each node sensors can be seen on Table 3.

Table 3. Address on each node sensor.

Node Sensor	IP Address
N1	192.168.1.102
N2	192.168.1.104
N3	192.168.1.103
N4	192.168.1.105
N5	192.168.1.101

IP Address and placement of node sensors was doing in randomly. Node Sensors was turned on at the same time then get the IP Address by randomly and sequentially along with the connection of node sensors. Network quality that generated from this experiment can be seen on Table 4. And shown on Figure 10. until Figure 12.

Table 4. Network quality data on star topology.

Node Sensor	Packet Loss (%)	Delay (ms)	Jitter (ms)
N1	3,8	7,026	0,25406
N2	0	5,556	0,23274
N3	9,2	6,61957	0,34254
N4	22,4	7,32232	0,3688
N5	2,4	6,77276	0,31191

When using star topology, packet loss value is not only affected by distance, but also the number of node sensors and the connected node sensor sequence proved by the packet loss of N5 is only 2,4% compared to the packet loss of N4 of 22, 4%. Based on that, Mi-Fi modem still using the SISO system (Single Input Single Output) it is proven by looking on IP Address the data services run in alternating ways and the first connected priority. However, the overall value of the resulting network quality is quite good that is marked with a packet loss value of not more than 25%.

4.3. Sensor accelerometer

After performing the connection test, a sensor test is performed which becomes an indicator of the system that has been made to work optimally on the bridge monitoring system. The result of comparison of sensor data output between raw data and online data using MQTT protocol is very much different which can be seen in figure 9 until 11.

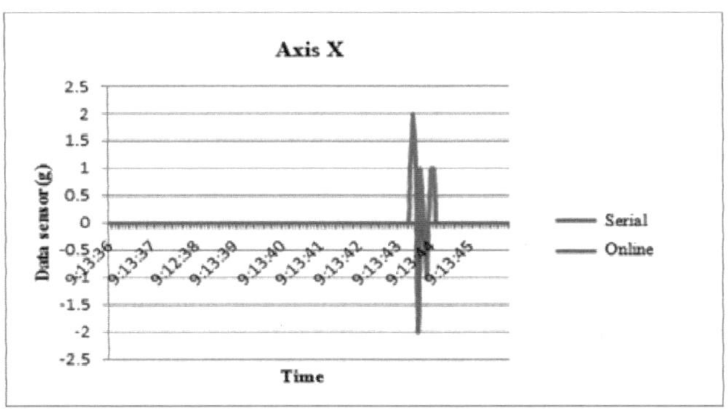

Figure 9. Output data sensor of accelerometer axis X.

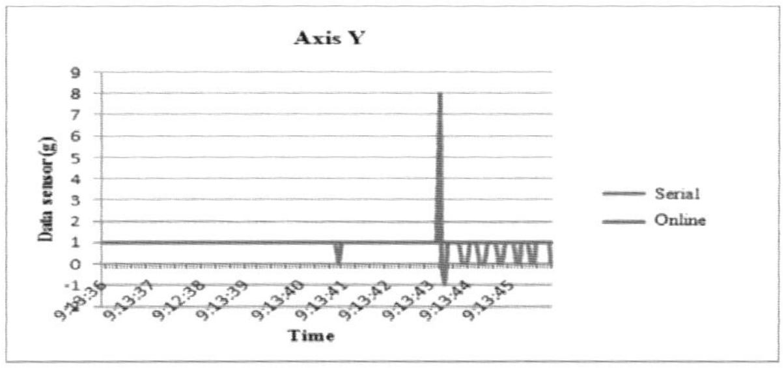

Figure 10. Output data sensor accelerometer axis Y.

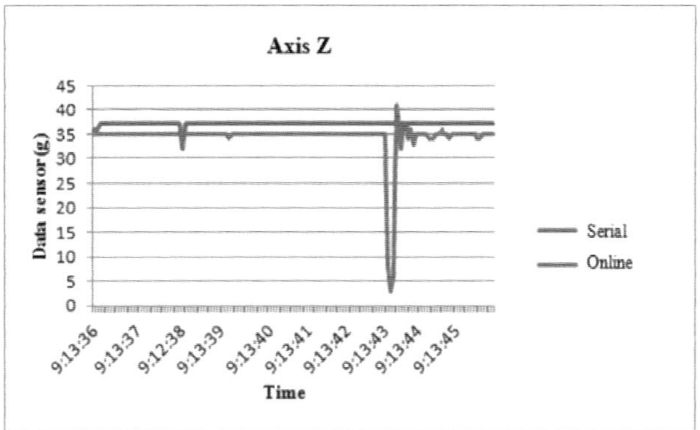

Figure 11. Output data sensor accelerometer axis Z.

Based on the test, MQTT protocol is very good at publishing and subscribing sensor data transmitted. But, MQTT protocol still unable to compensate for the very short vibration detection speed of the ADXL345 accelerometer sensor. There's a big gap between serial data and online data. With the same setting that node sensor can take more than 100 data in one second to improve the detection of vibration, in serial data all of the data was recorded alternately between X, Y and Z axis. However, transmitting and uploading data to the cloud will certainly make this data can be subscribed and seen by anyone anywhere connected to the internet but it still not efficient for a sensor that produces data very quickly and very much.

5. Conclusion

On protocol MQTT, transmitting and uploading needs a time for the process that caused delay at the time of reading online data on MQTT Dashboard. In addition, sensor data requires ACK (acknowledgment) or the correction of the sensor data transmission process that has effect on the transmitted data does not collide and there is no error data but it will impact on the time required in the transmission of sensor data that will become progressively longer. Accelerometer sensor can offset and represent vibrations that occur on the bridge in the event of a vibration caused by a vehicle engine that occurs very briefly. Anyway, to build a health monitoring system on the bridge there are still complex problems. This research still has many limitations such as this research is only based on the point of view of telecommunications engineering without comparing with civil engineering data so that it is still not perfect to called as SHM. Need further research that comparing when it LoS (Line of Sight) and nLoS (non-Line of Sight) and also vibration sensor data itself. Furthermore, tests the quality of the network from the MQTT broker itself and radiation pattern are needed to find out more about the anomalies that happen on the bridge.

Acknowledgment
Thanks to Mr Sudarmono from Politeknik Negeri Semarang, for supporting and helping by giving the bridge prototype as the object of this project.

References

[1] Soman R, Onoufriou T and Kyriakides M 2013 Damage assessment of bridges using compound SHM-signal processing and communication challenges in *Signal Processing Conference (EUSIPCO), 2013 Proceedings of the 21st European* 1-5

[2] Susanto T and Budipriyanto A 2014 Deteksi Kerusakan dan Penempatan Sensor Percepatan Berbasis Respons Dinamis Struktur: Kasus Jembatan KA Porong, Sidoarjo *Jurnal Aplikasi Teknik Sipil* **12** 1-12

[3] Rad H N, Sani S S and Rad M N 2015 A new inexpensive system for SHM of bridge decks using wireless sensor networks based on measurements of temperature and humidity in *Knowledge-Based Engineering and Innovation (KBEI), 2015 2nd International Conference on 2015* 998-1009

[4] 24_SE_M_2015 Pedoman Perencanaan Sistem Monitoring Kesehatan Struktur Jembatan

[5] Mahmud M A, Bates K, Wood T, Abdelgawad A and Yelamarthi K 2018 A complete Internet of Things (IoT) platform for Structural Health Monitoring (SHM) in *Internet of Things (WF-IoT), 2018 IEEE 4th World Forum on* 275-279

[6] Abdelgawad A and Yelamarthi K 2016 Structural health monitoring: Internet of things application in *Circuits and Systems (MWSCAS), 2016 IEEE 59th International Midwest Symposium on* 1-4

[7] Mahmud M A, Abdelgawad A, Yelamarthi K and Ismail Y A 2017 Signal processing techniques for IoT-based structural health monitoring in *Microelectronics (ICM), 2017 29th International Conference on* 1-5

[8] Amalina E N and Setijadi E 2013 Perbandingan Topologi WSN (Wireless Sensor Network) Untuk Sistem Pemantauan Jembatan *Prosiding Conference on Smart-Green Technology in Electrical and Information System*

[9] Satria G O, Satrya G B and Herutomo A 2015 Implementasi Protokol Mqtt Pada Smart Building Berbasis Openmtc *eProceedings of Engineering* **2** 2015

[10] Wulandari R 2016 Analisis QoS (Quality of Service) Pada Jaringan Internet (Studi Kasus: UPT Loka Uji Teknik Penambangan Jampang Kulon–LIPI) *Jurnal Teknik Informatika dan Sistem Informasi* **2** 2016

[11] Somani N A and Patel Y 2017 Zigbee: A low power wireless technology for industrial applications *International Journal of Control Theory and Computer Modelling (IJCTCM)* **2** 27-33

[12] Lee S, Kim H, Hong D K and Ju H 2013 Correlation analysis of MQTT loss and delay according to QoS level," in *Information Networking (ICOIN), 2013 International Conference on* 714-717

IOP Conf. Series: Materials Science and Engineering 1108 (2021) 012026 doi:10.1088/1757-899X/1108/1/012026

Accelerometer sensor data analysis of bridge structural health monitoring system

A A Hapsari[1,*], E Supriyanto[2], A Hasan[2] and A Suharjono[2]

[1] School of Graduate Studies, Management and Science University, Shah Alam, Malaysia
[2] Departement of Electrical engineering, Politeknik Negeri Semarang, Semarang, Indonesia

*012018020311@sgs.msu.edu.my

Abstract. The use of Wireless Sensor Network (WSN) for monitoring systems continues to grow, as it is required data analysis of a sensor node in which there is an accelerometer, use WSN system intended for structural health monitoring application. This monitoring system prototype, built by integrating software and hardware with device microcontroller and sensor. The node sensor produces data and processed with a microcontroller, the resulting data is then immediately transmitted to the cloud by internet network using MQTT. The system was built with a firmware-based Internet of Things (IoT) and this verified using accelerometer sensor node. The data has also been explored and analyzed in the time domain and frequency domain using FFT. Using MQTT, data can be sent directly to the internet in the form of raw data or graph data over time. And in this paper authors do an analysis of the raw data of the output sensor accelerometer ADXL345 and applied to monitor the bridge. From experiments, conducted output accelerometer sensor can vary depending on the sensitivity of the values.

1. Introduction

The bridge is one of the very important infrastructures for road transport. Along with the progress of the times, bridge construction built increasingly long and pretentious. However, the bridge looks strong, sturdy and stately could just become very fragile and dangerous for the users. Therefore, there are required monitoring system to find out the cause of the damage to the structure of the bridge which can cause the collapse [1]. Used of Structural Health Monitoring System (SHMS), structure analysis can be performed to detect and identify the presence of damage or degradation of the structure of the bridge so that it could prolong the service bridge because it can find out the cause of the damage [2]. \Factors cause damage and a reduction in the strength of the structure of the bridge such as the presence of overload, earthquake, vibrations, collisions, and the degradation of the quality of construction of the bridge. By placing a sensor node on the bridge can be known of the condition of the bridge and the observable parameters must not exceed the limits specified service (Serviceability Limit State) [3]. Several other studies on Structural Health Monitoring System to observe communication system and find out the most appropriate topology used for monitoring bridge structure when used on large bridges [4].

For the analysis of data generated by the sensor on the system can use a Fast Fourier Transform algorithm and wavelet transform to produce clearer data for further analysis [5].

IOP Conf. Series: Materials Science and Engineering 1108 (2021) 012026 doi:10.1088/1757-899X/1108/1/012026

Research on Structural Health Monitoring System using Wireless Sensor Network mode has been there one of them using Arduino, ATmega32, and Zigbee is quite expensive sensors are on each node. Because of it, we needed a more efficient alternative system.

Based on the reasons in the background. The author conducting research on Structural Health Monitoring System with the design of a simpler hardware and system with cheap and efficient technology. We propose Accelerometer Sensor Data Analysis of Prototype Bridge Structural Health Monitoring System Prototype with Internet of Things. And we use the internet of things based firmware that can display the data directly from the sensor to the internet with cheaper hardware.

2. Related works

On research about Structural Health Monitoring System on Single Degree of Freedom Bridge [6]. Deals with monitoring bridges with sensors and accelerometer sensors Wight in Motion. This research is already using the method of WSN. And the data is displayed in some form, namely time and frequency. But the accelerometer data is not yet in the show on the internet. Wireless Sensor Networks Technology for Bridge Health Monitoring [7] describes the research on monitoring the bridge with WSN. This research data further discusses the power consumption and power on the system. But data from the sensors are not shown in this study. And just discuss power supply is needed on WSN system.

Next research [8], already using WSN method. And use multiple sensors. This research uses a sensor device is relatively expensive and not yet explained how the data is analyzed. The sensor data is still in the form of graphs in the time domain. And not explained how the generated sensor data, and how to process data from the sensor.

Research about Structural Health Monitoring Measurements under Train Speed Changing [9]. In this research are described and illustrated the influence of train speeds towards the railway bridge structure with accelerometer sensor. The device can monitor displacement when trains pass. The data in this paper is a comparison of the data fast and slow trains train when passing the bridge. Already using WSN method and a few sensors deployed directly on the bridge but the data haven't been uploaded or displayed directly on the internet.

Based on related works that are already described in this research will be designed a system with sensor and accelerometer node MCU ESP8266 which is the firmware for the application of the technology of the Internet of Things on a prototype monitoring system health structures of the bridge. The data on this system can be seen anywhere by internet network as it uses MQTT. So it will be a frugal and efficient system. The resulting data will be uploaded directly to the cloud using the MQTT protocol, saved by the database using node .js and the data saved in the database. And data can be viewed by subscribers wherever and whenever.

3. System description

On this research built a prototype of health monitoring system bridge with wireless sensor network. This study uses several interconnected sensors node can be sending data from the accelerometer sensor reading of results. Draft and a general overview of the health monitoring system of the bridge structure is shown in Figure 1.

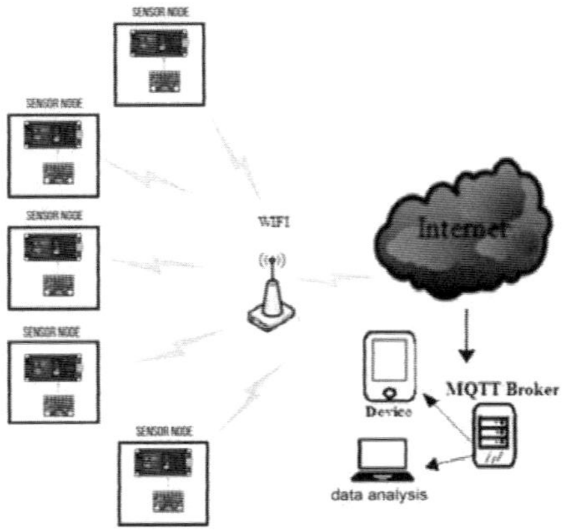

Figure 1. Architecture design of the SHM system.

The workings of this system, starting with the determination of the reference or indication that will be applied. The sensor used in these tools namely accelerometer and sensors will be installed in the body section of the prototype structure of the bridge. Data from the accelerometer sensor can be analyzed and compared the results. Several nodes comprising sensors, accelerometer and sensor node MCU will be installed at some point and can monitor the condition of the bridge simultaneously. With each node connected with Wi-Fi networks that are generated by the node MCU ESP 8266, connected with a Wi-Fi modem that is connected to the internet. Data can be displayed in the form of a real-time graph. The system is designed with a highly efficient. The creation of this system is divided into two parts, namely the manufacture of hardware and software.

3.1. Hardware

Wireless sensor nodes are composed of four components, a processing unit, a transceiver unit, a battery unit, and a sensor unit. This research uses the node MCU ESP 8266 as data processing. And node MCU ESP 8266 already there is a wireless module with Wi-Fi 802.11 b/ g/ n, so it can instantly transmit wireless data. Node MCU ESP 8266 working on the resolution 32 bit [10]. In this paper, the structural vibration is sampled by an accelerometer sensor. Accelerometer sensor used is the ADXL 345 type that has a pretty cheap price. Not only low-cost but also have low- power 3-axis accelerometer with a measurement range of ±16 g. Node MCU and accelerometer sensors work on the voltage of 3.3 volts. Sensor output data has ADXL 345 with the digital format. The minimum resolution that can be read by the accelerometer sensor is of 4mg/ LSB.

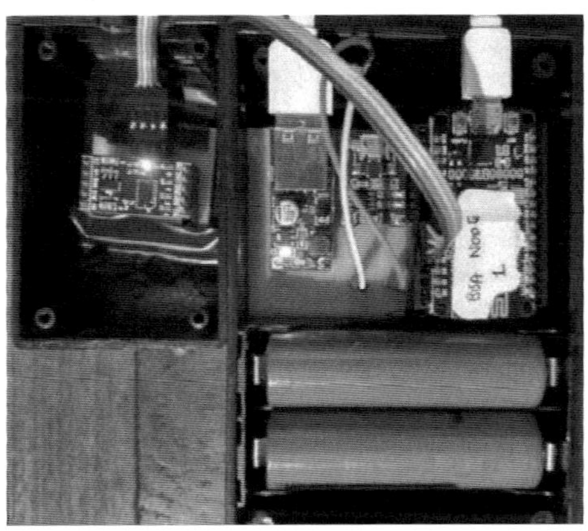

Figure 2. The hardware of wireless sensor node.

IOP Conf. Series: Materials Science and Engineering 1108 (2021) 012026 doi:10.1088/1757-899X/1108/1/012026

3.2. Software

The sensor nodes are configured with the program so that the sensor can work and produce data. Used the Arduino IDE software to upload programs to the node MCU ESP 8266. Upload the program to the node MCU esp8266 with the Arduino IDE that has been previously installed library from node MCU ESP8266 version.

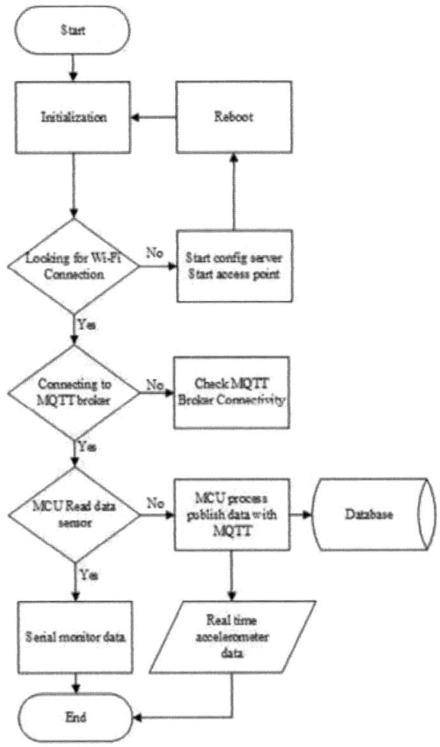

Figure 3. Flow diagram of the program for the node sensor.

The configuration for Wi-Fi and wireless connections, as well as on the program via Arduino IDE. And then configure the program for the connection with MQTT. The flow of the program can be seen in Figure 3. In this system using five sensors to sample the real data from the board. The system used MCU with a wireless module to send the data sampled from the bridge prototype. All of the node sensors sends the sampled data to the cloud using MQTT. The data that appear and then saved into a database using the node.js program. The data acquired and stored in the database is subsequently carried out further analysis. It also carried the serial data retrieval, to do an analysis of the data of the accelerometer sensor is still in the form of raw data.

3.3. Digital accelerometer sensor data

The experiment can be seen in the comparison of the data from the accelerometer sensor by performing different sensitivity settings and compare with the datasheet from the sensor being used. Accelerometer digital has raw data output data that can be converted to mg or g [11]. Sensitivity on digital sensors related to the value of the bit of the sensor. The value of g from digital accelerometer sensor can be determined based on the value of the LSB of the data that can be spelled out with formula one, and the value of the LSB can be searched with the formula two [12].

$$g = \frac{data\ LSB\ value}{1000} \tag{1}$$

It can be viewed and compared with the results of existing data on the datasheet the sensor to be used. The value of LSB vary depending on the value of bit sensors are used. Digital accelerometer has different bits of resolution depending on the type.

$$One\ LSB = \frac{Full\ g\text{-}range}{Number\ of\ counts} \tag{2}$$

4. Experiment results

This study observed how the data generated by the accelerometer sensors on a prototype bridge. previously seen how the data generated by the sensor node. Data over time can be seen live with MQTT applications and can also be stored in the database is next on to draw in the shape of the graph. The first experiment to see if the sensor has been successfully transmitted data directly with MQTT. And the result can be seen on the graph of fig 4. Displaying sensor data sent wireless and online using MQTT.

The next test is taking the sampling data resulting from a reading of the accelerometer sensor node in place on the body of the prototype model of the bridge. experiment with changing the values of the sensor sensitivity accelerometer ADXL 345 and compare with the datasheet. the raw data generated from the sensor output accelerometer sensitivity differently when in settings with different values. but from the resulting data would be worth approaching when computed using the formula (1).

By laying the same sensor that is horizontally aligned right at the Earth's surface and produce a value of 0 g on the X-axis, 0 g on the Y-axis, and 1 g on the Z axis, that can be seen in Table 1.

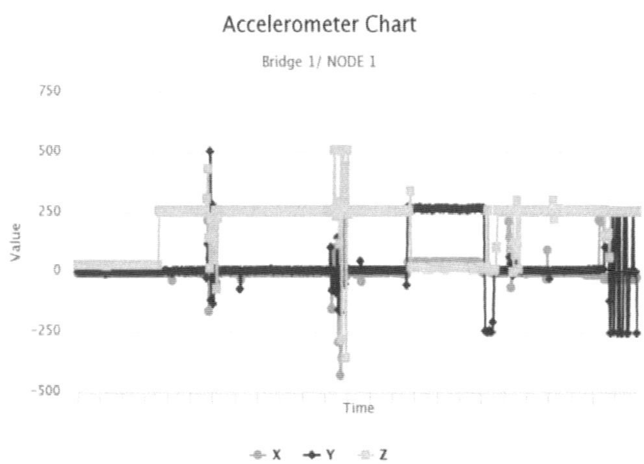

Figure 4. Live data sensor with X, Y, Z axis from accelerometer.

Figure 5. Experiment set up on bridge model.

Table 1. Data values to g accelerometer sensors with different sensitivity.

Sensitivity	x (g)	y (g)	z (g)	LSB
2	-0,0585	0,0507	1,0491	3,9
4	-0,0624	0,0546	1,0452	7,8
8	-0,0624	0,0468	1,0296	15,6
16	-0,0630	0,0315	1,0395	31,5

Then do the experiment by giving objects to move above the bridge, it generates data read out sensor as shown in Figure 6. It can be seen from the graph that there are changes to the reading of data sensors accelerometer in the time domain and then do the conversion with FFT.

On Figure 6 there are shows the graph of the experimental data, and describe the form of the output signal from the sensor is x, y, z-axis. Figure a shows the output signal which is read by a sensor node and later changed into the form of FFT on the graph b. And look a peak magnitude values on the graph of the FFT. from the resulting data shown graph FFT on each axis x, y, and z.

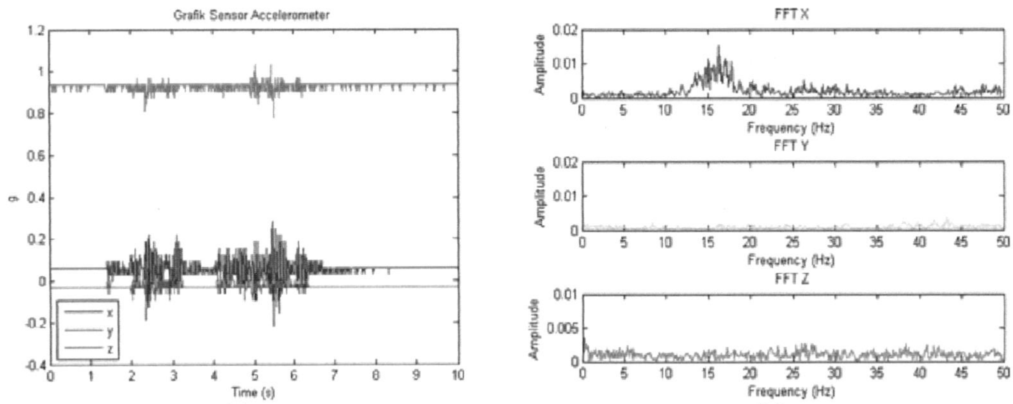

(a) (b)

Figure 6. Three-axis sensor node data response node 1 (a) graph time sensor node with FFT (b).

Test on the model of the bridge is done by taking the sensor data when given a load moving and not given a load moving. When the constituent components of the bridge like a bolt that is still attached to the full and is not complete. then carried out observations on the outcome of the reading of the sensor. Shown in figure 7 and 8.

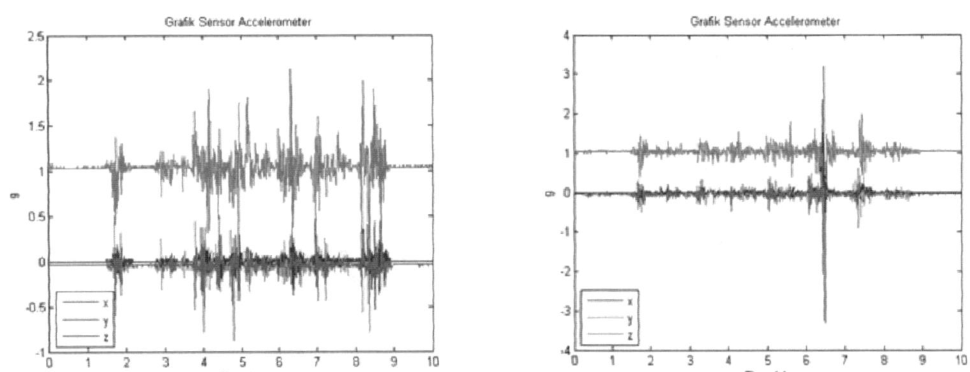

(a) (b)

Figure 7. Three-axis sensor node data response when mounting bolts tight (a) data response when the bolt loose (b).

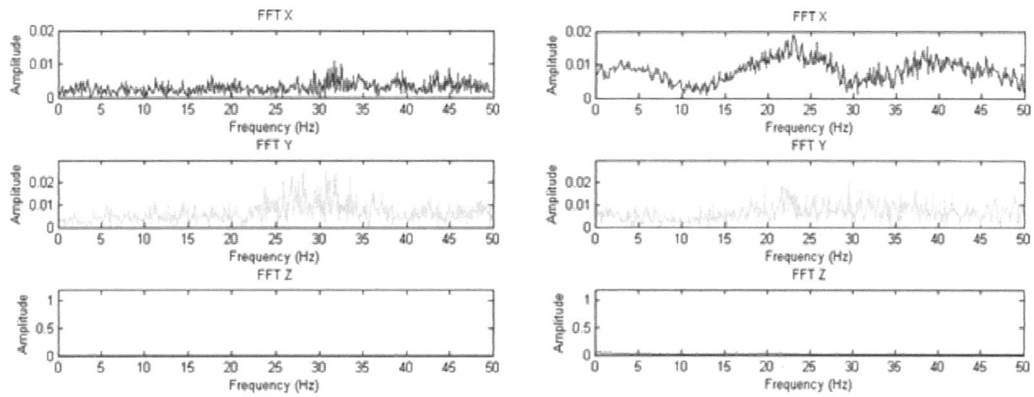

(c) (d)

Figure 8. FFT graph when bolts tight (c), FFT graph when bolt loose (d).

The accelerometer sensor node is set with a 10 ms delay and an analysis of 1000 data inputs is performed. The higher peak value difference is generated on the "x" and "y" node axes when there is a change in the bridge structure at the maximum value with the FFT computation of the node 4 sensor data which detects the state of the bridge at the time of the bolt nut condition removed, and with a complete bolt nut condition.

Node 4 is the sensor node that is placed closest to the location of the nut and bolt release on the bridge model. The maximum output value of sensor node 4 can be seen in Table II.

Table 2. Output maximum value of data sensor node 4.

Condition	LSB	X (g)	Y (g)	Z (g)
Complete bolt	31,2	0,5616	1,3416	2,496
Lost bolt	31,2	3,1824	1,0296	4,056

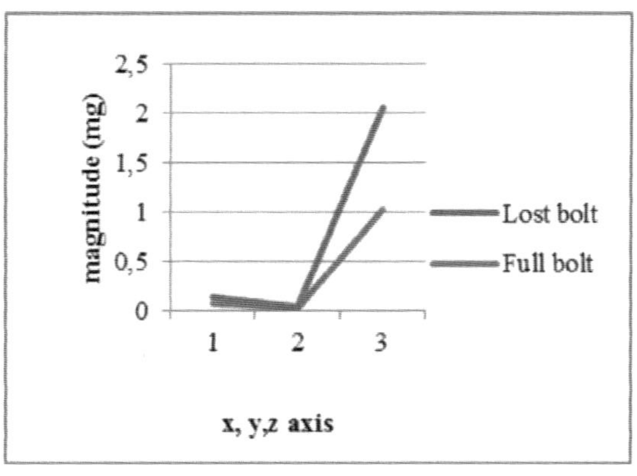

Figure 9. Graph of *amplitude* maximum FFT axis x,y,z data sensor when bolts tight and loose.

When the model of the bridge is passed through the moving load, the accelerometer sensor installed produces a different data change output similar to that of the bridge model body. The change is also obtained by reading the different sensor output values. The peak value of the resulting FFT is in the range of 0-1 Hz. And when the bridge model changes by removing the nuts and bolts, there is an increase in data readings as can be seen in the average FFT peak value graph in Figure 9.

5. Conclusion

From the research that has been done, it can be seen that the output data from the accelerometer sensor in the form of raw data, the analysis needs to be done first, and calibrate the sensor and match the datasheet of the sensor used. The raw data generated by the sensor is analyzed by converting it according to the sensor datasheet into units of "g" or "mg" so that it can be used further. And after observation, knows that the raw output value of the sensor data obtained still meets the value of g as found in the datasheet of the accelerometer sensor ADXL 345 when the sensor is still and the bridge is not passed through the load.

When there is a moving load across the bridge it appears that the accelerometer sensor readings are changing. And in the result seen at the maximum sensor reading value has a higher value when the bolt release.

From some tests that have been done known that FFT can be done on accelerometer sensor data. FFT is performed on each axis. And the faster of sensor readings are used the higher the sampling frequency. Accelerometer sensor can be used as vibration detector and also displacement at bridge monitoring system. The FFT process is performed on the accelerometer data to know the value in the real of frequency.

Acknowledgment

Thanks to Mr Sudarmono for support at Politeknik Negeri Semarang, for the bridge prototype.

References

[1] Kawchuk G N, Hartvigsen J, Edgecombe T, Prasad N and Van Dieen J H 2016 Structural health monitoring (vibration) as a tool for identifying structural alterations of the lumbar spine: a twin control study *Nat. Publ. Gr.* 6–11

[2] Bassett K and Bassett K 2010 Vibration Based Structural Health Monitoring for Utility Scale Wind Turbines By

[3] Yang J, Chang F and Derriso M M 2013 Structural Health Monitoring *Struct. Heal. Monit.*

[4] Setijadi E, Bp S and Suprobo P 2013 Design of Large Scale Structural Health Monitoring System for Long-Span Bridges Based on Wireless Sensor Network 169–174

[5] Shome S K, Sen S, Mondal K and Datta U 2016 Development and Performance Analysis of Wireless Sensor Node for Structural Health Monitoring using Fast Fourier and Wavelet Transform 491–495

[6] Putra S A, Trilaksono B R, Harsoyo A and Imam A 2016 Agent-based structural health monitoring system on single degree of freedom bridge: A preliminary study *2015 Int. Conf. Inf. Technol. Syst. Innov. ICITSI 2015 - Proc.*

[7] Zhou G D and Yi T H 2013 Recent developments on wireless sensor networks technology for bridge health monitoring *Math. Probl. Eng.*

[8] Niu J, Deng Z, Zhou F and Cao Z 2009 A Structural Health Monitoring System Using Wireless Sensor Network 2–5

[9] Kaloop M R, Hu J W and Sayed M A 2016 Yonjung High-Speed Railway Bridge Assessment Using Output-Only Structural Health Monitoring Measurements under Train Speed Changing

[10] Skraba A, Kolozvari A, Kofjac D, Stanovov V and Semen E 2016 Streaming Pulse Data to the Cloud with Bluetooth LE or NODEMCU ESP8266 428–431

[11] Setiawan I, Setiyono B and Susilo T B 2009 Hasil Uji Kalibrasi Sensor Accelerometer *e-Journalundip* 118–122

[12] Description G, Ideas A and Specifications K 2009 MMA7455 3-Axis Accelerometer Module (# 28526) Package and Pin Descriptions + Z + X **8333**(916) 1–7

IOP Conf. Series: Materials Science and Engineering 1108 (2021) 012027 doi:10.1088/1757-899X/1108/1/012027

Double security system with RFID and password based on Arduino Mega 2560 and PLC

S Syahid*, S Supriyatna, Y Badruzzaman, D Antono and A Santoso

Department of Electrical Engineering, Politeknik Negeri Semarang, Semarang, Indonesia

*syahid@polines.ac.id

Abstract. This research focuses on the use of RFID (Radio Frequency Identification) systems that are used as personal identification cards in the room access system. This system is intended to maintain the security and privacy of the room from someone who does not have the authority to enter the room. This security system uses a two-tier security system, namely through an RFID identity card and security code. The design and implementation of this system, an electric key work system is evaluated based on the solenoid component and the password used when accessing the room. RFID tags used in this system are in the form of "Mifare RFID Card 13.56 Mhz" cards and store unique codes used as personal identification. This code is read by the RFID reader in validating its authority with Arduino Mega 2560 to regulate the system of electric locks and passwords used to open the door. From the test results, 100% RFID tags can be recognized by the RFID reader, 100% RFID tag and password reading system is successfully used to access the room.

1. Introduction

The growing human mindset in developing science and technology, and supported by high curiosity makes science and technology grow rapidly over time. Technology contributes greatly to development in various fields. Especially in the area of security and practicality. In the field of security, with the existence of technology, safeguarding things such as safeguarding a space that can only be entered by certain people can be stronger. In the field of practicality, with the existence of technology, it is easier for humans to carry out their activities such as doors that can be opened and closed by themselves without being operated by humans.

The absence of safeguards against anyone who will go through the automatic door makes the room that can be entered through the automatic door becomes vulnerable to criminal activities such as theft, misuse and others.

Therefore, there is a need for innovation and development of this automatic door to overcome this problem, namely by adding RFID (Radio Frequency Identification) reader, RFID card, keypad, Arduino mega 2560 microcontroller and PLC as access to be able to enter the restricted space by the door this automatic [1,2].

2. Literature review

RFID or Radio Frequency Identification is a data telecommunication device using radio waves to exchange data between one reader with an electronic tag attached to a particular object [3,4]. Data communication technology between an RFID reader with electronic tags (RFID tags) in this system are contactless, real time and wireless. Data identification on RFID tags is carried out through radio frequencies that propagate through the air media at a certain range according to the features possessed by each RFID module (consisting of RFID readers and RFID tags) used. In general, RFID tag data that is unique is stored or embedded in a chip card so that the influence of natural conditions such as dust, dirt or air temperature will not reduce the quality of data communications that occur.

The features possessed by RFID technology are the advantages of RFID technology when compared to other identification systems such as barcodes and magnetic cards. But this advantage will be relative because it will depend on the use of an identification technology in an application that will be implemented. This technology has been used in various applications related to object identification systems in several previous studies, such as opening doors, accessing computers, turning on motorbikes, and controlling equipment in office spaces such as lights, computers and lighting [5,6].

Previous research that had been made was "the use of the RFID System as a Room Access Limit" [7] with a system design like Figure 1.

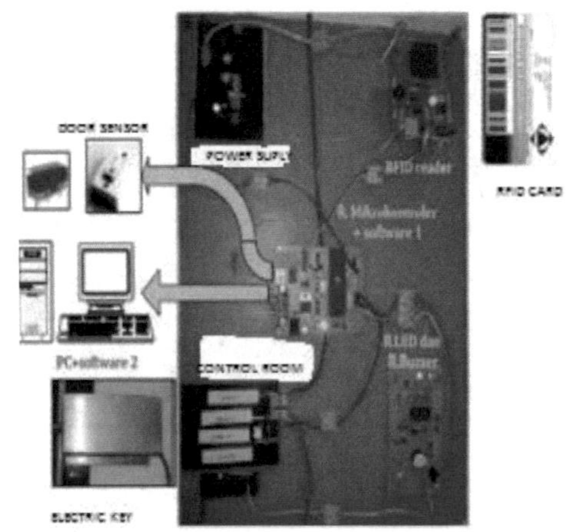

Figure 1. Design and realization of RFID systems as Room Access Limit.

Researchers have also conducted research in 2013 concerning RFID applications "the Development of Monitoring Software for Fertile Detection of Dairy Cows Using RFID Technology to Increase the Success of Artificial Insemination". An overview of previous research, as shown in Figure 2 as follows:

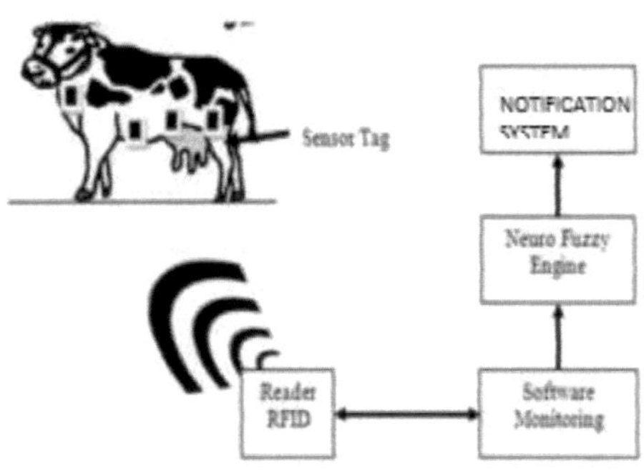

Figure 2. Monitoring diagram of fertile period detection of dairy cows using RFID technology [8].

Through this research the Double Security System with RFID and Password Based On Arduino Mega 2560. The road map is as follows:

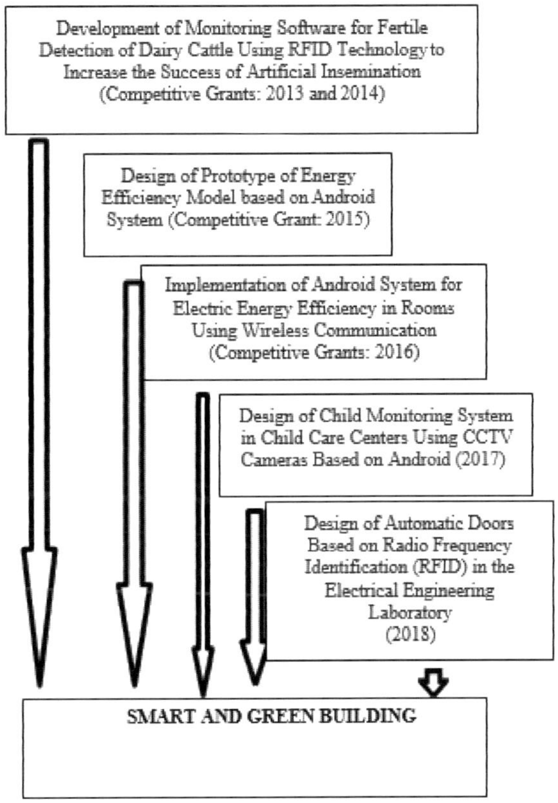

Figure 3. Roadmap.

3. Results and discussion

3.1. Work description

To enter the room, the new door can be opened after the RFID card is scanned by the RFID reader and enter the password according to the RFID card. If the password entered is incorrect, the door will not open. And if the password entered is correct, the 5 Vdc relay will be active and provide input to the PLC. The new door will open after the outer door sensor is active, and will close after the inside and outside door sensors are not active. If you have entered the password correctly, but the sensor door is not active for 10 seconds, the door will not open even if someone will enter the laboratory room. To open the door again, have to scan the RFID card and enter the card password correctly again. Whereas if someone wants to get out of the laboratory room, the door will open by itself without having to scan the card and enter the password correctly.

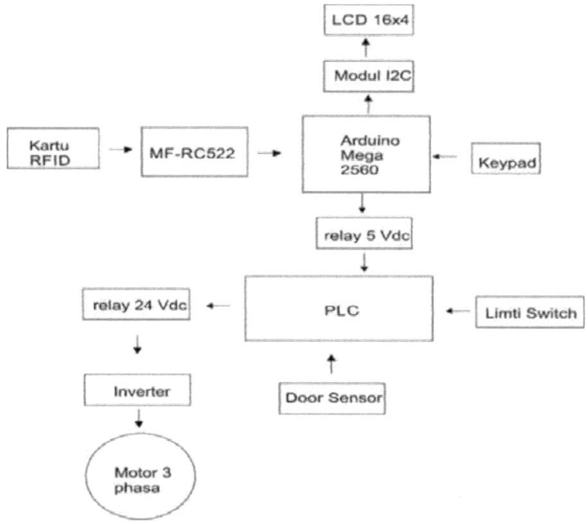

Figure 4. Block circuit diagram.

3.2. Circuit of tools

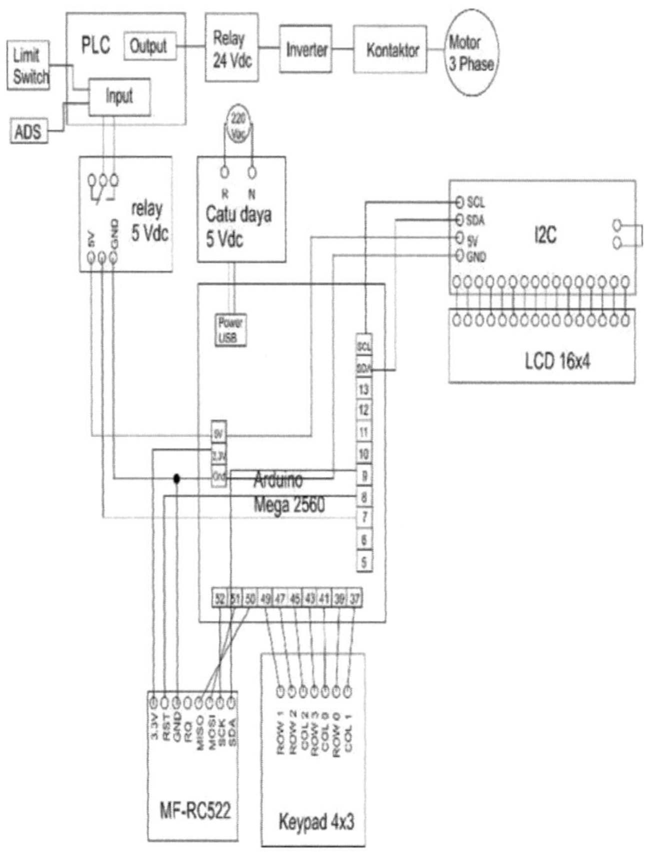

Figure 5. Circuit of tools.

3.3. Flowchart

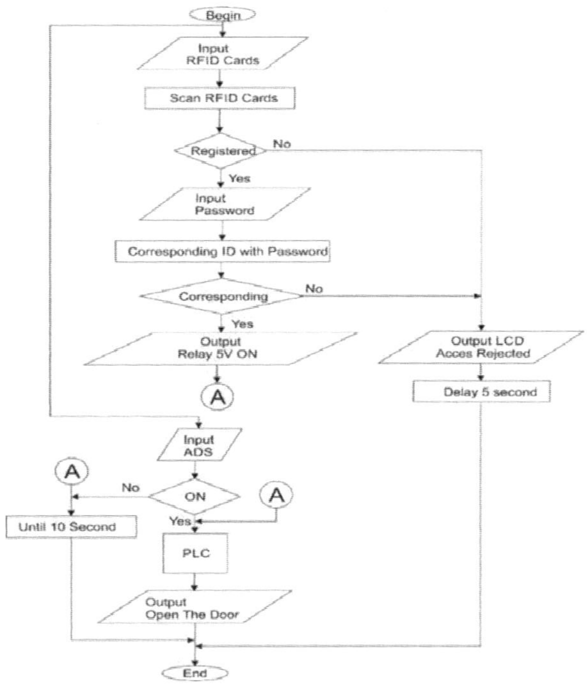

Figure 6. Flowchart.

3.4. Test of results

3.4.1. Initial LCD display

Figure 7. Initial LCD display.

3.4.2. RFID and the password

Table 1. ID number of RFID cards and password is used.

RFID	ID Number	Registered	True Password
RFID 1	632082054111	Yes	1111
RFID 2	14916016840181	Yes	2222
RFID 3	101193143403	Yes	3333
RFID 4	158202633275	Yes	4444
RFID 5	158152273261	Yes	5555
RFID 6	95194740178	No	
RFID 7	14113720145224	No	

Table 2. Result of RFID and password to open the door.

RFID	Password Input	Access	ADS ON
RFID 1	1234	Rejected	Before 10s
RFID 2	2222	Received	Before 10s
RFID 3	3333	Received	Before 10s
RFID 4	1582	Rejected	Before 10s
RFID 5	5555	Received	After 10 s
RFID 6	5454	Rejected	Before 10s
RFID 7	1411	Rejected	Before 10s

Figure 8. RFID card is registered and entry password.

Figure 9. RFID card is not registered.

Figure 10. The Password is true.

Figure 11. The Password is false.

4. Conclusion

The door will open after scanning the registered RFID card and entering the password according to the RFID card used. And the auto door sensor is active before 10 seconds, calculated from when entering the RFID password correctly.

The door will remain closed even though it has scanned and entered the password correctly, because the auto door sensor is not active in 10 seconds from the time of entering the password correctly. To reopen the door, it is necessary to scan and enter the password correctly again.

References

[1] Kadam M S S, Shinde M V S, Desai M M M and Deshmukh M P G 2015 RFID Based Car Parking Security System Using Microcontroller IC89c52

[2] Bolic M, Simplot-Ryl D and Stojmenovic I 2010 *RFID SYSTEMS, Research Trends and Challenges* (Wiley)

[3] Ahsan K, Shah H and Kingston P 2010 RFID Applications: An Introductory and Exploratory Study *International Journal of Computer Science Issues* **7**(3)

[4] Daniel H, Albert P and Mike P 2007 *RFID A Guide to Radio Frequency Identification* (John Wiley & Sons)

[5] Prabhu B S, Su X, Ramamurthy H, Chu C C and Gadh R 2006 WinRFID: a middleware for the enablement of radiofrequency identification (RFID)-based applications *Mobile, wireless, and*

sensor networks: Technology, applications, and future directions, **313**
[6] Graafstra A 2007 *How Radio-Frequency Identification and I Got Personal* [online] retrieved from http://spectrum.ieee.org/computing/hardware/hands-on
[7] Rachmat H H and Hutabarat G A 2014 Pemanfaatan Sistem RFID sebagai Pembatas Akses Ruangan *Jurnal ELKOMIKA* **2**(1)
[8] Syahid B S 2013 Pengembangan Software Monitoring Pendeteksi Masa Subur Sapi Perah Menggunakan Teknologi RFID Untuk Meningkatkan Keberhasilan Inseminasi Buatan *Seminar Nasional Politeknik Negeri Semarang*

Power saving of IP over WDM: Coherent taxonomy, mechanism and technique: A review

L P Pratama[1,*], S N Saud[2], E D Wardihani[3] and N M Lehat[2]

[1] Faculty of School of Graduated Studies, Management and Science University, Shah Alam, Malaysia
[2] Faculty of Information Sciences and Engineering, Management and Science University, Shah Alam, Malaysia
[3] Faculty of Telecommunication engineering, Politeknik Negeri Semarang, Semarang, Indonesia

*legend92pp@gmail.com

Abstract. In this survey, the comparison of highlights on the importance of energy saving in IP over WDM (IPoWDM) is presented and followed by the review on the existing related article by using systematically data review based inclusion and exclusion criteria. The adaptive clock frequency, sleeping mode, and hybrid grooming mechanism protection is the latest breakthrough in the IPoWDM mechanism. The researcher identifies trend of survivability is divided into three protections: link scheme, path protection and restoration. This paper focuses on the techniques and methods employed, and analyze the difference from the latest article review. The analysis based on the literature review indicates the development in the research trend and estimates the energy saving based on the techniques, mechanism and modeling and protection scheme by identifying the advantages and benefits for the next research. Results of obtained, we viewed from the ease and the great impact on power saving, the technology of sleep mode on the standby network is possible. The scheme that is carried with a protection mechanism is set to sleep and wake quickly. The easiness is a principal aspect that encourage to look into best protection strategies to save power in IPoWDM networks. Nevertheless, these does not prevent the possibility of development another mechanism. Finally, the potential for future research directions are also discussed.

1. Introduction

Nowadays, a router identifies the total capacity of traffic which is much smaller than the total capacity of an OXC carrier. The mechanism of IP router accumulates low-speed traffic flows and transfers the traffic over light paths provided by the optical layer. The IP node capacity reflects the largest share of the total power consumption as high as at 60% [1]. A power-efficient of IP over WDM network with network configuration provides higher energy savings with a slight increase in the consumption of power at optimization of light. This can be happening by using the implementation of the hybrid grooming [2] or by using intelligent traffic engineering and routing by utilizing sleep mode operation [3]. The goal for reducing electricity invoice due to energy consumption has become a crucial issue for all industries especially in the Information and Communication Technology (ICT) sector [4].

The objective of the studies is to minimize the total energy consumption by assigning corresponding technique base on topology for each router at different time based on traffic work prediction from the existing research. This topic has been discussed from previous studies and manage to determine that the techniques and limitations of each implementation for each topology that has been done. They attempt meet in finding for similarities of both protection and restoration mechanisms for each layer in order to get a common technique strategy applied by combining network consumption of optical network infrastructures [5].

Through the information in current existence of simulations and research, the percentage of energy saving can be obtained from the diverse mechanism of energy functions by assuming at different levels considering the three areas configuration that allowed network equipment to save energy. These level are identifying as networks hardware components, network design and protocols design [6]. These is due to the perspectives in minimization the energy IP over WDM network which is a necessity to research problem. Some papers have research problem for energy efficiency to get the energy savings from the IP over WDM network. These approaches based on virtual-topology experiment and traffic grooming design to reduce network energy wasted.

The explanation of general IP over WDM architecture are related in section II. In section III, the principle of network energy consumption. The solution of power saving several of mechanism in previous articles in section IV. In section V resume to get classification and discussion of reviews paper. Finally, the conclusion the papers in VI.

2. IP over WDM network architecture

This area is a regulator IP over WDM architecture. In Figure1 below introduces linear models of IP over WDM network [7]. The example using three nodes that are directly connected to the optic fibers. The basic structure on CIR (Core IP routers) preparation are from low-level traffic ingression of the package core network from other level access device routers. The CIR communicates through the Ethernet port to the muxponder OXC interface on the optical layer. The routers can specifically have divided into two structure layers, (i) the Basic Structure Device (BSD) and (ii) the Physical Structure Device (PSD). In this study it will be focus solely on the BSD. The BSD are a layer comprises of chassis device, switch fabric or control processor component, power supply device, fan devices, and Ethernet. BSD underpins the card in bidirectional transferring manner in between Line Card (LCard) / Slot Card (S-Card) / Port Card (P-Card) which are in steady state capacity. It can be more than one L-Card/SCard/P-Card installed in BSD. All type of cards can be in a fully utilized slots or empty slots. Each slot could be added or upgraded with L-Cards. There are combination numbers of SCards in each L-Card. S-Card is use to integrate media modules, which can support diverse capacity of port pluggable with difference light beam and interface (e.g., LAN port, 16/64/256 STM, 1 GE, 10 GE and 100 GE) [8]. The LCard, S-Card and P-Card are combined structure known as a physical interface.

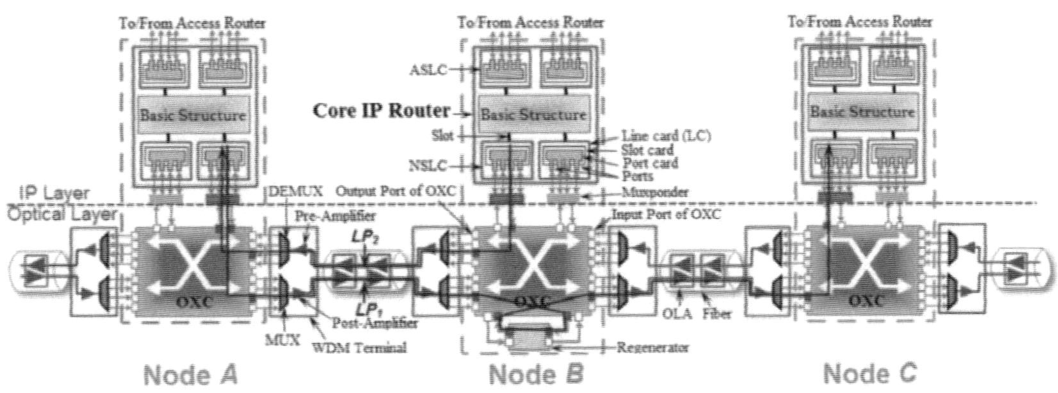

Figure 1. The linear architecture of IP over WDM network [7].

The Figure 1 shows the linear architecture of IP over WDM network. The characteristics of this linear architectures are about direct connection of Long Reach (LR) interface port, which communicates to

Extended Long Haul (ELH) Muxponder devices. The principle uses in this architecture are a light path that need one Network-Side LC (NSLC) connecting in one Muxponder at every port to communicate with one port to another which is Optical-Electrical-Optical (OEO) conversion. Generally, the OEO conversion provides the signal restoration to performed before the signal quality reach the amount of degrading below the threshold limit which known as the 3R bidirectional structure (retiming, reshaping and re-amplification). The regenerator synchronously regenerates raw signals to WDM tunnel and fiber optic. These WDM tunnel and fiber optic interface connector are also known as Muxponder. This Muxponder, then combined with gain booster to become the Multiplexer. The Multiplexer will transfer to WDM terminal directly. In the other side, the De-multiplexer will have received a signaling and regenerated it in the preamplifier to become demuxponder [8,9]. At the same time, some additional electronics signals are simultaneously providing necessary gain to all wavelengths in both directions shown as flow "a" and "b" in the Figure 2. These "a" and "b" flow are also the optical transmission line which connecting two unidirectional fibers to the optical amplifiers [7].

The EO conversion are processed by traffic data over the optical transport until the destination. The other hand, the OE conversion is reassembled by merging the data become the electrical phase which is translated by the IP router. The transparent mode (linear line "b", Figure 2) shows the OEO conversion in the intermediate router are incomplete stage which then will be filtered during electronic signal conversion at the destination router. The weakness of this basic technique is, if there is a physical damage to the signaling layer, this will interfere the signals transmission routing and automatically will proceed the transmission signal to the alternate routing which is longer en-route to the destination route. This will consume more time than the original design method en-route [10]. The opaque mode of OEO conversion processing in each transit node (linear line "a", Figure2.), are allowing grooming to recover and control the packet delivery in each hop of optic wavelength. However, the process will take a lot of additional processing time, power consumption, and latency [4]. When compared to previous modes, the translucent mode is more saving for cost and energy in the IP domain on traffic regeneration in getting the effective bandwidth usability and decrease network fee [2].

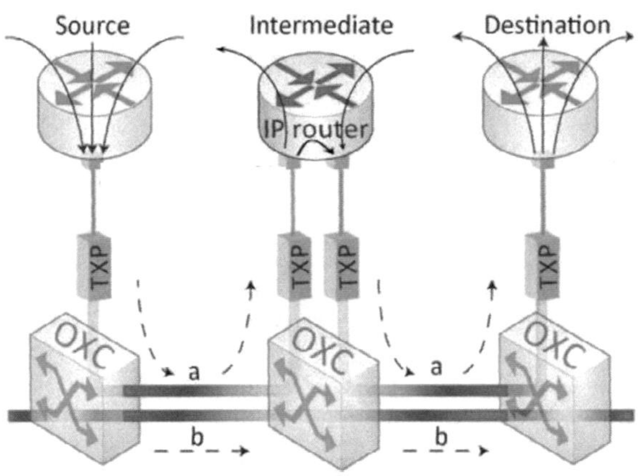

Figure 2. Transformations modes in network layers: a: opaque, b: transparent [10].

3. Principles of network energy consumption

Each development's attempt was constrained to major limited factor on the expenses of network device equipment. From the data in research [11], the amount of energy consumption on the Internet estimating to grow up by 1-2% per years. In 2017, the global power consumption prediction increase in telecommunication networks sector reach the amount of 220% compared to 2007. In according, the energy saving in the telecommunication sector is reasonable a challenge for the researcher reducing energy consumption become effective and efficient. The principal of light path technical scheme in backbone IP Over WDM networking divided into two strategies, either 1) making lightpath non-bypass or 2) lightpath bypass in the connection between routers. In Zao et al [12], the connection of the router is depending on the strategies using a non-bypass network, shall there is any incident or network failure in the lightpath between the node networks it must be discontinued. Then, all data packets which carry by the lightpath will be stop and the router must be rerouting of the path. By this time, the data packet will be stuck on the IP layer intermediate routers. However, these strategies do not need the OEO exchange in the node, which passed by the routing path wherein there are additions of energy consumption used in optic ports layer. On the other side, if the bypass network strategies are held, then all data, which carried on the lightpaths forwarded to the optical, transmit in intermediate nodes until the packet arrives in the destination.

Beletsioti et al [13], presented the results in three comparisons of heuristic algorithms (ISLB, GF, and OVTC) obtaining power consumption on the IPoWDM optic by minimizing on the count of optical paths in the virtual architecture. The example of a small network topology consists of 11 links, which are connected to eight nodes virtually and the topology of Deutsche Telekom of 14 nodes and 23 links are compared. In Figure3, the result of the experiment presented the lower limit level of the total power consumption by using ISLB, GF and OVTC algorithm. The number of random traffic sequentially from 20 Gbps to 120 Gbps, which the IP router is specifically as in need supporting device for power consumption and the smallest is optical amplifiers, asper indicated in Figure4. Furthermore, the percentage details are 93% of energy consumption in the IP router, 6% in the WDM and 1% in the optical amplifiers (EDFAs).

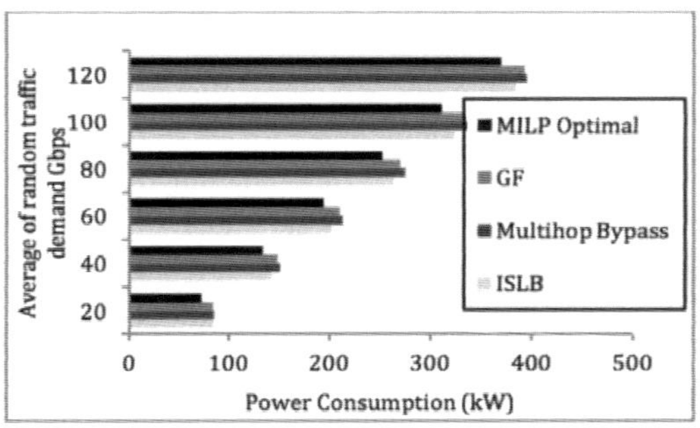

(a)

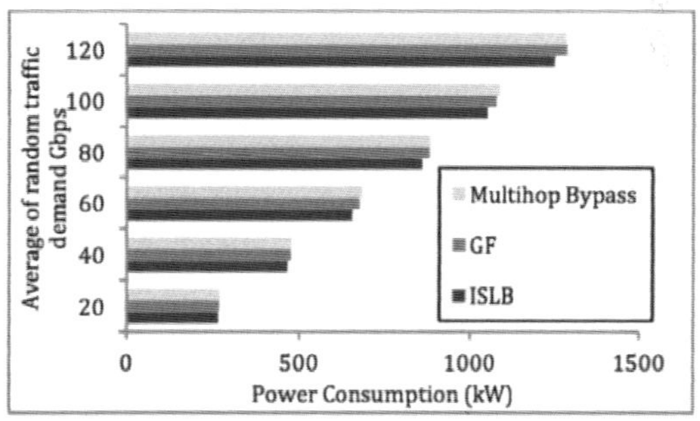

(b)

Figure 3. Power consumption in (a) Snode-net and (b) Deutsche Telekom [13].

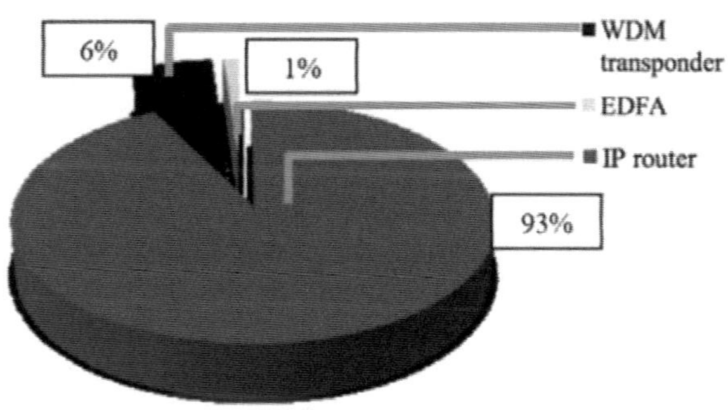

Figure 4. Energy consumption distribution under the ISLB implementation [13].

4. The solution of survivable in IP over WDM management

The efficiency in designing has saving many collections of different aspects of technology. The functionalities and requirements of telecommunication networking must be replied overall standardization aspects: (1) Scalability; (2) Resilience; (3) Quality of service; and (4) Service management [4]. The main advantage of energy efficient integrated routing in backbone IPoWDM optical networks for improving bandwidth utilization or minimizing power consumption.

4.1. Requirement adaptive clock frequency mode protection scheme

In Zhao et al [12], it is suggested for energy efficiency by the management of clock frequency strategies on WDM networks. The adaptive clock frequency that merged with the lightpath bypass technique to concurrently producing number total consumption more efficiency in every of the router cards energize by an IP over WDM networking. The function as a lower limit implementation, the clock frequency and the power consumption of a router device can systematically be transformed and keep a direct connect with overhead steady. The global optimization design network based on formulating of the decomposed models wherein ideally determine the number of needed router devices and the clock frequency embraced by every router board in a time slot. The consequence of energy utilization situation is mentioned on Figure5. The management fore see that without considering the adaptive clock frequency strategy, it can have looked without of remembering the adaptive clock frequency scheme and the number of energy utilization is doubled of that with clock frequency, moreover 14% higher than that the situation with the busy hour traffic request.

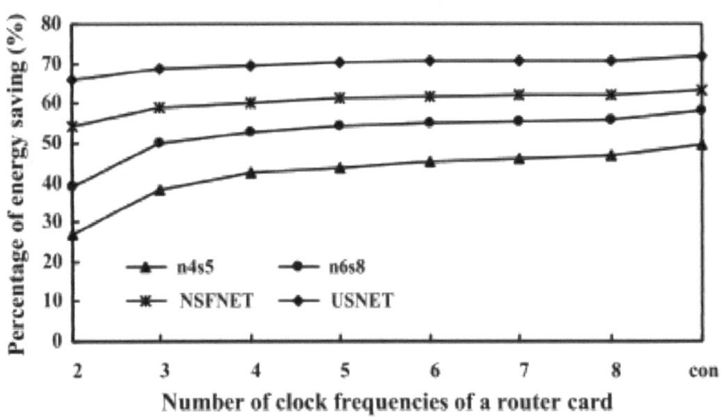

Figure 5. Results of energy saving using by the adaptive router card clock frequency strategy in any topology.

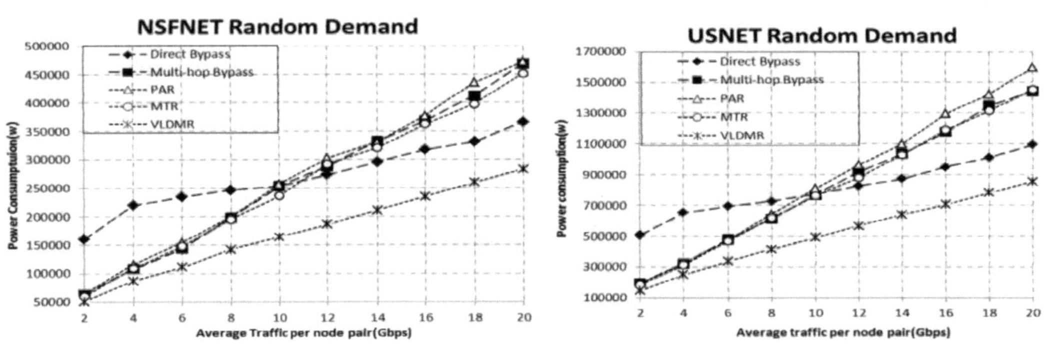

Figure 6. Comparison of total power consumption between different designs by using sleeping method in (a) NSFNET and (b) USNET [7].

4.2. Requirement sleeping mode protection scheme

Protection mechanisms are based on additional equipment to include backup resources in case of failure accident. The protection can be shared in three kind of protection: 1) life path switching, 2) span switching and 3) ring switching. The protection using scheme for backup resources are a standby scheme or in a state of sleep mode for allowing recovery of traffic when a failure appears in the resource end-to-end. In Ren et al [14], the IPoWDM networking planner, with some device of ports and IP router might need, wherein provided by various router devices. Therefore, due to the traffic activity requests among nodes various at different times, the condition can happen where every port on router device in low traffic situation. An essential status on router device sleep is not anyone device of the router ports contained the carry`s traffic. Sleep mode is explained as hibernation mode, where the resource can automatic active if the main link has failure in a short time. Main link of part are active mode and backup link are standby mode.

In Heddeghem et al [7], the summary on a various formulation method for degrading the energy utilization in core network is identified and discussed. Numerous factors have affected and provides an outline of the discrete energy saving methodologies. For ideal port on router designation, there is proposed of a blended router port distribution scheme called "mixed mode". As shown Figure6, the interleaving similarity to get well step of the composite network consumption at the distinctive proportions are to choose that the lower utilization and the result relating to energy consumption and scheme can be downgraded the power consumption by over 40% compared with the non-sleeping mode under the lightpath bypass technique.

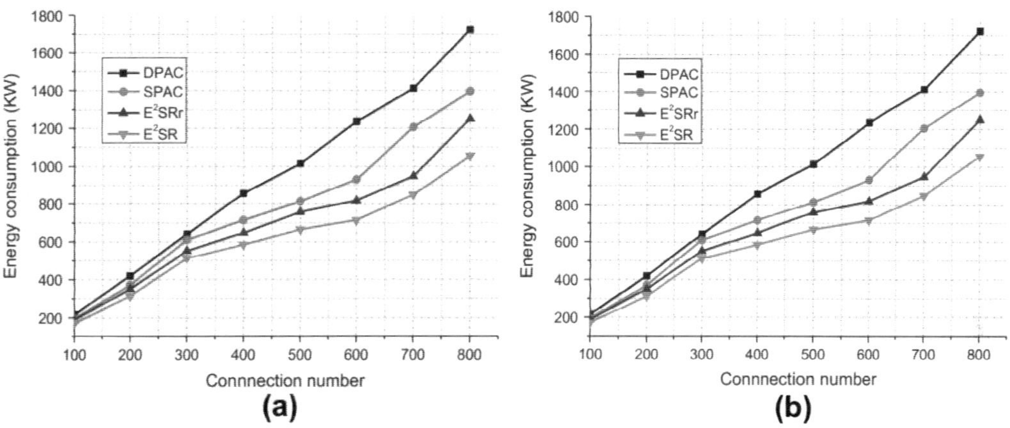

Figure 7. (a) Power consumption of NSFNET and (b) power consumption of USNET [15].

4.3. Requirement hybrid grooming mechanism protection scheme

The important solution for traffic grooming in the optical cross-connect allows full access to routing and switching. However, it cannot give a transparent mechanism for transmitting and receiving by time division multiplexing/ demultiplexing scheme. In Hou et al [2], another mechanism called the hybrid grooming system. The system has two important concepts: (a) the IP request in multiple layers by using the traffic grooming which is manipulated by lightpath in a high-capacity to get electrical mitigation on Transmission Ports (TPs). (b) The lightpaths on optical bypass guarantees the forming of an all-optical (OOO) single scheme and will not change Optical-Electrical-Optical (OEO) conversions on intermediate connection. As result, the consumption of power in each router ports connection to OEO operations is greatly reduced. In Chen et al [16], the switch node can be classified as an optical multi-hop partial grooming optical cross-connect (OXC) in any specific network scenario irrespectively. The interconnection of the optical layer, allows the pass through physical platforms (PP) to Reconfigurable Optical Add-Drop Multiplexers (ROADM). The ROADMs are responsible on physical fiber for fulfilling and delimiting traffic schematic on the optic links and permitting in the intermediate transit node to bypass lightpath in physical platforms optic.

In Guo et al [17], the power efficiency on grooming survivable are optimized using the metric referred by power ratio. The power ratios have an estimation of energy consumption to establish rapport lightpath over degreasing with grooming as requirements to mitigate intemperate power utilization. Therefore, the incorporated multi-hop grooming will merge with the power survivable hybrid which more efficiency on the waveband port as energy saving. In Zhang et al [15], from the results of experiments indicate that the power ratio is comparison between the total power consumption and hybrid grooming methods in Figure7. The power efficient of hybrid grooming method is recommended to upgrade network source advantage. From the energy model of IPoWDM hybrid, the decline of electrically lightpath conversion can be degraded because of power consumption.

5. Clarification and discussion

This paper reviews the classification and discussion of the feature on protection and reconfiguration in the related article. The general context of taxonomy under category was facilitated by being group grouping in reading material that has been done by previous researchers to develop algorithms or learning the techniques of power saving in IP-over-WDM apply. In recent years, many studies and researches have been performed for effort a power efficient in networking optimization, which can be referred in [4] and [18] as two comprehensive of survey standard. Most of the research in the literature focus on model to decrease the total power consumption or reduce the capacity of traffic density in optical transmission WDM and IP networks. We grouping at least in three large groups in the use of mechanism: based on switching on-off scheme [19,20] in components of the network or we call protection in link, the standby or sleeping mode [9] to protection the path routing of the traffic [21].

A traffic model of power consumption in routers is presented in three groups: 1. dynamic model [3,11], 2. static model [22,23] and 3. combined model [8,24]. The application developing on approach an ILP optimization International Technical Conference 2018 ITC2018 (4-6 Oct 2018, Semarang, Indonesia) 5 model [2,25] or heuristic based on algorithm [5,26] to minimize the power of the network, or practical in router board [1] and base on routing configuration scheme [27,28]. The mentioned three aspects of the problem survivability of power saving in this related paper is protected, restorated and mathematical predicted model (as the performance metric and formulating). According to these protection aspects, we classify into two ways. In the first group, the protection in Link performance hop failure tolerance requirement. The second group, protection in path end-to-end traffic.

5.1. Performance measures

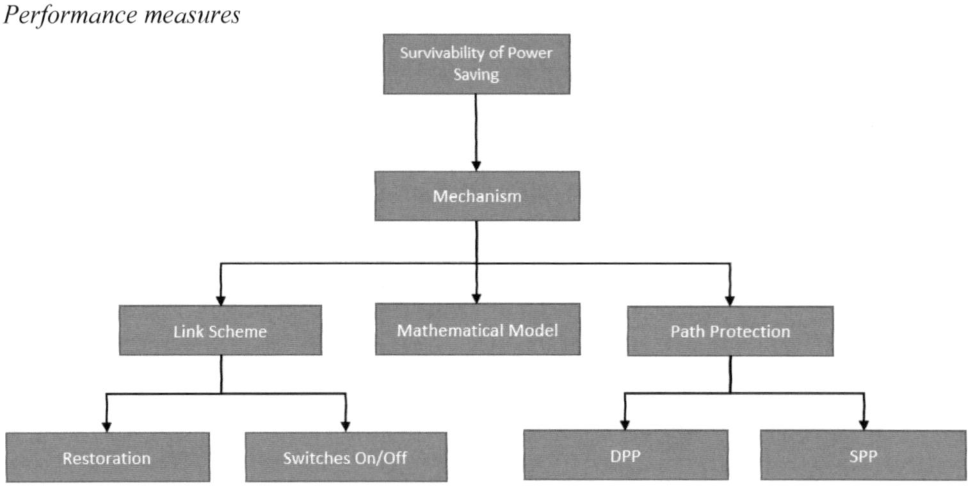

Figure 8. Taxonomy of research literature on survivability power saving in IP over WDM.

The approach of different validation and adoption of performance evaluation in the limited various studies have been tried to reviewed. The motivations of this work to help efficient performance in the same whereby the challenges and recommendations of the research have been in place to describe and innovate for better development of concepts. In this section, all of these parameters are discussed in Table 1 comparison of related work used in the reviewed papers. Table 1 presents various studies on IP over WDM to prove the energy saving classification, as well as a comprehensive survey related of various criteria and sub-criteria for evaluation. In Table 1, as illustrated by variations in the percentage saving of criteria among various method.

Table 1. Summaries of previous articles approaches and methods.

Reference	Focus Mechanism			Modeling of Approach				Protection/Scheme			
	Link Scheme	Path protection	Math-model	ILP	Heuristic	Routing base	Math-model	Switches On-Off	Restoration	DPP	SPP
[2]	•			•					•		
[3]		•		•							•
[5]		•			•						•
[8]		•		•	•				•		
[9]		•		•						•	•
[10]			•				•				
[11]	•				•				•		
[12]	•			•	•			•			
[13]	•			•	•				•		
[14]		•		•					•		
[17]	•				•				•		
[19]	•			•				•			
[20]	•			•				•			
[21]	•				•				•		
[22]	•			•				•			
[23]		•		•	•					•	
[28]	•			•	•				•		
[24]		•		•						•	
[25]	•				•						•
[26]	•	•				•		•	•		
[27]		•				•			•		
[29]		•				•			•		
[30]	•			•				•			
[31]			•				•				
[32]			•				•				
[33]		•				•					•

Our review to perform evaluation of various classifiers of method in queries studies involved in; this aspect measures the effectiveness of focus mechanism, modeling approach, and protection scheme. In this section, all of these parameters are discussed in every chart comparison of related work used in the reviewed papers. The variance in the use of these criteria indicates an achievement in giving summary the power and capacity to specific next researcher.

5.2. Clarification: Objective fiction and traffic model

This study works to summarize the modern classification base Objective Fuction and Traffic Model in IPoWDM to know that how far the development of methods to get power saving. The focus of technique and method to compile the relevant for the literature is set as a basis of the indicate taxonomy. The taxonomy from literature is review the development in trend the research in specific area to show the advantage and benefit for the next research. Some of different article apply the subject from a basic aspect, while other develop the technique and method efficiently. Some make contributes for advance and complexity to the real solution on the research.

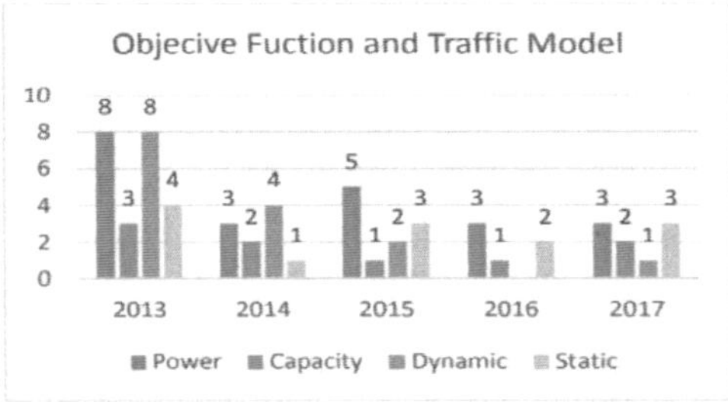

Figure 9. Summaries of previous articles Objective fuction and traffic model from reference.

From the chart in Figure 9, the review of previous articles divided in two focus objective function: 1) focus on efficiency power and 2) focus for optimization in capacity. It is differenting aspect but analogous to get a minimizing power saving. Some of work show the result of attempt that the power effective relies on the total of active device. the disabling unused ports on a line card reduces the device power consumption. Developing of strategy on an IPoWDM architecture may require to discover the routing that reducing the quantity of utilized connections while optimizing all the demands. On the other hand, utilizing higher capacity of connection offers speed and capability more over optical capacity in every node. Furthermore, prediction of traffic models makes each research very interesting when they use or comparing the dynamic or static model. For example, in [21], traffic model indirectly effects the dynamic threshold on network power consumption by control of high DHT (Dynamic High Thresholds) and allowing for more efficient utilization of lightpath capacity. Therefore, the algorithm inspected is more evaluate of sampling in 100 experimentations that display diverse traffic based on the initial scramble.

5.3. Focus mechanism

The IPoWDM networks are referred to give guarantee for further generation networks of core architecture on resilience and the adjustments ability by proposing the controlling on IP protocols and feature at optical layers e.g. substantial bandwidth provision, reasonable latency, minimum error of bit rate, and others [29]. The challenging aspect of researcher for reconfiguring designing efficient power-saving strategies as a new paradigm of resource energy saving has been proposed for more proportional on the energy consumption and amount of the traffic executed.

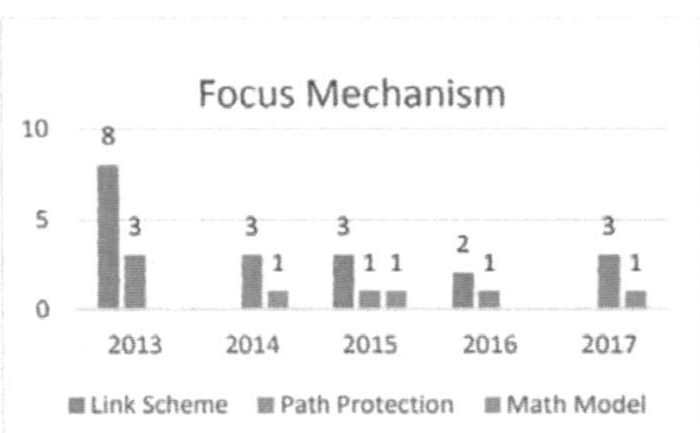

Figure 10. Summaries of previous articles in focus mechanism.

Based on the graph in Figure10, the focus mechanism to get the best power saving changes from the link scheme to the path protection. This is felt good where the path protection will produce a balance in transferring data from source to destination. There is two group in survivability mechanisms: 1. focus in path protection and 2. focus in link scheme. Path protection is improvement scheme for planned failure, whilst link scheme is manual improvement restoration carried out after or during a failure. According to research [30] the weakness of survivability based on restoration scheme is less then dependable and require more time to restored, then the path protection is more normally utilized by administrators.

5.4. Modeling approach

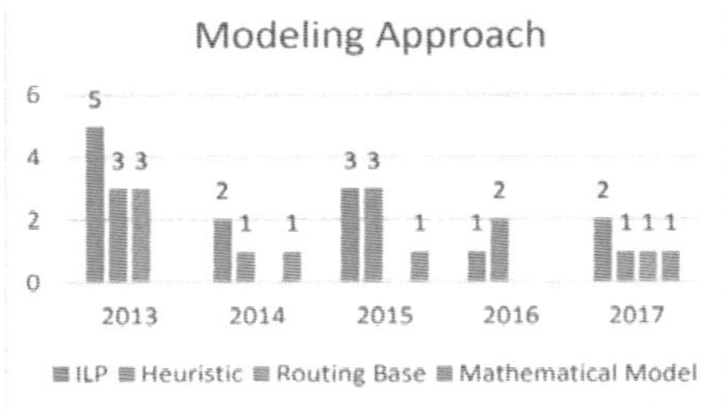

Figure 11. Summaries of previous articles in modelling approach.

In this systematically review, there are four approaches adopted: 1) approaches for design in MILP, 2) approaches for design heuristic method, 3) approaches for design routing base, and 4) Math-model. All approaches adopted aim to getting the power saving in the IP Over WDM network.

In modeling approaches, the ILP approaches and Heuristic approaches are always used from year to year. ILP approach is intended as a linear mathematical calculation method in obtaining optimal solutions to the scheme by maximizing or minimizing objective functions. Whereas the Heuristic Method approach aims to solve simple problems based on samples. This technique can be further investigated into more complex problems.

5.5. Protection / schame

Finally, these are comparison between protection and scheme. Some of research that has been done, focuses on these two fields. on the 5.c, it has been alluded that the protection in this case tends to prioritize the concept of survivability (DPP or SPP) and migration (restoration metrics) while the link scheme is more inclined to physical concepts.

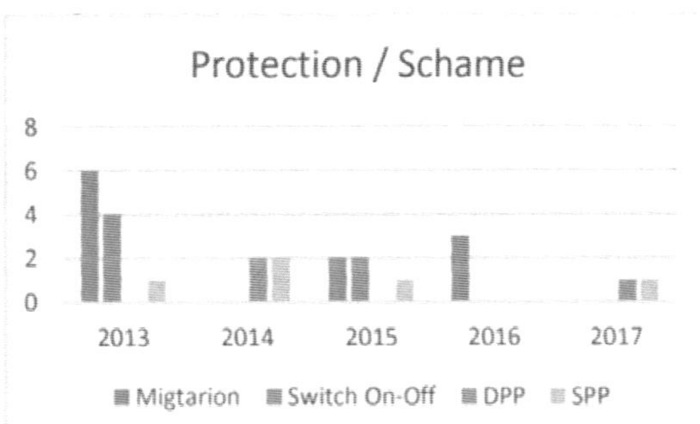

Figure 12. Summaries of previous articles in protection/schame.

In Figure12, Migration techniques that are often used of researcher. These technique has two ways in general, namely Migration (upgrading devices / relocation) or setting routing metrics (link failure protection). In the subjective author, we have a tendency of view researching in the sleep mode scheme to reduce power consumption on network devices. We notice that the internal factors of the existing scheme cannot be changed pragmatically and sleep mode can follow the existing scheme.

6. Conclusion

In this paper review the general view in terms of techniques research to get the energy saving estimates, which is one of methods to approximation the total of energy consumption. The increase of the energy consumption is substantial responsibility and a lot on strategies for energy saving in telecommunication aspect. We group the metadata for comparing techniques on reducing energy consumption from previous work in the development of IP over WDM technology. This is due to the development of various telecommunication technologies by researchers with diverse background to achieve knowledge requisite at diverse phase of implementation. Finally, the mechanism, protection scheme, capacity and power values appropriate is solution survivability technologies as well as energy saving by researcher as a global issue in the telecommunication networking.

Acknowledgment

The author (s) would like to thanks in Management and Science University (MSU) for providing to support and facilities to conduct this research.

References

[1] Meo M, Le Rouzic E, Cuevas R and Guerrero C 2014 Research challenges on energy-efficient networking design *Comput. Commun.* **50** 187–195

[2] Hou W, Guo L, Gong X and Sun Z 2013 Multi-domain integrated grooming algorithm for green IP over WDM network *Comput. Commun.* **36**(3) 342– 350

[3] Jalalinia S S and Cavdar C 2017 Green and resilient design of telecom networks with shared backup resources *Opt. Switch. Netw.* **23** 97–107

[4] Maaloul R, Chaari L and Cousin B 2018 Energy saving in carrier-grade networks: A survey *Computer Standards and Interfaces* **55** 8–26

[5] Mayoral A, Lopez V, Gerstel O, Palkopoulou E, Gonzalez De Dios O and Fernandez-Palacios J P 2014 Minimizing resource protection in IP over WDM networks: Multi-layer Shared Backup Router in *Conference on Optical Fiber Communication, Technical Digest Series* **7**(3) 440–446

[6] Galán-Jiménez J and Gazo-Cervero A 2014 Using bioinspired algorithms for energy levels assessment in energy efficient wired communication networks *J. Netw. Comput. Appl.* **37**(1) 171–185

[7] Van Heddeghem W, Lannoo B, Colle D, Pickavet M and Demeester P 2014 A Quantitative Survey of the Power Saving Potential in IP-over-WDM Backbone Networks *IEEE Commun. Surv. Tutorials* 1–1

[8] Dey S K and Adhya A 2015 IP-Over-WDM Network Design Methodology to Improve Efficiency in Overall Expenditure Due to Cost and Energy Consumption *J. Opt. Commun. Netw.* **7**(6) 563

[9] Wu G and Mohan G 2014 Power-efficient integrated routing of sublambda connection requests with traffic splitting in IP over WDM networks *Comput. Networks* **70** 16–29

[10] Yevsyeyeva O and Ilyashenko Y 2014 Energy effective Routing and Wavelength Assignment in IP-overDWDM Networks *First Int. Sci. Conf.* (2) 26– 28

[11] Beletsioti G A, Kyriakopoulos C A, Papadimitriou G I, Nicopolitidis P and Varvarigos E 2015 Green networks: An energy-oriented model for IP over WDM optical networks *Int. Congr. Ultra Mod. Telecommun. Control Syst. Work.* 1–6

[12] Zhao X, Shen G, Shao W and Peng L 2015 EnergyMinimized Design and Operation of IP over WDM Networks with Traffic-Aware Adaptive Router Card Clock Frequency *IEEE J. Sel. Areas Commun.* **33**(12) 2847–2862

[13] Beletsioti G A, Papadimitriou G I and Nicopolitidis P 2016 Energy-Aware Algorithms for IP over WDM Optical Networks *J. Light. Technol.* **34**(11) 2856–2866

[14] Ren C, Wang S, Ren J, Qian W, Zhang X and Duan J 2017 Energy-efficient virtual topology design in IP over WDM mesh networks *Comput. Networks* **112** 223–236

[15] Zhang X, Wang H and Zhang Z 2014 Survivable green IP over WDM networks against double-

link failures *Comput. Networks* **59** 62–76
[16] Chen X and Phillips C 2014 An evolutionary based dynamic energy management framework for IP-overDWDM networks *Sustain. Comput. Informatics Syst.* **4**(2) 94–105
[17] Hou W, Guo L and Gong X 2013 Survivable power efficiency oriented integrated grooming in green networks *J. Netw. Comput. Appl.* **36**(1) 420–428
[18] Pyoung C K and Baek S J 2018 Joint load balancing and energy saving algorithm for virtual network embedding in infrastructure providers *Comput. Commun.* **121** 1–18
[19] Rizzelli G, Morea A and Tornatore M 2013 An Analysis of Daily Power Consumption under different On-Off IP-over-WDM Translucent Design Approaches *Opt. Fiber Commun. Conf. Fiber Opt. Eng. Conf.*
[20] Rizzelli G, Morea A, Tornatore M and Pattavina A 2013 Reach-related energy consumption in IP-Over-WDM 100G translucent networks *J. Light. Technol.,* **31**(11) 1828–1834
[21] Melidis P, Nicopolitidis P and Papadimitriou G 2016 Dynamic Threshold Reconfiguration Mechanisms for Green IP-Over-WDM Networks **34**(18) 4354–4363
[22] Cercós S S, Resendo L C, Ribeiro M R N, Fagertun A M and Monroy I T 2015 PowerAware Rationale for Using Coarse-Grained Transponders in IP-Over-WDM Networks *J. Opt. Commun. Netw.* **7**(9) 825
[23] Musa M, Elgorashi T and Elmirghani J 2017 Energy efficient survivable IP-Over-WDM networks with network coding *J. Opt. Commun. Netw.* **9**(3) 207–217
[24] Lui Y, Shen G and Shao W 2013 Design for EnergyEfficient IP Over WDM Networks With Joint Lightpath Bypass and Router-Card Sleeping Strategies *J. Opt. Commun. Netw.* **5**(11) 1122
[25] Kuperman G, Modiano E and Narula-Tam A 2014 Network protection with multiple availability guarantees *Comput. Networks* **74**(PA) 1-12
[26] Xu Z, Huang J, Zhou Z, Ding Z, Ma T and Wang J 2013 A novel grooming algorithm with the adaptive weight and load balancing for dynamic holding-time-aware traffic in optical networks *Opt. Fiber Technol.* **19**(5) 392–399
[27] Coiro A, Listanti M, Valenti A and Matera F 2018 Energy-aware traffic engineering: A routing-based distributed solution for connection-oriented IP networks *Comput. Networks* **57**(9) 2004
[28] Yu X, Xiao G and Cheng T H 2013 Historical data learning based dynamic LSP routing for overlay IP/MPLS over WDM networks *Opt. Fiber Technol.* **19**(4) 309–318
[29] Yang S, Wieder P, Yahyapour R and Fu X 2017 EnergyAware Provisioning in Optical Cloud Networks *Comput. Networks* **118** 78–95
[30] López Vizcaíno J, Ye Y, López V, Jiménez F, Musumeci F, Tornatore M, Pattavina A and Krummrich P M 2014 Protection in optical transport networks with fixed and flexible grid: Cost and energy efficiency evaluation *Opt. Switch. Netw.* **11** 55–71

Wind speed monitoring and alert system using sensor and weather forecast

M Mardiyono*, R E Sari and O N Dini

Electrical Engineering Department, Politeknik Negeri Semarang, Semarang, Indonesia

*mardiyono@polines.ac.id

Abstract. Strong wind allows uproot the trees and causes human victims to be injured or dead. Unfortunately, the strong wind comes suddenly to the trees on the street and many people and traffic are around there. It needs wind speed monitoring and strong wind early warning to avoid the victims due to fallen tree. This paper explains the development of wind monitoring and early warning system using wind speed sensor and weather forecasting. The steps involve investigating the wind speed, designing the system, developing the system, testing, and validating. The result shows that the system can measure the wind speed on the spot in real time and trigger the alert to the people around if the strong wind was occurred. On the other hand, it also informs the forecast of wind speed according to weather web services data few hours before. Applying this system will help the government to prevent the victims of fallen tree due to strong wind.

1. Introduction

Fallen tree caused by strong wind is able to cause human casualties. During 1995 up to 2007 there were 407 death tolls where 44% of victims (180 people) are fallen tree on the street in US [1]. The winds that are caused the deadly fallen tree in US involve thunderstorm, non-convective high winds, tropical cyclones, and tornadoes. In Indonesia, the strong wind occurs in transition time from dry to rainy season while the thunderstorms often occur in January or February. There is no detail information related to human victims due to fallen tree in Indonesia but it was occurred in 28 March 2017 that 12 trees uprooted in Jakarta during two hours' heavy rain. To prevent or decrease the victims of deadly fallen tree, the government should provide early warning system of dangerous strong wind which may cause deadly fallen trees.

Several researchers deal the problem with developing wind speed warning system to inform the wind speed level to the people surroundings the neighborhoods. Wind speed Detection system is developed using agile method, wind sensor, microcontroller and software to calculate the wind speed level and display the warning light or message when it identifies the strong wind [2]. Other cross wind warning system utilizes rotary encoder proximity anemometer, microcontroller, alert Red LED and alarm which are ON if the wind speed is greater than 22 knots [3]. The next approach is using weather data mining collected by sensors of rain fall, temperature, humidity, wind speed and direction. The data are transmitted to the server and used to predict flood disaster early warning system [4]. Based on the approaches mentioned in previous works, it is likely that the wind speed warning system still can be improved adequately by combining both local wind speed sensor and weather data forecasting available in the weather web services. This weather forecast data can be used to give early warning system before the extreme wind strikes.

This paper explains the investigation of wind speed level that can uproot the trees in Indonesia and development of wind speed early warning system using wind speed local sensor and weather forecast. This work was motivated by the facts that many news reporting the fallen tree death victims related to strong wind on the road. This system hopefully informs the dangerous level of wind speed in real time and warns people to get away from the trees. Weather forecast data was also carefully processed such that it can be used to inform people several hours in advanced, before the strong wind will come to strike the suspected locations. The paper is organized as follows. Section 2 describes the overview of related works followed by section 3 that explains the research method. Section 4 discusses the result and discussion and finally it is concluded in section 5.

2. Related works

Recently natural disaster management and mitigation has been equipped by early warning system to reduce or prevent human victims. The natural hazards overspread in the forms of earth quake, tsunami, flood, landslide, storm, snow, lightning, fallen tree, and etc. Warning system is enhanced with multi sensors, data processing/computing, and informs alerts to authorities and communities. Researches emphasizing to earth quake warning system include earthquake alarm systems (ElarmS) in California [5], earthquake early warning using accelerometer and seismometers applying in Italy [6], and intelligent building damage index alert system due to earthquake in Malaysia [7]. On the other hand, the topics of early warning system regarding to tsunami has been studied by Chatfield and Brajawidagda [8] in Indonesia using social media Twitter to inform whether the earthquake effects the tsunami or not, Kanazawa in the project of Japan Trench earthquake and tsunami monitoring network covers the sea regions involving Hokkaido, Tohoku, and Kanto [9]. Furthermore, Pinto and Cipriano develop Tsunami Early Warning System (TEWS) comprising sensors, communication, processing, dissemination, and decision support system [10]. Other researchers concern to the flood detection and monitoring system including Mousa et al. [11], Putra et al. [4], Rahadianto et al. [12], and Febrianto et. al. [13]. Early warning systems related to lanslides and debris flows are also studied and developed by Baum and Godt [14] that group the warning levels (null, outlook, watch, and warning), Jian et al. [15] that propose 3D WebGIS platform utilizing several sub-systems (wireless sensor network (WSN), analyser, criteria decision, and alert), Kebaili et al. [16] that enhance the prototype using web of thing and MongoDB Big Data database, and Sattele et al. [17] who study the reliability and effectiveness of debris flow warning system.

The contributions of this work are proposing new framework of natural disaster early warning system relating to dangerous wind that is potentially uprooted the trees. The framework involves existing modules of previous works [2-4] and proposed modules. The existing modules include wind sensor/anemometer, microcontroller, display, and alarm. In this research, the framework is enhanced by adding the module of weather forecast harvesting from internet and monitoring module to show the information in computer screen. This module allows the system to inform few hours in advanced, before the arrival of the strong wind.

3. Methods

The method of this research utilizes prototype model including requirement communication, quick plan, modeling quick design, construction of prototype, deployment delivery and feedback. This section emphasizes on discussion of modeling quick design and construction of prototype as part of the methods.

3.1. The proposed framework

The goal of proposed framework was designed to warn the people around the strong wind disaster area in text alert display and alarm notification in real time and few hours before using weather forecast data. The Wewarn framework is illustrated in Figure 1.

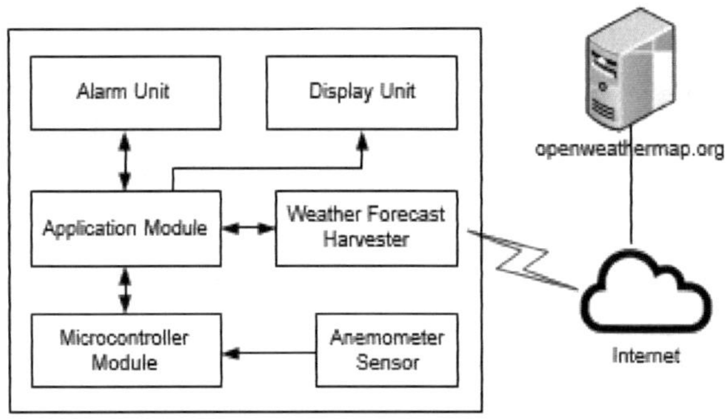

Figure 1. Wewarn framework.

The framework consists of hardware unit consisting of anemometer sensor and microprocessor module, software unit involving application unit and weather forecast harvester, notification unit comprising of alarm unit and display unit. This framework also requires weather forecast web services data from openweather.com. Each module is described as follows:

- Anemometer sensor is used to convert wind speed to electric signal to be processed by microprocessor module. This research utilized anemometer sensor HS- FL2 that converts blade revolution to its respective voltage.
- Microcontroller module as processing unit. It converts the analog signal of anemometer sensor to serial data format. The wind speed data can be read via USB port by application unit. Microprocessor was used in this research is Arduino Uno. Small script of C Arduino was embedded to convert analog data to wind speed value in m/s and then transfer to USB port.
- Weather forecast harvester is small client program using API weather forecast harvester from openweathermap.com. The data is presented in XML format and provided every 3 hours. The weather data used in this research involve in wind speed, temperature, and date-time in specific location (Semarang Indonesia).
- Application unit is a software developed in VB.Net that converts wind speed value from m/s to km/h and determines the warning thresholds in tree conditions such as safe (< 18 km/h), beware (18 - =<60 km/h), danger (>60km/h). The alert of tree conditions refers to wind speed Beaufort scale.
- Alarm unit is a sound media to warn the people around the sensor that actives if the wind speed was categorized in beware and danger level.
- Display unit shows the wind speed on location, level of warning, and weather forecast data in next three hours. Display unit is a LCD monitor. It also displays the early warning notification in text format few hours before based on wind speed data from weather forecast web services.

3.2. Wind speed measurement circuit design

To measure the wind speed, this system needs anemometer sensor, microprocessor, and embedded program developed in C Arduino. This design utilizes analogue anemometer sensor HS- FL2. The measurement range is from 3.6 km/h up to 108 km/h represented in voltage output of 0.17 - 5 volts with tolerance error ± 0.3%. The wind sensor HS-FL2 is depicted in Figure 2.

Figure 2. HS-FL2 wind sensor.

The circuit of wind sensor and microprocessor (Arduino Uno) is illustrated in Figure 3.

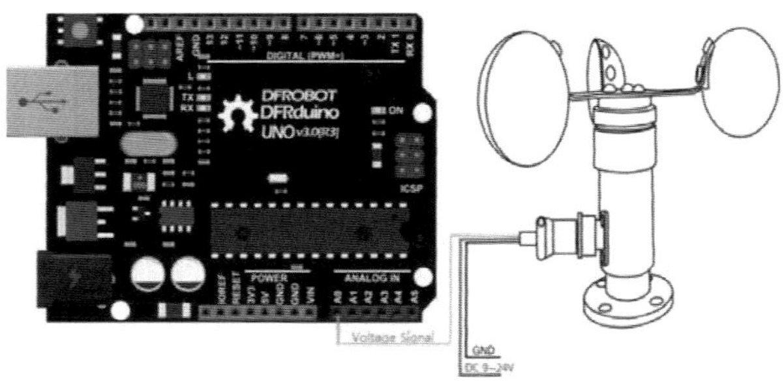

Figure 3. Circuit of wind sensor-microprocessor.

The sensor should be connected to DC power supply 6-24V for output voltage 0-5 V. Signal output of wind sensor is connected to analog input of Arduino A0. Small program written in C Arduino is used to convert analog signal to wind speed value in km per hour. The pseducode of such program is described as follows:

```
void loop
sensorValue ← read analog voltage A0 outvoltage ← sensorValue *
5.0/1023 Windspeed ← 6 * outVoltage * 3600/1000 Serial.println
← Windspeed
```

The program reads analog voltage of wind sensor via port A0 and it is assigned to variable sensor Value. Furthermore, the value of sensorValue is calculated by multiplying 5.0 and dividing 1023 due to decimal data of A0 is integer ranging from 0 to 1023 and the output voltage of wind sensor is 0 to 5 V. To convert outvoltage to wind speed in km/h the value is computed by the equation 6 * outVoltage * 3600/1000. Last step is transferring Windspeed data to serial port (USB).

3.3. Software design

The software running in a computer involves application unit and weather forecast harvester. Both of software are developed using VB.Net. Application unit reads serial data on USB port connected to Arduino board. Wind speed data in km/h are categorized into tree levels namely safe, beware, and danger. If the status is beware and danger, the application will trigger the alarm unit and send notification to display unit (screen). Thus the people around the location will be warned to go away from the trees. On the other hand, weather forecast harvester requests weather forecast data according to specific location (Semarang) to openweathermap.org. The response contains several weather data including date, time, temperature, wind speed, air pressure, precipitation, and weather condition.

In this design, data of wind speed, temperature, and time are used and displayed on the screen. If the next hour wind speed forecast data status is beware or danger the notification will be displayed to the screen as the early warning information. The steps of program are illustrated in a flowchart in Figure 4.

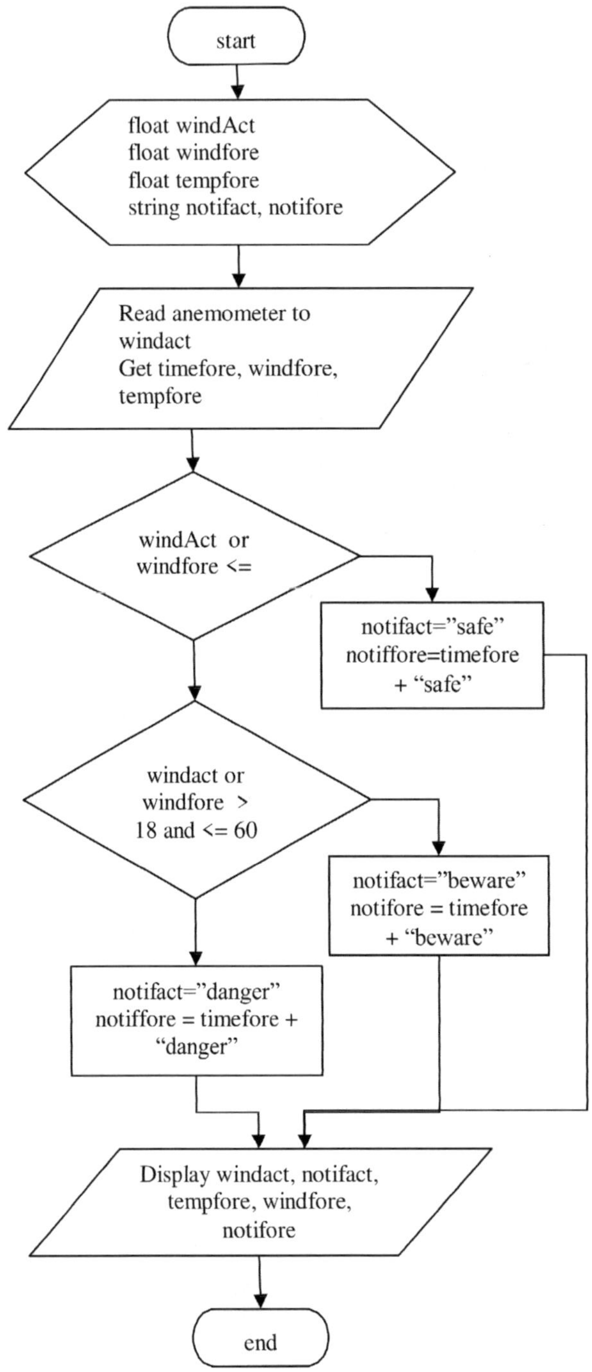

Figure 4. Flowchart of Wewarn.

Figure 4 describes the flowchart of Wewarn system. It obtains data from two sources including anemometer data and forecast data of weather web services. Actual wind speed of anemometer is assigned to windact variable whereas the forecast data (date time, wind speed, and temperature) are assigned to timefore, windfore, and tempfore variable. Each value of windact and windfore is checked wether the value is less than or equal 18 Km/h or not. If the result is yes, then the notifact variable is set to "safe" and notifore variable is assigned to the value timefore and "safe". Otherwise both value of windact and winfore are examined if the values are greater than 18 km/h and less than 60 km/h. If the result is yes then the message "beware" is set to windact and windfore variable, else the massage "danger" is assigned. At the end, all of the data are displayed to monitor including windact, notifact, timefore, windfore, and notifore value.

3.4. Wind speed threshold design

Threshold of wind speed warning refer to beufort wind scale described in Table I. Safe category is set ranging from scale 0 to 3 (gentle breeze) that represents wind speed 10 knots or 18 km/h. This speed will cause leaves and small twigs constantly moving and light flight extended. The second category is beware which refers to scale 4 to 7 having the speed between 11 to 33 knots or 19.8 to 59.4 (rounded off to 60) km/h. The last category is danger which refers to scale 8 to 12. This category will cause from twigs breaking off trees until trees broken or uprooted and structural damage.

Table 1. Beufort scale.

Scale	Wind speed		Desc
	Knots	Km/h	
0	< 1	<1	Calm
1	1-3	1.8-5,4	Slight a ir
2	4-6	7.2-10.8	Slight breeze
3	7-10	12.6-18	Gentle breeze
4	11-16	19.8-28.8	Modera te breeze
5	17-21	30.6-37.8	Fresh breeze
6	22-27	39.6-48.6	Strong breeze
7	28-33	50.4-59.4	Near gale
8	34-40	61.2-72	Gale
9	41-47	73.8-84,6	Strong gale
10	48-55	86.4-99	Storm
11	56-63	100.8-113.4	Violent storm
12	>63	>113.4	Hurricane

3.5. Prototype development

Wewarn system has three components involving anemometer sensor, microprocessor, and desktop application. Two software are developed and installed to Arduino microprocessor and desktop computer. Figure 5 illustrate the prototype design.

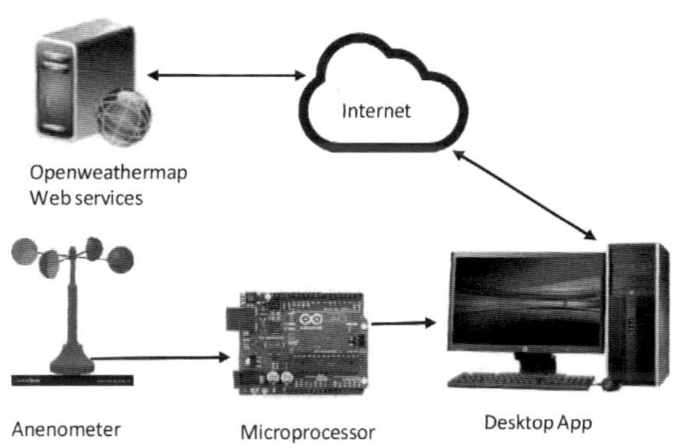

Figure 5. Prototype design.

Threshold of wind speed warning refer to Beufort wind scale described in Table I. Safe category is set ranging from scale 0 to 3 (gentle breeze) that has wind speed 10 knots or 18 km/h. Beware category is set from greater than 18 Km/h to 60 Km/h (4-7 scale). The over 60 Km/h wind speed is a danger level.

4. Results and discussion

The testing of Wewarn prototype is performed in Laboratory of Electrical Engineering Department including wind speed measurement, alert system, weather data harvesting, and display. The results are compared to previous researches in comparative analysis views.

4.1. Wind speed measurement standard error

The measurement data of wind speed are compared to the other digital anemometer measurement data. Wind source is got from fan having three speeds. The sensor measures the wind speed ranging from the distance 0.5m, 1m, 1.5m, 2m, and 2,5, These data are analyzed using standard error equation. Based on the data of Table II, Standard Error (SE) is obtained 4.4%. It means that the measurement is acceptable and applicable.

Table 2. Measurement data.

		Measurements (Km/h)	
Speed	**Distance (m)**	**Other digital anemometer**	**This device**
1	0.5	7.92	8.1
1	1	6.89	6.8
1	1.5	5.38	5.4
1	2	0.00	0.1
1	2.5	0.00	0.5
2	0.5	8.34	8.6
2	1	7.79	7.3
2	1.5	6.33	6.3
2	2	2.68	2.9
2	2.5	0.00	0.5
3	0.5	10.3	10.4
3	1	8.02	8.0
3	1.5	6.64	6.5
3	2	3.91	3.8
3	2.5	0.34	0.5

$$\text{Standard Deviation} = \sqrt{\frac{\sum(x_1-\bar{x})^2}{n-1}}$$

$$= \sqrt{\frac{0.41}{14}}$$

$$= 0.17$$

$$\text{Standard Error (\%)} = \frac{Standar\ Deviation}{\sqrt{N}}\ x\ 100\%$$

$$= \frac{0.17}{\sqrt{15}}\ x\ 100\%$$

$$= 4.4\ \%$$

4.2. Wind speed monitoring and alert

Information of wind speed measurement and weather forecast data are shown in LCD monitor. Application built in VB processes the input data of anemometer sensor and weather forecast of openweather.org. It also notifies the alert in text warning and alarm if the wind speed data are in beware and danger level. The result of wind speed monitoring and alert is illustrated in Fig 6.

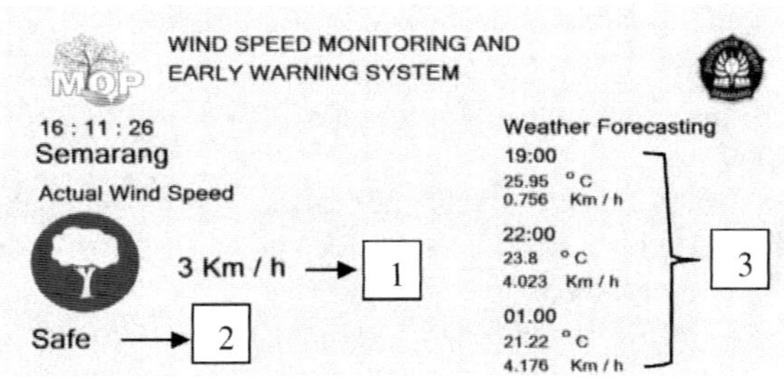

Figure 6. Wind speed monitoring and alert.

Figure 6. shows the information of wind speed measurement, weather forecast data, and alert. The notation 1 describes the current actual wind speed of 3 Km/h whereas the notation 2 explains the alert of wind speed level in safe category. Number 3 illustrates the weather forecast every three hours later including time, temperature, and wind speed.

This research has successfully developed the monitoring and alert system of wind speed. It can be implemented in public area to notify the warning if the level of wind speed is not safe. The system also informs early warning to public if weather forecast data is in level of beware and danger. Related to wind speed monitoring research, this work has similar topics with [2] and [3]. Both of previous researches measure the actual wind speed measurement and group in several levels that this work is similar with this research. This research has innovation feature by combining the actual measurement and weather forecast data for notifying community around the system about the level of wind speed in advanced. Regarding to natural disaster warning, this research has similar topic with early warning of tsunami [8- 10], earthquake [5-7,9], flood [4,11-13,18], land sliding [16], and wind [2,3].

5. Conclusion

This paper describes the development of wind speed monitoring and early warning system using wind sensor and weather forecast data. All of the functionalities including wind speed measurement, weather data harvesting, and warning system have successfully worked based on the testing result. The measurement data is analyzed using standard error equation with other digital anemometer and it is obtained the SE of 4.4%. This work can be enhanced by displaying of actual and weather forecast data in running text display as an outdoor device and informing the warning to social media.

References

[1] Schmidlin T W 2009 Human fatalities from wind-related tree failures in the United States, 1995-2007 *Nat. Hazards* **50**(1) 13–25

[2] Nasir H M, Nur W and Dallina F 2016 Wind Speed Detection System : Framework and Implementation in *International Conference on Information and Communication Technology (ICICTM)* 61– 64

[3] Aulia R, Ervianto E and Lysbetti N 2016 Rancang Bangun Catu Daya dan Software dalam Apikasi Anemometer untuk Alat Peringatan Angin Kencang *J. Online Mhs. Bid. Tek. dan Sains* **3**(1) 1–11

[4] Putra R, Zaini Z, Madona E and Nasution A 2016 Desain dan Implementasi Peringatan Dini Banjir Menggunakan Data Mining dengan Wireless Sensor Network *J. Nas. Tek. Elektro* **5**(2) 181–191

[5] Gasparini P 2007 *The ElarmS Earthquake Early Warning Methodology and Application across California in Earthquake Early Warning systems*, P. Gasparini, G. Manfredi, and J. Zschau, Eds (Berlin: Springer) 45–63

[6] Zollo A and Lancieri M 2007 Real-time Estimation of Earthquake Magnitude for Seismic Early Warning *in Earthquake Early Warning systems*
[7] Mardiyono M and Suryanita R 2011 Prediction of Building Condition after Earthquake Using Artificial Neural Network in *Seminar Nasional Teknologi Informasi III* 47–55
[8] Chatfield A T and Brajawidagda U 2013 Twitter early tsunami warning system: A case study in Indonesia's natural disaster management in *46th Hawaii International Conference on System Sciences* 2050–2060
[9] Kanazawa T 2013 Japan Trench earthquake and tsunami monitoring network of cable-linked 150 ocean bottom observatories and its impact to earth disaster science *in IEEE International Underwater Technology Symposium* 1–5
[10] Pinto D and Cipriano A 2014 Tsunami early warning system architecture based on automation technologies *in 1st International Conference on Information and Communication Technologies for Disaster Management ICT-DM 2014* 1–5
[11] Mousa M, Zhang X and Claudel C 2016 Flash Flood Detection in Urban Cities Using Ultrasonic and Infrared Sensors *IEEE Sens. J.* **16**(19) 7204–7216
[12] Case S and Java E 2015 Risk-Level Assessment System on Bengawan Solo River Basin Flood Prone Areas Using Analytic Hierarchy Process and Natural Breaks *in International Conference on Data and Software Engineering* 195–200
[13] Surabaya C S and Java E 2016 Urban Flood Risk Mapping Using Analytic Hierarchy Process and Natural Break Classification *in International Conference on Knowledge Creation and Intelligent Computing (KCIC)* 148–154
[14] Baum R L and Godt J W 2010 Early warning of rainfall-induced shallow landslides and debris flows in the USA *Landslides* **7**(3) 259–272
[15] Huang J, Huang R, Ju N, Xu Q and He C 2015 3D WebGIS-based platform for debris flow early warning: A case study *Eng. Geol.* **197** 57–66
[16] Kebaili M O, Foughali K, Fathallah K, Frihida A, Ezzeddine T and Claramunt C 2016 Landsliding Early Warning Prototype Using MongoDB and Web of Things Technologies *Procedia Comput. Sci.* **98** 578–583
[17] Sättele M, Bründl M and Straub D 2015 Reliability and effectiveness of early warning systems for natural hazards: concept and application to debris flow warning *Reliab. Eng. Syst. Saf.* **142** 192–202
[18] Dersingh A 2016 Design and Development of a Flood Warning System via Mobile and Computer Networks in *2016 International Conference on Electronics, Information, and Communications (ICEIC)* 1–4

IOP Conf. Series: Materials Science and Engineering 1108 (2021) 012030 doi:10.1088/1757-899X/1108/1/012030

Victim identification application with face recognition based on android

L Triyono*, S Handoko and K Y Puspawaty

Politeknik Negeri Semarang, Semarang, Indonesia

*liliek.triyono@polines.ac.id

Abstract. The increase in the population causes the problems faced are also increasing. Problems that arise include that, the number of accident victims who have recently often occurred on the highway, where the victims of the accident suffered minor injuries to severe and some even have to take lives due to accidents on the highway. And the second is crime, like a murder case that can take someone's life. Often victims do not carry identities, making it difficult for the police or related parties to identify victims of accidents and killings. This system is made in the hope that it can help the police or related parties to be able to recognize the face of the victim without an identity just by using the android application. So that it can save more time in identifying the identity of the victim. Victim face recognition algorithm using Facial Landmark Detector and for its implementation using the Dlib library. To do facial recognition, it is done by taking Figures of victims using an application on android which is then sent to the server for the recognition process. After the image is sent to the server, the recognition process will be carried out, the results of the introduction process are sent back to android to display the results. The level of accuracy of the victim's face recognition with a wound if using a tolerance of 0.35 is 100%, but if using tolerance 0.2 the level of accuracy is only 25%.

1. Introduction

Increasing population, the problems faced were also increasing. Two problems include the first, the number of accident victims who have often happened on the highway lately, where victims of the accident suffered minor injuries to severe and some even have to take lives due to accidents on the highway. The second problem is criminal acts such as murder cases that can take someone's life. Often the victim does not carry an identity so that it is difficult for the police or related parties to identify the victim's identity if the victim is unconscious or dies.

Currently there are many applications that use face detection features. Like applications used for door security systems using face recognition technology [1]. However, this application has not used facial extraction feature, which is a feature based on the shape and placement of facial attributes, such as eyes, eyebrows, nose, lips, and chin, and the relationship between attributes, so that the level of accuracy is still less than optimal. Then there are also mobile applications as monitors and controllers on embedded building security systems [2]. This application uses Eclipse IDE to build applications and java as the programming language. But this application uses IP Camera as a device to take Figures. The disadvantage of this device is that if the network connection is not stable, then this device cannot be used optimally, so an internet connection is needed that is fast and stable if using an IP camera as a medium to take facial Figures. Next is the Student Attendance application with Face Recognition [3]. This application can identify student faces and store the data as attendance data in Excel files so that it can facilitate relevant parties to recap student attendance data. But there are a few weaknesses in this application, that is the photos taken must be in the same lighting conditions as the condition of the face photos stored in the database, this will be a little difficult if the student is absent in the afternoon or when the weather conditions are minimal and not in accordance with the light conditions photos in the database, so that the face cannot be recognized by the system. Subsequent research is the application of face recognition and gender information which is a computer application to identify faces automatically or verify someone's face from the camera to capture someone's face [4]. This is usually used in access control systems and can be compared to other biometrics such as fingerprint recognition systems or iris. This application can help the process of increasing security in obtaining information on someone's identity by detecting faces. But in this application can only display the results of the introduction in the form of name and gender only, and for data storage has not used a database so that the data stored is limited. And the last is a computer vision based application with the aim of detecting visual hacking [5]. The application created can recognize the user's face and will issue a warning window if a peeping face is detected. However, the weakness of this application is that it cannot detect faces facing the right or the left and the face with the head tilted even though the face position is still facing forward, so the accuracy level will be reduced.

The purpose of this study is to build an Android-based face recognition application. This application is used to identify the face of the victim and determine the identity of the face in accordance with the data already stored in the database on the server. Based on several existing face recognition applications, many still use desktop or PC media to run the application. To make the application more practical, the Face Recognition application will run on the android platform. Because at this time most people already use Mobile with an Android system so that it will be easier and more practical if this application can be run on Android. And this application not only displays the identity in the form of a name, but this application will also display the victim's identity in full which includes name, address, place of birth date, and religion.

2. Research methods

The method used in this study is the Waterfall method which includes analysis, design, writing program code, and testing. The stages described in more detail in this section are the design process. The design process for the development of android-based face recognition applications includes a general system description, context diagram, and ERD.

2.1. System overview

"Face recognition" Victim Identification Application with Android-based Face Recognition is an Android-based application system where the interface and android functions are built using Java, and Python, while for data storage, the system uses a MySQL database. The working principle of this system is to recognize the face of the victim without an identity and display the results of the identification of the identity of the person. This application runs on the Android operating system.

The stage of face recognition includes taking Figures / photos from the Android phone through the face recognition application to then be sent to the server then the server processes the data to be able to recognize and display the results of identification of the victim's face in the form of the victim's identity to the android application that was used to take previous Figures. But this application has restrictions that can only recognize faces that are still intact, if there are accident victims who suffered severe injuries in the face, then this application cannot accurately identify the face.

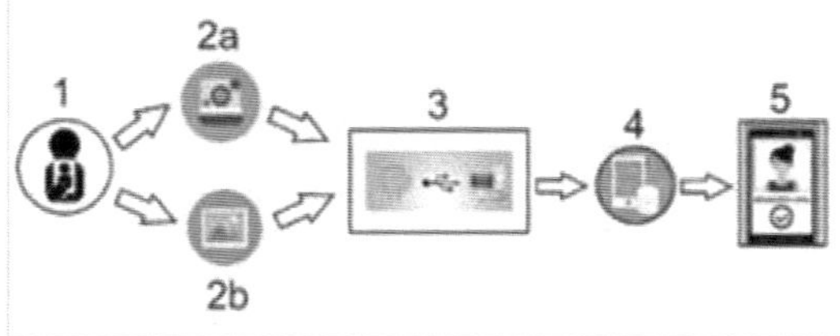

Figure 1. Application overview.

Figure 1 is an overview of the Face Recognition System for Victim Face Recognition Applications. From the Figure, the process starts from number 1, that is the user captures images from the android application, then goes to processes 2a and 2b where the user can choose to take Figures from the camera or from the gallery. Then on the 3rd process the image is sent from the android application to the server, then on the server the face recognition process uses the Dlib library. Then the server also updates the process data in the MySql database in process number 4. Then on the 5th process, the face has been identified and the recognition results are displayed in the Face Recognition application on android.

This system has 2 entities involved, namely the user (user) who will run and use the Face Recognition application and the admin who is tasked to enter the identity data of people who are recognized into the database. The introduction data includes data on Photo, Name, Address, Place of Birth Date and Religion. Figure 2 will describe the tasks of the admin:

Figure 2. Admin enters data.

Then for the flow of the process of running the application the author describes using a flowchart as follows:

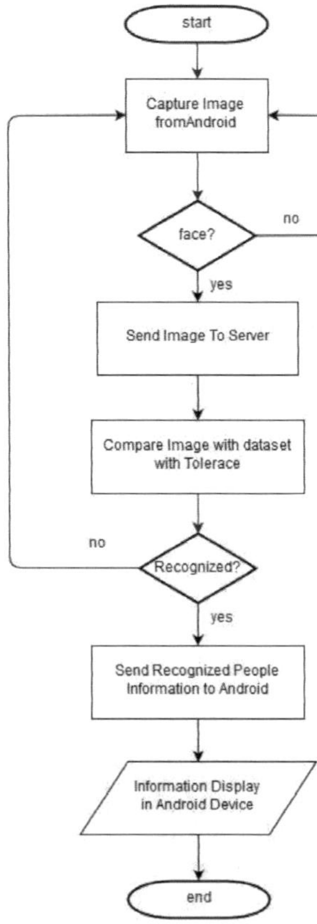

Figure 3. Flowchart system.

From the Flowchart system in Figure 3 it can be explained that the face recognition process begins with capturing images from an Android application, then face detection is done through the application, if the face is detected, a box image appears on the face that can then be captured. This face detection facility is used so that users do not come from sending random images and there are no drawn faces to be sent to the server, so the purpose of this face detection facility is to streamline the face recognition process for a person's face image. Furthermore, if the face is not detected, the process will return to the previous stage until a face is detected. Then after the face image is captured, the image will be sent to the server, then this image data will be compared with the training data that the admin has entered into the face recognition database. After being compared and processed using a certain tolerance, if the face can already be identified, the face recognition result will be sent to Android and then displayed to the android application used to capture the image. But if the face is not recognized, then the process will be repeated to the image capture process for later recognition process (Face Recognition) again. Tolerance is a server-defined limit value to limit the maximum distance value on facial recognition results. The smaller the tolerance, the smaller the distance generated, the greater the level of accuracy. And vice versa if the tolerance is greater, the distance generated is greater so that the level of accuracy will decrease. Tolerance values range from 0.1-0.7 and can be set on the server.

2.2. ERD

Database relations or ERD indicate a mechanism used to manage or organize data physically. The linkages between tables used in system design can be seen in Figure 4.

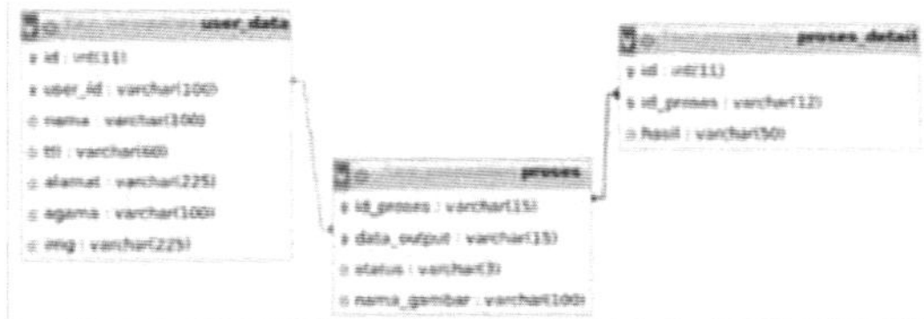

Figure 4. ERD of face recognition application.

The process table relates to the parent tables whose data will be retrieved. The user_id field in the user_data table is related to the id_product field in the order table, then the data_output field in the process table. And the id_proses field in the process table is related to the id_proses field in the detail_detail table. This relation functions at the time of determining the identity data of the victim that is recognized by the system to be displayed to the Face Recognition application.

3. Results and discussion

3.1. Functional testing

Functional testing is done on android and server applications. Functional testing on the features of the android application is shown in Table I.

Table 1. Server functional testing.

No	Testing	Status
1.	Receive photos from the Android Application	Succeed
2.	Insert process data into the Database	Succeed
3.	Recognize the face of the victim without injury	Succeed
4.	Recognize the face of the victim with a wound	Succeed
5.	Choose the recognition results with the smallest distance	Succeed
6.	Update status of process in the Database	Succeed
7.	Send introduction results to the Android application	Succeed

Table I shows the results of server functionality testing from the Face Recognition application. From the test results there are 7 functions that have been tested and succeeded. The server can already recognize the victim's face without the wound and face of the victim with a wound (on a certain scale). It can be concluded that the server from the Face Recognition application has been running well to recognize the faces of victims.

Then functional testing on the android application is shown in Table 2.

Table 2. Android functional testing.

No	Testing	Status
1.	Access the camera	Succeed
2.	Take photos from the camera	Succeed
3.	Access photo storage (gallery)	Succeed
4.	Take photos from the gallery	Succeed
5.	Upload photos to the server	Succeed
6.	Showing the results of facial recognition	Succeed

The table above shows the results of testing the functionality of the Face Recognition application and there are 6 functions that have been tested and succeeded. Testing of Android-based Face Recognition applications includes taking Figures from the gallery or camera, sending images to the server and displaying facial recognition results. It can be concluded that the andorid-based Face Recognition application has gone well.

Some functional tests on the Face Recognition application will be shown in more detail in some images. Tests that will show details of the Face Recognition application include taking Figures from the camera, taking Figures from the gallery, uploading images to the server, displaying the results of recognition. Then for the server test display includes receiving images from android, processing facial recognition and displaying facial recognition results.

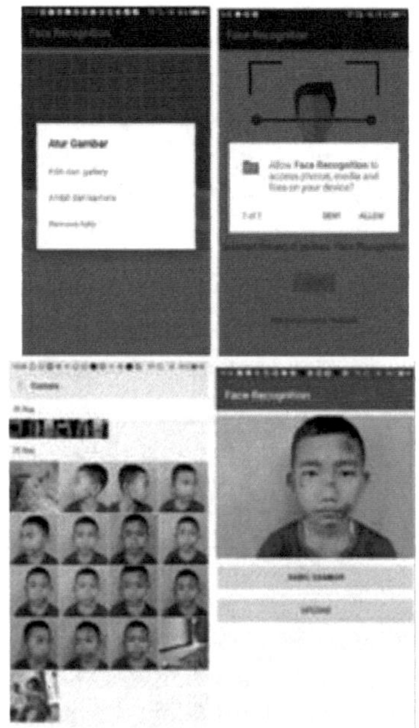

Figure 5. Take a Figure from the gallery.

Figure 5 is a Figure capture detail from the gallery. On the 3 menu options, click "Select from gallery" then the application will display the permission window to access the gallery and take Figures from the gallery. If successful, it will look like the lower right Figure.

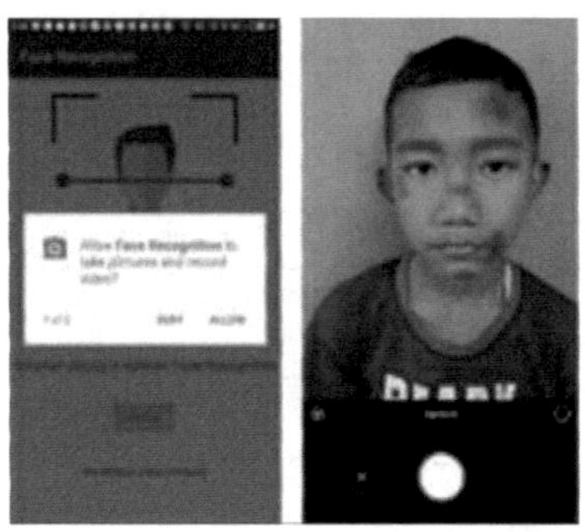

Figure 6. Take a Figure from the camera.

Figure 6 is a Figure capture detail from the camera. On the 3 menu options, click "Take from kamrea" then the application will display the permission window to access kamrea and take a Figure of kamrea. If successful, it will appear as shown in Figure 8 on the right.

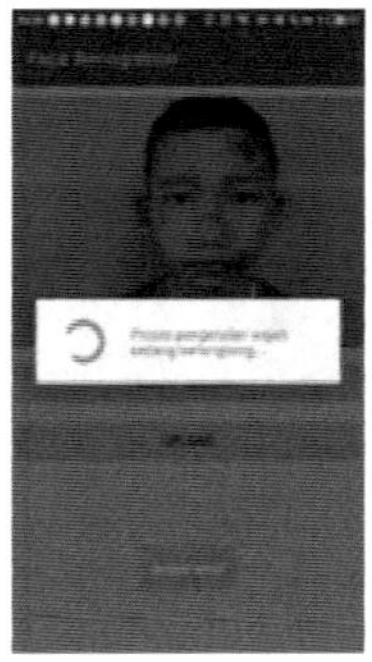

Figure 7. The process of uploading images to the server & the face recognition process.

Figure 7 shows the process of uploading images to the server, if the toast appears that contains "upload successfully" then this indicates that the image from android was successfully uploaded to the server.

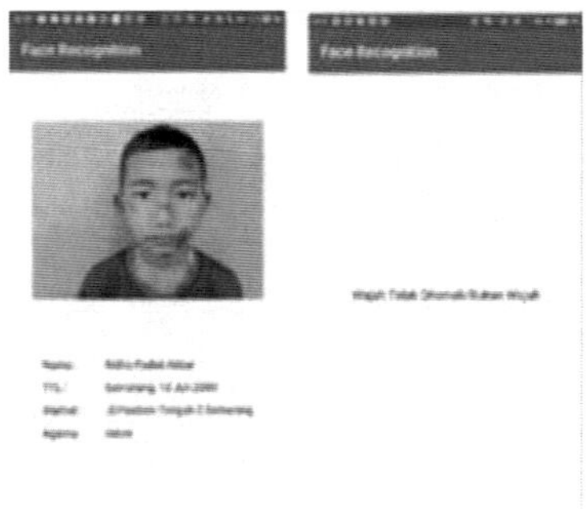

Figure 8. Face recognition results appear on android.

Figure 8 will appear when the server is able to recognize the victim's face. And the identity displayed includes name, place of birth date, address and religion. If the server cannot recognize the image uploaded from the Face Recognition application, it will display results like the right part 10 image.

Figure 9. The process of receiving images from Android.

Figure 9 shows that the server successfully received the images uploaded from the Face Recognition application. the server also managed to add process data to the database. Which is used to distinguish the process of drawing one with the other drawing process.

Figure 10. The introduction process and the results of the introduction.

Figure 10 shows the victim's face recognition process. After getting the recognition result, the server will update the process data that the server has added to the previous database. Then the victim's identity will also appear on the server in accordance with the results of his introduction.

3.2. Testing with several conditions

In testing the Face Recognition application, there are 3 types of testing, namely testing the face with wounds, facial recognition with minimal light conditions, and face recognition taken directly using an android smartphone camera. The results of facial testing with wounds that use a tolerance of 0.35 are all image can detect and recognized.

The following are the results of facial testing with a wound that uses a tolerance of 0.2 are any image can recognized successfully.

Based on the results of the test, the percentage obtained to recognize the victim's face with a wound if using a tolerance of 0.2 is 25%. And if using tolerance 0.35 the success rate reaches 100% (based on 4 data that has been tested).

Then test with photos with less brightness. On the server, the server can detect faces in the image but the server cannot recognize the face.

Then test with photos taken directly from the camera. Recognition results taken directly from the camera can be recognized but use a tolerance of 0.4.

Based on testing with 5 data set can be seen that the results of testing image produce using photos taken directly with the camera, successfully identified the 5 people with a tolerance of 0.4. When using a tolerance of 0.2, the five people cannot be recognized by the server, and if using tolerance is 0.3, there are 3 of 5 people who are known, and the remaining 2 are not recognized by the server.

4. Conclusion

Based on the results of testing "Android-based Face Recognition Application", it can be concluded that:

- The victim identification application with Android-based face recognition has been successfully created on Android devices and has gone through the process of testing functionality with the results of all functions running properly.
- The success rate for testing the victim's face with a wound that uses 0.2 tolerance is 25% of the 4 experimental data. For the success rate of the victim's face with a wound that uses a tolerance of 0.35 reaches 100% of the 4 trial data.
- The number of pixels in the image that is known to greatly affect the length of the introduction process, in this test, there are 50 known images from 10 different people. And the pixel of each image is 2048 x 1536px. Then the average time of recognition is approximately 2 minutes
- Lighting when taking Figures that will be recognized also greatly influences the recognition process. If the light is lacking or dark, the system cannot recognize the face.
- For testing with a face with a wound, it is recognized that the tolerance is set to 0.35. When using a tolerance of 0.2, there are several faces with wounds that the system cannot recognize.

- Introduction using photos taken directly with the camera, successfully identified 5 people with a tolerance of 0.4. When using a tolerance of 0.2, the five people cannot be recognized by the server, and if using tolerance is 0.3, there are 3 of 5 people who are known, and the remaining 2 are not recognized by the server.

References

[1] Hermawan D, Setiawan A B, Prasetya D A and Rabi A 2017 Sistem Pengaman Pintu Gudang Senjata Rudal Arhanud TNI AD dengan Identifikasi Wajah *Semin. Nas. Sist. Inf.* 887–897

[2] Asmara S B 2016 Perancangan Dan Implementasi Aplikasi Mobile Sebagai Pemantau Dan Pengendali Pada Sistem Keamanan Gedung Berbasis Embedded *in e-Proceeding of Engineering* **3**(1) 674–680

[3] Yusuf M, Ginardi R V H and A Solichah A 2016 Rancang Bangun Aplikasi Absensi Perkuliahan Mahasiswa dengan Pengenalan Wajah *J. Tek. ITS* **5**(2) 2–6

[4] Nurmala P, Gazali W and Budiharto W 2015 Sistem Kontrol Akses Berbasis Real Time Face Recognition dan Gender Information *ComTech* **6**(2) 198–207

[5] Alexander L W, Sentinuwo S R and Sambul A M 2017 Implementasi Algoritma Pengenalan Wajah untuk Mendeteksi Visual Hacking *E-Journal Tek. Inform.*, **11**(1)

[6] Putra A T 2016 Face Detection Menggunakan Javascript sebagai Filter Awal Presensi Berbasis Web *KOMPUTAKI* **2**(1) 1–9

[7] Nugraha M 2017 Sistem Pengenal Wajah Manusia untuk Personalisasi Perintah pada Robot *J. Ilm. KOMPUTASI* **16** 139–143

[8] Saputra D I S, Ramadhan K A N, Anjar W S and Pamungkas R A 2017 Pelacakan Dan Deteksi Wajah Menggunakan Video Langsung pada Webcam *Telematika* **10**(1) 50–59

[9] Perdana R N, Irawan B and Setianingsih C 2017 Perancangan Pintu Pintar Untuk Deteksi Wajah Nyata Berbasis Pengolahan Citra Digital *in e-Proceeding of Engineering* **4**(3) 4169–4176

[10] Permana N, Wijayanto I and Rusdinar A 2017 Perancangan dan Implementasi Prototipe Sistem Keamanan Mobil Berbasis Pengenalan Wajah dan Sistem Operasi Android sebagai Kontrol in *e-Proceeding of Engineering* **4**(1) 436– 446

IOP Conf. Series: Materials Science and Engineering 1108 (2021) 012031 doi:10.1088/1757-899X/1108/1/012031

Performance analysis of shape, color and texture features on tracking information face based on CBIR

J Jumi*, A Zaenuddin and T Mulyono

Politeknik Negeri Semarang, Semarang, Indonesia

*jumi_mail06@gmail.com

Abstract. The image has the features of shape, color and texture that are vary. Each feature has a different performance in supporting the accuracy of information retrieval using a process approach to CBIR (Content-Based Image Retrieval). On the image with different objects different performance will be generated on each feature. For example, that the performance features of the form of the more dominant compared features color and texture on the image with the face, while the object on the image with the object of interest feature is more dominant than the features of texture and shape. In this research was conducted on the analysis of the performance features of the shape, color and texture in supporting the accuracy of a search using the approach of CBIR (Content Based Image Retrieval). The method used are invariant moment, color moment and GLCM (Grey Level Co-occurrence Matrix). The results showed that the best search accuracy is 95%, where the features of shape has a performance by 50%, 30% color feature and texture feature by 20% with 600test image with object database face.

1. Introduction

Image is a data that describes the physical form of an object. Imagery through its feature content has uniqueness or characteristic that can distinguish between object one with other object. This allows the image to be an alternative key to information retrieval other than using a text field key. The search model using the image as a key field can be done through the CBIR (Content Based Image Retrieval) approach using the features owned by each image.

In the development of research on content-based retrieval image or called Content Based Image Retrieval (CBIR) has done some retrieval method based on image features. Research on CBIR using texture features as a similarity analysis method or similarity as a basis for retrieval has been done by producing 54% similarity precision value [1]. The use of combination method of color and shape features has been done that applied to the medical database [2]. CBIR research by combining more than two image features is also done using the features of shapes, colors, textures and edges with a value of retrieval accuracy of more than 75% [3].

2. Related works

Several CBIR studies using a combination of 3 features yielded a precision value of more than 60% [4], subsequently similar studies have resulted in a recall value of test results using Wang data being over 85% [5], followed by similar studies with value results precision and recall more than 90% [6]. Research using the combination and weighting of image features on indoor asset imagery has also been done by generating retrieval accuracy varies according to the percentage of weighting of the three features, in which the shape features are more dominant than the color features and texture features [7].

From some research above can be concluded that combination of some image feature will yield level of accuracy of resemblance higher than without doing combination. Thus each feature has different performance in its support for improving the accuracy of the search by CBIR. The extraction method used also affects the retrieval accuracy by the CBIR approach.

In this research, the performance analysis of each feature is a feature of shape, color and texture to support the improvement of retrieval accuracy. Each feature has a different percentage of support for CBIR-based information retrieval. In the image of different objects will be different percentage of performance of each feature generated.

3. Research methods

Some research that has been done have not done performance analysis of shape, color and texture in support of CBIR image search accuracy improvement. In this research have done retrieval image using three features with research step shown by Figure 1.

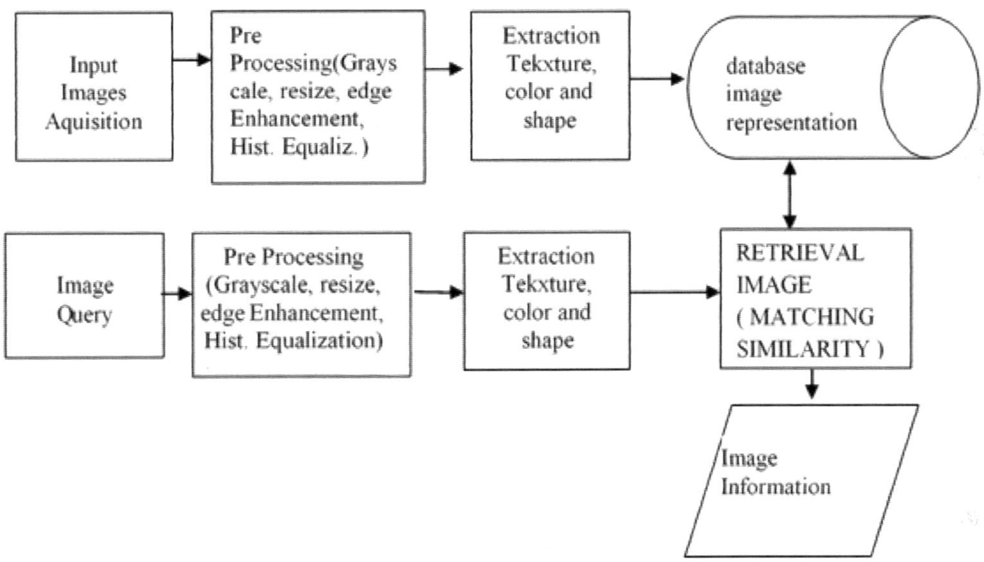

Figure 1. An overview of CBIR retrieval stages using CBIR approach using texture, color and shape features.

Details of the stages of this study are as follows:

3.1. Preprocessing

Preprocessing is the improvement of image quality prior to feature extraction with the aim of improving image feature performance in support of retrieval accuracy. There is a preprocessing difference in feature form extraction with color and texture feature extraction. The preprocessing difference is to get a quality image before feature extraction.

3.1.1. Preprocessing features of the texture and color features

- Resize: The larger of image size so time extraction is also longer, so it needs resize stages to speed up the computation process. At this stage the image resized to 300 x 300 pixels.
- Grayscale: Preprocessing grayscale is done for texture feature extraction to make texture more visible, whereas in color feature extraction not done because the color value to be calculated.

- Histogram Equalization: At this stage is done histogram alignment so that the image quality becomes more contrast to get the value of color features and texture quality. The result of histogram alignment can be seen in Figure 2.

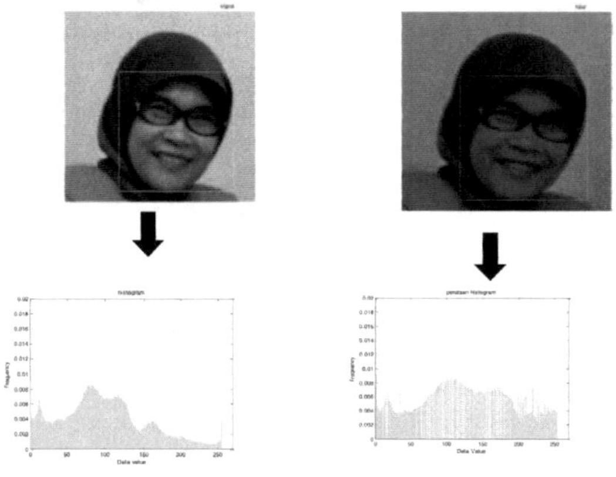

(a) (b)

Figure 2. Comparison of images before and after the histogram process, (a) Original Images, (b) Image of Histogram Equalization (HE).

3.1.2. Preprocessing of the shape feature

- Grayscale: Especially on the calculation of the features of the color element form is not taken into account, so that the change of image to grayscale for faster computation process.
- Resize: Image size will affect the length of feature extraction time, the smaller the size, the faster the computation of its feature extraction. At this stage the image resized to 300 x 300 pixels.
- Edge Enhancement dan Histogram Equalization (HE): The edge enhancement stage will produce a new image with a clearer or sharper edge of the object, so that the shape of the image will be clearer. This stage uses convolution method with Sobel operator [8]. The edge sharpening result can be seen in Figure 3.

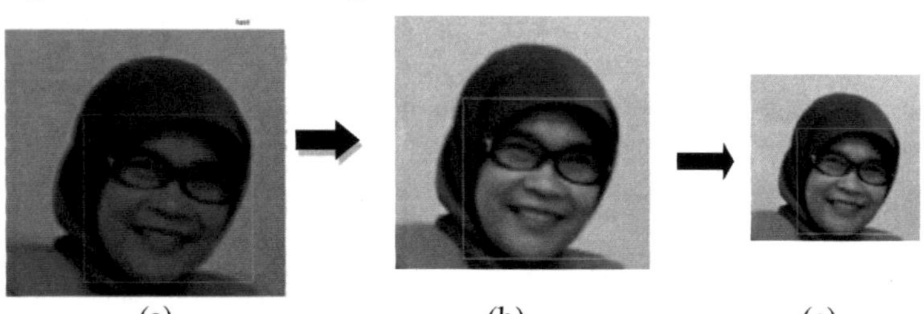

(a) (b) (c)

Figure 3. Preprocessing stages of shape features, (a) Histogram equalization image, (b) Grayscale, (c) Resize.

3.2. Extraction of the texture feature

Texture is the regularity of certain patterns formed from the arrangement of pixels on an image. Texture value can be used as one of the variables for measuring the similarity of image. There are three algorithms for texture processing in image that is structural, spectral and statistical [8]. The first and second-order stats commonly referred to as Gray Level Co-Occurrence Matrix (GLCM) are one of the most commonly used feature extraction methods. First-order feature extraction is a feature retrieval method based on the characteristics of an image histogram. The hogrogram shows the probability of occurrence of the pixel gray value of an image. From the value of the resulting histogram then calculated several dimensions of the first order feature i.e. mean, skewness, variance, kurtosis and entrophy.

In some cases, the use of first order cannot be used to distinguish between objects one with other objects, so it is necessary to use GLCM on feature extraction feature. Haralick et al have proposed GLCM through extraction using a kookurency matrix which is an intermediate matrix that represents the neighboring relationship between pixels in images in different orientation directions and spacetime spacing as in Figure 4 [9].

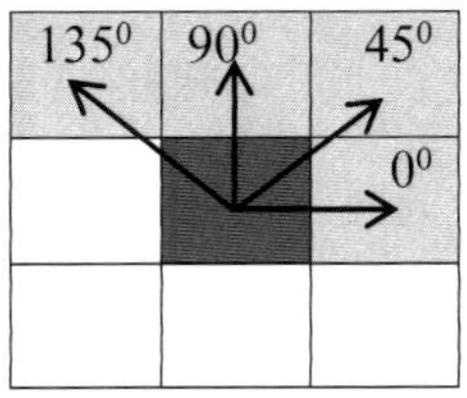

Figure 4. Neighboring relationships between pixels.

One technique that can be used to get a second-order statistical feature is to calculate the probability of an adjacency relationship between two pixels at a certain distance and angle orientation [8]. The pixel connectivity relationship used in this study is using angle orientation in four directions that is 0^0, 45^0, 90^0, 135^0. While the kookurensi matrix is a square matrix with the number of elements as much as the square of the number of pixel intensity levels in the image.

After obtaining the kookurency matrix, it can be calculated second order statistic feature which represents image and subsequently extracted texture feature using kookurensi matrix [9]. The second-order statistical features calculated in this study are energy, homogeneity, contrast and correlation. The energy dimension is calculated using equations 1, 2, 3.

$$P(i,j|d,\theta) = \frac{P(i,j|d,\theta)}{\sum_i \sum_j P(i,j|d,\theta)} \quad (1)$$

$$Energy = \sum_i \sum_j P(i,j)^2 \quad (2)$$

Energy to measure concentration of intensity pairs on the GLCM matrix.

$$Homogeneity = \sum_i \sum_j \frac{P(i,j)}{1+|i-j|} \quad (3)$$

While homogeneity indicates homogeneity of variation in image intensity. The Contrast is calculated using equation 4.

$$Contrast = \sum_{i,j} P(i,j)|i-j|^2 \quad (4)$$

Contrast shows the size of the spread of matrix elements. If the location is far from the main diagonal, then the value of contrast will be great. Next Correlation is calculated using equation 5.

$$Correlation = \frac{(1-\mu_i)(1-\mu_j)P(i,j)}{\sigma_i \sigma_j} \quad (5)$$

Correlation shows the size of the gray-level linear dependence of the neighboring pixels in a gray image.

The stages of image search based on texture features are:

- Read the query image
- Convert RGB image to Grayscale
- Calculate 4-way GLCM matrix at 0^0, 45^0, 90^0, 135^0
- Calculate in each direction for energy value, homogeneity, contrast and correlation.
- Make matching similarity between the image of the query with the image on the database
- Select the minimum distance value as the most similar image.

3.3. Extraction of the shape feature

Feature feature extraction is done using invariant moment method. This method is used because it is not vulnerable to image changes caused by Rotation, Scale and Translation (RST) [10].

Invariant Moment to result seven constant invariant moment values against RST [11]. In the invariant moment can also be analyzed form the set of the moment of a function f (x, y) of two variables defined as follows: [8]

Moment order (p q) for the continuous function f (x, y) with p, q = 0, 1, 2, ... are defined as follows:

$$m_{pq} = \int_{-\infty}^{\infty}\int_{-\infty}^{\infty} x^p x^q f(x,y)dxdy \quad (6)$$

For the momentsentral formula is given as below:

$$\mu_{pq} = \int_{-\infty}^{\infty}\int_{-\infty}^{\infty} (x-\bar{x}))^p (y-\bar{y})^q f(x,y)dxdy \quad (7)$$

with

$$\bar{x} = \frac{m_{10}}{m_{00}} \text{ dan } \bar{y} = \frac{m_{01}}{m_{00}} \quad (8)$$

And for digital image

$$\mu_{pq} = \sum_x \sum_y (x-\bar{x}))^p (y-\bar{y})^q f(x,y) \quad (9)$$

The region is represented by seven invariant moments which can be searched from the central moments: μ00, μ10, μ01, μ20, μ11, μ02, μ30, μ12, μ21, μ03, and by defining the normalized central moments i.e. ηpq = μpq / μγ00, where γ = 0.5 (p q) 1 for p q = 2, 3, and so on, which finally obtained seven invariant moments, the seven invariant moments values are shown in equations 10 to equations 16.

$$M_1 = \eta_{20} + \eta_{02} \quad (10)$$

$$M_2 = (\eta_{20} + \eta_{02})^2 + 4\eta^2_{11} \quad (11)$$

$$M_3 = (\eta_{30} - \eta_{12})^2 + (3\eta_{21} - \eta_{03})^2 \quad (12)$$

$$M_4 = (\eta_{30} + \eta_{12})^2 + (\eta_{21} + \eta_{03})^2 \quad (13)$$

$$M_5 = (\eta_{30} - 3\eta_{12})(\eta_{30} + \eta_{12})[(\eta_{30} + \eta_{12})^2 - 3(\eta_{21} + \eta_{03})^2] + (3\eta_{21} - \eta_{03})(\eta_{21} + \eta_{03})[3(\eta_{30} + \eta_{12})^2 - (\eta_{21} - \eta_{03})^2] \quad (14)$$

$$M_6 = (\eta_{20} - \eta_{02})[(\eta_{30} + \eta_{12})^2 - (\eta_{21} + \eta_{03})^2] + 4\eta_{11}(\eta_{30} + \eta_{12})(\eta_{21} + \eta_{03}) \quad (15)$$

$$M_7 = (3\eta_{21} - \eta_{30})(\eta_{30} + \eta_{12})[(\eta_{30} + \eta_{12})^2 - 3(\eta_{21} + \eta_{03})^2] + (3\eta_{12} - \eta_{30})(\eta_{21} - \eta_{03})[3(\eta_{30} + \eta_{12})^2 - (\eta_{21} + \eta_{03})^2] \quad (16)$$

The values of these invariant moments do not change with respect to rotation, translation and scale [11]. The stage of feature extraction features are as follows:

- Read the image
- Convert the image to Gray Scale
- Calculate seven invariant moment values
- Calculate the distance value through the seven values of the form feature between the query image and the database image
- The most similar image distance value is the searched image

3.4. Extraction of color feature

Preprocessing with histogram equalization in Figure 2 produces a more contrasting image. The image is an input image on the color feature extraction. HSV color space (Hue, Saturation and Value) can sort colors in images like human vision so as to recognize images well [8]. HSV is one way to define colors based on color wheel. Hue to measure the angle around the color wheel (0 degrees in red, 120 degrees in green and 240 degrees in blue). Saturation shows the color wheel radius so that the proportion between dark colors to light or pure white or declared the degree of color purity that indicates how much white color is given on the color. Value attribute that states the amount of light received by the eye regardless of the color or declared the brightness of color or brightness.

The HSV model is a derivative of RGB so it needs conversion from RGB to HSV. RGB or Red, Green and Blue are the colors formed by the color model which is a mixture of primary colors of red with green and blue compositions. Stages of image search based on the color features are as follows:

- Read the image
- Convert RGB color image to HSV
- Quantize each pixel in 256 histogram bins
- Perform normalization of the histogram by dividing the number of pixels
- Save 256 values of color features in the database vector
- Calculate the image similarity using ueclidean distance
- The image that has the smallest resemblance distance is the image sought.

3.5. Similarity measurement

Measurement of similarity is used Euclidean distance, where in images that have a close similarity distance zero can be categorized the most similar. Calculating the value of the similarity distance using the equation 17.

$$D(Q, M) = \sqrt{\sum_{n=1}^{k} (Q_n - M_n)^2} \quad (17)$$

Where Q and M refers to the features of query images and images in the database in the dimension number n.

Results of the similarity calculation were sorted in such a way that the similarity whose value is the closest to zero has the highest level of similarity.

3.6. Performance measurement of each feature in supporting retrieval accuracy

Image data search is done by using CBIR (Content Based Image Retrieval) method by using three image features to match the similarity of image that will be the key of image data search. Each feature has different performance in favor of improving retrieval or search results. In this research, K-Fold Cross

Validation and Confusion matrix methods are used to measure feature performance in support of retrieval accuracy.

The data sharing of training data and test data for accuracy, precision and recall testing is varied from K = 3 to k = 10. For example, if K = 3, then the first part starts from the 1st data up to the 180th data as training data, the second part starts from the 181th data up to the 360th data as the validation data and the third part starts from the data to -361 up to 480th data as training data and 481th to 600th data are test data.

4. Results and analysis

Tests in this study using 5 types of face images with the number of images as much as 600 measuring 300 x 300 pixels.

4.1. Retrieval results

Retrieval testing is done by varying the number of clusters and the percentage variation of feature weights. Test results with weighted variations of 50% for feature weight, 30% color feature weight and 20% texture feature weight with variation of 5 clusters with number of displayed images shown in Figure 5.

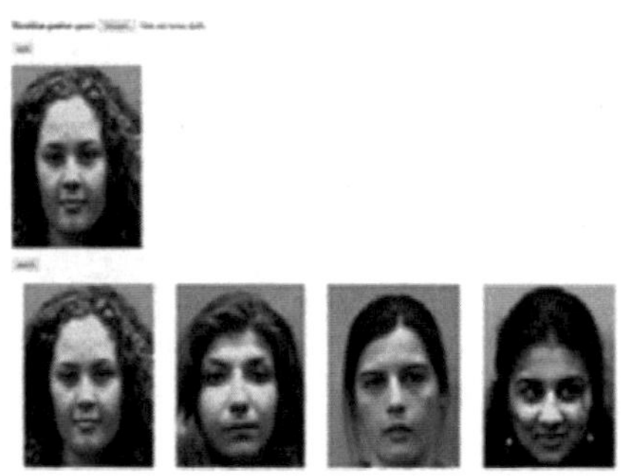

Figure 5. Image of search results.

In Figure 5. The retrieval result is shown with a similarity ranking of 1 to 24 from the face image database. Tests with weighted variations and cluster number variations will result in different retrieval outputs in the same data.

4.2. Discussion

At this stage, the analysis of facial image akurasiretrieval based on the features of shape, color and texture using face image database before and after clustering. In Table 1 it shows the percentage of retrieval accuracy using different feature weights on the image database before clustering.

4.2.1. Before clustering. The result of retrieval accuracy analysis on face image database before clustering with weighted variation on shape, color and texture features can be seen in Table 1.

Table 1. The percentage retrieval accuracy with variation of feature weighted scheme prior to clustering.

Img	Equal	S1	S2	S3	S4
F 1	75.23%	78.99%	87.52%	83.69%	68.03%
F 2	74.13%	77.65%	85.52%	83.69%	67.33%
F 3	75.20%	78.99%	87.52%	83.69%	68.03%
F 4	75.15%	78.99%	85.52%	82.00%	67.33%
F 5	74.13%	76.40%	87.52%	83.69%	68.03%

Based on Table 3.1, it is shown that the retrieval accuracy with the S2 scheme yields the highest accuracy of over 85% and the S4 weight variation yields the lowest accuracy of less than 69%. The S2 scheme is a feature weighting where 50% feature shapes, 30% color features and 20% texture features. From Table 1 it can be seen that performance of texture features is more dominant in face image retrieval. The retrieval accuracy graph with the same feature weight and feature weight varies before clustering is shown in Figure 6.

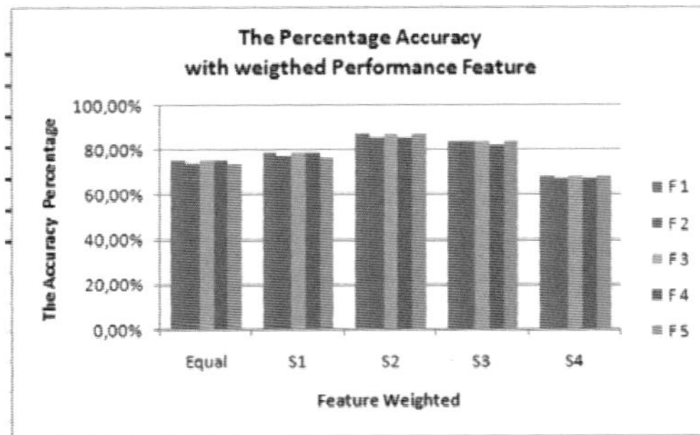

Figure 6. The graph of retrieval accuracy with weighted of performance feature before clustering.

In Figure 5.6, it shows that S2 weight of performance feature variation yields the highest retrieval accuracy compared to other weight variations.

4.2.2. After clustering. Clustering stages and variation of weighting schemes affect the speed of the retrieval process and the level of retrieval accuracy. Percentage of retrieval accuracy with difference of feature weight scheme and number of clusters can be seen in Table 2.

Table 2. Percentage of retrieval accuracy with weighted feature schemes and different of number clusters.

K	Equal	S1	S2	S3	S4
3	72,13%	75,59%	92,67%	82,38%	63,33%
5	75,23%	77,45%	95,60%	86,36%	65,40%
7	76,40%	77,75%	95,60%	86,38%	65,67%
9	76,40%	77,45%	95,67%	86,87%	65,66%
10	76,40%	77,65%	95,67%	86,67%	65,69%
15	74,1%	75,15%	93,55%	84%	64,2%
20	73,1%	74,25%	93%	83,3	64,67%

Based on Table 2 shows that the variation of S2 weighting scheme with a 50% weight form feature, 30% color features and 20% texture features as well as cluster number variations between 5 to 10 clusters yields a retrieval average accuracy of more than 95% higher than the cluster others. While S4 weighted variation with 30% feature weight form, 50% color feature weight and 20% texture feature weight with variation of cluster number 3 yields the lowest accuracy of less than 64%.

The retrieval accuracy graph on the Face image database uses the same and varied feature weights that are S1, S2, S3 and S4 with variations of cluster numbers ranging from 3 to 20 can be seen in Figure 6.

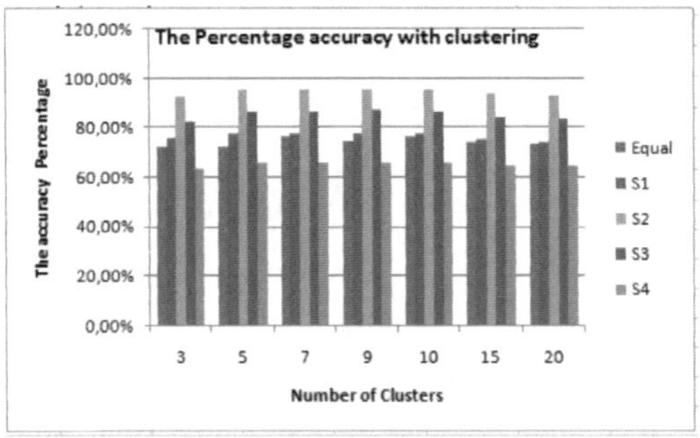

Figure 7. Graph of retrieval accuracy with weighted of performance feature after clustering.

Figure 7 shows that retrieval accuracy tends to rise to a variation in the number of clusters ranging from 5 to 10 and tends to decrease in the number of clusters greater than 10. This test is performed on 600 face image data with 480 training images and 120 test images. the test was taken randomly with 24 data displayed.

4.2.3. Performance feature test. To know the performance of image features in supporting the accuracy of the system conducted testing. Test method used in this study is K-fold Cross Validation. In this test uses 3 feature values that have 7 dimensions for shape features, 9 dimensions for color features and 16 dimensions for texture features. The total data used in this test is 600 image data that is 120 face1 image (F1), 120 face2 image (F2), 120 face3 images (F3), 120 face4 image (F4) and 120 image face5 (F5).

Image performance test feature using K-fold with value 2, 3, 6 and 9 with SMO classification. Variations of K-fold values to determine the magnitude of K-fold values that produce the percentage accuracy, precision and recall reach the highest level. Table comparison of accuracy, precision and recall value can be seen in Table 3 and confusion matrix value on K-fold with k = 3 value shown by Table 4.

Table 3. Values of accuracy, precision and recall of image features.

K-Fold	Accuracy (%)	Prcision (%)		*Recall (%)*	
		F1	100	F1	100
		F2	97.2	F2	85.1
2	94.99	F3	100	F3	100
		F4	99.2	F4	99.6
		F5	100	F5	95.0
		F1	100	F1	100
		F2	98.1	F2	85.1
3	95.62	F3	100	F3	100
		F4	96.8	F4	100
		F5	88.8	F5	97.5
		F1	100	F1	100
		F2	98.1	F2	85.1
6	95.62	F3	100	F3	100
		F4	96.8	F4	100
		F5	88.8	F5	97.5
		F1	100	F1	100
		F2	98.1	F2	85.1
9	95.62	F3	100	F3	100
		F4	96.8	F4	100
		F5	88.8	F5	97.5

Table 4. Confusion matrix of face image feature with K-Fold = 3.

Actual	Prediction					
	F1	F2	F3	F4	F5	Total
F1	119	0	0	0	0	119
F2	15	103	0	3	0	121
F3	0	0	119	0	1	120
F4	0	0	0	120	0	120
F5	0	2	0	1	117	120
Total	134	105	119	124	118	600

Table 4 it shows that starting at K-fold with values of 3, 6 and 9 has a high accuracy value of more than 95% compared to K-fold of 2. At K-fold with value 3 obtains a mean value of precision equal to 95.7% and the average recall value of 95.5%. Furthermore, accuracy, precision and recall on K-fold 6 and 9 also has the same precision, precision and recall value.

4.2.4. Computation time. In testing the face tracking time is done by the variation of the number of clusters of the face image database. The test result is the average time of the face tracking which can be seen in Table 5.

Table 5. The average face time trial test results with variation of cluster number.

Number of cluster	The tracking of time (milisecond)				
	Equal	W1	W2	W3	W4
3	93	93	93	93	93
4	62	62	62	62	62
5	56	56	56	56	56
6	55	55	55	55	55
7	54	54	54	54	54
8	53	53	53	53	53
9	51	51	51	51	51
10	49	49	49	49	49
15	47	47	47	47	47
20	45	45	45	45	45

Based on table 5 shows that on the same cluster, it tends to have the same average.

5. Conclusion

Each feature performs differently in support of CBIR-based retrieval accuracy. In this study proves that the use of different feature-weighted schemes on each normalized feature value and cluster determination affect the increase of face retrieval image accuracy.

The research results show that the shape feature in the face image is the dominant factor to determine the level of similarity in the retrieval process, while the color features and texture features are the complementary features. The average computational time required for a retrieval is 5 mili-second.

References

[1] Sani 2007 Aplikasi Image Retrieval berdasarkan tekstur dengan menggunakan transformasi Haar Wavelet *Seminar Nasional Sistem dan Informasi, Bali*

[2] Vijay K 2008 A Content Based Approach to Image Database Retrieval *Journal of Computer Applications* **1**(4)

[3] Herry S 2009 Sistem Pencarian Citra Digital Menggunakan Content Based Image Retrieval *Proceeding Seminasif*

[4] Hiremath and Jagadesh 2012 Content Based Image Retrieval Based on Color, Texture and Shape

Feature Using Image and Its Complement *International Journal of Computer Science and Security* **1**(4)

[5] Sumana K and Remesh A 2015 Efficient Technique Using Color, Texture and Shape Feature In Skecthes *IJAETAE* **5**

[6] Lingadalli K R and Ramesh N 2015 Content Based Image Retrieval Using Color, Shape and Texture *International Advanced Research Journal in Science, Engineering and Technology*

[7] Jumi, Harjoko A and Ashari A 2015 Content Based Image Retrieval for Asset Management based on Weighted Feature and K-Means Clustering *JATIT* **77**(1)

[8] Gonzales R C and Woods R E 2008 *Digital Image Processing, Third Edition* (New Jersey: Pearson Prentice Hall)

[9] Harralick R M, Shanmugam K and Dinstein I 1973 Texture Features for Images Classification *IEEE Transaction on Systems, Man and Cybernectics*

[10] Acharya T and Ray A K 2005 *Image Processing Principle and Applications* (USA: John Willey & Sons)

[11] Castleman K R 1996 *Digital Image Processing* (New Jersey: Prentice Hall Inc.)

IOP Conf. Series: Materials Science and Engineering 1108 (2021) 012032 doi:10.1088/1757-899X/1108/1/012032

Analysis of Deep Water Culture (DWC) hydroponic nutrient solution level control systems

A Nursyahid*, T A Setyawan, K Sa'diyah, E D Wardihani, H Helmy and A Hasan

Department of Electrical Engineering, Politeknik Negeri Semarang, Semarang, Indonesia

*arif.nursyahid@polines.ac.id

Abstract. Hydroponics is the cultivation of plants that use water as a planting medium without using soil. One of the simplest hydroponics is Deep Water Culture (DWC). DWC is a hydroponic technique that allows plant roots to always be submerged in water containing nutrients. The purpose of the application of controlling the level of hydroponic nutrient solution in the type of Deep Water Culture (DWC) in the box is to ensure that the roots of the plant are always submerged in nutrient solution so that the nutrients are still fulfilled. The results obtained from controlling the level of hydroponic nutrient solution in the type of Deep Water Culture (DWC) in the box is when the level of nutrient solution is less than the specified threshold then the 12VDC pump relay will live to drain the source water to the nutrient solution reservoir. The HC-SR04 sensor is used for reading the level of nutrient solution, so it has an effect on determining the pump life time. The HC-SR04 sensor has a reading error of 0.87%. The duration of the pump life is determined based on the linear regression equation. Controlling the level of nutrient solution based on linear regression linear calculation has a good accuracy of 88.6%.

1. Introduction

Agriculture is a very important sector for Indonesian people. But along with technological developments, the industrial sector is also increasing. The increase has an impact on the narrowness of agricultural land. To overcome these problems, the solution that can be done is to use a hydroponic farming system. Hydroponics is a method of cultivating plants that use water as a planting medium without using soil, but using water given the nutrient solution needed by plants.

One of the hydroponic techniques used by the community is Deep Water Culture (DWC). DWC is a hydroponic technique that supplies plants with nutrient solutions directly to plant roots. The DWC technique will ensure that the root of the plant is always submerged in a nutrient solution [1]. This Deep Water Culture type hydroponics allows planting indoors or in a box. Planting done in the box can use TL lamps as a substitute for sunlight.

In this DWC hydroponic technique, nutrient solutions need to be aerated to add oxygen dissolved in water because lack of oxygen will interfere with nutrient absorption by plant roots. Plants need a continuous supply of nutrient solutions so that the plants remain nutritionally fulfilled. In hydroponics, nutrition plays an important role in the productivity and quality of lettuce. With balanced absorption of nutrients, it is very important in determining the quality of lettuce products [2]. Sometimes the level of nutrient solution is less than the specified limit which causes the root of the plant not to be submerged in nutrient solution. Likewise, vice versa if the level of nutrient solution exceeds the threshold then the stem of a small plant will also be submerged in nutrient solution, so that the plant will wither even die.

To overcome these problems, a system is needed to control the level of nutrient solution. One method that can be used to adjust the level of nutrient solution is the linear regression method. With this control system, it is expected that the roots of plants can always be submerged in nutrient solutions so that the nutrients remain fulfilled.

2. Related works

In the study "Automated system developed to control pH and concentration of nutrient solution evaluated in hydroponic lettuce production" [3] carried out control of pH and the concentration of nutrient solution in lettuce plants. Selenoid valve will automatically drain acid / base solution and drain nutrients when the pH value and nutrient solution density do not match the threshold. But in this study did not control the level of nutrient solution. The level of nutrient solution in the DWC hydroponic plants needs to be controlled so that the plant roots are always submerged in nutrient solution so the plant does not die.

In the study "A Development of an Automatic Microcontroller System for Deep Water Culture (DWC" [1], pH control on DWC hydroponics uses a pH sensor as a reader of the pH value in solution. In addition, in this study also paid attention to the level of nutrient solution by flowing water into and out of the reservoir. In this study, monitoring the level of nutrient solution is important to maintain the level of nutrient solution that is appropriate to the reservoir. But monitoring still uses LCDs and remote monitoring and control cannot be carried out.

In the study "Electrical Conductivity and pH Adjusting System for Hydroponics by using Linear Regression" [4] carried out control of the concentration of nutrient solution and pH using linear regression method on red lettuce and green lettuce. This research was carried out by taking sample data by adding AB solution using a solenoid valve. But this research is still done in a simulation and has not been implemented in the real system.

In this study a new system will be designed to control the level of nutrient solutions in hydroponic plants of the DWC type using linear regression methods. With the level control system of nutrient solutions using this linear regression method, it is expected that there will be no more hydroponic plants from DWC that die because the roots of the plant are not submerged in nutrient solution.

3. Proposed methods

3.1. Hardware configuration

In the control system of the level of DWC hydroponic nutrient solution uses two microcontrollers as the program brain, namely NodeMCU and Arduino Nano. NodeMCU in the solution level control system is used to upload the results of monitoring by the sensor to the database server and access the data threshold level of nutrient solution that is regulated via android. The threshold data is then sent to Arduino Nano serially. Monitoring the level of nutrient solution in DWC hydroponics using HC-SR04 ultrasonic sensor to determine the height of the nutrient solution in the reservoir.

The nutrient solution level control system works as a back response to the results of monitoring nutrient levels in the DWC hydroponic plants in the box. The results of monitoring the level of nutrient solution by the HC-SR04 sensor will then be compared with the predetermined threshold. When the monitoring results are not in accordance with the specified threshold, the controller will respond to Arduino Nano to turn on the source water pump relay. To regulate the duration of the active pump linear regression method is used. The pump will be active with time delay according to the linear regression equation. The results of the linear regression equation were obtained from the experimental sample data processed using simple linear regression. Then the pump will pump water from the source reservoir to the nutrient solution reservoir so that the level of nutrient solution is in accordance with the specified measuring limit. Figure 1 shows the architecture of the control system for the level of nutrient solution for hydroponic plants of the DWC type in the box.

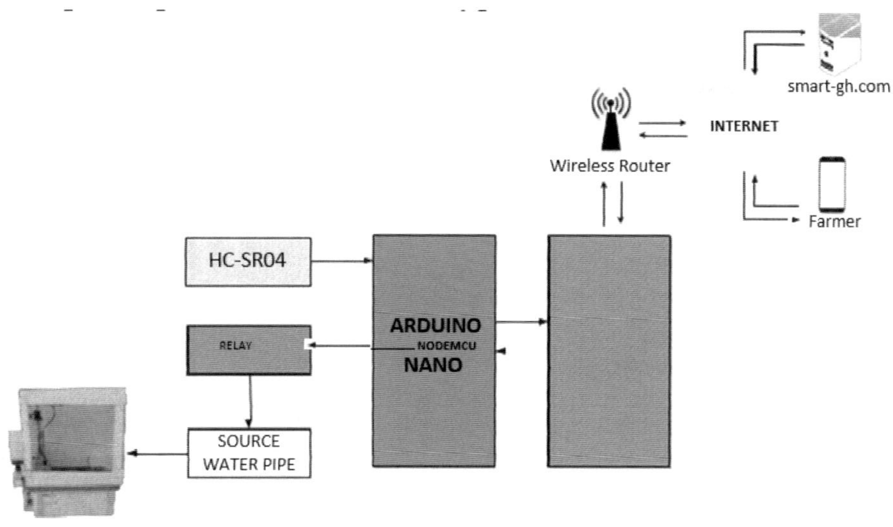

Figure 1. DWC hydroponic nutrient solution level control system architecture.

3.2. Software configuration

Designing software in this system includes designing programs on Arduino IDE to run systems and designing Android interfaces for remote control. The design of the software on the Arduino IDE is used to write the program to NodeMCU and Arduino Nano to execute the commands given. While the software design on Android is used for monitoring and controlling the level of nutrient solution remotely. Figure 2 is a flow diagram of the nutrient solution level control programming in the Arduino IDE.

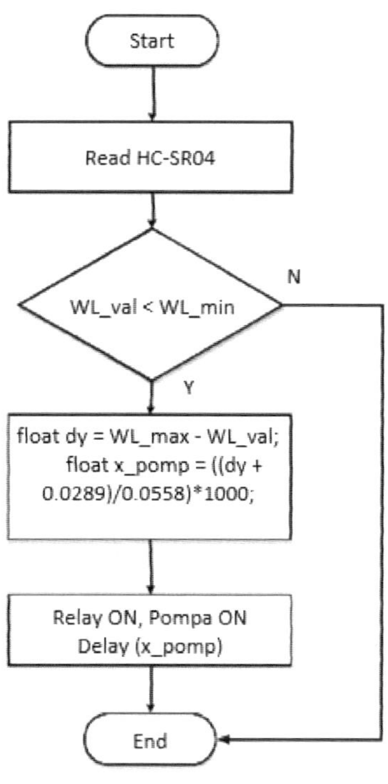

Figure 2. Flowchart of the DWC hydroponic nutrient solution level.

IOP Conf. Series: Materials Science and Engineering 1108 (2021) 012032 doi:10.1088/1757-899X/1108/1/012032

3.3. HC-SR04 calibration sensor

The HC-SR04 sensor is used to measure the level of nutrient solution in the reservoir. The HC-SR04 sensor test is done by aligning the position of the HC-SR04 sensor with a ruler measuring instrument in the nutrient solution reservoir to determine the level of nutrient solution. The HC-SR04 sensor calibration is done by comparing the measurement results of the HC-SR04 sensor with the measurement results with a ruler measuring instrument. Figure 3 shows the calibration of the HC-SR04 ultrasonic sensor with a ruler measuring instrument.

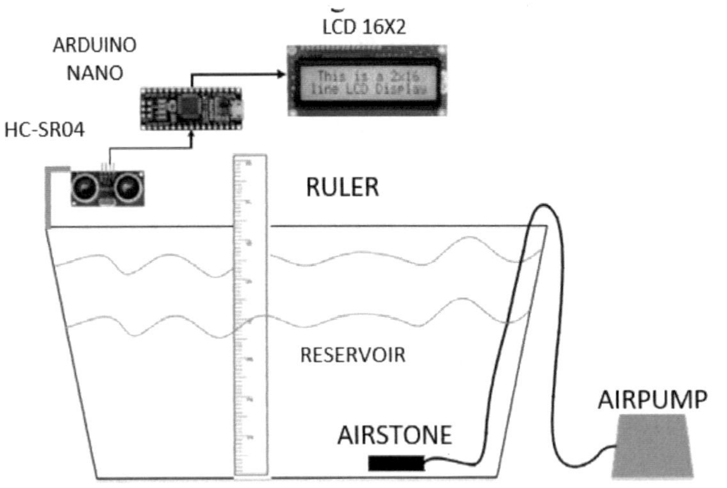

Figure 3. Calibration of HC-SR04 sensor using a ruler measuring tool.

3.4. Linear regression equations

Linear regression is an analysis to build a mathematical model that is used to explain the form of the relationship between X variables and Y variables. Y variable is also called the dependent variable or response variable whose values depend on variable X, which is called the independent variable or free variable. Linear regression design as a mathematical calculation of nutrient solution level begins with sample data experiments. The sample data testing aims to get the linear equation level of the nutrient solution by determining the dependent variable X and the independent variable Y. The variable X in the control system of the nutrient solution level is the length of time the pump lives in seconds. While the variable Y is the value of the increase in the level of nutrient solution. In testing the sample data, the addition of the nutrient solution level using a pump is given by giving different pump life times. Equation 1 is a simple linear regression equation.

$$Y = a + bX \quad (1)$$

In linear regression testing, sample data was searched. The sample data is then processed using equation (1) so as to obtain the equation for the level of nutrient solution. Then testing the equation is done to get its accuracy.

4. Experimental results and discussion

4.1. Experiment of HC-SR04 sensor

The HC-SR04 sensor experiment was carried out by providing water in a reservoir with different water levels. The HC-SR04 sensor is placed together with a ruler measuring instrument in the same time and condition. Table 1 shows the comparison of measurement results using HC-SR04 sensor with a ruler measuring instrument.

Table 1. Comparison of HC-SR04 sensor measurement with ruler.

No	HC-SR04 Sensor (cm)	Ruler (cm)	Difference (Absolut) (cm)	Accuracy (%)
1	2	2	0	100
2	2,07	2	0,07	96,5
3	4	4	0	100
4	6,01	6	0,01	99,8
5	6,06	6	0,06	99
6	7,05	7	0,05	99,29
7	8,88	9	0,12	98,67
8	8,98	9	0,02	99,78
Average Accuracy (%)				99,13

From the test results using HC-SR04 sensor and ruler measuring instrument, the measurement results using HC-SR04 sensor have the biggest error of 0.12 cm. Based on sensor specifications, HC-SR04 sensor has an error value of ± 0.3 cm. While the average error of nutrient solution level is 0.041. In this test, the HC-SR04 sensor can work properly because the accuracy value is still in the sensor safe range according to the datasheet. Based on table 1, comparison of measurement results using HC-SR04 sensor with a ruler measuring instrument can be described in graphical form as shown in Figure 4.

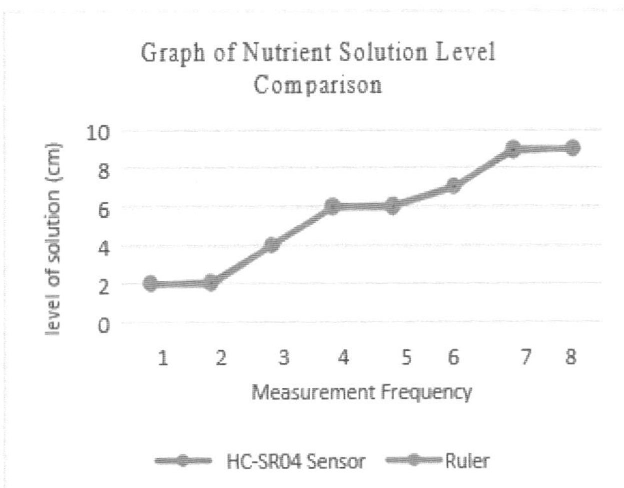

Figure 4. Graph of Level Comparison of HC-SR04 sensor solution with Ruler.

4.2. Results of the Android interface

The threshold level of nutrient solution can be done remotely using an Android-based cellphone. The monitoring page and threshold level of nutrition solution settings on Android are shown in Figure 5.

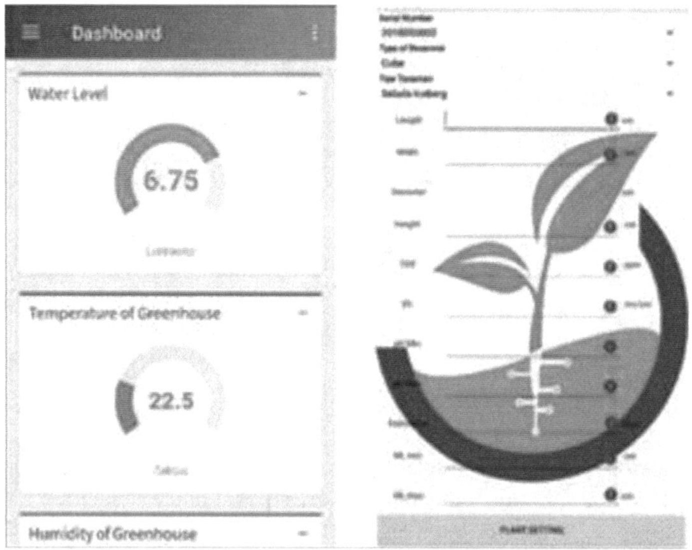

Figure 5. Monitoring and setting threshold levels of nutrient solutions on Android.

4.3. Experiment of linear regression equations

Sample data for the nutrient solution level control system is carried out by giving the pump time to live 1 second, 2 seconds, 3 seconds, 4 seconds, 5 seconds, and 6 seconds. Each t-value was tested for 50 times, so that the sample data collected were 300 sample data. Linear regression equations from the sample data obtained can be calculated by equation 1. Graph data of the hydroponic nutrient solution DWC in the box can be seen in Figure 6 and a graph of the increase in the level of nutrient solution can be seen in Figure 7.

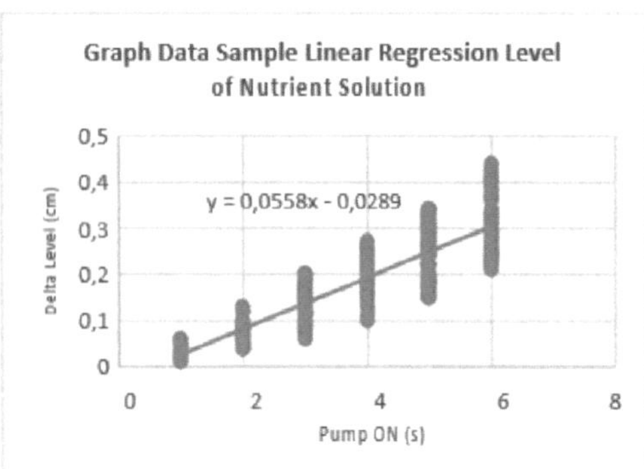

Figure 6. Chart of linear regression levels of nutrient solution levels.

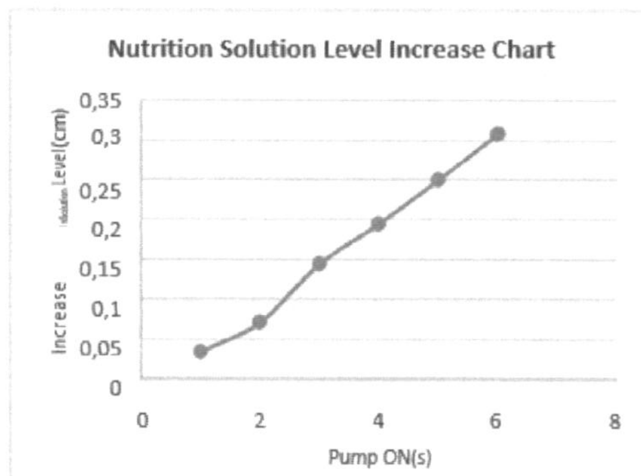

Figure 7. Increased nutrient solution level graph.

After the equation for the level of nutrient solution is obtained, then the trial of the equation is carried out. To get a live pump time which is an X variable, the equation used is:

$$X = \frac{Y + 0{,}02895}{0{,}0558} \tag{2}$$

The results of the linear regression accuracy experiments are shown in the following table:

Table 2. Linear regression accuracy data level of nutrient solution Y = 0.5 cm.

No	Start level (cm)	Set poin t	End level (cm)	Delay (second)	delta level (cm)	delta setpoint (cm)
1	1,88	2,38	2,41	9478.50	0,53	0,03
2	2,39	2,89	2,8	9478.51	0,41	0,09
3	2,8	3,3	3,18	9478.52	0,38	0,12
4	3,18	3,68	3,57	9478.53	0,39	0,11
5	3,58	4,08	3,99	9478.54	0,41	0,09
6	3,99	4,49	4,3	9478.55	0,31	0,19
7	4,4	4,9	4,76	9478.56	0,36	0,14
8	4,66	5,16	5,06	9478.57	0,4	0,1
9	5,17	5,67	5,49	9478.58	0,32	0,18
10	5,57	6,07	6,18	9478.59	0,61	0,11
Average						0,116
Accuracy Percentage (%)						76,8

The results of the experiment with Y = 0.5 cm have an average value of accuracy of the test of 76.8%.

Table 3. Linear regression accuracy data level of nutrient solution Y = 1 cm.

No	Start level (cm)	Set point	End level (cm)	Delay (second)	delta level (cm)	delta setpoint (cm)
1	6,44	7,44	7,32	18439.07	0,88	0,12
2	7,32	8,32	8,24	18439.07	0,92	0,08
3	7,94	8,94	8,8	18439.07	0,86	0,14
4	8,24	9,24	9,36	18439.07	1,12	0,12
5	9,36	10,36	10,15	18439.07	0,79	0,21
6	2,26	3,26	3,2	18439.07	0,94	0,06
7	3,64	4,64	4,6	18439.07	0,96	0,04
8	5,28	6,28	6,15	18439.07	0,87	0,13
9	5,62	6,62	6,56	18439.07	0,94	0,06
10	6,81	7,81	7,68	18439.07	0,87	0,13
Average						0,109
Accuracy Percentage (%)						89,1

The results of the experiment with Y = 1 cm have an average value of accuracy of the test of 89.1%.

Table 4. Linear regression accuracy data level of nutrient solution Y = 1.5 cm.

No	Start level (cm)	Set point	End level (cm)	Delay (second)	delta level (cm)	delta setpoint (cm)
1	7,68	9,68	9,34	36360.21	1,66	0,34
2	1,63	3,63	3,65	36360.21	2,02	0,02
3	3,79	5,79	5,9	36360.21	2,11	0,11
4	6,08	8,08	8	36360.21	1,92	0,08
5	7,63	9,63	9,53	36360.21	1,9	0,1
6	2,05	4,05	4,04	36360.21	1,99	0,01
7	4,3	6,3	6,34	36360.21	2,04	0,04
8	5,95	7,95	7,89	36360.21	1,94	0,06
9	7,78	9,78	9,67	36360.21	1,89	0,11
10	9,21	11,21	11,09	36360.21	1,88	0,12
Average						0,099
Accuracy Percentage (%)						93,3

The results of the experiment with Y = 1.5 cm have an average value of accuracy of the test of 93.3%.

Table 5. Linear regression accuracy data level of nutrient solution Y = 02 cm.

No	Start level (cm)	Set point	End level (cm)	Delay (second)	Delta level (cm)	Delta setpoint (cm)
1	7,68	9,68	9,34	36360.21	1,66	0,34
2	1,63	3,63	3,65	36360.21	2,02	0,02
3	3,79	5,79	5,9	36360.21	2,11	0,11
4	6,08	8,08	8	36360.21	1,92	0,08
5	7,63	9,63	9,53	36360.21	1,9	0,1
6	2,05	4,05	4,04	36360.21	1,99	0,01
7	4,3	6,3	6,34	36360.21	2,04	0,04
8	5,95	7,95	7,89	36360.21	1,94	0,06
9	7,78	9,78	9,67	36360.21	1,89	0,11
10	9,21	11,21	11,09	36360.21	1,88	0,12
Average						0,099
Accuracy Percentage (%)						95,05

The results of the experiment with Y = 2 cm have an average value of accuracy of 95.05%.

From the results of testing the linear regression accuracy of the above nutrient solution level, the average accuracy of testing was 88.6%. From these tests have a high degree of accuracy, meaning that equation (2) can be applied well to the DWC hydroponic nutrient level control system.

In the overall system test, equation (2) is applied to the whole DWC hydroponic system. This test is carried out to determine the performance of the system when combined with other DWC hydroponic systems. Then the results of the overall system testing are compared with accuracy testing without being integrated with other hydroponic systems. In this test 10 tests were carried out. The overall system accuracy results are shown in table 6.

Table 6. System trial data overall minimum level = 6.5 cm and maximum level = 9 cm.

No	Start Level (cm)	End Level (cm)	Y (cm)	Delta Setpoint (cm)	Accuracy (%)
1	6,15	8,46	2,85	0,54	81,05
2	5,85	8,41	3,15	0,59	81,27
3	6,1	9,23	2,9	0,23	92,07
4	6,46	8,41	2,54	0,59	76,77
5	6,2	8,34	2,8	0,66	76,43
6	6,08	8,87	2,92	0,13	95,55
7	5,9	9,11	3,1	0,11	96,45
8	6,13	8,75	2,87	0,25	91,29
9	6,3	8,64	2,7	0,36	86,67
10	5,79	8,97	3,21	0,03	99,07
Average Accuracy Percentage (%)					87,66

5. Conclusion and recommendations

5.1. Conclusion

- Sensor HC-SR04 can be used as an indicator for the process of controlling the level of DWC hydroponic nutrient solution with an average accuracy percentage of 99.13%
- Control of nutrient solution level using linear regression method has an average accuracy of 88.6%.

- The control system for nutrient solution levels combined with the overall system has an accuracy of 87.66%.
- The greater the increase in the level of nutrient solution (variable Y), the better the accuracy of the control with the linear regression method.

5.2. Recommendations

- There needs to be a serial connection test between NodeMCU and Arduino Nano so that the sending and receiving of threshold data is more stable.
- Control methods should use multiple linear regression so that it can be used in different volume volumes.
- Control methods used not only to increase the level of solution, but also reduce the level of solution.

References

[1] Saaid M F, Yahya N A M, Noor M Z H and Ali M S A M 2013 A development of an automatic microcontroller system for Deep Water Culture (DWC) *Proc. - 2013 IEEE 9th Int. Colloq. Signal Process. its Appl. CSPA 2013* 328–332

[2] Abou-Hadid A F, Abd-Elmoniem E M, El-Shinawy M Z and Abou-Elsoud M 1996 Electrical conductivity effect on growth and mineral composition of lettuce plants in hydroponic system *Acta Hortic.* **434** 59–66

[3] Domingues D S, Takahashi H W, Camara C A P and Nixdorf S L 2012 Automated system developed to control pH and concentration of nutrient solution evaluated in hydroponic lettuce production *Comput. Electron. Agric.* **84** 53–61

[4] Kaewwiset T and Yooyativong T 2017 Electrical Conductivity and pH Adjusting System for Hydroponics by using Linear Regression in *2017 14th International Conference on Electrical Engineering/Electronics, Computer, Telecommunications and Information Technology (ECTI-CON) Electrical* 761– 764

Nutrient Film Technique (NFT) hydroponic nutrition controlling system using linear regression method

A Nursyahid*, H Helmy, A I Karimah and T A Setiawan

Department of Electrical Engineering, Politeknik Negeri Semarang, Semarang, Indonesia

*arif.nursyahid@polines.ac.id

Abstract. Nutrient Film Technique (NFT) hydroponic cultivation could be an alternative to overcome the increasing food need. Nutritional needs supply of hydroponic plant are one of parameters that need to be considered. NFT hydroponic nutrient controlling system aims to facilitate farmers to maintain the amount of nutrition according to plant needs. This system uses linear regression method to control the amount of nutrients. The accuracy results obtained from controlling system experiment are 87.84%.

1. Introduction

Recently, the number of human population was increasing rapidly. The increasing of human population will also causes increasing of food demand. But the availability of agricultural land is decreasing. In addition, the climate change can cause crop failure. Hydroponic can be an alternative way to solve this problem.

Hydroponics is a method of cultivating plants without using soil as a planting medium. Hydroponic systems use water which is given a nutrient solution as a plant need [1]. One of the hydroponic techniques that used is Nutrient Film Technique (NFT). In NFT technique, plant roots are placed in a layer of shallow water that circulated and contains nutrients. The advantage of using NFT technique is plant growth is easier to control.

In the hydroponic plants cultivation, some parameters need to be considered. One of the most important parameters is TDS/EC nutrient solution. Each plants require a different balance of composition and amount of nutrients. The amount of nutrients that given must be according to the plant's needs, such as lettuce requires TDS value of 400-600 ppm [2]. The value of TDS or EC level is effected on plant growth, development, dan yields [3]. Giving too little or excessive amount of nutrients, can cause crop yields not optimal, or can even cause crop failure. Therefore, a system that able to control nutrient solution is needed.

Hydroponic nutrition controlling system is using TDS sensor to measure value of nutrition. This system can maintain TDS values to suit plants nutrition needs by using linear regression method.

2. Literature review

The research of Electrical Conductivity and pH Adjusting System for Hydroponics by using Linear Regression [4], control the micro-controller to adjust EC dan pH values in reservoir by using linear regression method. EC and pH are measured without using sensors but using EC and pH measurement tool, so the input system is given manually.

In the research Estimation of Electrical Conductivity and pH in Hydroponic Nutrient Mixing System using Linear Regression Algorithm [5], presents relation between A&B solution with EC, amount of nitric acid with pH, relation between EC dan pH in hydroponic nutrient system to get the adjusting equation value of the nutrient solution using linear regression analysis.

In research Nutrient Film Technique (NFT) Hydroponic Monitoring System Based on Wireless Sensor Network [6], system is made to ease farmer to monitor hydroponic cultivation in real time using WSN. This system using EC sensor to measure value of nutrient solution.

This paper aims to propose automatic hydroponic nutrient controlling system with linear regression method by using analog TDS sensor. This system can be controlled in real time.

3. System design and methods

3.1. System design

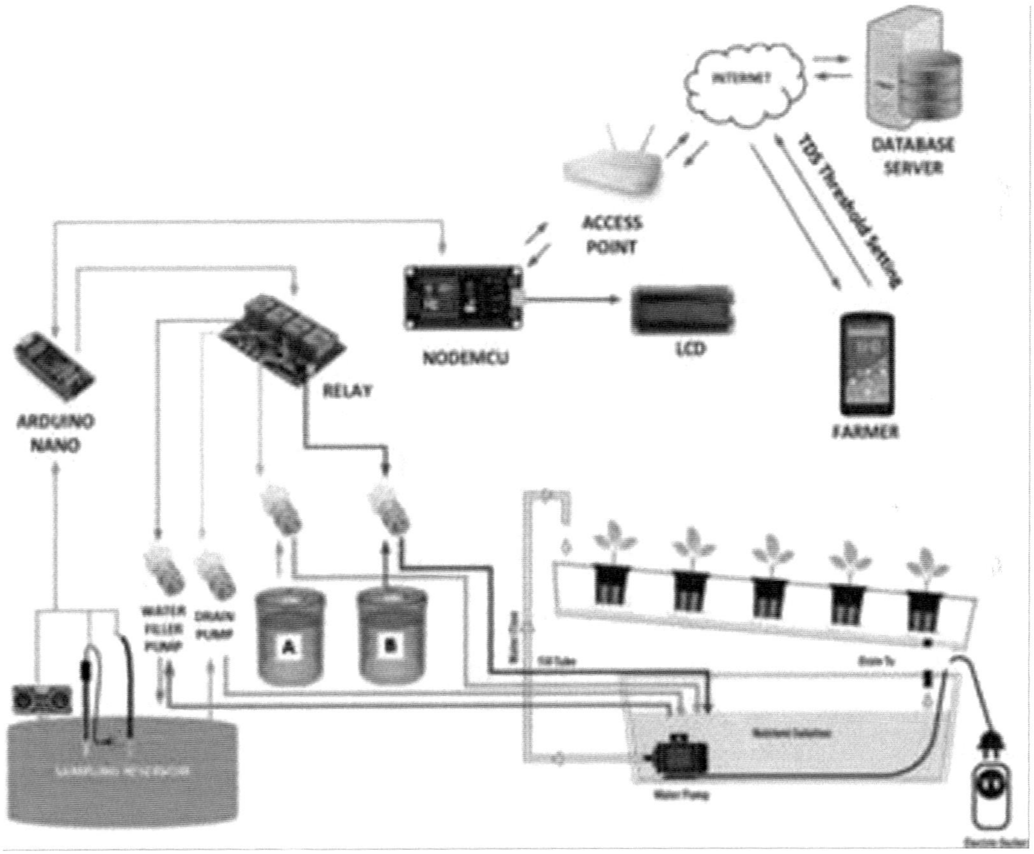

Figure 1. Controlling system of NFT hydroponic plant.

The nutrient hydroponic controlling system consist of NFT hydroponic module, Nodemcu and Arduino Nano microcontrollers, relay, 12V DC pump, Analog TDS sensor, DS18B20 temperature sensor, WLAN internet connection that connected to database server and web, and media access information. Fig. 1 shows the design of nutrient hydroponic controlling system.

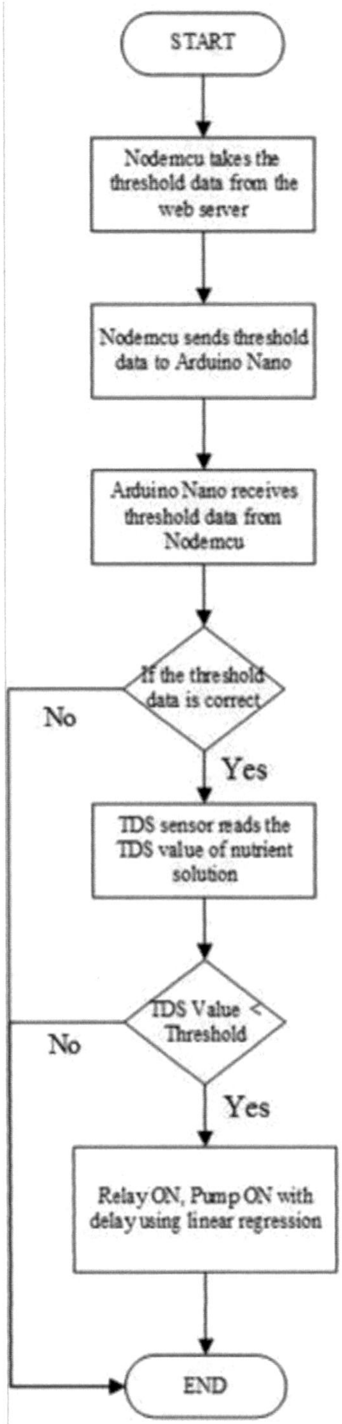

Figure 2. Nutrient hydroponic controlling system flowchart.

Based on Fig. 1, the system works as feedback from monitoring TDS values. Nodemcu micro-controller will retrieve the threshold data from the web server and then send it to Arduino Nano. Nodemcu will give an order to the Arduino Nano to monitor the nutrient value of the nutrient solution. Monitoring using analog TDS sensor, DS18B20 temperature sensor that placed in the sampling reservoir. Furthermore, Arduino Nano will compare the nutrient value from the results of monitoring with a predetermined threshold. Fig. 2 shows the flowchart of nutrient hydroponic controlling system.

When the value of nutrient monitoring results is less than the predetermined threshold, then the controller will give a back response. The response is an order to the Arduino Nano microcontroller to activate the relay which connected to the actuator. In this system the actuator is 12 Volt DC pump. This pump will pull nutrients to the nutrient solution reservoir so that the nutrient value suits the plant's needs. In this system linear regression is used to adjust the activate time of relay that connected to 12V DC pump. The system will set the activate pump delay according to the linear regression equation. The equation is obtained from the sample data collection experiment which is then processed using simple linear regression

3.2. Linear regression methods

This research using linear regression's equation to control pump delay of AB mix nutrient solution. The general equation of linear regression is:

$$Y = a + bX \tag{1}$$

Where
Y is dependent variable
X is independent variable
a is y-intercept
b is regression coefficient

3.2.1. b can be find by using:

$$b = \frac{[n \, \Sigma \, XY] - [(\Sigma X)(\Sigma Y)]}{[n \, \Sigma \, X^2 - (\Sigma X)^2]} \tag{2}$$

Where
n is number of data

3.2.2. a can be find by using:

$$a = \bar{Y} - b\bar{X} \tag{3}$$

Where
$\bar{Y}$ *is mean Y*
$\dot{X}$ *is mean X*

The implementation of linear regression method in this research is by taking sample data to determine the values of a and b, so that the equation can be obtained to be applied in the system.

The sample data collection experiment performed 20 times trial of each A solution and B solution. to determine the increase in TDS value of nutrient solution. This experiment is using pump delay 0.5s, 1s, 1.5s, and 2s, and carried out on a nutrient solution tank with a volume of 30 liters. The result of sample data test for A solution shown through graph in Fig.3.

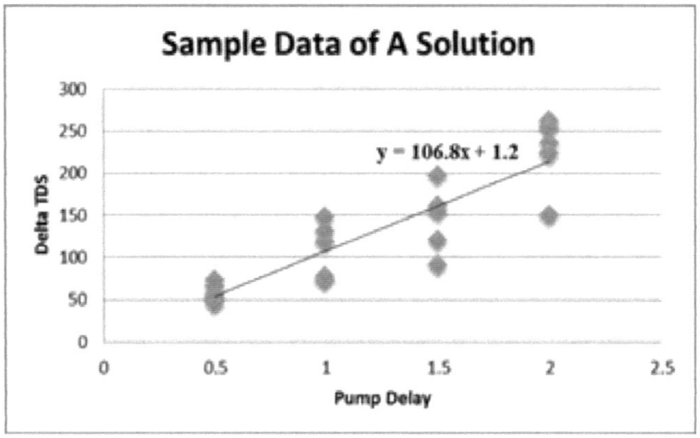

Figure 3. Sample data of A solution's test result.

From data sample collecting experiment of A solution obtained the value of a=1.2 and b=106.8. The equation for nutrient solution A as follow:

$$Y = 1.2 + 106.8\,X \tag{4}$$

The result of sample data test for B solution shown through graph in Fig.4.

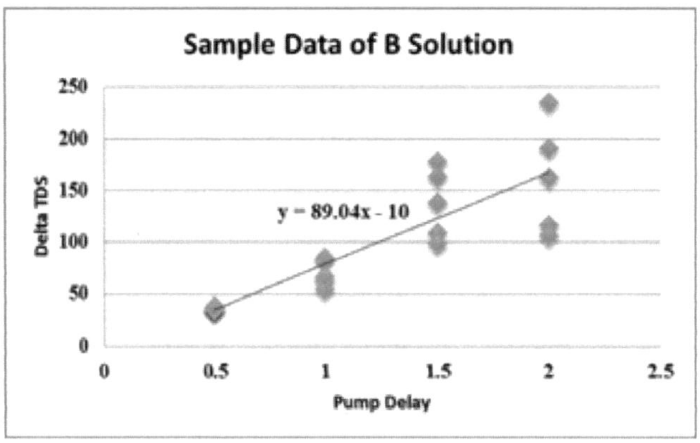

Figure 4. Sample data of B solution's test result.

And from data sample collecting experiment of B solution obtained the value of a=-10 and b=89.04. The equation for nutrient solution B as follow:

$$Y = 89.04X - 10 \tag{5}$$

4. Experimental results and discussion

4.1. Experiment of controlling hydroponic nutrient

Hydroponic nutrient controlling experiment aim to find out system performance using linear regression equation. Threshold is set with delta set point value of 50 ppm, 100 ppm, 200 ppm, 300 ppm, and 400 ppm. The results of experiment are shown in table I – table V:

IOP Conf. Series: Materials Science and Engineering 1108 (2021) 012033 doi:10.1088/1757-899X/1108/1/012033

Table 1. Results data of nutrient controlling with delta set point 50 ppm.

No	Set Point (ppm)	Initial TDS (ppm)	Final TDS (ppm)	Delta set point (ppm)
1	388	338	349	50
2	399	349	378	50
3	428	378	402	50
4	452	402	423	50
5	473	423	454	50
6	504	454	472	50
7	522	472	491	50
8	541	491	510	50
9	560	510	531	50
10	581	531	551	50

Table 2. Results data of nutrient controlling with delta set point 100 ppm.

No	Set Point (ppm)	Initial TDS (ppm)	Final TDS (ppm)	Delta set point (ppm)
1	369	269	329	100
2	429	329	386	100
3	486	386	451	100
4	346	246	302	100
5	402	302	355	100
6	455	355	393	100
7	493	393	431	100
8	697	597	650	100
9	750	650	710	100
10	369	269	329	100

Table 3. Results data of nutrient controlling with delta set point 200 ppm.

No	Set Point (ppm)	Initial TDS (ppm)	Final TDS (ppm)	Delta set point (ppm)
1	457	257	371	200
2	571	371	490	200
3	690	490	621	200
4	821	621	728	200
5	451	251	362	200
6	562	362	493	200
7	693	493	595	200
8	446	246	369	200
9	569	369	477	200
10	677	477	587	200

Table 4. Results data of nutrient controlling with delta set point 300 ppm.

No	Set Point (ppm)	Initial TDS (ppm)	Final TDS (ppm)	Delta set point (ppm)
1	568	268	442	300
2	742	442	653	300
3	953	653	814	300
4	571	271	497	300
5	807	507	704	300
6	537	237	426	300
7	726	426	652	300
8	952	652	851	300
9	687	387	557	300
10	857	557	753	300

Table 5. Results data of nutrient controlling with delta set point 400 ppm.

No	Set Point (ppm)	Initial TDS (ppm)	Final TDS (ppm)	Delta set point (ppm)
1	632	232	444	400
2	844	444	782	400
3	673	273	546	400
4	946	546	856	400
5	647	247	533	400
6	933	533	831	400
7	643	243	521	400
8	921	521	845	400
9	646	246	499	400
10	899	499	825	400

In this experiment, delta set point is results of set point minus initial TDS. The initial TDS is TDS value before the system controls, while the final TDS is TDS value after the system controls.

4.2. Accuracy of nutrient controlling system

Based on the result of nutrient controlling system, calculated accuracy of the controlling system by comparing the value of Final TDS with set point. The calculation of error data and accuracy of nutrient controlling system is given in Table VI – Table X below:

Table 6. The calculation error and accuracy of controlling with delta set point 50 ppm.

No	Set Point (ppm)	Final TDS (ppm)	Error rate (%)	Accuracy (%)
1	632	444	10.05	89.95
2	844	782	5.26	94.74
3	673	546	6.07	93.93
4	946	856	6.42	93.58
5	647	533	4.02	95.98
6	933	831	6.35	93.65
7	643	521	5.94	94.06
8	921	845	5.73	94.27
9	646	499	5.18	94.82
10	899	825	5.16	94.84
	Average (%)		6.02	93.98

Hydroponic nutrient controlling experiment with delta set point 50 ppm has the biggest accuracy of 95.98%, and an average accuracy of 93.98%.

Table 7. The calculation error and accuracy of controlling with delta set point 100 ppm.

No	Set Point (ppm)	Final TDS (ppm)	Error rate (%)	Accuracy (%)
1	369	329	10.84	89.16
2	429	386	10.022	89.98
3	486	451	7.20	92.80
4	346	302	12.72	87.28
5	402	355	11.69	88.31
6	455	393	13.63	86.37
7	493	431	12.58	87.42
8	697	650	6.74	93.26
9	750	710	5.33	94.67
10	369	329	10.84	89.16
	Average (%)		10.16	89.84

Hydroponic nutrient controlling experiment with delta set point 100 ppm has the biggest accuracy of 94.67%, and an average accuracy of 98.84%.

Table 8. The calculation error and accuracy of controlling with delta set point 200 ppm.

No	Set Point (ppm)	Final TDS (ppm)	Error rate (%)	Accuracy (%)
1	457	371	18.82	81.18
2	571	490	14.19	85.81
3	690	621	10	90.00
4	821	728	11.33	88.67
5	451	362	19.73	80.27
6	562	493	12.28	87.72
7	693	595	14.14	85.86
8	446	369	17.26	82.74
9	569	477	16.17	83.83
10	677	587	13.29	86.71
	Average (%)		14.72	85.28

Hydroponic nutrient controlling experiment with delta set point 200 ppm has the biggest accuracy of 90%, and an average accuracy of 85.28%.

Table 9. The calculation error and accuracy of controlling with delta set point 300 ppm.

No	Set Point (ppm)	Final TDS (ppm)	Error rate (%)	Accuracy (%)
1	568	442	22.2	77.8
2	742	653	12.0	88.0
3	953	814	14.6	85.4
4	571	497	13.0	87.0
5	807	704	12.8	87.2
6	537	426	20.7	79.3
7	726	652	10.2	89.8
8	952	851	10.6	89.4
9	687	557	18.9	81.1
10	857	753	12.1	87.9
	Average (%)		14.7	85.3

Hydroponic nutrient controlling experiment with delta set point 300 ppm has the biggest accuracy of 89.8%, and an average accuracy of 85.3%.

Table 10. The calculation error and accuracy of controlling with delta set point 400 ppm.

No	Set Point (ppm)	Final TDS (ppm)	Error rate (%)	Accuracy (%)
1	632	444	29.7	70.3
2	844	782	7.3	92.7
3	673	546	18.9	81.1
4	946	856	9.5	90.5
5	647	533	17.6	82.4
6	933	831	10.9	89.1
7	643	521	19.0	81.0
8	921	845	8.3	91.7
9	646	499	22.8	77.2
10	899	825	8.2	91.8
Average (%)			15.2	84.8

Hydroponic nutrient controlling experiment with delta set point 400 ppm has the biggest accuracy of 92.7%, and an average accuracy of 84.8%.

Based on the data on table VI to table X, the nutrient controlling system using linear regression method has the biggest average accuracy of 93.98% on the 50 ppm point set and the smallest average accuracy is 84.8% on the 400 ppm point set. Figure 4 shown the graphic of the accuracy calculation from controlling experiment.

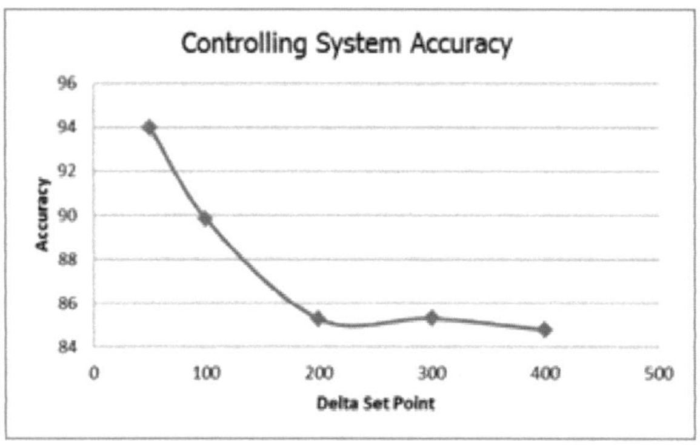

Figure 5. Graphic of controlling system accuracy.

Based on the graph, the bigger differences between set point and final TDS value, the bigger error percentage, and the lower accuracy percentage.

This is due to testing of sample data carried out on 30 liter reservoirs which have not been applied to NFT hydroponics. While the control system testing is done on the NFT hydroponic greenhouse. In NFT hydroponic greenhouse, the volume of nutrient solution is greater because of there is nutrient solution that is flowed to irrigate plant roots. The difference in the amount of water volume of ± 5 liters causes the control of nutrient density solution to be less than optimal.

5. Conclusion and future works

The application of analog TDS sensor can be used as indicator and input of controlling system. Sample data collection using 30 liter tank, with pump delay 0,5s to 2s gets equation results $Y= 1.2 + 106.8\ X$ for A solution, and $Y=89.04\ X - 10$ for B solution. Based on the results of the research, controlling system with linear regression method has an average accuracy of 87.84%. The bigger delta set point value, the bigger average of error, and the lower average of accuracy. The use of linear regression method needs to be observed by adding more sample data and using more variables to get the equation in order to make better controlling system.

References

[1] Saaid M F, Yahya N A M, Noor M Z H and Ali M S A M 2013 A development of an automatic microcontroller system for Deep Water Culture (DWC) *Proc. - 2013 IEEE 9th Int. Colloq. Signal Process. its Appl. CSPA 2013* 328–332

[2] Jones J B J 1997 *A Guide for the Hydroponics & Soilless Grower, First Edit* (Delray Beach: St. Lucie Press)

[3] Alexander T 2000 *Best of the Growing Edge, First Edit* (USA: New Moon Publishing, Inc.)

[4] Kaewwiset T and Yooyativong T 2017 Electrical Conductivity and pH Adjusting System for Hydroponics by using Linear Regression in *2017 14th International Conference on Electrical Engineering/Electronics, Computer, Telecommunications and Information Technology (ECTI-CON) Electrical* 761–764

[5] Kaewwiset T 2017 Estimation of Electrical Conductivity and pH in Hydroponic Nutrient Mixing System using Linear Regression Algorithm

[6] Nursyahid A, Setyawan T A, Hasan A and Mahaidayu M G 2016 Nutrient Film Technique (NFT) Hydroponic Monitoring System *J. Appl. Inf. Commun. Technol.* **1**(1) 1–6

Fall detecting clothes in realtime based seniors full body motion capture system using multiple inertial sensors

T R Yudantoro*, F Pramuditya, R Apriantoro and S Jum'atun

Dept. of Electrical Engineering, Politeknik Negeri Semarang, Semarang, Indonesia

*tryudan@gmail.com

Abstract. The physical ability of the elderly will decrease along with the aging process, thus increasing the potential for falls. An elderly person who has experienced a fall and is not detected for a long time will bring many possible consequences. Early detection of an elderly fall will help to minimize this possibility by reducing the time between the occurrence of an event and the arrival of medical assistance The purpose of this research is to build a Fall Detecting Clothes in Realtime Based Art Full Body Motion Capture System Using Multiple Inertial Sensors. This system builds by Waterfall methodology where every research step is done in sequence, start from analysis, design, coding, testing. Testing methodology is black box, which is a test every feature function. The Fall Detecting Clothes in Real Time Based Systems Full Body Motion Capture System Using Multiple Inertial Sensors has been tested using black testing box with the results of all feature functions working properly. The system has been tested with several conditions with the results of all controls and output data works properly as well.

1. Introduction

Worldwide, the growth of the elderly population is very fast compared to other age groups. Elderly belongs to the fastest growing population group in the world. The population is expected to double from 606 million to 1.2 billion in the next half century [1]. According to the UN report in 2011, in 2000 the elderly population reached 7% of the world's population and is expected to increase to 28.68% in 2045 [2].

The physical ability of the elderly will decrease along with the aging process, thus increasing the potential for falls. An elderly person who has experienced a fall and is not detected for a long time will bring many possible consequences. Early detection of an elderly fall will help to minimize this possibility by reducing the time between medical events and the arrival of help [3]. Incidence of falls in the elderly can result in physical impact (most frequent head injury) and physiological effects (fear of falling). If emergency treatment arrives late, falling injuries can result in disability, paralysis, and even death [4]. Based on a survey in the United States, around 30% of elderly people over 65 years of age experience falls. half of these numbers occur repeatedly. Falling incidence experienced by Americans in the age group of more than 65 years was recorded as many as 1800 incidents per year which caused death [5].

In the current era of globalization requires humans to be busy working to fulfill their needs. As a result, time for supervision and assistance to the elderly is very minimal. Fall detection technology is currently being developed with a variety of methods, both using a camera, and using sensors that can detect movement. The use of cameras for monitoring is a good method with high precision, but often the use of the camera causes discomfort in the elderly because how old do they feel their movements are curbed and always monitored by the camera. Therefore, it is necessary to have a system that is able to carry out remote monitoring well without giving a sense of oversight to the activities of the elderly with early warning to accelerate the provision of medical help to reduce the risk of death in the elderly who experience a falling incident. The prototype that we will design uses full-body motion capture technology which provides a solution to remotely monitor the movement of the elderly in real time by reducing discomfort due to excessive monitoring as done by the camera. The combination of the gyroscope sensor and accelerometer is used to detect every movement of the elderly. The results of the sensor readings were sent by NodeMCU via the wifi network. If detected falls, the system will send a notification to the family along with the coordinates of the scene to be followed up to contact the nearest medical officer. This protoype is in the form of a special shirt that has 10 IMU sensors installed with a power bank power supply that can be used flexibly and easily to recharge the power with a 4G LTE connection to support data transfer speeds, so notifications can be sent quickly and accurately.

2. Methods

This system using waterfall methods. Start from analysis, design, coding, testing.

MySQL database still the better to choose when it comes to data reliability and secure assurance of performed transactions. Relational database management systems are more popular and offer both free and paid support. In case any issue arises, it can be solved easier than NoSQL databases. Relational databases are the go to solution for difficult querying and data keeping requirements. They are more competent and outclass in this domain.

2.1. Analysis

The first step taken at this stage of the analysis is to validate by ensuring that the system is made in accordance with the needs of the user and able to handle the problems that exist in the user, as well as analyzing the needs of both hardware and software to build the system. Figure 1 is system overview of Fall Detection Clothes. This system uses hardware consisting of 10 IMU sensors (MPU-6050) that are connected to each other in a specially designed shirt. The prototype that we designed uses full-body motion capture technology which provides a solution to remotely monitor the movement of the elderly in real time by reducing discomfort due to excessive surveillance as the camera does. The combination of gyroscope sensor and accelerometer is used to detect every movement of the elderly based on the acceleration of angle changes. Results from sensor readings are sent by NodeMCU via the wifi network. If detected falls, the system will send a notification to the family along with the coordinates of the scene to be followed up to contact the nearest medical officer. The design power supply of this tool uses a power bank that can be used flexibly and easily to recharge the power with a 4G LTE connection to support data transfer speeds, so notifications can be sent quickly and accurately.

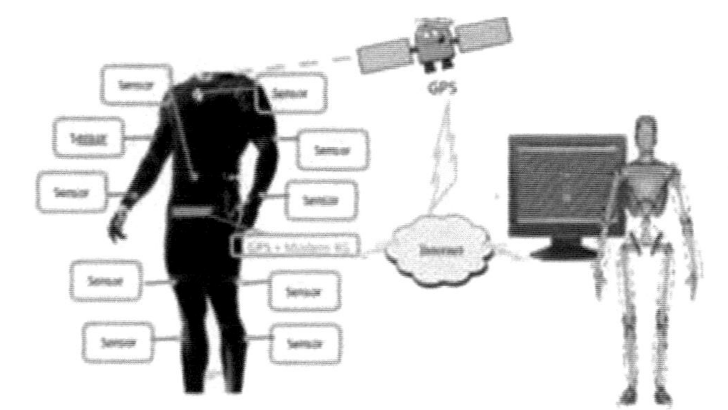

Figure 1. System overview of fall detection clothes.

2.2. Design
Design stage produces block diagram which shows the relationship between inter blocks in the system. Figure 2 is block diagram of Fall Detection Clothes. Elderly clothes design falls there is an MPU6050 element and the power supply that enters the Lolin V3 MCU Node element which is connected to the wifi network so as to produce output data stored in the database, the data is processed to generate android push notifications when the elderly is dropped.

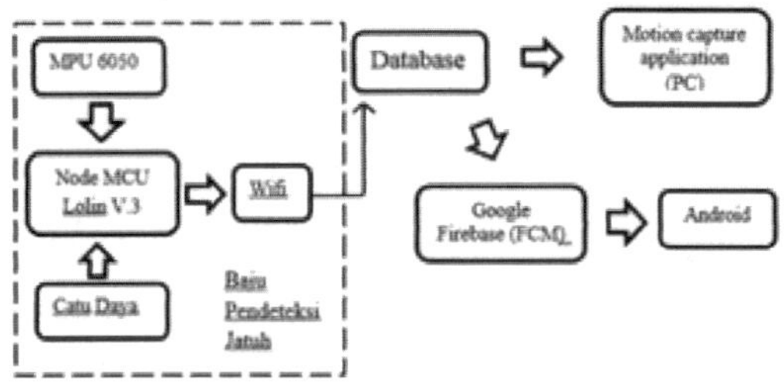

Figure 2. System overview of fall detection clothes.

2.3. Coding
The language used in making Android applications of Fall Detection Clothes is java with the addition of a retrofit library, Picasso, Recycle View, and Firebase Cloud Messaging. Encoding on MCU nodes uses C language where the MCU node is used as a client http.

2.4. Testing
The types testing to test functionality is blackbox that focuses on functional of system.

3. Results
The results obtained from this system are in the form of old clothes falling and the push notification application for the elderly in a state of falling. Figure 3 shows the appearance of clothes for elderly in falling conditions. Figure 4 shows the splash screen display of the Android application.

Figure 3. the appearance of clothes for elderly in falling conditions.

Figure 4. The splash screen display of the Android application.

Figure 5 shows the experiment when the elderly falls and Figure 6 shows the push notification given by the Android application when the elderly is in a state of falling.

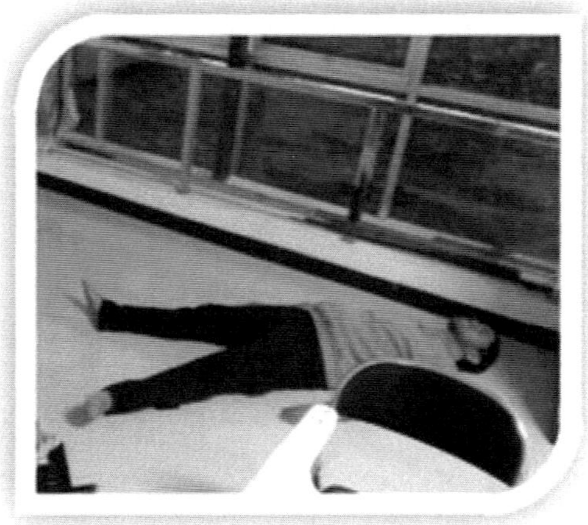

Figure 5. Fall experiment.

Figure 6. The push notification given by the Android application when the elderly is in a state of falling.

3.1. *Blackbox testing*

This test is based on main function that already plan in previous time before create an application. Table 1 displays whether application work properly as expectation or not. And it shows the result black box testing of this system.

Table 1. Black box testing of fall detection clothes.

Data	Expected results	Results	Conclusion
Retrieve gyro and accelerometer data	Store gyro and accelerometer data in database	All gyro and accelerometer data can store in database	In accordance with system specifications
Retrieve Notification based fall detection condition	Display notification on android application	Display notification on android application	In accordance with system specifications
Call telephone number that integrated in Android application where elderly in fall condition	Call telephone number that integrated in android application Where elderly in fall condition	Can make a call with The telephone number that integrated when elderly in fall condition	In accordance with system specifications

4. Conclusion

Based on the results of this system, it concludes that the Fall Detecting Clothes in Realtime Based Seniors Full Body Motion Capture System Using Multiple Inertial Sensors has been tested using black box testing with the results of all feature functions work properly. The system has been tested with several conditions with the results of all controls and output data work properly as well.

References

[1] Firmansyah M 2016 Rancang Bangun Sistem Fall Detection untuk Orang Lanjut Usia Berbasis Inertial Measurement Unit *Jurnal Penelitian dan Pegembangan Telekomunikasi, Kendali, Komputer, Elektrik, dan Elektronika (TEKTRIKA)* **1**(2)

[2] Mateska A, Pavloski M and Gavrilovska L 2011 RFID and sensors enabled in-home elderly care

IOP Conf. Series: Materials Science and Engineering 1108 (2021) 012034 doi:10.1088/1757-899X/1108/1/012034

In 2011 Proceedings of the 34th International Convention MIPRO 285-290

[3] Riduwan 2004 *Metode Riset* (Jakarta: Rineka Cipta Halaman) p 104

[4] Sugiyono 2013 *Metode Penelitian Kuantitatif, Kualitatif dan R&D* (Bandung: Alfabeta) p 244

[5] Viet V and Choi D 2011 Fall Detection with Smart Phone Sensor *The 3rd International Conference on Internet (ICONI)* 27–31

Study method accordance for presstool design learning between POLINES and PT Mekar Armada Jaya in the framework of link & match

R M Wattimena*, A S Ardjo, H Hartono and A S Alfauzi

Politeknik Negeri Semarang, Jl. Prof Sudarto S. H., Tembalang, Semarang 50275, Indonesia

*riles.wattimesna@polines.ac.id

Abstract. PT Mekar Armada Jaya, located in Cibitung, Bekasi, West Java, is a national-scale industry engaged in manufacturing various vehicle parts, especially those made of steel sheets. The process of making various parts of this vehicle requires a press tool. This company designs, makes, studies, and develops all the required press tools, therefore human resources (HR) are needed who have the competency to design and make presstools. Based on this, PT Mekar Armada Jaya held a partnership to produce the intended HR needs, namely graduates of the D3 Mechanical Engineering study program who were ready to work both at this company and at other similar companies. Along with the link between the Semarang State Polytechnic and PT Mekar Armada Jaya above, it is necessary to study the alignment of methods and learning strategies for the presstool design. The results of this study are in the form of design learning methods that will be carried out in the D3 Mechanical Engineering study program and learning handouts for the topic of presstool design.

1. Introduction

Profile of graduates of Semarang State Polytechnic (Polines) of three years diploma in Mechanical Engineering Study Program, one of which is the ability to design presstool. Presstoolsor dies are equipments or tools to form sheet metal (steel, stainless steel, copper, aluminum) into certain forms such as vehicle components, electronic appliance components, household appliances, home appliances. Figure 1. shows the atmosphere of automotive component production at PT Mekar Armada Jaya, Figure 2. shows the components produced from sheet metal with a presstoolfound at PT Mekar Armada Jaya.

The main courses that form the competency of designing Prestool are Design of Production Assistance, the topics in this course are the design of jigs, fixture, press tool and mold. This course has been supported by 3 or 4 lecturers since 1983.

Support for this course is in the form of reference books published before 1990, design aids (CAD) software. The problems faced are:

Figure 1. Presstool construction.

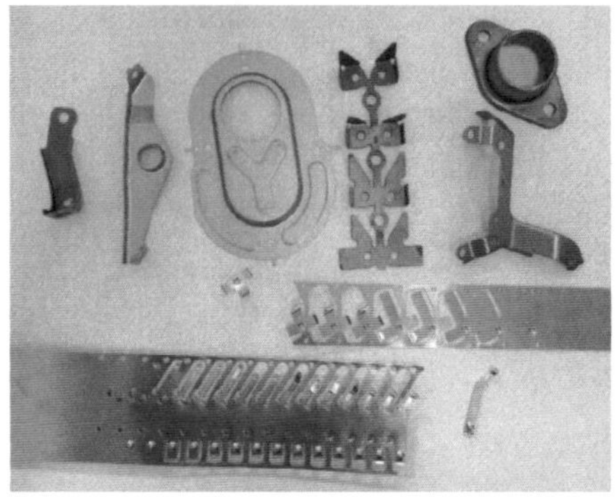

Figure 2. Automotive components produced by the press tool.

- The three lecturers referred to are over 60 years old, 1 lecturer is 57 years old so that in the near future they will retire,
- Among the four Lecturers who have had experience working at a company as a presstool designer are only 1 person, so that practical expertise is not yet possessed,
- The four lecturers referred to as masters of CAD are very weak,
- the four lecturers have not yet developed the learning methods of designing the tool that is in accordance with design procedures in the industry,
- the development of design technology support software in the industry is very rapid so that the three years diploma in the study program has lagged far behind,
- the price of software supporting design technology used in the industry is very expensive.

As a result of the above problems that can occur are: • Competency in designing the design of Prestool Graduates of the study program is no longer suitable to the needs of the industry • the study program graduates become unable to compete when applying to companies that need the ability to design Prestool.

PT Mekar Armada Jaya, located in Cibitung, Bekasi, West Java, is a national-scale industry engaged in manufacturing various vehicle parts, especially those made of steel sheets. The process of making various parts of this vehicle requires a press tool. This company designs, makes, studies, develops all the required press tools, therefore human resources (HR) are required who have the competency to design and produce Press Tools. Based on this, PT Mekar Armada Jaya held a partnership with the study program to produce the intended HR by graduteswho were ready to work both at this company and at other similar companies. Along with the link between the Semarang State Polytechnic and PT Mekar Armada Jaya above, it is necessary to study the accordance of methods in designing presstool. The results of this study are expected to teach learning methods for presstool in the study program to be matched with those carried out in industry. Based on the background above, the following problems can be identified:

- Presstool design learning methods in the study program it's time to be reviewed to fit the presstool design methods applicable in the industry (PT Mekar Armada Jaya)
- It is necessary to immediately formulate the presstool design learning method of the results of a study at PT Mekar Armada Jaya, which can be applied to the study program.
- Graduates in the study program must be able to compete with similar study program graduates from other tertiary institutions when they are recruiting for industries that require competence in presstool expertise.

Based on the description on the background and identification of the problems above, it can be formulated that the problems that must be overcome are: learning methods for designing tools such as what is implemented at PT Mekar Armada Jaya which can be adopted as methods for designing presstool in the study program can have competencies that are in line with what is needed by industry?

The purpose of this study is to conduct a study of the accordance of presstool design methods Polines and PT Mekar Armada Jaya in the Link & Match Framework. The benefit of this research is the existence of standard design methods that are appropriate to what is carried out in the industry, so that both lecturers and students have high design competencies

2. Literature review

2.1. Machine design method

Budynas-Nisbet [1] proposes design phases in the field of mechanical engineering known as the Sighley's Mechanical Design Method as shown in figure 3. Identification of needs is often a very creative act, because the need may only be in the form of unclear dissatisfaction, an uncomfortable feeling, or a feeling that something is not right. Needs are often not clear at all. For example, the need to do something about food packaging machines can be indicated by noise levels, by variations in package weight, and by slight variations but clearly in the quality of the packaging or wrapping.

The problem definition is more specific and must include all specifications for the object to be designed. Specifications are the number of inputs and outputs, the characteristics and dimensions of the space that objects must occupy, and all restrictions on this quantity. We can think of an object as something designed in a black box, which must determine inputs and outputs along with their characteristics and constraints.

The specifications determine the cost, the amount to be produced, the expected age, range, operating temperature, and reliability. Specified characteristics can include speed, feed, temperature limits, maximum range, expected variations in variables, dimensions, and weight limitations, etc. Available labor skills and competitive situations are also implicit obstacles. Anything that limits designer freedom of choice is an obstacle. Many materials and sizes are listed in the supplier catalog, for example, but these are not all easily available and deficiencies occur frequently.

Synthesis of schemes that connect system elements which may sometimes be called concept discovery or concept design. This is the first and most important step in the synthesis task. Various schemes must be proposed, investigated and measured in terms of predetermined metrics. Analysis must be carried out to assess whether the system's performance is satisfactory or better, and, if it is satisfying, how well it performs. System schemes that do not survive analysis are revised, corrected, or discarded.

Design is an iterative process in which we proceed through several steps, evaluate the results, and then return to the previous procedure phase. Thus, we can synthesize some system components, analyze and optimize them, and return to synthesis to see what effect they have on the remaining parts of the system. For example, the design of a system for transmitting power requires attention to the design and selection of each component (e.g., gears, bearings, shafts). Evaluation is a significant phase of the total design process. Evaluation is the final proof of a successful design and usually involves testing a prototype in a laboratory. In this phase we want to find out if the design really satisfies the needs. Is this reliable? Will it be able to compete with similar products? Is it economical to produce and use? Is it easy to maintain and adjust? Can profits be from selling or using them? How likely is it to produce product liability lawsuits?

Presentation is selling work. Engineers, when presenting new solutions for administration, management, or supervisors, are trying to sell or prove to them that this solution is better. Unless this can be done successfully, the time and effort spent to get a solution is largely wasted. When designers sell new ideas, they also sell themselves. If they are repeatedly successful in selling new ideas, designs, and solutions for management, they begin to receive raises and promotions; in fact, this is how a person succeeds in his profession.

2.2. *DIN-VDI 2221 design method*

In 1993 the VDI 2222 guidelines were replaced by the systematic approach of the VDI 2221 guidelines for the development and design of systems and technical products [VDI-Richtlinie 2221 1993]. This guideline is still based on VDI 2222. The aim of this guideline is to propose a general methodology to design systems and technical products and to support methodical and systematic design, so as to produce a more efficient work style. This guideline does not depend on branches and the content and aspects of the organization's design. This guide emphasizes broad applications in engineering mechanics, precision mechanics, switches and software development and engineering planning. This guide is based on system engineering and problem solving.

One major aspect of it is integral data processing and application of CAD. This aspect takes up one third of the description in Weißdruck. The remaining suggestions address the iteration of the process and adaptation of the list of requirements throughout the entire process. The advice in this Wehrdruck guideline contains more research results than practical instructions for thinking and acting. This means that the inspection question step has changed to the requirements of a good problem solver. Thus, Weißdruck contains paragraphs with the results of successful designer research [VDI-Richtlinie 2221, 1993].

Examination questions such as those in VDI 2222 as well as suggestions regarding thought processes were eliminated. The instructions of VDI 2221 compared to VDI 2222 were reduced to "defining functions and structures, looking for the principle of solutions and their structure and structure in the module." Terminology is more abstract than VDI 2221. Especially the terminology and module structure were introduced. Weißdruck covers 42 pages Figure 4 shows instructions and work in designing.

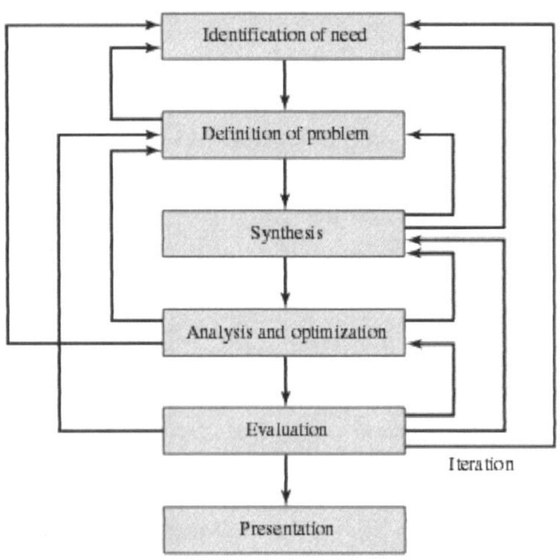

Figure 3. Design phases in mechanical design.

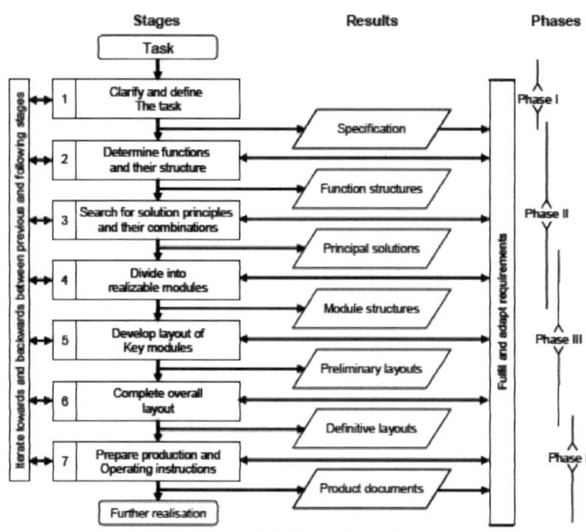

Figure 4. DIN-VDI 2221 design method [1] Guidline [2].

3. Data and discussion

3.1. *Data collection results*

The observations of the design method used at PT MAJ and its comparison from the German Engineers Association (VDI) and Bandung State Polytechnic Manufacturing (Polman) can be seen in Table 1.

Table 1. Comparison of design methods.

No	PT MAJ	VDI 2222	VDI 2222 POLMAN	POLINES
1	Receiving orders	Clarification of assignments		Receive/choose assignments
2	Drawing parts	Conceptual Design	Pre Design	Product Analysis
3	Generate CAD data	Shape Design	Concetual	Product Data Acquistion and Simulation
4	Production Planning Schedulling (PPS)	Pre Layout	Design	Punch Desaig
5	CAE Analysis	Layout	Settlement	Dies Design
6	Dies Design	Documentation		Determine the DieSet
7	Simulation			Creating Work Drawing
8				Die set Simulation
9				Presentation
10				Documentation

Presstool Designing facilities obtained from PT MAJ and compared to those owned by Polines can be seen in table 2.

Table 2. Comparison of Support Facilities for Presstool Design.

Facillition	PT MAJ	POLINES
Computer	PC 64 bit Monitor 21"	`PC 64 bit Monitor 17"
Software	Auto CAD, CAM, CAE, CG Press PAM Stamp	Auto CAD, Inventor, Solidwork, Master Cam (pirated)
Reference	n.a	Fundamentals of Tool Design, Metals Handbook
Standard	Standard Componens for Press Die, Punch CAM Unit, Sankyo, Mitsumi Dayton, NT tool	Leinco Component Standart
Machines	Large Capacity CNC Machines	Conventional machinies, CNC Machines
Laboratory	3- D Scanner	n.a

Comparison of the use of design methods between PT MAJ, Polman Bandung, and Polines can be seen in table 3.

Table 3. Comparison of use of design methode.

Design Parameters	Polines	Polman Bandung	PT MAJ
Dept of analysis	Very Good	Good	Fairly
Dimension Determination	Good	Good	Very Good
Simulation	Good	Good	Very Good
CAD Implementation	Very Good	Very Good	Very Good
CAE Implementation	Bad	Very Good	Very Good
Level of Difficulty	Good	Good	Very Good
Presstool Dimension	Good	Good	Very Good

Based on studies conducted based on table 1 and table 2, the method of planning carried out in the D3 Study Program in Mechanical Engineering Polines is recommended:

- Keep using existing methodsb. At some stages it can be adopted that is carried out at PT MAJ and Polman Bandung, such as simulations with PAM Stamp software.
- Polines should provide CAE software facilities such as those used by PT MAJ, so that the design results will be more accurate

In connection with the implementation of learning presstool design in POlines can be suggested as follows:

- Integrate bench worksheets and CNC practice worksheets with material / assignments on presstool designers so that the competency in designing precision equipment (jigs, fixtrures, presstools, molds) will be even higher.
- Provides more complete design software such as PAM Stamp, CG Press, and CAE

4. Conclusion

Based on studies conducted based on table 1 and table 2, the method of planning carried out in the D3 Study Program in Mechanical Engineering Polines is recommended:

- Keep using existing methods
- At some stages it can be adopted that is carried out at PT MAJ and Polman Bandung, such as simulations with PAM Stamp software.
- Polines should provide CAE software facilities such as those used by PT MAJ, so that the design results will be more accurate

In connection with the implementation of learning presstool design in Polines can be suggested as follows:

- Integrate bench worksheets and CNC practice worksheets with material / assignments on presstool designers so that the competency in designing precision equipment (jigs, fixtrures, presstools, molds) will be even higher.
- Provides more complete design software such as PAM Stamp, CG Press, and CAE

References

[1] Budynas-Nisbet 2006 *Mechanical Engineering: Shigley's Mechanical Engineering Design, Eighth Edition* (McGraw–Hill Primis

[2] Jansch J and Birkhofer H 2006 The Development of Guideline VDI Direction *International Design Conference – Design*

Biaxial pressure pack carburizing method to modification local low carbon steel's mechanical properties

A Satito*, H Hariyanto and S Supandi

Department of Mechanical Engineering, Politeknik Negeri Semarang, Semarang, Indonesia

*aryosatito@gmail.com

Abstract. The absolute requirement for steel material can be hardened directly is the carbon content of this material is more than or equals to 0.3 %. This steel material is recognized as medium carbon steel and above. Carburizing process is an effort to increase low carbon content in soft steel material which have less than 0.3 % it can be done by adding carbon content up to 0.3 % and above. The tool commonly used for this process called pack carburizing device and the process occur in furnace atmospheric pressure. In this research, carbon content adding process held on pressurized pack carburizing container, and the pressure was given mechanically from two axial direction named biaxial pressing carburizing process. The speciment material used in this research is local low carbon steel (ST.37) and coconut shell active charcoal powder. Process parameter The process parameter are pressure, temperature, and hoding time. The best result is achieving at a process temperature of 1000 °C with 3 hour holding time and biaxial pressure of 2 MPa with 115.4 HRB, 823,6 N/mm2, and 0,41 % the incensement of carbon content from 0,27 % to 0,40 %.

1. Introduction

Local low carbon steel, usually known as ST.37, has approximately carbon content average of 0.27 %. This condition becomes an obstacle if the low carbon steel want to be harden directly, because the steel must have carbon content ≥ 0.3%. to be able to harden the speciment's surface directly on heat treatment process. The process of adding carbon content on ST.37 steel will be carried out by the biaxial pressure case carburizing method. By this method, a heat treatment process to increase carbon content by "burried" the soft steel specimens in active coconut charcoal powder containers and pressured from two axial directions, then heated to specified temperature and holding time. The pressure on carbon powder is predicted will distributed more evenly the absorption of carbon to the specimen than pressureless pack carburizing which used in previous research.

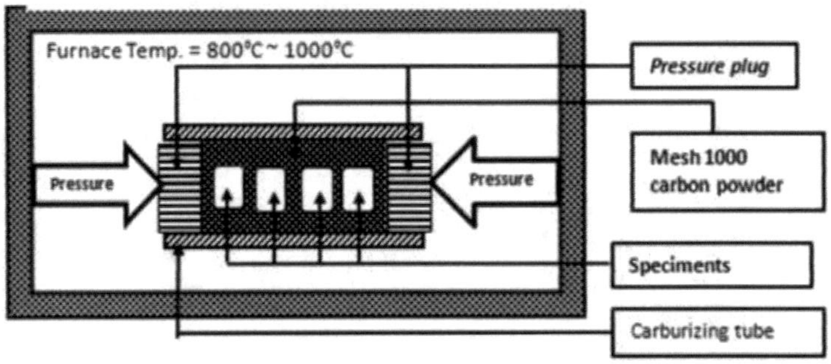

Figure 1. Biaxial pressure pack carburizing process scheme.

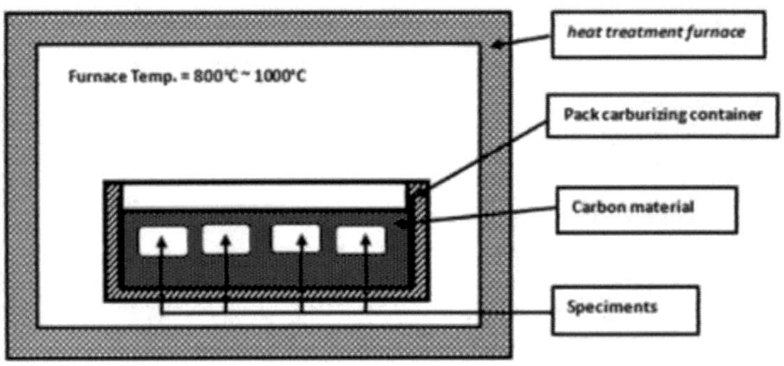

Figure 2. Pressureless pack carburizing process scheme used in previous research.

This biaxial pressure pack carburizing's method is predicted that the diffusion and penetration of carbon atoms into the surface of the specimen will be deeper. This method is believed will increase more carbon content at the surface of the specimen than the open pressureless pack carburizing method used in previous studies.

Steel is an alloy of iron (Fe) and carbon (C) with a carbon element content of between 0.2% up to 2.1%. And steel is classified into 3 main groups, namely soft steel is steel with carbon content ≤ 0.20% and does not contain other elements besides Si and Mn, low carbon steel is steel with carbon content of 0.10% ~ 0.25 % and high carbon steel is with a carbon content of 0.56% ~ 1.70% with the addition of other elements to the composition [1]. Meanwhile, soft steel and low carbon steel is very difficult to increase the value of hardness if the heat treatment process is carried out directly [2]. To be able to increase the value of steel hardness through the heat treatment process (heat treatment) is determined by the carbon content in the steel itself. If the carbon content in steel > 0.30% [3], the steel will not be able to increase its hardness value by means of direct heat treatment process.

One of the ways to increase carbon content in low carbon steel is to increase the carbon content by wrapping low carbon steel specimens with carburized material which is generally carbon powder, graphite powder, or activated charcoal powder in a heat-resistant container to be heated to temperatures between 800 ° C ~ 1050 ° C. If the process temperature has been reached, this temperature must be maintained for 2 hours to 3 hours (holding time). The next process is to cool specimens that have been heated for a certain period of time suddenly, this cooling process is known as a quenching process [4]. To increase carbon content's in steel composition, a carburizing process on low carbon steel must be conducted using carburized material in the form of a mixture of coconut shell charcoal powder and calcium carbonate powder (CaCo3) at 900 ºC and 1000° C [5]. The result is the carbon content in low carbon steel increases up to 70% greater than the initial carbon content. The carburizing process is an effort to add carbon content to the specimen by wrapping the specimen with active charcoal powder from a coconut shell [6]. The result achieved is an increase in the hardness value of the speciments up to 80% with a process temperature of 950 ° C with a holding time of 2 hours. And an effort to improve mechanical properties on low carbon steel surfaces (0.19% C) by wrapping specimens using carburized material mixture. The results achieved increased carbon composition on the surface of the specimen by 80% from 0.19% C to 0.35% C as thick as 1.2 mm [7]. Some of the research carried out above is very in line with the objectives of this study, namely modifying the material of soft steel and low carbon steel into materials that meet the requirements of making tensile link components that must have a certain hardness value and adequate wear resistance.

Biaxial pressurized pack carburizing use in this research is a method for distributing pressure which applied from two opposite directions axially in a pressing tube as shown in the image below. The purpose of this compression method is that the density of the object being pressed in the pressure tube is more evenly distributed when compared to the pressure in the pressure tube only given from one direction which is commonly called uniaxial pressure method.

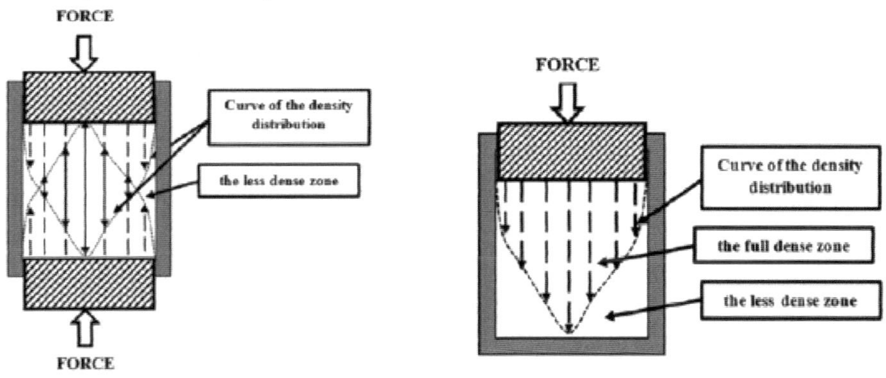

Figure Biaxial pressing process

Figure Uniaxial pressing process

Figure 3. Biaxial pressure distribution and uniaxial pressure distribution scheme.

1.1. Test of chemical composition of specimens

Emission spectrometer is one of the chemical analysis tools for the determination of metal elements in a massive metal and metal alloys, both qualitatively and quantitatively [8]. Spectrometer tests were performed to determine the chemical composition of genuine tensile links, specifically the carbon content of C so that it can be known the type of material used. In addition to genuine tensile link, ST.37 material was also tested by a spectrometer to prove that ST.37 material is low carbon steel with carbon C content below 0.3%. To determine the mechanical properties of metals and non-metals one of the tests used is a tensile test. Tensile testing is a method used to test the strength of a material by giving the force load in the opposite direction [8]. And hardness testing is one of several types of testing used because it can be carried out on small test specimens without difficulty regarding specifications. The most widely ysed test is to press the penetrator or indenter with the shape and load according to the predetermined standard of the object and to measure the dimension of the former stressor indenter formed above it [9]. The value of the tensile strength test results can be converted to the value of violence in the Brinell scale with the following equation [10]:

$$TS, MPa = 3{,}45 \ x \ HB \qquad (1)$$

with : *TS = Tensile Strength, N/mm² ; and*
HB = Hardness Brinell Number, HB

To solve the problem of adding carbon content on a local low carbon steel (ST.37), several researchers do some research as mentioned above and no one use the carburizing process which enclose the pressure in the process. The suggested strategies in pack carburizing process with pressure enclosed are effective and can be implemented in the material hardening shop, because the super fine coconut active carbon charcoal size 500 mesh ASTM have a character like fluids [11].

2. Methods

Research implementation in the form of Ishikawa diagram

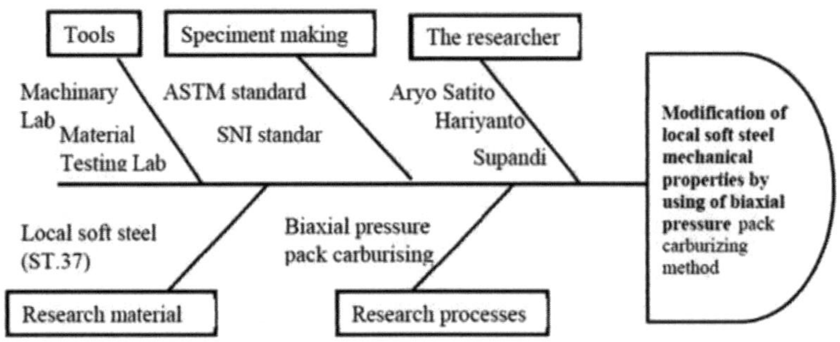

Figure 4. Ishikawa research's diagram.

2.1. Experiment's material

The material used in this experiment is local soft steel that is easily found on the local market, namely ST.37 which has a maximum carbon content up to 0.27% with an average tensile strength of 500 N / mm2 and a surface hardness of 79 HRB. The carburiser material is active coconut shell charcoal powder of 500 mesh ASTM.

2.2. Research equipment

Research equipment used in this experiment is placed in the Lab. Of Machining and Lab. of Materials Testing at Politeknik Negeri Semarang. Whereas the carbon content test is carried out in the Material Test laboratory Ceper Polytechnic of Manufactur.

2.3. The experimental variables

The variables carried out in this study are the pressure on the carburising tube, the carburizing process temperature, and the carburizing process time. All of these variables will affect the mechanical properties, especially the hardness of the specimen.

2.4. The experimental steps

The experimental steps carried out in this study are as follows:

- Low carbon steel type ST.37 is formed according to ASTM E8/8M-15a standard
- Perform carburizing process by means of case hardening methods using biaxial pressure pack carburising equipment units with the specified variables, namely temperature, pressure, and processing time.
- Test the composition of carbon on the surface of the specimen.
- Test the hardness of specimens by the Rockwell B testing method
- Analyze data from all specimen hardness test results.

2.5. Research output

The annual output of the study is the data of specimen hardness values and recommendations of the carburizing process that are most suitable for ST soft steel material. 37 as material for making machine components.

2.6. Outcome indicator

The expected outcome indicator of this research is the hardness value of the specimen reaches 80 to 110 HRB such as the value of the hardness of the machine component material in general.

3. Results and discussion

3.1. Initial material's data

3.1.1. Spectrographic test data

Table 1. Chemical composition of local soft steel material (ST. 37) "as received".

Element	Fe	C	Si	Mn	P	S	Cr
%	98.8	0.266	0.147	0.428	0.0485	0.0285	0.0116

Source: Politeknik Manufaktur Ceper

3.1.2. Tensile strength test data

Table 2. Tensile strength values of local soft steel materials (ST. 37) "as received".

Type of test	Utimate Tensile Strength (N/mm^2)				Average UTS (N/mm^2)
	1	2	3	4	
Tensile test	505	480	520	495	500

3.1.3. Surface hardness test data

Table 3. Surface Hardness values of local soft steel materials (ST. 37) "as received".

Type of hardness Test	Rockwell B Hardness Number (HR_B)				Hardness Number Average (HR_B)
	1	2	3	4	
Rockwell "B"	76	75	78	78	76,75

3.2. Results

3.2.1. Carbon content test results. The initial grade value of local ST. 37 soft steel material is 0.26%, while the carbon content of local ST. 37 steel material after undergoing the process of biaxial pressure pack carburizing is highest at 0.45% at a temperature variation of 800°C with a holding time of 3 hours at pressure biaxial of 2 MPa and the lowest carbon content value of 0.35% is at a temperature variation of 1000°C with a holding time of 2 hours with biaxial pressure of 3 MPa.

3.2.2. The tensile test results

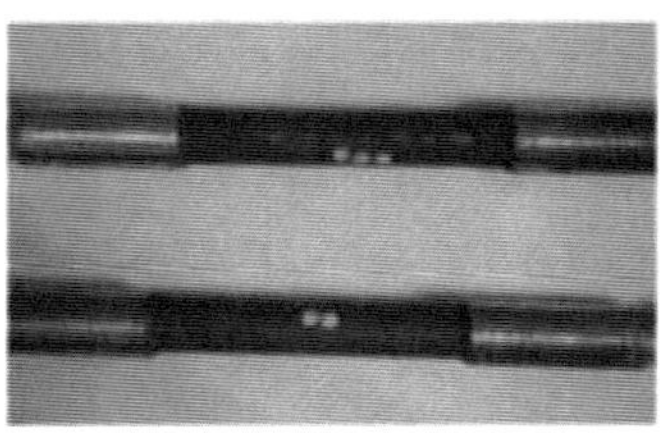

Figure 5. Tensile test specimens according to ASTM E8 / E8M-15a standards.

Table 3. Hardness and tensile strength data of local soft steel (ST. 37) after biaxial pressure pack carburizing process and carbon content, hardness value and tensile strength of local soft steel (ST. 37) after the carburising process at a pressure of 2 MPa.

Temperature process (°C)	Holding time (h)	Carbon %	UTS (N/mm^2)	Hardness (HR_B)
800	2	0.42	817.9	102.1
	3	0.45	767.1	103.1
900	2	0.38	720.1	105.4
	3	0.41	942.3	108.1
1000	2	0.38	793.8	112.1
	3	0.41	889.0	112.3

The initial tensile strength value of local ST. 37 soft steel material is 80 HRB, while the highest tensile strength value of local ST. 37 steel material after undergoing the biaxial pressure pack carburizing process is at a temperature variation of 900°C with a holding time of 3 hours that is 942 N / mm2 and the lowest tensile strength value of 720 N / mm2 is at a temperature variation of 900°C with a holding time of 2 hours and biaxial pressure of 2 MPa.

There was a decrease in the value of the level of tensile strength in the results of the process with biaxial pressure parameters of 3 MPa, carburising temperature of 1000°C and holding time of 2 hours. This case is happened because the fireretardant insulation on the equipment fails to function as a seal of CO gas leakage generated during the experiment.

3.3. Discussion

3.3.1. Carbon content test results. The initial grade value of local ST. 37 soft steel material is 0.266 % carbon content, while the carbon content of local ST. 37 steel material after undergoing the process of biaxial pressure pack carburizing is highest at 0.45% at a temperature variation of 800°C with a holding time of 3 hours at pressure biaxial of 2 MPa and the lowest carbon content value of 0.35% is at a temperature variation of 1000°C with a holding time of 2 hours with biaxial pressure of 3 MPa.

3.3.2. Tensile test results. The initial tensile strength value of local ST. 37 soft steel material is 80 HRB, while the highest tensile strength value of local ST. 37 steel material after undergoing the biaxial pressure pack carburizing process is at a temperature variation of 900°C with a holding time of 3 hours that is 942 N / mm2 and the lowest tensile strength value of 720 N / mm2 is at a temperature variation of 900°C with a holding time of 2 hours and biaxial pressure of 2 MPa.

3.3.3. Hardness test results. Testing the hardness of the local ST. 37 Soft Steel Material specimens through the carburizing pack process using the Rockwell B scale (HRB) method. The initial hardness value of local ST. 37 soft steel material is 80 HRB, while the local hardness value of ST. 37 steel material after undergoing the highest carburizing pressure pack carburizing process is at a temperature variation of 1000°C with a holding time of 3 hours i.e. 144 HRB at biaxial pressure of 3 The lowest MPa and hardness value of 102 HRB is at a temperature variation of 800°C with a holding time of 2 hours with a biaxial pressure of 2 MPa.

4. Conclusion

- The value of carbon content, tensile strength value, and hardness value which is the mechanical properties of local soft steel (ST. 37) can be increased by the biaxial pressure pack carburising method with active medium sized 500 mesh and 1000 mesh mesh according to ASTM standards. As a parameter of the carburising process temperature used 800 ° C, 900 ° C and 1000 ° C.

While the biaxial pressure chosen is 1 MPa, 2 MPa and 3 MPa. The holding time is determined by 2 hours and 3 hours.
- Biaxial pressured pack carburizing on variations of the process parameters as mentioned above affects the mechanical properties of local soft steel material (ST. 37) in each variation.
- The highest increase in carbon content of the specimen by 0.45% C and also the tensile strength value of 950 N / mm2 was obtained when the specimen was processed at 900 °C, holding time for 3 hours and biaxial pressure of 3 MPa. While an increase in the hardness of the specimen of 144 HRB was obtained when the specimen was processed at a temperature of 1000 °C, holding time for 3 hours and biaxial pressure of 3 MPa.

4.1. Recommendation
The recommendations of biaxial pressured pack carburizing parameters for local soft steel material (ST. 37) if will be used as replacement material for machine's components making is a temperature of 900 °C and holding time 3 hours which has a hardness value of 105 HRB and a tensile strength of 720 N / mm2. Because at these mechanical properties the processed materials can still be easily carried out.

Acknowledgement
This 2019's competitive applied research team would like to express its gratitude and appreciation to:
- Politeknik Negeri Semarang which has funded this activity.
- The director of Politeknik Negeri Semarang which has facilitated this activity
- The leader of P3M of Politeknik Negeri Semarang as the facilitator of this activity.

References
[1] Abdulrazzaq M A, Dr. 2016 Investigation the Mechanical Properties of Carburized Low Carbon Steel *Journal of Engineering Research and Application* **6** 59-64
[2] Nitha N, Bontong Y, Syam R and Asmal H S 2017 Behavior of Low Carbon Steel Mechanical Properties Due Pack Carburizing Media Charcoal Buffalo Bone *International Journal of Research Science and Management* **4** 38-47
[3] Nwoke V U, Nnuka E E, Odo J U and Obiorah S M O 2014 Effect of Process Variables on Mechanic Quenched In Different Media *International Journal of Scientific and Technology Research* 3 4 388-398
[4] Elzanaty H 2014 The Effect of Carburizationon Hardness and Wear Properties of The Mild Steel Samples *International Journal of Innovation and Applied Studies* **6** 995-100
[5] Omunakwe R 2017 Effect of Carburization With Palm Kernel / Coconut Shell Mixture on The Tencil, e Properties and Case Hardness of Low Carbon Steel *Fouye Journal of Engineering and Technology* 2 1 101-105
[6] Oluwafemi O M, Oke S R, Otunniyi I O and Aramide F O 2015 Effect of Carburizing Temperature and Time on Mechanical Properties os AISI / SAE 1020 Steel Using Carbonized Palm Kernel Shell Leonardo Electronic *Journal of Practice and Technologies* **27** 41-56
[7] Alias S K, Abdullah B, Jaffar A, Latip S A, Kasolang S, Izham M F and Abd Ghani M A 2005 Mechanical Properties of Paste Carburized ASTM A516 Steel *Procedia Engineering Science Direct, Elsevier* **68** 525-530
[8] Anggraini D and Nugroho A 2005 Aplikasi Spektrometer Emisi Pada Analisis Unsur-unsur Bahan Paduan Alumunium AlMgSi-1 *Jurnal Teknik Bahan dan Nuklir* 1, 2.
[9] Vikas V and Choudhary R 2017 Mechanical and Wear Properties of Carburized Mildsteel Samples *International Journal of Modern Trends in Engineering and Science* 4 6
[10] Callister W D 2007 *Material Science and Engineering* 7th (Canada John Wiley & Sons, Inc.)
[11] German M R 1994 *Powder Metallurgy Science* (New Jersey Metal Powder Industries Federation)

IOP Conf. Series: Materials Science and Engineering 1108 (2021) 012037 doi:10.1088/1757-899X/1108/1/012037

Analysis of vibration due to misalignment in the clutch cluster installation of centrifugal pump

I G Widodo*, A Khoryanton and A Pramono

Departement of mechanical engineering, Politeknik Negeri Semarang, Jl. Prof. H. Soedarto, S.H., Tembalang Semarang 50275, Indonesia

*Ignatius.gunawan.widodo@polines.ac.id

Abstract. Knowledge of vibration and the resulting data is very important for maintenance and troubleshooting where this ability can help companies reduce the occurrence of downtime and can increase profits both in terms of production and from a longer machine life. One of the causes of high vibrations is a misalignment in a centrifugal pump installation that is driven by an electric motor and connected by a fixed clutch. The method used is to do differences in height in the horizontal, vertical and angular directions by providing variations in the rotation of the electric motor and pump, then checking at four bearing positions on the electric motor and pump. The results showed an increase in vibration at each addition of the sim on the electric motor fastener bolts, the greatest vibration occurred in the MOV (indoor motor axiall) with the addition of a 1.5mm sim of 3.01 mm / sec an average increase of 29% and at the largest pump vibration in POA (outdoor axial pump) with the addition of a 1.5mm sim of 9.88mm / sec, increased by an average of 58%.

1. Introduction

Vibration is widely used as a tool for analyzing damage to industrial machines [1]. Knowledge of vibration and the resulting data is very important for maintenance and troubleshooting. This capability can help companies reduce downtime and can increase profits both in terms of production and longer engine life. The vibrations that arise due to cyclic force through the elements of the machine, where these elements act with each other and energy is determined through the structure in the form of vibrations. The impact of vibration is a noise, a decrease in performance and pump performance and can damage components at the pump, especially on the shaft and bearings [2].

The purpose of this study is to modify the practicum for centrifugal pump installation that can be used to test the characteristics of machine vibrations caused by the presence of misalignmet. So from the results of the analysis of vibration characteristics that can be predicted damage to its components. In this study want to prove that one of the causes of vibration is the occurrence of misalignment on the shaft connected with a fixed coupling in a centrifugal pump installation.

The outline of disability (*misalignment)* is divided into two conditions including:

- ***Angular misalignment*** where the position of the two axes of both shafts forms an angle, as in figure 1 below this. Shows that the difference between the shaft of the motor shaft and the axis of the pump shaft forms an angle where the clutch gap has a difference, the axial vibration shows the amplitude 1x rotation per minute in the axial direction, and will decrease at 2x revolutions per minute, also 3x revolutions per minute.

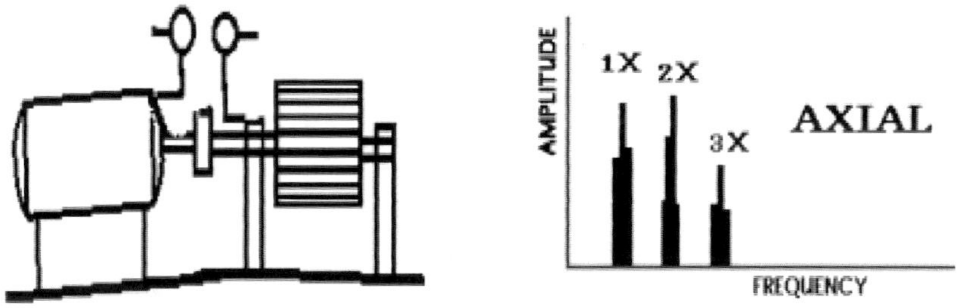

Figure 1. Angular misalignment [3].

- *Parallel misalignment* in which the position of the axes of the two shafts are parallel but have different height which is shown on the surface of the two parallel couplings but the height is different. The axial plane that occurs shows 2x revolutions per minute in the radial direction. Parallel misalignment has the same vibration symptoms compared to angular misalignment, but shows higher radial vibrations approaching 180 phases. When angular or parallel misalignment becomes severe, it can produce a much higher harmonic amplitude 8 times or even the whole series of high frequency harmonics.

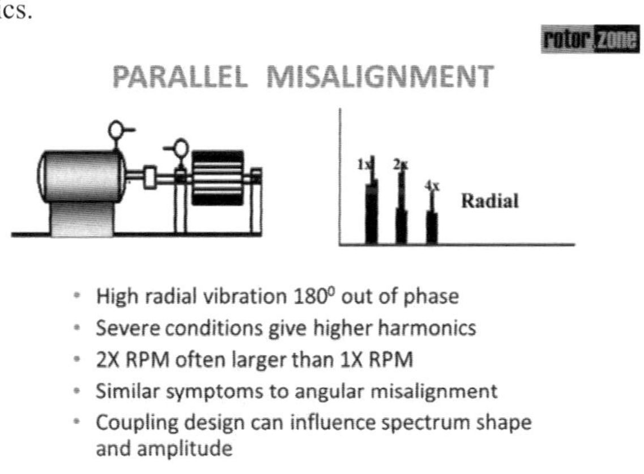

Figure 2. Parallel misalignment.

Based on ISO 10816 standard table classification machine vibrations in general are taken based on empirical experience with testing various machines and equipment in the industry shown in figure 3 below.

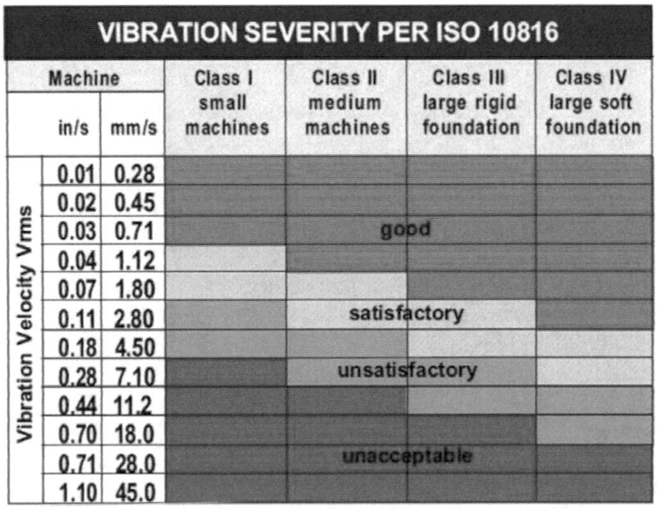

Figure 3. Standard vibration machine according to ISO10816.

Misalignment occurs as a result of loosening of the fastening bolts on the electric motor or pump holder, errors during initial setup or due to the continuous workload on the pump installation.

2. Research methods

The methodology in this study follows the research flow diagram as can be explained as follows:

The first stage in this research is planning a centrifugal pump installation as one of the machines that are widely used in industry, by planning laboratory-scale pump installations. In this plan considered in determining the method of testing pump vibrations with the condition of the equipment and materials available in the maintenance and repair laboratory.

The second stage in this research is to make a vibration characteristic test device due to misalignment of the pump installation using materials according to plan and using existing equipment in the maintenance and repair laboratory.

The third step in this research is to test the vibration of the centrifugal pump under conditions of shaft misalignment with several variables such as the relationship between the pump coupling and the motor during Unfault, the position of the bolts on the prototype of the pump base and the motor, all the motor base bolts and all the pump base bolts in loose / loose state, all motor base bolts are loose / loose, while all pump base bolts are firmly mounted. From the data of the test results then the vibration characteristics are analyzed as a basis for determining damage predictions for centrifugal pump components.

3. Results and discussion

The results of the modification of the vibration test equipment on the pump installation with the following data:

1). IRD 808 Vibration meter vibration test equipment

2). 3 Phase electric motor, 3 HP

3). Centrifugal pump type

• Brand: Torisma Pump

IOP Conf. Series: Materials Science and Engineering 1108 (2021) 012037 doi:10.1088/1757-899X/1108/1/012037

- Head: 70 m
- Capacity: 30 liters / sec
- Power: 3700 watts
- Voltage: 380 volts
- Rotation: 2950 rpm

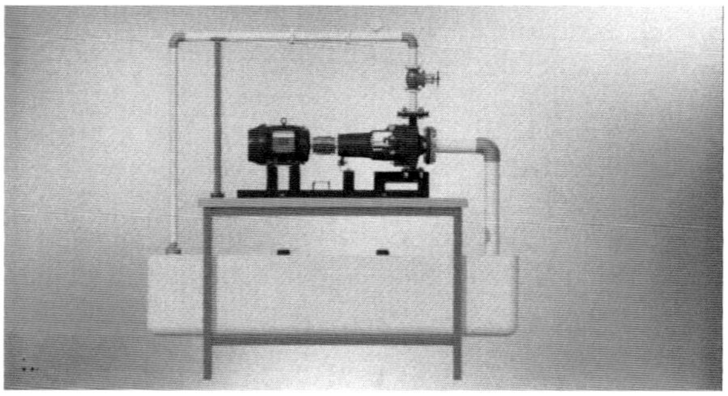

Figure 4. Centrifugal pump installation.

Determination of the vibration measurement point is used to obtain optimal results, in this study vibration signal retrieval is done both at the motor and at the pump, as in Figure 5. While in table 1 and table 2 displayed the results of vibration measurements in the pump section and on the motor part, the value of vibration increase uniformly occurs in the pump area.

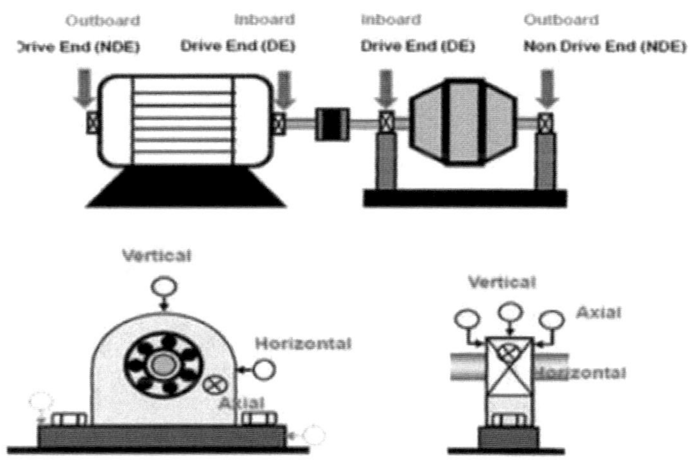

Figure 5. Vibration measurement point [4].

IOP Conf. Series: Materials Science and Engineering 1108 (2021) 012037 doi:10.1088/1757-899X/1108/1/012037

Table 1. Results of vibration measurements on the motor.

Result of measurement of angular misalignment vibration(mm/sec)							
Thickness shim (mm)		**MOH**	**MOV**	**MOA**	**MIH**	**MIV**	**MIA**
0,5	1480	0,9	0,86	1,18	1,11	0,97	1,43
	1200	0,7	0,66	0,98	0,91	0,77	1,23
	1000	0,5	0,46	0,78	0,71	0,57	1,03
0,75	1480	1,24	1,51	1,62	1,67	1,43	2,04
	1200	1,54	1,91	2,02	2,07	1,83	2,44
	1000	1,84	2,31	2,42	2,47	2,23	2,84
1	1480	1,45	2,65	1,98	1,05	1,81	2,48
	1200	1,2	1,4	1,73	0,8	1,56	2,23
	1000	0,95	1,15	1,48	0,55	1,31	1,98
1,5	1480	1,75	1,75	2,09	1,19	1,63	3,01
	1200	1,55	1,55	1,89	0,99	1,43	2,81
	1000	1,35	1,35	1,69	0,79	1,23	2,61

Table 2. Results of vibration measurements at the pump.

Result of measurement of angular misalignment vibration(mm/sec)							
Thickness shim (mm)		**POH**	**POV**	**POA**	**PIH**	**PIV**	**PIA**
0,5	1480	0,93	1,02	2,57	1,08	0,98	1,58
	1200	0,73	0,82	2,37	0,88	0,78	1,38
	1000	0,53	0,62	2,17	0,68	0,58	1,18
0,75	1480	1,56	1,03	3,5	1,74	1,69	2,44
	1200	1,96	1,43	3,9	2,14	2,09	2,84
	1000	2,36	1,83	4,3	2,54	2,24	3,24
1	1480	1,42	1,84	6,44	1,24	1,65	3,22
	1200	1,17	1,59	6,19	1	1,4	2,97
	1000	0,92	1,34	5,94	0,75	1,15	2,72
1,5	1480	1,92	1,97	9,88	1,92	1,41	5,04
	1200	1,72	1,77	9,68	1,72	1,21	4,84
	1000	1,52	1,57	9,48	1,52	1,01	4,64

From the results of the above vibration tests can be made in the form of figure below.

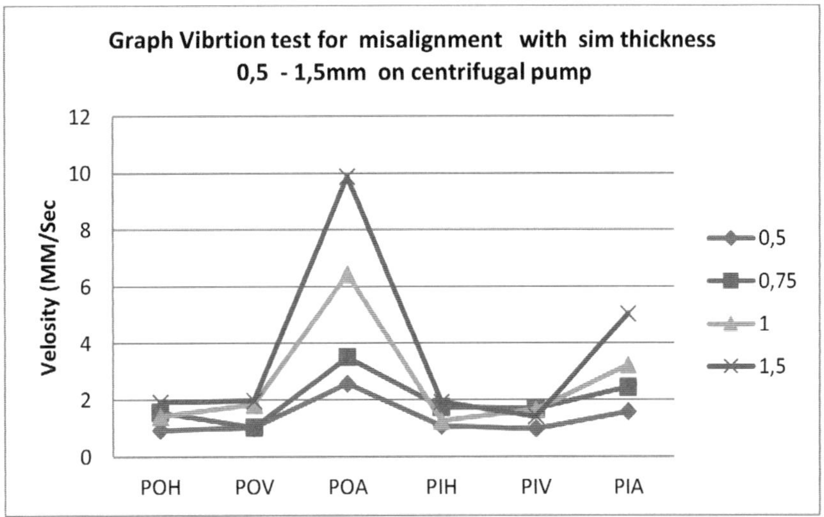

Figure 6. Graph Vibration test for misalignment conditions.

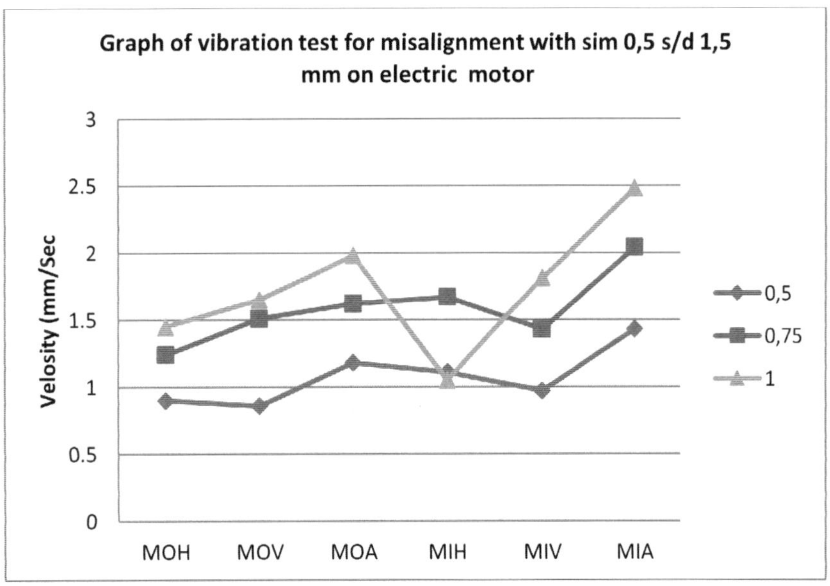

Figure 7. Graph of vibration test on an electric motor.

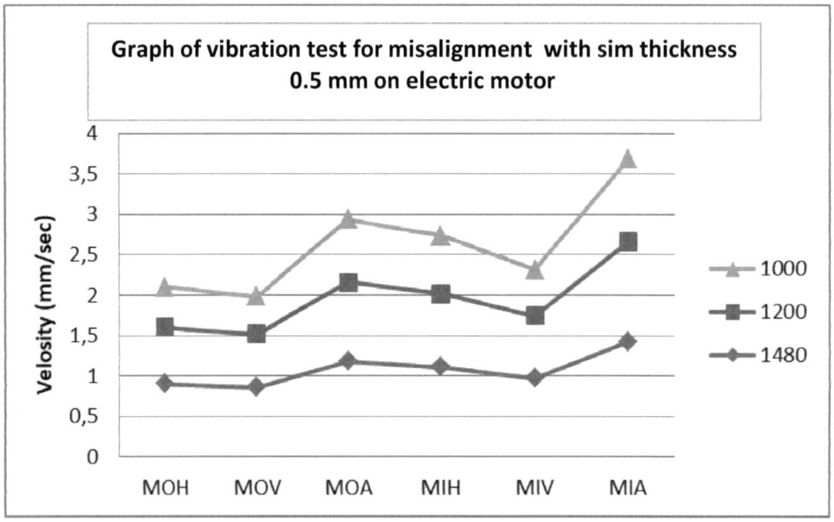

Figure 8. Graph of Vibration test vs rotation on an electric motor bearing.

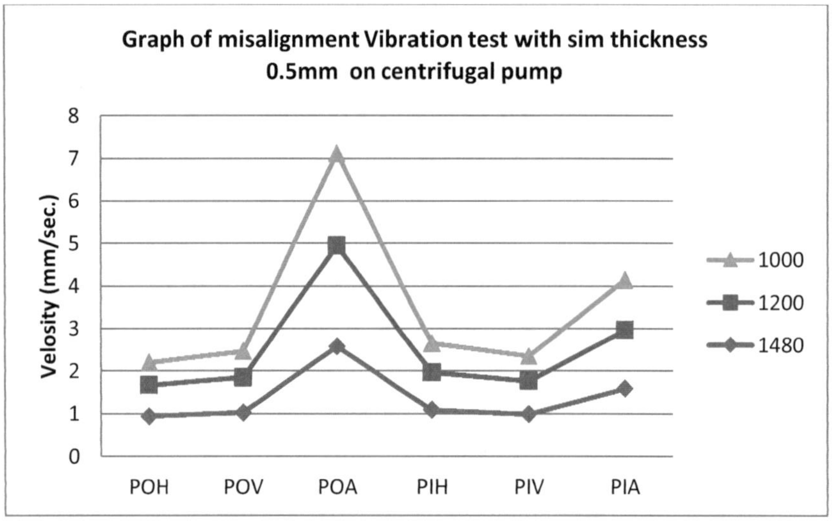

Figure 9. Graph of vibration test vs rotation on pump bearing.

From the results of the discussion above it is known that the vibration test with a variety of misalignment shows that the greater the misalignment on both shafts the greater the vibration and causes damage to the indoor axial motor bearings and outdoor axial pumps, the greatest vibrations occur in axial indoor motors (MIA) at 5.04mm / sec at 1.5mm shim thickness and at outdoor axial (POA) pumps at 9.66mm/sec at 1.5mm shim thickness.

4. Conclusions

- One method to see the performance of the pump is to check the vibration test on the electric motor bearings and pump bearings. The test results show the misalignment with a difference in shim height of 0.5 mm of vibration at the MIA point shows a magnitude of 1.43 mm / sec with a difference of shim height of 1.5 mm of vibration increased by 5.02 mm / sec. Whereas at the pump changes occur at the POA point to the depth of 0.5mm shim of vibration at 2.57mm / sec., At the depth of the 1.5mm shim the vibration becomes 9.88mm / sec.
- One of the causes of high vibrations in pump installations is the occurrence of impulse on both the pump shaft and the motor (misalignment) which can damage the bearings.
- Vibration checks on pumps and motors can be done every 4-5 hours to prevent damage to the pump and to maintain pump readiness.

Acknowledgments to the Politeknik Negeri Semarang (Polines) DIPA funds that have funded this research, would be able to provide benefits for the Maintenance and Repair Laboratory and students of the Department of Mechanical Engineering as a medium for the Performance Test Practicum.

Acknowledgment

Thanks to the Politeknik Negeri Semarang through applied research funds, which provided us with the opportunity to complete this research.

Reference

[1] Aji K 2007 Deteksi Kerusakan Bantalan Gelinding Pada Pompa Sentrifugal Dengan Analisa Sinyal Getaran (Doctoral dissertation, Universitas Sebelas Maret)

[2] Suwandi A, Wahono D R and Hermawanto D 2009 Analisis karakteristik getaran pada kereta api rel listrik dan kereta api rel diesel *Jurnal Standardisasi* **11**(2) 98-105

[3] Scheffer C 2005 *Associate Professor"Pump Condition Monitoring Through Vibration Analysis"* Department of Mechanical and Mechatronic Engineering: Stellenbosch University

[4] Marscher W D, Olson E J, Onari M M and Boyadjis P A 2016 Vibration Problems & Solutions for Pumps and Other Turbomachines In *Proceedings of the 32nd International Pump Users Symposium. Turbomachinery Laboratories, Texas A&M Engineering Experiment Station*

Model for identification of rice type using combination of shape and color features

J Jumi*, A Zaenuddin and T Mulyono

Politeknik Negeri Semarang, Semarang, Indonesia

*jumi_mail06@gmail.com

Abstract. Rice is an agricultural commodity that is a staple food in Indonesia with hundreds of types of rice that have different characteristics. The type of rice can be distinguished from color and shape. The main feature that is dominant and can distinguish each type of rice is the color and shape. This feature is the main key in identifying types of rice. Identification is done by comparing the similarity of rice images using the value of color and shape features. The similarity can be determined through the difference in feature values between the query image and the database image. The closer the difference is to zero, the higher the level of similarity. The degree of similarity will affect the accuracy of image recognition at the time of identification. In this study, an analysis of the accuracy of image identification and measurement of computation time was carried out. Improved identification accuracy using the weighting of color and shape feature values. Extraction of the two value features using the invariant moment and color moment. Preprocessing before extraction using Grayscale, resize, edge Enhancement, Histogram Equalization. Clustering of rice image data using K-Means clustering. The results showed that the accuracy of identification with 400 rice image test data, reached more than 95% in the weighting scheme Ws (weighted Shape) = 40% and Wc (weighted color) = 60% with an average computing time of 5 milliseconds at 10 the cluster.

1. Introduction

Rice is one of the agricultural commodities that are needed as the main food ingredient. Rice has a variety of types with different characteristics-different. The type of rice has different quality and economic value so that it can influence the perception of customers who consume it [1]. The variety of rice types of rice necessary database for easy identification and tracking information about the types of rice. Identification is done through the introduction of rice images by using the image feature as a key field.

The value of the feature extracted from the object image affects the accuracy of the identification of the object which in this study is rice as the object. Identification through the introduction of image features is influenced by various variables including accuracy and accuracy in the process of calculating the value of features. One other variable that can affect accuracy is the use of feature value weighting schemes.

Generally, the image on an object dominates different features so that a feature weighting scheme is needed by the object. The image of an object in rice has a more dominant color feature to distinguish the similarity between one image of rice with another image of rice. The shape and texture features of rice images are as supporting features for image similarity measurement [2]. Image similarity can be grouped in three main groups in the measurement namely the similarity of shape, color similarity and the last is the similarity of texture [3]. This similarity is used as a key introduction to image-based data identification.

If more nearly zero, then have a higher level of similarity [3]. Data that are grouped based on the value of features that have been weighted is a process to improve accuracy and shorten the time for information retrieval.

The stages of feature weighting scheme with different percentages and clustering were carried out in this study. The stages are carried out to increase the accuracy value and speed up image identification time. Identification is done by comparing the feature values between the query image with the image in the database [4].

Previous research that discusses image similarity search based on feature or color feature extraction. The research has been carried out without regard to feature weights and clustering to produce a similarity accuracy value of more than 75% [5]. A computer vision-based approach is used to classify six types of rice using twenty-one features of rice grains. The classification uses seven features based on color and fourteen morphological features. The method used is a neural network that reaches an accuracy of 88.30% [6]. Support multi-class vector machines (SVM) are also used to classify rice into premium classes, grade A, grade B, grade C and achieve 86% accuracy [7]. The neural network method in 4 groups of a total of 52 types of Philippine rice grains can achieve an average accuracy of 70%, by classifying rice groups from rice images [8]. Here, images of individual rice grains are obtained by placing rice grains one at a time on the scanner. After scanning it continues using the color and morphological features to classify the six types of rice once again using neural networks. Their model achieves an average accuracy of 84: 8%. In Silva et al. [9] Singh et al. [10] classifying 9 different types of rice using thirteen morphologies, fifteen textures, and six color features and achieving 92% accuracy.

This study aims to analyze the identification of image-based rice types using feature value weighting schemes and the calculation of search computing time after clustering using the K-Means technique. The K-Means algorithm is used to determine the cluster position in each image by first calculating the image distance with all centroids using the Euclidean distance method.

2. Methods

This research uses a method by combining color and shape features. The method used consists of three stages, namely data acquisition, recognition, and identification shown in Figure 1. This study uses a merging method that was not found in previous studies, namely the method of weighting features and clustering based on color and shape features for the process of identification of rice types.

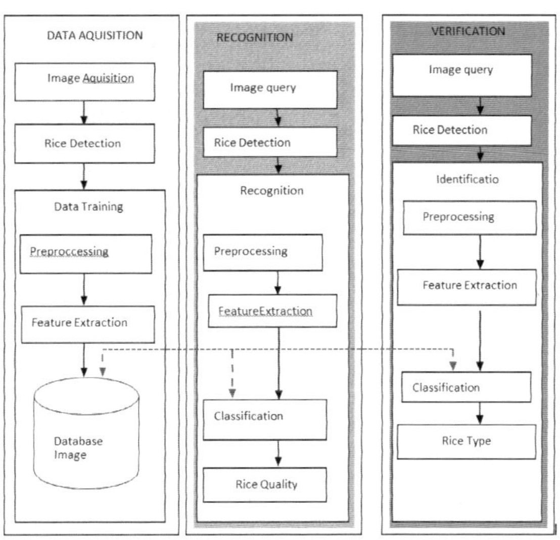

Figure 1. Overview of image search stages.

The details of the stages of this research are as follows:

2.1. Preprocessing
Preprocessing is the stage of improving image quality before feature extraction to improve the accuracy of image feature extraction results. There is a difference between preprocessing in color feature extraction and shape feature extraction. The difference in preprocessing is to get a quality image before feature extraction.

2.1.1. Preprocessing color features
a. *Resize*
   The larger the image size, the longer the extraction time, so that the resize step is needed to speed up the computation process. At this stage, the image is resized to 200 x 200 pixels.
b. *Histogram Equalization*
   At this stage, histogram leveling is done so that the image quality becomes more contrasted to get a quality color feature value.
   Histogram leveling results can be seen in Figure 2 and Figure 3.

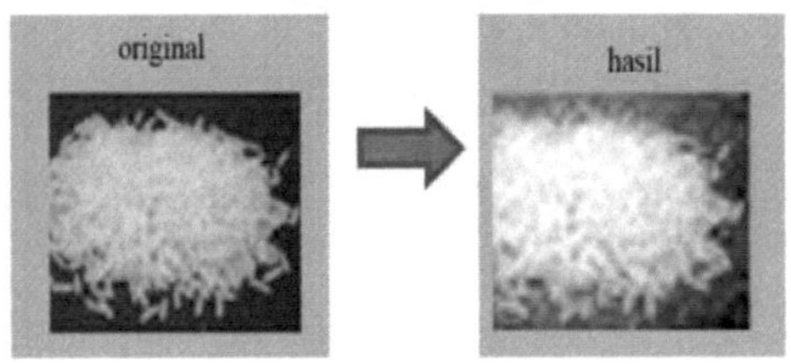

(a) (b)

Figure 2. Comparison of the image before and after the histogram process, (a) Original Image, (b) Image of Histogram Equalization (HE).

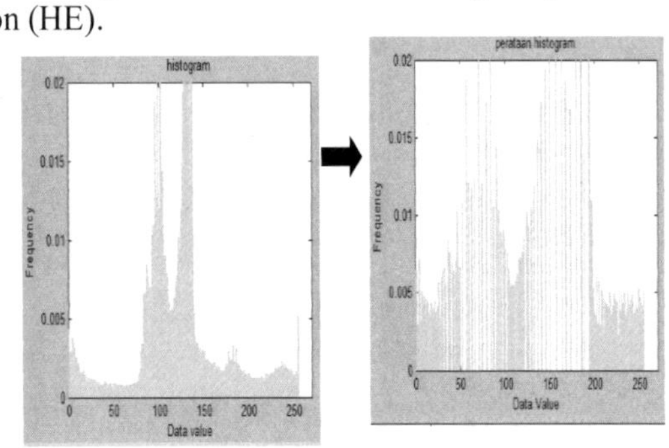

(a) (b)

Figure 3. Comparison of histograms, (a) Original image histogram, (b) Image histogram of HE results.

2.1.2. Preprocessing shape features
a. *Resizing*
   The size of the image will affect the length of feature extraction time, the smaller the size, the faster the computation of feature extraction. At this stage, the image is resized to 200 x 200 pixels.
b. *Grayscale*
   Especially in calculating the features of color elements are not counted so that the image is changed to grayscale so that the computation process is faster.
c. *Edge Enhancement dan Histogram Equalization (HE)*
   The edge enhancement stage will produce a new image with clearer or sharper edges of the object, so the shape of the image will be clearer. This stage uses the convolution method with the Sobel operator [3]. The edge sharpening results can be seen in Figure 4.

Figure 4. Preprocessing Stages of Shape Features, (a)Original Image (b) Grayscale Image, (c) Resize, (d) Histogram Image Equalization.

2.1.3. Feature extraction. Color and shape feature extraction is performed on the image after going through the preprocessing stage. Color feature extraction using color moment and form feature extraction using invariant moment.

The color moment uses three main moments of the color distribution of the image, namely the mean, standard deviation, and skewness, so this method produces three values for each color component [11]. Color Moments method is a method used to distinguish images based on color features [12]. The color moment assumes that the color distribution of an image can be expressed as a probability distribution.

The invariant moment produces feature values that are not susceptible to image changes caused by Rotation, Scale, and Translation (RST) [11]. Analysis of the set of moments of a function f (x, y) of two variables can also be done with [2].

2.1.4. Weighting of features. This study uses the weighting of different feature values for identification accuracy analysis. Weighting varying feature values will result in different levels of identification accuracy. Weighting is carried out with four variations namely the W1 scheme (0.25, 0.75) is a 25% weighting scheme for the color feature weights and 75% the shape feature weights. The W2 scheme (0.6, 0.4) is a 40% weighting scheme for color feature weights and 60% for shape feature weights. The W3 scheme (0.75, 0.25) is 75% the color feature weight and 25% the shape feature weight. And in the W4 scheme (0.4, 0.6) is a 60% weighting scheme for the color feature weights and 40% the shape feature weights.

2.1.5. Clustering. Data clustering to speed up the identification process. In this study, the K-Means Clustering and Euclidean distance methods are used as a measurement of image similarity. The similarity measurement equation uses equation 1.

$$D(Q,M) = \sqrt{\sum_{n=1}^{k} (Q_n - M_n)^2} \quad (1)$$

Which Q_n and M_n are features of the query image and database image in the nth dimension. Image similarity is determined by the difference in the value of the identified image features with the image in the database. The difference in feature values close to zero has the highest level of similarity.

3. Results and analysis

Tests in this study used 5 types of rice there are IR-35, IR-50, IR-64, Sticky Rice, and Menthik with a total of 400 rice images measuring 200 x 200 pixels.

IOP Conf. Series: Materials Science and Engineering 1108 (2021) 012038 doi:10.1088/1757-899X/1108/1/012038

Table 1. Data on rice image acquisition.

No	Image of Rice	Name	Grain Type	Level
1		IR 35	Whole seed	2
2		IR 35	Whole seed	1
3	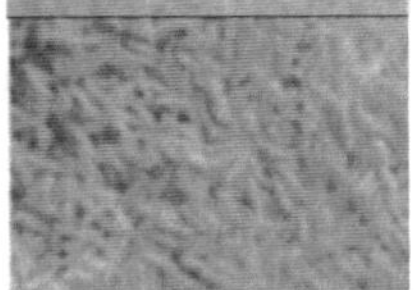	IR 50	Whole seed	2
4		IR 50	Whole seed	1
5	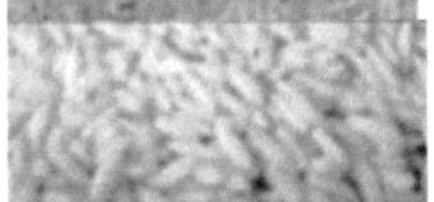	Ketan	Whole seed	3
6		Ketan Merah	Whole seed	3
7		IR 64	Whole seed	3
8		IR 64	Whole seed	5
9		Menthik	Whole seed	2
10	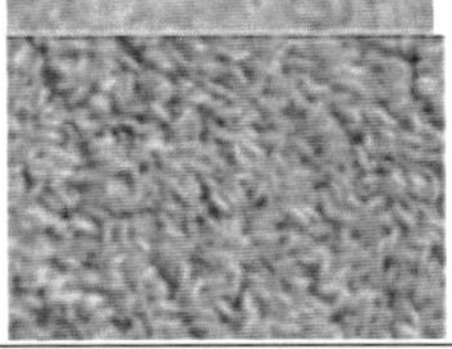	Menthik	Whole seed	1

In Table 1 an example of rice image data with names, grain types, and quality levels as the identification of rice quality. Quality at level 1 is poor quality rice, level 2 is medium quality rice, level 3 is pretty good quality, level 4 is good quality rice, and level 5 is premium quality rice.

3.1. Identification results

Identification testing is done by varying the number of clusters and variations in the percentage of feature weights. Test results with 60% weighting variation for color feature weights and 40% form feature weights with 10 cluster variations are shown in Figure 5.

Figure 5. The results of rice tracking the identity of rice (0.6 and 0.4).

In Figure 5. the results of identification of the quality and type of rice are shown using similarity values with the Euclidean distance method. Retrieval ranks similarity 1 to 10 from the rice image database and testing with weighting variations and variations in the number of clusters will produce different retrieval outputs on the same data.

3.2. Discussion

At this stage, retrieval accuracy is calculated in the rice image database. Furthermore, a retrieval accuracy analysis was carried out on the rice image database before and after clustering with the number of clusters that varied from 3 to 15 clusters.

3.2.1. Before clustering. Retrieval accuracy analysis at this stage uses a database of rice images before clustering with weighting variations in shape, color, and texture features. The accuracy value of the retrieval results with the same feature weights and varying feature weights can be seen in Table 2.

Table 2. Percentage of retrieval accuracy with the same feature weight and variable feature weight in the rice image database.

Image of Name	Equal Weight	W1	W2	W3	W4
I1	75,55%	78,59%	86,75%	84,38%	61,33%
I2	74,67%	79%	87%	83,87%	60,67%
I3	75,05%	79,45%	87,10%	83,33%	60%
I4	74,69%	79,15%	87,50%	86,67%	59,33%
I5	75,65%	78,75%	86,67%	84,38%	60,67%
I6	75%	78,50%	86,36%	84%	60%
I7	75,76%	79,45%	86,67%	83,87%	59,33%
I8	74,89%	78,65%	87,33%	83,33%	60,67%
I9	74,87%	79,55%	86%	84%	61,33%
I10	75,35%	79,25%	87,33%	84,38%	60,67%

Table 2. shows that the retrieval accuracy with the variation of W2 produces the highest average accuracy and the variation of the weight W4 produces the lowest accuracy.

3.2.2. After clustering. Grouping is done on the image that is based on similarity, the results will narrow the search space for image data information. The process is very beneficial because it will shorten the retrieval time and increase the accuracy of the retrieval. Retrieval accuracy with variations in weight and number of clusters can be seen in Table 3.

Table 3. Percentage of retrieval accuracy with variations in weight of features and variations in number of clusters.

Number of clusters	Equal Weight	W1	W2	W3	W4
3	72,55%	75,59%	92,67%	82,38%	63,33%
4	72,67%	75%	92,33%	82,33%	63,67%
5	73,05%	76,45%	93,33%	83,36%	64,00%
6	73,69%	76,15%	93,67%	83,67%	64,33%
7	74,65%	77,75%	94,33%	84,38%	65,67%
8	74%	77,50%	94%	84%	65,00%
9	74,76%	77,45%	94,67%	83,87%	64,33%
10	75,89%	79,65%	95,67%	86,67%	66,93%
11	74,47%	79,15%	93,55%	84%	66,23%
12	74,15%	77,25%	93%	85,33%	65,67%
13	74,20%	76,15%	92,23%	82,63%	64,63%
14	73,25%	75,59%	92,13%	82,44%	64,55%
15	73,13%	75,29%	92,09%	82,23%	63,43%

Table 3 shows that the weighting variation of W2 with a weight of 60% of color features, and 40% of color features and variations in the number of clusters 10 produces the highest retrieval accuracy of more than 95%.

While the weighting variation of W4 with 40% weight of color features and 60% weight of color features with variations in the number of clusters 3 produces the lowest accuracy which is less than 64%.

Retrieval accuracy graphs in the rice image database using the same and varied feature weights namely W1, W2, W3, and W4 with variations in the number of clusters ranging from 3 to 15 are shown in Figure 6.

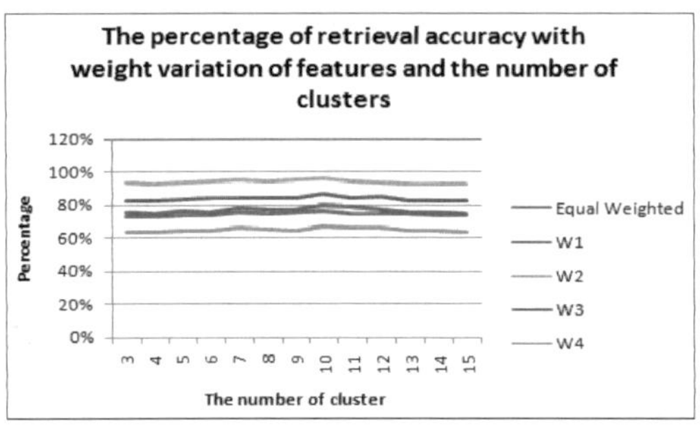

Figure 6. Retrieval accuracy graph with feature weight variation and cluster number variation.

Figure 6. is the retrieval accuracy that tends to rise until reaching variations in the number of clusters of 10 and tends to decrease after the number of clusters is greater than 10.

3.3. Computation time

The computational time that takes place to retrieve an image in a rice image database is influenced by the size of the image database and its database management techniques. The average computational time for searching in an image database before clustering is with the variation in the number of clusters shown in Figure 7.

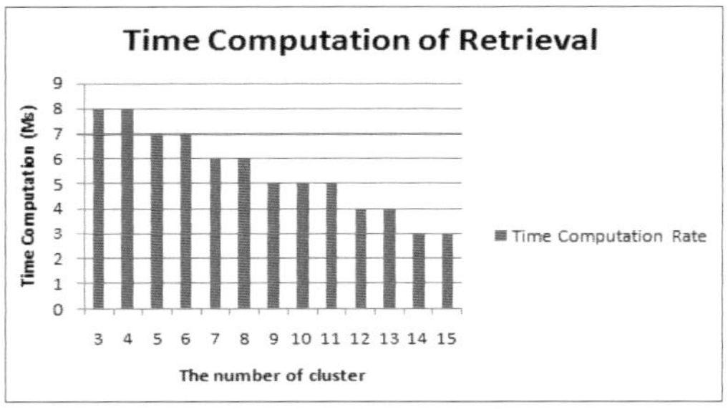

Figure 7. Time computation of retrieval.

Figure 7 explains that the greater the number of clusters, the faster the computing time.

4. Conclusion

The level of accuracy in the process of retrieval of rice images is influenced by several variables including image acquisition, preprocessing, feature extraction methods, and identification techniques in the database. This study proves that the use of different feature weighting schemes for each normalized feature value and using a variety of cluster numbers will affect the increase in accuracy and retrieval speed in the process of identifying rice types. The highest level of accuracy obtained in retrieval is the variation of the number of clusters 10 with a weighting variation of W2 that is 60% color features and 30% shape features. From this study, it was found that the color features of rice images had a more dominant factor in determining the level of similarity in the process of identifying types of rice. As for the form features as a complementary feature. The average computing time required for retrieval is 5 milliseconds.

References

[1] Vadivel A, Majumdar AK and Shamik S 2004 Characteristics Of Weighted Feature Vector In Content-Based Image Retrieval Applications *International Conference on Intelligent Sensing and Information Processing* IEEE.

[2] Gonzales R C and Woods R E 2008 *Digital Image Processing* ThirdEdition (New Jersey: Pearson Prentice Hall)

[3] Castleman K R 1996 *Digital Image Processing* (New Jersey: Prentice Hall Inc.)

[4] Jumi and Harjoko A 2012 Image Similarity Analysis Based on Shape, Color and Texture Feature of Asset Image *International Conference on Computer Science Electronics and Instrumentation*, Yogyakarta, Indonesia.

[5] Xing-yi Huang, Jian Li and Song Jiang 2004 Study on identification of rice varieties using computer vision [J]" In: *Journal of Jiangsu University (National Science Edition)* 2.003

[6] Harpreet Kaur and Baljit Singh 2013 Classification and grading rice using multi-class SVM" In: *International Journal of Scientific and Research Publications* **3** 1–5

[7] Guzman J D and Peralta E K 2008 Classification of Philippine rice grains using machine vision and artificial neural networks In: *World conference on agricultural information and IT, IAALD AFITA WCCA 2008*, Tokyo University of Agriculture, Tokyo, Japan, 24-27 August, 2008. Tokyo University of Agriculture. 2008, pp. 41–48.

[8] Zhao-yan Liu, Fang Cheng, Yi-bin Ying, and Xiu-qin Rao 2005 Identification of rice seed varieties using neural network". In: *Journal of Zhejiang University. Science*. B 6.11 1095

[9] Chathurika Sewwandi Silva and Upul Sonnadara 2013 Classification of Rice Grains Using Neural Networks" In: *Proceedings of Technical Sessions*. **29** 9–14

[10] Singh S K, Bejagam K K, An Y and Deshmukh S A 2019 Machine-learning based stacked ensemble model for accurate analysis of molecular dynamics simulations The *Journal of Physical Chemistry*

[11] Acharya T, Ray A K 2005 *Image Processing Principle and Applications* (USA: John Willey &

Sons)
[12] Susilo A 2007 Web Image Retrieval untuk Identifikasi bunga dengan Pengelompokan Content Warna, Institut Teknologi Sepuluh November, Surabaya.

IOP Conf. Series: Materials Science and Engineering 1108 (2021) 012039 doi:10.1088/1757-899X/1108/1/012039

Design models of automatic humidity control system in flash off painting line room to reduce defect products at PT. Kubota Indonesia

A Khoryanton*, G Giyanto, A Satito, S Harmanto, H Haryanto and T Tegar

Department of Mechanical Engineering, Semarang State of Polytechnic, Semarang, Indonesia

*ampala.khoryanton@polines.ac.id

Abstract. The main challenge often faced by PT. Kubota Indonesia in connection with the quality of production is still a lot of level of disability in the production process, especially in the painting process. The process of painting is highly required according to the standard of both the process and condition of the room so that the results of the painting do not occur defects. Based on production data in the period from October 2018 to January 2019 there were 80% defective products with blistering defects of the total defective products, this needs a deeper analysis of the causes of product defects. Blistering defects in the form of bubbles on the surface of the painting result occur because the standard of air humidity is not achieved when the flash off process is caused because there is no tool to condition the air in the flash off the chamber. The purpose of this research is to design an automatic humidity control system model that can be used as an alternative to PT. Kubota Indonesia as a method for reducing humidity so that it can reduce the potential causes of blistering defects. Research methods include problem identification, literature study, design, and manufacture of models and testing. Problem identification is done by using the 5M + 1E method to find the root of the problem. The process of designing and making models is divided into 3 important activities namely designing 3D design concepts using solid works software, designing a control system software with Arduino Uno and designing hardware models. Tool testing is done by testing the automatic humidity control system model on the prepared test chamber model. The results obtained from this study are models of automatic humidity control systems that can condition the humidity in the test room and maintain the humidity according to the setpoint.

1. Introduction

The main challenge often faced by PT. Kubota Indonesia in connection with the quality of production is still a lot of level of disability in the production process, especially in the painting process. Based on production data in the period September 2018 to February 2019, there are 5% defective products, this needs a deeper analysis of the causes of these product defects.

One of the departments with the most stringent quality control standards is the painting department. This painting department is part of the production department which is responsible for the process of painting components before they are assembled on the assembling line and the painting process for spare parts that are not included in the assembling process. The process of painting is highly required according to the standard of both the process and the condition of the room so that the painting results do not occur defects. But the condition of the painting process room (flash off process) still does not meet the standards. Ideally, the best condition for spraying paint is when the relative humidity is between 40-50% [1]. Humidity in the room is still dependent on the conditions of the surrounding environment and is still frequently encountered the humidity in the room is above the specified standard. This is certainly not in accordance with the work standards at PT Kubota Indonesia where every job must follow the specified standards. Humidity in the flash off room that is not suitable also results in defects in the material of painting [2]. The problem in this study is how to control the relative humidity in the painting room of the flash off a process in order to obtain a standard condition that is the relative humidity value between 40-50 %.

In order to overcome the problem in the flash off space, the purpose of this study is to design a model of a relatively automatic humidity control system in the flash off the room to reduce product defects. With the availability of this system, the painting process is expected to run according to standards and be able to reduce the product defect ratio painting.

2. Literature review

Humidity is the concentration of water vapor in the air [3]. This concentration can be expressed in absolute humidity, specific humidity or relative humidity. A tool for measuring humidity is called a hygrometer. A humidistat is used to regulate the level of humidity in a building.

Absolute humidity is analogous if all the water in one m3 is condensed into a container, the container can be a scale of absolute humidity. Absolute humidity has values that range from 0 gram / m3 when the air is dry to 30 grams / m3 when the water vapor becomes saturated at 30 ° C. Relative humidity is very important in estimating weather [4].

The relative humidity is the ratio between the actual humidity and the capacity of the air to hold water vapor. When the actual humidity is expressed by the actual vapor pressure, the air capacity to collect water vapor is saturated vapor pressure. So that relative humidity can be written with percent [5].

Relative Humidity is the ratio between the actual water vapor pressure at a certain temperature and the saturated water vapor pressure at that temperature. Another understanding of RH is the comparison between the amount of water vapor contained in the air at a certain time with the maximum amount of water vapor that can be accommodated by the air at the same pressure and temperature [5].

The humidity of the air in a closed room can be adjusted as desired. The regulation of the humidity of the air is based on the principle of equal water potential between the air and solution or with certain solid materials. The high moisture content in the air can cause various kinds of problems both for humans and for the material in surrounding areas, including making physiological pressure and discomfort, causing disease, accelerating mold growth and increasing insect population, accelerating metal corrosion, reducing electrical resistance to insulators, damaging the surface finishing process, causing structural failure, causing paint failure and wall fouling in buildings [6].

3. Research methods

3.1. Problem analysis

Problem identification is the first step of the research carried out with the aim of knowing and understanding the problems that occur. Identification is done through field observations and direct brainstorming at the study site with the operators and leaders of the relevant departments.

3.2. Literature review

Literature studies are needed to look for clues to solving problems and provide a theoretical basis by referring to scientific journals, handbooks, and manual books.

3.3. Design concept

The design concept is an outline of the design plan that will be created. This design concept is further adjusted to the component to be selected according to specifications. The design concepts and components that have been selected are then carried out in a three-dimensional (3D) tool design process using Solidworks.

3.4. *Hardware and software design*
Hardware and software design is used to determine the components needed in the manufacture of humidity control equipment as well as programs to control it.

3.5. *Testing*
In the testing phase, a series of tests are performed on the equipment that has been made.

- Simulation Tools. Automatic humidity control system simulation is done by placing the device in a flash off a room or another room measuring 6m x 6m then the device is set to the value of the humidity in accordance with existing standards. Then the device will read the actual humidity in the room and will start to condition the humidity to the determined humidity value and keep the humidity in the room to remain constant.
- Moisture Value Validation. The automatic humidity control device is installed with a temperature and humidity sensor to read the actual humidity value in the room, then the humidity value is displayed via the LCD. From the humidity value displayed on the LCD compared to the humidity value displayed by the temperature and humidity reader in the room.

4. Results and discussion

4.1. *Identification of problems*
Problems found in research conducted at PT. Kubota Indonesia is the absence of air conditioning systems in the flash off space painting process so that the moisture in the flash off space does not match the working standard chart (WSC) which results in the final product being painted there is a product defect in the form of blistering defects. Here are the data that support the identification of problems in PT. Kubota Indonesia.

4.2. *Data defect product in the painting department*
Painting department of PT. Kubota Indonesia is divided into 2 painting lines, each of which goes through stages of the work process starting from Parkerizing, Oven, Loading, Under Coating, Flash Off, Top Coating, Oven, Unloading and Nishugata. Each process must be carried out in accordance with a working standard chart (WSC) to avoid unexpected product defects. Actually, after checking during the final painting process, product defects are often encountered. Here are the results of defect product on Table 1 Data Defect Product Department of Painting, October 2018 - January 2019.

Table 1. Data defect product department of painting.

Month	Defective type (pcs)			
	Blistering	*Sagging*	*Mottle*	*Etc*
October	2459	487	2092	29
November	2194	809	1042	123
December	1517	585	917	508
January	1808	145	1132	64

4.3. *Test room model data and components of the automatic humidity control system model*
In this study, a trial room model and a component of the humidity control system were made in order to determine the effectiveness of the system to be made before it was applied directly to the flash off space of PT. Kubota Indonesia. The trial room model will be made using acrylic material with dimensions of

1500 mm × 500 mm × 800 mm with the acrylic thickness of 5 mm. The following is a description of the model of the trial room which can be seen in Figure 1:

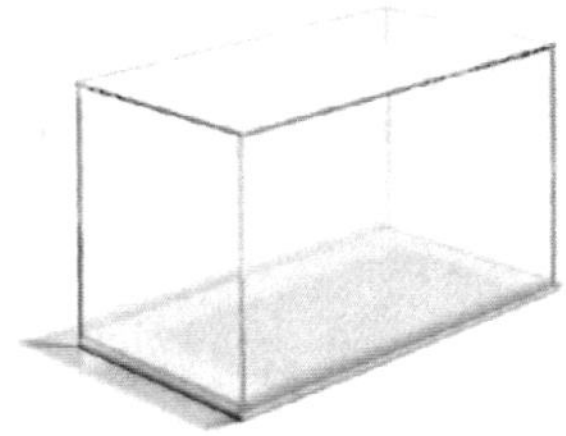

Figure 1. Testing room model.

The design of the automatic humidity control system model was made by adjusting the test chamber model, but still using the same work system that should be applied to the flash off space of PT. Kubota Indonesia. Following is the design of the automatic humidity control system model that will be created as shown in figure 2.

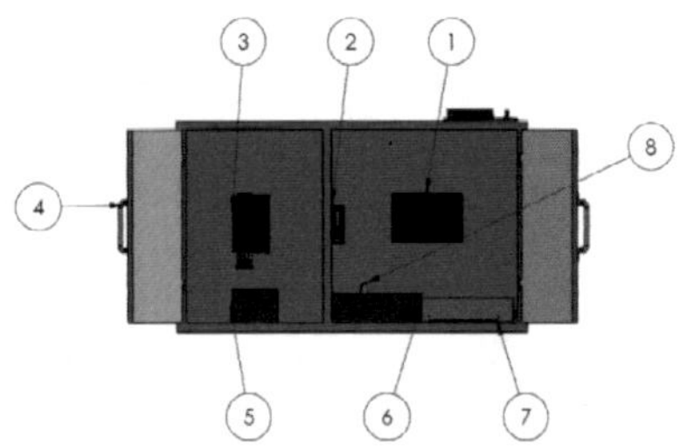

Figure 2. Humidity controller.

4.4. Test and analysis results
Calculation of cooling load is needed to analyze the factors that influence heat transfer that occurs in the test room so that we know the amount of heat energy that must be removed in the room. Cooling loads that are affected are, among others, external heat loads caused by heat entering through conduction (walls, ceilings, glass, partitions, floors), radiation (glass), and convection (ventilation and infiltration) and internal heat loads caused by heat that arises because of people/occupants, lights, and equipment/machinery.

Table 2. Recapitulation of cooling load analysis results.

No.	Source of Cooling Load	Total Cooling Load (Watts) (Watt)
1	External Factor	
	Cooling load from the wall	31.51
	Cooling load from the roof	11.59
	Cooling load from infiltration	7.33
2	Internal Factors	
	Cooling load of LCD I2C	5
	Q total	55.43
	Q safety (10%)	5.543
	Q	60.973

4.5. Calculation of thermoelectric cooler performance
The specifications of the thermoelectric cooler used in this study are as follows:

$\mathbf{Q_{max}}$ = 55.74 W $\mathbf{\Delta T_{max}}$ = 73.38 °C

I_{max} = 6.4 A **V_{max}** = 16.04 V
R =2.24**Ω**

Next, the Seebeck coefficient of the module when operated at $T_h = 28°C$ is as follows:

$$\alpha = \frac{V_{max}}{T_h} = \frac{16.04\ V}{(28\ °C + 273\ °C)} = 0{,}053\ ^{V}/_{K}$$

So, the coefficient produced by the module at the average temperature is 0.053 V/K.
Calculate thermal resistance

$$\theta = \frac{\Delta T_{max}}{I_{max} \cdot V_{max}} \times \frac{2T_h}{(T_h - \Delta T_{max})}$$

$$= \frac{73.38\ °C}{6.4\ A \cdot 16.04\ V} \times \frac{2(301\ K)}{((28°C + 273°C) - 73.38°C)} = 1{,}89\ ^{K}/_{W}$$

Calculate electrical resistance

$$R = \frac{V_{max}}{I_{max}} \times \frac{(T_h - \Delta T_{max})}{T_h}$$

$$= \frac{16.04\ V}{6.4\ A} X \frac{((28°C + 273°C) - 73.38°C)}{(28°C + 273°C)} = 1.89\ \Omega$$

Calculate the amount of current
Using a thermoelectric cooler performance graph with a T_h data of 28 °C is as follows:

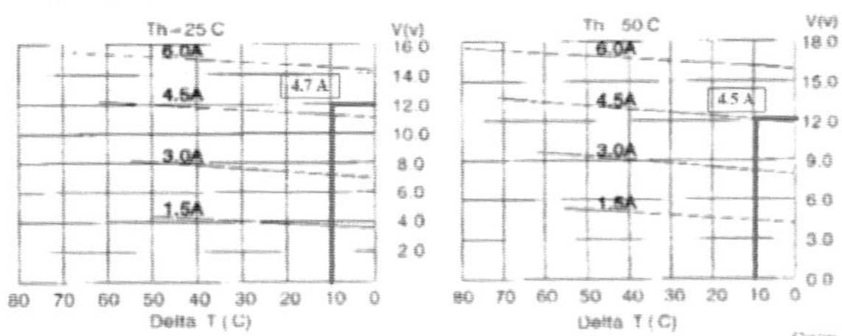

Figure 3. Curve flow.

The known T_h from the figure 3 above is aimed at temperatures of 25°C and 50°C. Whereas the average T_h the test is 28°C so the current sought must be interpolated. The ΔT value is obtained from the reduction between T_h and T_c by 3°C. Thus, the result of the interpolation of the current value at $T_h =$ 28°C is 4.69 A.

Calculate the voltage: $V = IR + \propto \Delta T$

$$= 4.69\ A \cdot 1.89\ \Omega + 0.053\ V/K \cdot (301\ K - 298\ K) = 9.02\ V$$

Calculates the heat released $q_{em} = \propto IT_h - \frac{\Delta T}{\theta} + \frac{I^2R}{2}$

$$= 0.053\ ^{V}/_{K} \cdot 4.69\ A \cdot (273°C + 28°C)$$

$$- \frac{3\ °C}{1.89\ ^{K}/_{W}} + \frac{(4.69\ A)^2 \cdot 1.89\ \Omega}{2} = 94.02\ W$$

Calculates the heat absorbed

$$q_{abs} = \propto IT_c - \frac{\Delta T}{\theta} - \frac{I^2R}{2}$$

$$= 0.053\ ^{V}/_{K} \cdot 4.69\ A \cdot (273°C + 25°C)$$

$$-\frac{3\ ^{\circ}\mathrm{C}}{1.89\ ^{\mathrm{K}}/_{\mathrm{W}}}-\frac{(4.69\ \mathrm{A})^{2}\cdot 1.89\ \Omega}{2}=93.27\ \mathrm{W}$$

So, the heat absorbed by the thermoelectric cooler is 93.27 W. While the heat released by the module into the environment is 94.02W.

5. Conclusion

The automatic humidity control system created and tested in the test chamber model can be used as a new alternative method in the air conditioning process replacing conventional methods using conventional air conditioners. From the results of the analysis of the calculation of the cooling load and the performance of the thermoelectrical device, the system is able to transfer heat according to the expected target. The cooling load that must be transferred is 60,973 Watts, while the performance of the thermoelectrical device is able to transfer heat reaching 93.27 watts of heat absorbed and 94.02 watts of heat released.

References

[1] Sipilä E, Virkki J, Sydänheimo L and Ukkonen L 2016 Experimental study on brush-painted passive RFID-based humidity sensors embedded into plywood structures *International Journal of Antennas and Propagation*

[2] Khumgorn S and Osothsilp N 2015 Improvement of Paint Thickness in Electro Deposition Painting Process of Truck Manufacturing In *Advanced Materials Research* **1077** 270-275 Trans Tech Publications.

[3] Hall A E 2017 Humidity and plant productivity. In *Handbook of agricultural productivity* 23-40 CRC Press.

[4] Khairunnisa F and Harmadi H 2017 Rancang Bangun Alat Ukur Kelembaban Udara Berbasis Mikrokontroler ATMega328 dengan Sensor Serat Optik Evanescent Menggunakan Film Gelatin *Jurnal Fisika Unand* **6** 217-224

[5] Riyanto S R 2017 *Rancang Bangun Alat Kontrol Suhu Dan Kelembaban Pada Fermentasi Tempe Kedelai Berbasis Mikrokontroler* (Doctoral dissertation, Politeknik Negeri Sriwijaya)

[6] Wahyono R E 2016 Rancang Bangun Sistem Kendali Otomatis Temperatur Dan Kelembaban Kumbung Jamur Tiram (Pleurotus Sp) Berbasis Mikrokontroler.

Effect of Vibrating Fluidized Bed (VFBD) and continuous microwave drying on drying characteristic of green tea

S U Handayani[1,*], M E Yulianto[1], S Sutrisno[1], D Ariwibowo[1], R Amalia[1] and I Hartati[2]

[1]Industrial Technology Department, Diponegoro University, Semarang, Indonesia
[2]Chemical Engineering Dept., Wahid Hasyim University, Semarang, Indonesia

*sriutamihandayani@lecturer.undip.ac.id

Abstract. In industrial green tea processing, people usually use endless chain pressure (ECP) dryers and fluidized bed dryers. ECP dryers have disadvantages including case hardening, uneven dryness, high energy consumption and long drying time, while fluidized bed dryers still have disadvantages of long drying time and high hot air capacity to ensure tea leaves can be fluidized perfectly. Therefore, another alternative dryer is needed that can overcome the problems in the ECP dryer and fluidized bed dryer. This study aims to determine the comparison of drying characteristics between vibrating fluidized bed dryer (VFBD) that have been developed previously with microwave dryers. Drying condition on both dryer was set at 80oC, the airflow velocity at VFBD was 1.09 m/s. Result shows the time needed by the microwave at temperature to dry 500 grams of tea leaves to reach a constant weight is 24 minutes, whereas if using VFBD it takes 85 minute. In VFBD, the constant rate period takes 25% of total drying time, and the rest is critical and falling rate period. While in microwave drying, around 50% of total drying time is constant rate period and the rest is critical and falling rate period. The falling rate period in VFBD requires longer time, which is up to 75% of the total drying time. Highest drying rate in VFBD dryer was 0.08 gr water/gr solid.min, while microwave dryer was 0.14 gr water/gr solid.min.

1. Introduction

Tea (Camellia sinensis) is the most common beverage in the world. Deped on how it is processed, there are several types of tea such as black tea, white tea, green tea, oolong tea, and others. Tea contains active ingredients such as: polyphenols, alkaloids, amino acids, proteins, volatile compounds, minerals and small amount of trace elements [1]. Polyphenols in green tea are dominated by Epigallocatechin (EGC), Flavanolscatechin (C), epicatechin (EC), Gallocatechin (GC), Gallocatechin 3-O-Gallate (GCG), Catechin 3-O-Gallate (CG), Epicatechin 3-O-Gallate (ECG) and Epigallocatechin 3-O-Gallate (EGCG) [2-6].

White tea has the highest antioxidant, followed by green tea and then black tea. Drying rate is the change in the moisture content of the dried product in unit time. The catechins content in the tea can be affected by which leaves are harvested, how the leaves are processed, and how the tea is prepared [4].

Polyphenols are quickly oxidized after harvesting due to the enzyme polyphenol oxidase. For the production of green tea, freshly harvested leaves are rapidly steamed or pan-fried to inactivate enzyme polyphenol oxidase [6]. This process approximately 10-15 minute in a process known as fixing [7]. The enzymes in leaf shoots will be deactivate in order to prevent oxidation and fermentation and to maintain a green color. The main relevant enzymes in tea plants include polyphenol oxidase, peroxidase, catalase, and ascorbic acid oxidase. Fast, even and high-temperature fixing is a feature of green tea processing, and is important for high quality tea [7]. Green tea processing is then followed by rolling, shaping and drying. Rolling aims to obstruct their cell walls, release leaf moisture and form the final product and reduce the volume.

Drying in green tea processing is intended to reduce water content, extend the shelf life by decelerate or stopping microorganisms' growth [8,9], prevent biochemical change during storage [9] and stop the enzymatic oxidation process of polyphenols if there are still active enzymes. The catechins content in green tea is determined during the fixing and drying process, so the selection of its methods is important to determine the quality of green tea. There are various drying methods including spray drying, rotary drying, fluid-bed drying, infrared heating, microwave heating and others [10]. Drying treatments can affect nutritional and qualitative characteristics of green tea such as color, vitamins, total flavonoid, total phenolic content, chlorophyll and antioxidant activity of final product [8].

In the green industry, the dryer used for processing green tea is an endless chain pressure (ECP) dryer or fluidized bed drier. ECP dryers have disadvantages including case hardening, uneven dryness, high energy consumption and long drying time. Because fluidized bed dryers have more advantages, many tea industries are turning to fluidized bed dryers [11], but fluidized bed dryers still have disadvantages of long drying time and high hot air capacity to ensure tea leaves can be fluidized perfectly. To our knowledge, research on the development of continuous dryers using microwaves for drying green tea has never been done.

Microwave heating is directly related to dielectric material loss. If put in a high-frequency electrical field, the substance can absorb energy. As a consequence, electrical dipole polarization and conducting will be produced within a dielectric material composed of polar molecules with positive and negative poles. Such ordered scattered polar molecules vibrate instantaneously and vigorously in reaction to the alternate high-frequency electrical field of the microwave as shown in Fig. 1. Resistance to molecular resistance and motion must be resolved. The temperature of the substance rises as heat is produced by friction [12].

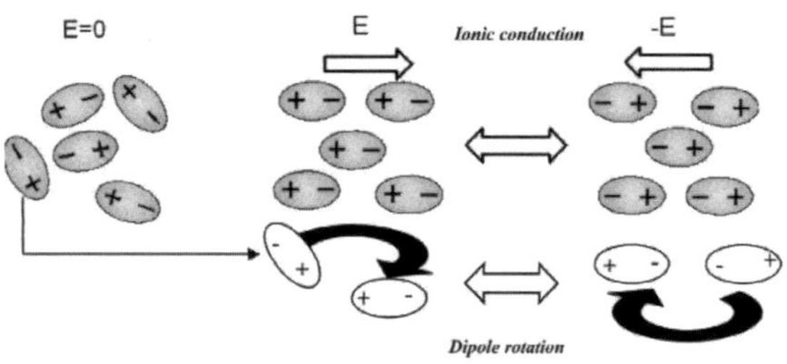

Figure 1. Mechanism of microwave heating [12].

The microwave heating rate and energy consumption affected by the dielectric property of a material to be dried. The main factors that determine the dielectric behavior of dielectric materials include the frequency and temperature of microwave heating and the moisture content of the material [12]. The dielectric property also depend on density, composition and structure of material [13].

Moisture content can effect the dielectric constant and dielectric loss. Regardless of the temperature, materials with higher moisture contents consistently exhibit higher values of the dielectric constant and dielectric loss, because the microwave absorption and dissipation increase with increasing free water in the material [14].

In hot air drying, the material is heated by an external heat source (from outside the material to be heated) and the heat is transferred from the outside to the inside material. Microwave heating, on the other hand, is based on the quick polarization and depolarization of the charged groups as the material is exposed to a microwave field, resulting in simultaneous internal heat generation [12]. Microwave has the ability to penetrate and heat the substance volumetrically due to the interaction between the electrical field and the water molecules [15,16]. A large vapor pressure formed at the central region of a material facilitates quick moisture removal [17].

Microwave for drying application has many advantages compared to other types of dryers include [12,17-20]:

- Shorter drying time.
- Lower power consumtion due to shorter process time with respect to traditional drying.
- Deep penetration of microwave energy which enables heat to be produced effectively without directly contacting the materials to be dried.
- Better product quality.
- Less space and easy control.
- In combination with vacuum drying, high product quality can be achieved by low temperatures and the rapid transfer of energy from microwave heating.
- Fast drying in the microwave will produce a product with a better antioxidant impact [15,21], color and appearance, sensory and textural attributes [18].

Research on microwaves for drying has been done in mackerel [18], strawberries [22], potato slices [19,23], ginseng slices [24], cassia alata [25], strobilanthes crispus [15], shitake mushroom [26], agaricus bisporus [27], blueberry leaves [28], green asparagus [29], celery [30], lemongrass [31], etc.

2. Material and methods

2.1. Material

Fresh tea leaves (*Camelia sinensis* var. *assamica*) are supplied by PT Rumpunsari Medini, Kendal, Indonesia, clone TRI 2025.

2.2. Apparatus of vibrating fluidized bed drying

The vibrating fluidized bed dryer was a square continuous vibrating fluidized bed dryer with 2000 x 200 mm drying area [32] and consist of blower, LPG burner, mixing chamber, vibration generator, bed, hopper, panel and cyclone. The air is heated by an LPG burner and then sprayed into the mixing chamber and flows vertically to the bed. The air flow rate can be varied by changing the rotation of the blower using an inverter, while the temperature can be controlled automatically from the panel using thermocontrol. The VFBD was equipped with thermometer and hygrometer in the inlet, along the bed and outlet side. The schematic of the vibrating fluidized bed dyer can be seen in Fig.2 [32].

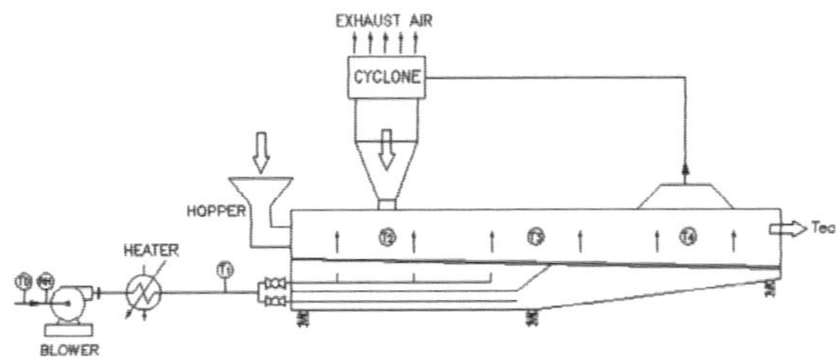

Figure 2. Schematic of vibrating fluidized bed dryer.

2.3. Apparatus of microwave-based drying machine

The microwave-based drying equipment was designed for laboratory scale, consist of hopper, microwave, fan, conveyor, electric motor, agitator and control panel, as presented in Fig.3. The dimension of the equipment is 3300mm(L) x 550mm(W) x 600mm(H). There are four commercial microve oven, which has 2450 MHz frequency, placed directly after the hopper. Each microwave equipped with agitator to stir the tea leaves in order to be exposed to the waves evenly. A thermocontrol is used to turn off the microwave power if the temperature is in accordance with the desired setting.

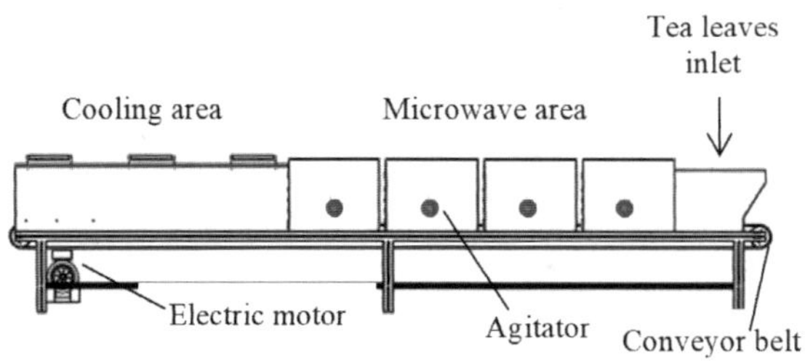

Figure 3. Schematic of continuous microwave dryer.

2.4. Method

Tea leaves that have been plucked are sorted until the third leaf from the shoots, then followed by fixing process in the microwave for 2 minutes. Tea leaves that have undergone fixing process are then rolled manually and ready to be dried.

VFBD was set at a temperature of 70°C and an air flow velocity of 1.09 m / s at a blower frequency of 50 Hz after the temperature and velocity have been reached then 500 gr tea leaves were fed into the hopper. The tea leaves were weighed every 15 minutes when they come out of the equipment [32]. The process were repeated until a constant weight was reached.

The microwave was set to 70°C, after the temperature has been reached, then 500 gr tea leaves were fed through the hopper. The conveyor will carry the tea leaves across the microwave and then out into the cooling area. This cooling area is only used during the fixing process. The tea leaves were weighed every 2 minutes when they come out of the equipment. The process were repeated until a constant weight was reached.

Moisture content of tea leaves was determined in dry basis, calculated as the rate of water amount in the product to its dry weight [33]. Moisture content (g water/g dry matter) values were calculated using Equation (1).

$$M_t = \frac{m_w}{m_w - m_s} \quad (1)$$

In this equation, m_w: water mass of tea leaves (gr), ms: dry mass of tea leaves (gr).

Moisture ratio was calculated using this equation [30]:

$$M_t = \frac{m_w}{m_w - m_s} \quad (2)$$

where MR is the moisture ratio, M is the moisture content at a given time (gr water per gr dry solids), Mo is the initial moisture content (gr water per gr dry solids), Me is the equilibrium moisture content (gr water per gr dry solids).

The drying rate is the change in the moisture content of the dried product by unit time [33]. Drying rate was determined as follows [34]:

$$DR = \frac{M_{t+dt} - M_t}{dt} \quad (3)$$

whereas, M_t and M_{t+dt} are the moisture content at t and t+dt respectively and t is the drying time (min).

3. Results and discussion

3.1. Drying time and moisture content

The time needed by the microwave to dry 500 grams of tea leaves to reach a constant weight is 24 minutes, whereas if using VFBD it takes 85 minutes (Fig.4). Significant time reduction is the main advantage of microwave dryers. Comparison between microwave dryers and spouted beds, ovens, also shows a eduction in time between 25% to 90% [16,35]. Shorter drying time will increase productivity and reduce drying costs. Based on the work published in the open literature by various researchers, the key factors affecting the time required for the drying of biological products are mainly affected by the drying mechanism and the conditions as well as the inherent properties of the material [19]. After fixing process, tea leaves containing high moisture contents, water accounts for the bulk of the dielectric component and is very intense in its reaction to microwave energy[19]. The same results were obtained for drying mint leaves [36], cassia alata [25], persimmon chips [37], chamomile leaves [38], sorbus fruits [39] and shitake mushrooms [26].

Comparison between the decrease in moisture content (dry basis) of tea leaves which are dried with a fluidized bed dryer and microwave can be seen in Fig.4. In VFBD, the constant rate period takes 25% of total drying time, and the rest is critical and falling rate period. While in microwave drying, around 50% of total drying time is constant rate period and the rest is critical and falling rate period. The falling rate period in VFBD requires longer time, which is up to 75% of the total drying time. Throughout the falling rate period, drying is inefficient as the dry product surface yields a high heat and mass transfer resistance layer, and the temperature gradient may be in the opposite direction of the moisture gradient.[16]. In addition, during the falling rate period, the moisture content is low, the water molecules therefore have a higher evaporation enthalpy, and the removal of these molecules by evaporation requires higher energy input. When drying foods and agricultural products with conventional hot-air drying methods, this low heat and mass transfer efficiency will results in prolonged drying time and hence a severe quality degradation [16].

High energy efficiency can be achieved in microwave drying during the falling rate period. It is because the energy directly penetrates the central area of the leaves, comes into contact with the moisture, so there is no need to transfer heat from the low moisture surface to the high moisture area [16].

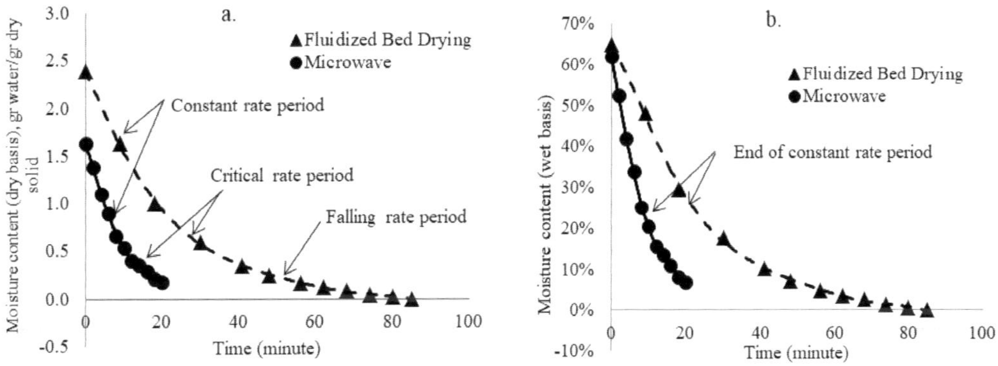

Figure 4. Comparison of changes in moisture content based on time of tea leaves in fluidized bed drying and microwave drying, a. dry basis, b. wet basis.

3.2. Moisture ratio

The dimensionless moisture ratio of green tea leaves was obtained using equation 1 and the relation of drying time and moisture ratio can be seen on Fig.5.

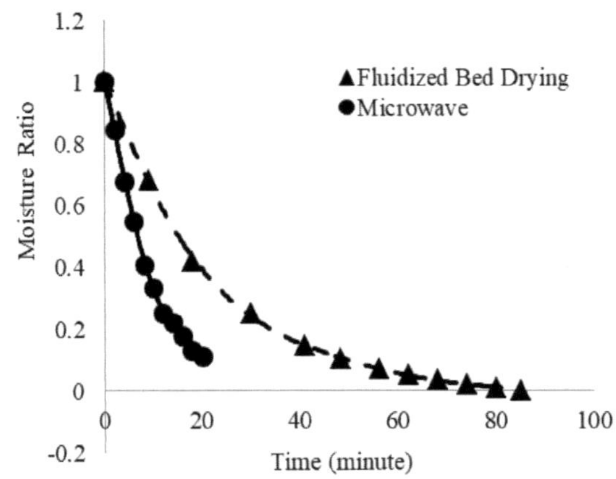

Figure 5. Moisture ratio.

3.3. Drying rate

Figure 6 shows the drying rate (gr. water/gr solid.min) against drying time for vibrating fluidized bed dryer and microwave. The drying rate of tea leaves in the continuous microwave drying was higher than vibrating fluidized bed drying, due to the shorter drying time.

In microwave drying, the initial phase of the drying, the drying rate increased quickly because more energy of the microwave was absorbed and the moisture content of the tea leaves was very high. As the drying process continues, drying rate drops rapidly due to the solids surface moisture content decreases.

In VFBD, the longer the drying time, the removal of water from tea leave due to the decreasing of drying rate. The drying rate decreased because the temperature difference between the tea and the air was very small and even roughly constant. As a result, the driving force of water to evaporate therefore decreases significantly [9].

Highest drying rate in VFBD dryer was 0.08 gr water/gr solid.min, while microwave dryer was 0.14 gr water/gr solid.min.

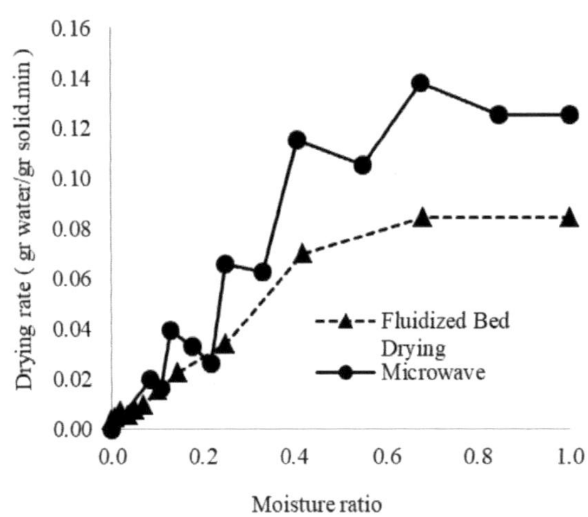

Figure 6. Drying rate.

4. Conclusion

The time needed by the microwave to dry 500 grams of tea leaves to reach a constant weight is 24 minutes, whereas if using VFBD it takes 85. In VFBD, the constant rate period takes 25% of total drying time, and the rest is critical and falling rate period. While in microwave drying, around 50% of total drying time is constant rate period and the rest is critical and falling rate period. The falling rate period in VFBD requires longer time, which is up to 75% of the total drying time. Highest drying rate in VFBD dryer was 0.08 gr water/gr solid.min, while microwave dryer was 0.14 gr water/gr solid.min.

Acknowledgement
This work was supported by Scientific Reserch Grant no 257-118/UN7.P4.3/PP/2019 from Ministry of Research, Technology and Higher Education of Indonesia.

References

[1] Yang X R, Ye C X, Xu J K, and Jiang Y M 2007 Simultaneous analysis of purine alkaloids and catechins in Camellia sinensis, Camellia ptilophylla and Camellia assamica var kucha by HPLC *Food Chem* **100** 1132–1136
[2] Ahmad M, Baba W N, Shah U, Gani A, Gani A and Masoodi F A 2014 Nutraceutical properties of the green tea polyphenols *Journal of Food Processing & Technology* **5** 1
[3] Pastoriza S, Mesías M, Cabrera C, and Rufián-Henares J A 2017 Healthy properties of green and white teas: An update *Food Funct* **8** 2650–2662
[4] Reygaert W 2017 An Update on the Health Benefits of Green Tea *Beverages* **3** 6
[5] Sharma V, Bhattacharya A, Kumar A, and Sharma H 2007 Health Benefits of Tea Consumption *Trop J Pharm Res* **6** 785–792
[6] Hayat K, Iqbal H, Malik U, Bilal U, and Mushtaq S 2015 Tea and Its Consumption: Benefits and Risks *Crit Rev Food Sci Nutr* **55** 939–954
[7] Ahmed S and Stepp J R 2012 Green Tea : Plants, Processing, Manufacturing and," no January, 2012
[8] Roshanak S, Rahimmalek M, Amir S, and Goli H 2016 Evaluation of seven different drying treatments in respect to total flavonoid , phenolic , vitamin C content , chlorophyll , antioxidant activity and color of green tea (Camellia sinensis or C assamica) leaves **53** 721–729
[9] Handayani S U, Yulianto M E, and Paramita V 2015 Efficacy of Zeolite Adsorption on the Green Tea Production by Fluidized Bed Dryer *Res J Appl Sci Eng Technol* **9** 1128–1131
[10] Langat N, Thoruwa T, Wanyoko J, Kiplagat J, Plourde B, and Abraham J 2014 Models and experiments for energy consumption and quality of green tea drying 43–50
[11] Temple S J and van Boxtel a J B 1999 Fluidization of Tea *J Agric Eng Res* **74** 5–11
[12] Rattanadecho P and Makul N 2016 Microwave-assisted drying: A review of the state-of-the-art *Drying Technology* **34** 1-38
[13] M Dekker *et al*, *No Title*
[14] Metaxas R C and Meredith R J *Industrial Microwave Heating Industrial Microwave Heating*
[15] Chua L Y W, Chua B L, Figiel A, Chong C H, Wojdyło A, Szumny A and Choong T S Y 2019 Antioxidant activity, and volatile and phytosterol contents of Strobilanthes crispus dehydrated using conventional and vacuum microwave drying methods *Molecules* **24** 1397
[16] Feng H, Yin Y, and Tang J 2012 Microwave Drying of Food and Agricultural Materials : Basics and Heat and Mass Transfer Modeling
[17] Zhang M, Tang J, Mujumdar A S, and Wang S 2006 Trends in microwave- related drying of fruits and vegetables **17** 524–534
[18] Viji P, Sai K S, Debbarma J, Das P D, Rao B M and Ravishankar C N 2019 Evaluation of physicochemical characteristics of microwave vacuum dried mackerel and inhibition of oxidation by essential oils *Journal of food science and technology* **56** 1890-1898
[19] Azimi-Nejadian H and Hoseini S S 2019 Study the effect of microwave power and slices thickness on drying characteristics of potato *Heat and Mass Transfer* **55** 2921-2930
[20] Arballo J R, Campañone L A and Mascheroni R H 2010 Modeling of microwave drying of fruits *Drying Technology* **28** 1178-1184
[21] Calín-Sánchez Á, Lech K, Szumny A, Figiel A and Carbonell-Barrachina Á A 2012 Volatile

composition of sweet basil essential oil (Ocimum basilicum L) as affected by drying method *Food Research International* **48** 217-225

[22] Asioli D, Rocha C, Wongprawmas R, Popa M, Gogus F and Almli V L 2019 Microwave-dried or air-dried? Consumers' stated preferences and attitudes for organic dried strawberries A multi-country investigation in Europe *Food Research International* **120** 763-775

[23] Rezaei S, Behroozi-Khazaei N and Darvishi H 2019 Microwave power adjusting during potato slice drying process using machine vision *Computers and Electronics in Agriculture* **160** 40-50

[24] Ning X, Feng Y, Gong Y, Chen Y, Qin J and Wang D 2019 Drying features of microwave and far-infrared combination drying on white ginseng slices *Food science and biotechnology* **28** 1065-1072

[25] Chua L Y W, Chua B L, Figiel A, Chong C H, Wojdyło A, Szumny A and Lech K 2019 Characterisation of the convective hot-air drying and vacuum microwave drying of Cassia alata: Antioxidant activity, essential oil volatile composition and quality studies *Molecules* **24** 1625

[26] Wang Q, Li S, Han X, Ni Y, Zhao D and Hao J 2019 Quality evaluation and drying kinetics of shitake mushrooms dried by hot air, infrared and intermittent microwave–assisted drying methods *LWT* **107** 236-242

[27] Jiang N, Lagnika C, Liu C, Li D, Yu Z, Wang L and Zhang M 2020 Dielectric properties of thermosonically treated Agaricus bisporus slices during microwave vacuum drying and correlation with the water state *Drying Technology* **38**(4) 448-459

[28] Borda-Yepes V H, Chejne F, Granados D A, Rojano B and Raghavan V S G 2019 Mathematical particle model for microwave drying of leaves *Heat and Mass Transfer* **55** 2959-2974

[29] Nguyen T V L, Tran T Y N, Lam D T, Bach L G and Nguyen D C 2019 Effects of microwave blanching conditions on the quality of green asparagus (Asparagus officinalis L) butt segment *Food Science & Nutrition* **7** 3513-3519

[30] Karabacak A Ö, Suna S, Tamer C E and Çopur Ö U 2018 Effects of oven, microwave and vacuum drying on drying characteristics, colour, total phenolic content and antioxidant capacity of celery slices *Quality Assurance and Safety of Crops & Foods* **10** 193-205

[31] Mabai P, Omolola A and Jideani A I 2018 Effect of drying on quality and sensory attributes of lemongrass (Cymbopogon citratus) tea *Journal of Food Research* **7** 68-76

[32] Handayani S U, Utomo M S K and Yulianto M E 2017 Performance Evaluation of Continuous Vibrating Fluidized Bed Dryer on Green Tea Production *Advanced Science Letters* **23** 2530-2532

[33] Çelen S 2019 Effect of Microwave Drying on the Drying 7–9

[34] Beigi M 2019 Drying of Mint Leaves : Influence of the Process Temperature on Dehydration Parameters, Quality Attributes, and Energy Consumption **21** 77–88

[35] Schubert H and Regier M 2006 Novel and traditional microwave applications in the food industry *Adv Microw Radio Freq Process - Rep from 8th Int Conf Microw High Freq Heat* 259–270

[36] Kripanand S M, Guruguntla S, and Korra S 2015 Effect of Various Drying Methods on Quality and Flavor Characteristics of Mint Leaves (Mentha spicata L) **3** 5

[37] Jia Y, Khalifa I, Hu L, Zhu W, Li J, Li K and Li C 2019 Influence of three different drying techniques on persimmon chips' characteristics: A comparison study among hot-air, combined hot-air-microwave, and vacuum-freeze drying techniques *Food and Bioproducts Processing* **118** 67-76

[38] Motevali A and Minaei S 2016 Energy analyses and drying kinetics of chamomile leaves in microwave-convective dryer *J Saudi Soc Agric Sci* **15** 179–187

[39] Lüle F and Koyuncu T 2015 Convective and Microwave Drying Characteristics of Sorbus Fruits (Sorbus domestica L) *Procedia - Soc Behav Sci* **195** 2634–2643

IOP Conf. Series: Materials Science and Engineering 1108 (2021) 012041 doi:10.1088/1757-899X/1108/1/012041

Improved Savonius double blade performance using modified blade shaped with variations of the wind flow ratios on the blade inlet and outlet side

Y D Herlambang*, S Supriyo and B Prasetiyo

Department of Mechanical Engineering, Politeknik Negeri Semarang, Jl. Prof. H. Sudarto, SH Tembalang, Semarang 50275, Indonesia

*masyusufdh@polines.ac.id

Abstract. The Savonius turbine is a wind turbine that has a vertical shaft parallel to the turbine blade and has advantages that can save wind energy from all directions at low speeds. The purpose of this study is to examine the Savonius turbine model with variation in the ratio of the blade inlet side and the blade outlet side to improve turbine performance. The research method used is turbine research; conduct an experimental test by varying the ratio of the inlet and outlet sides of the blade (1:1 and 1:2) to the speed of the blade's wind flow. Then conduct turbine performance analysis to get the best efficiency between the ratio of the inlet and outlet sides of the blade. The test was carried out at wind speeds of 5 m/s, 7 m/s, 9 m/s, 10 m/s, and 12 m/s with lamp loads distributed between 5-40 Watts. Analysis of turbine performance test The test results show that Savonius single blade wind turbine has the best efficiency of 8.43% at 5 m/s wind speed and is able to produce electricity of 10 Watts and then the best efficiency released by a 1:2 double blade wind turbine that has efficiency by 6, 9% at a load of 20 Watt and a 1:1 double blade wind turbine which has an efficiency 4.37% at a load 10 Watt.

1. Introduction

Electricity is the most important aspect supporting the activities of human life. Lots of alternative energy in Indonesia that can be utilized to produce electricity. One example of alternative energy that can be chosen is wind [1,2]. Wind energy is energy that is environmentally friendly and is renewable, for that it has the potential to be developed. The condition of Indonesia's wind speed has low wind speeds ranging from 2 m/s - 6 m/s. So it is necessary to design a wind turbine that is suitable for low wind speed areas and can meet electricity needs. Indonesia has a huge wind energy potential of around 9.3 GW and the total installed capacity is currently around 0.5 MW [3-5]. Wind is the air that moves due to the rotation of the earth and the difference in air pressure around it.

Betz Element's momentum theory based on the two-dimensional flow model of wind on the rotor explains the principle of conversion of wind energy in wind turbines [5,6]. The kinetic energy of an air mass m moving at velocity v can be expressed as [7-9]:

$$E = \frac{1}{2} m v^2$$

Considering a certain area, A, where air passes with velocity v, the flow rate of the volume of air V through over a given unit of time is [7-9]:

$$V = v\,A$$

The potential of the wind energy used in moving the wind turbine with a cross section A (m^2) is exhaled by the wind in the area with velocity v (m/s^2), then the rate of mass air flow passing through a place is [7-9]:

$$\dot{m} = \rho\,A\,v$$

So that the maximum kinetic energy produced by the union of time or the maximum kinetic power of the wind that passes through the turbine rotor [7-9]:

$$P_{kin} = \frac{1}{2}\,\rho\,A\,v^3$$

$$C_p = \frac{P}{P_0} = \frac{1}{2}\left|1 - \left(\frac{v_2}{v_1}^2\right)\right|\left|1 + \frac{v_2}{v_1}\right|,$$

So we get a value,

$$C_p = \frac{P_{mech}}{P_{kin}}$$

Tip speed ratio λ is the ratio between the speed of the tip of the blade and the speed of the wind passing through it. If $\lambda > 1$ means more blades experience lift. If $\lambda < 1$ means that more parts of the blade have a thrust [10-12].

$$\lambda = \frac{\pi D n}{60 v}$$

Where D is the diameter of the rotor, n is the rotor rotation, and v is the wind speed. Power coefficient and torque coefficient are different from tip speed ratio. The optimum tip speed ratio is determined by the rotor where the most efficient energy transfer and $C_{p,\,max}$ are for the maximum power coefficient [7].

Turbine power or mechanical power [13-15], expressed by the formula as follows:

$$P_{mek} = \frac{2\,\pi\,n\,T}{60}$$

Where T is the torque; n is the turbine shaft rotation.

Subsequently, to calculate the turbine efficiency, which is the ratio between the power generated by the shaft with kinetic power, namely [15,16]:

$$\eta_t = \frac{P_m}{P_k} \times 100\%$$

Where η_t is the coefficient of power or efficiency of the turbine, P_m is the mechanical power of the turbine, P_k is the kinetic power of the wind.

1.1. Generator power

Generator power is the electrical power produced by the generator [15,16].

$$P_g = V_p \times I_p$$

Where V_p is the generator output voltage, and I_p is the generator output current. So that the efficiency of the system can be found by the following equation:

$$\eta_s = \frac{P_g}{P_k} \times 100\%$$

2. Methods

The methods of this research include material selection, workmanship and assembly of Savonius rotor components including turbine shafts, blade flanges, turbine blades, pulleys, generators, timing belts, pillow blocks, gears, wheels, frames and loads. The next step is testing the performance of single blade, double blade (1:1) and double blade (1:2). The final method is analysis and discussion. The results of the processing are then displayed in the form of graphs of the efficiency characteristics of the system against tip speed ratio, voltage-current-electric power, and the relationship between generator power and rotation. The performance of each component is reviewed and analyzed. So it will produce a Savonius double blade rotor model with variations in the ratio of wind flow on the inlet and outlet sides of the blade that has the best performance. The test stages are presented in the figure below, where there are 3 tests of the Savonius rotor model namely single blade, 1:1 double blade, and 1:2 double blade with various wind speed test parameters namely 5 m/s, 7 m/s, 9 m/s, 10 m/s and 12 m/s.

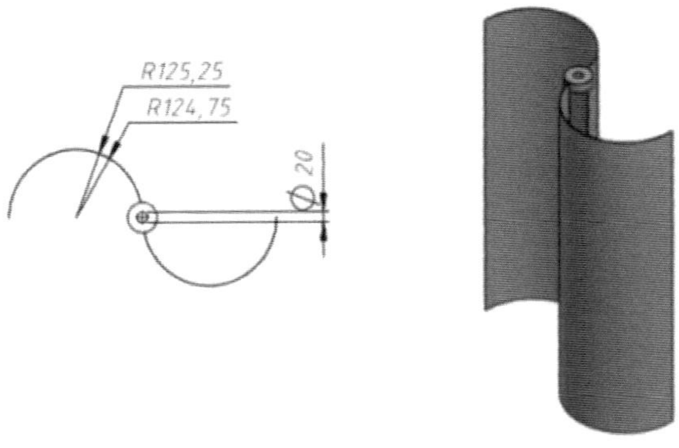

Figure 1. Savonius single blade wind turbines.

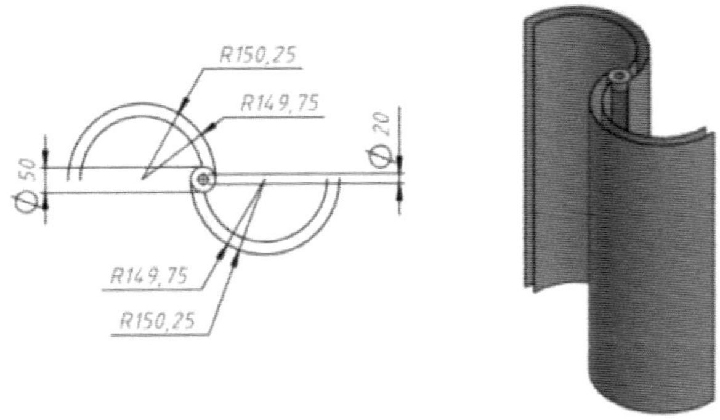

Figure 2. Savonius double blade wind turbine with a ratio of wind flow at the inlet side of the blade and the outlet side of the blade is 1: 1.

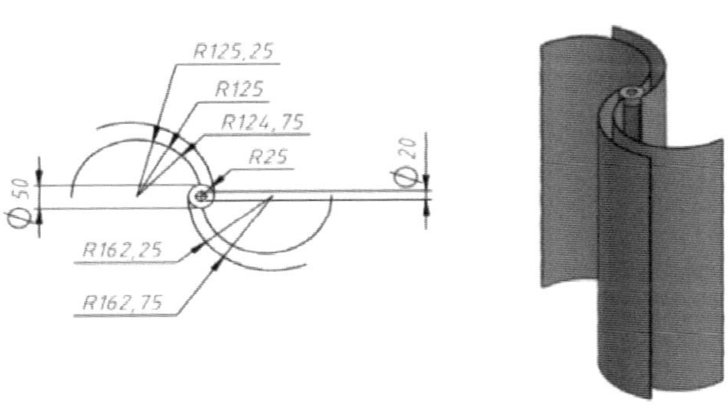

Figure 3. Savonius double blade wind turbine with a ratio of wind flow at the inlet side of the blade and the outlet side of the blade is 1: 2.

The Savonius turbine test circuit below is used when testing to obtain test data in the form of currents and voltages on 3 Savonius rotor models at various wind speeds. Furthermore, Savonius turbine performance was analyzed to get the turbine blade model that has the best characteristics.

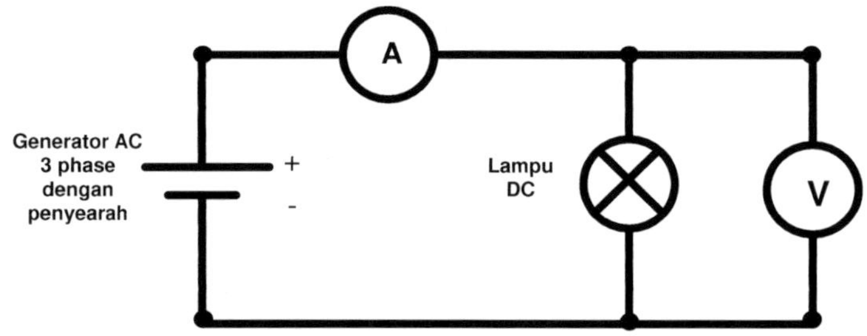

Figure 4. Wind turbine system testing circuit.

3. Results and discussion

The results of Savonius single blade, 1:1 double blade and 1:2 double blade wind turbines with wind speeds of 5 m/s, 7 m/s, 9 m/s, 10 m/s, and 12 m/s. Furthermore, it is processed into a graph of the characteristics of the relationship between system efficiency and speed ratio (η_s-*u/v*), graphs of generator power and turbine rotation characteristics (P_g-*n*), and graphs of current voltage characteristics (*V-I*) and electric current (*P-I*). The results of the study are Savonius sturbin models shown in the figure below. This step is done by designing a Savonius Wind Turbine model single blades and double blades with variations the ratio in the inlet blade and outlet side blade (1:1, 1:2, 1:3, 1:4) to the wind flow to improve performance.

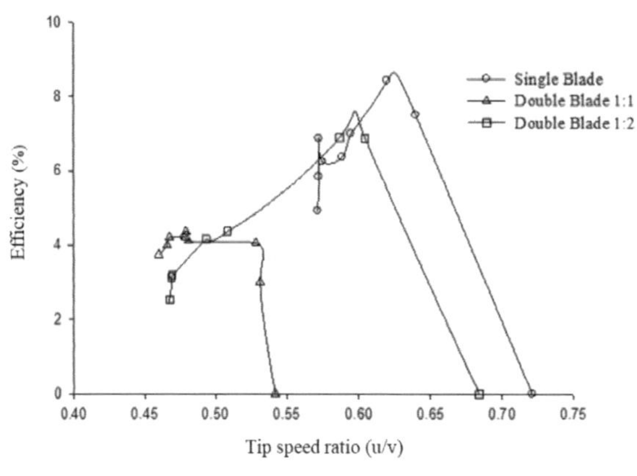

Figure 5. Savonius wind turbine performance: efficiency to tip speed ratio (η_s-u/v).

Figure 5 shows that efficiency tends to increase with increasing tip speed ratio values. In certain conditions the efficiency will decrease after reaching the saturation point. It happens because of changes in the load given to wind turbines, so that it will reduce the efficiency of the turbine. The test results in Figure 5 show that the highest system efficiency is owned by a single blade wind turbine at a wind speed of 5 m/s at 8.43% with a tip speed ratio (*u/v*) ratio of 0.619 at a 10 W load. Whereas the double blade ratio 1:2 wind turbine has the highest efficiency of 6.90% with a speed ratio (*u/v*) of 0.587 at a load of 10 W. This is followed by a decrease in system efficiency that occurs in a Savonius double blade 1:1 rotor of 4.37% with a speed ratio (*u/v*) of 0.479 at a load of 20 W.

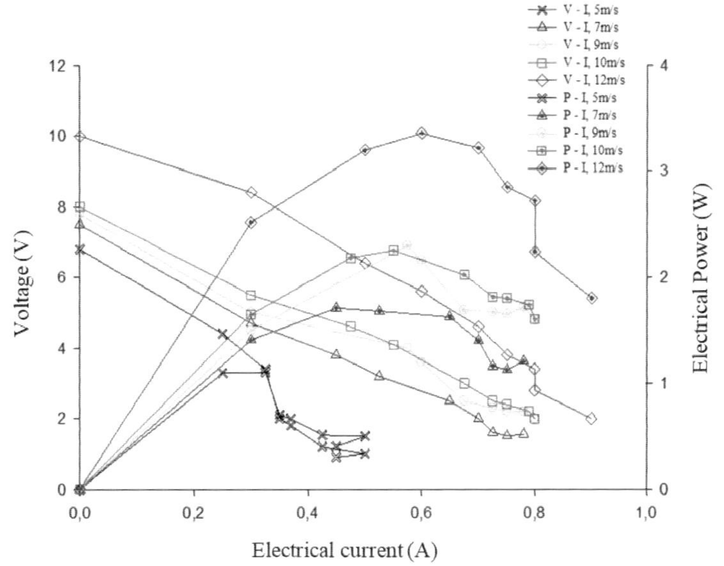

Figure 6. Performance characteristics of voltage-current-power of the Savonius wind turbines.

Figure 6 shows the characteristics of the relationship between voltage and current and electric power to the current in a 1:2. double blades wind turbine. The electrical power tends to increase with increasing current values. In certain conditions the efficiency will decrease after reaching the saturation point. It happens because of changes in the load given to wind turbines, so that it will reduce the efficiency of the turbine. The largest electrical power produced by wind turbines compilation of wind speeds reaches 12 m/s, which is 3.36 W at a measured load of 15 W. However, after achieving optimal electrical power values, there is a trend of decreasing the effectiveness of electric current loading. At this time, the maximum capacity of the Savonius turbine has reached its maximum limit.

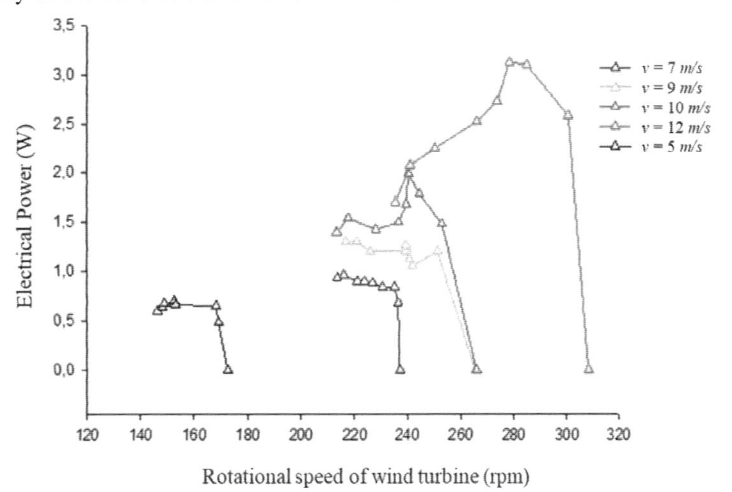

Figure 7. Performance of wind turbine between electrical power versus rotational speed at the ratio double blade 1:1.

The characteristic relationship between electrical power and rotation at the ratio double blade 1:1 is shown in Figure 7. At wind speed 5 m/s it can be seen that the minimum power of the generator is 0.48 W when the load reaches 5 W at 169.2 rpm. While the maximum power obtained is 0.7 Watt when the load reaches 20 W at 152.64 rpm rotation. At a wind speed of 12 m/s it can be seen that the minimum power of the generator is 1.7 W when the load reaches 40 W at 235.44 rpm. The generator power will increase as the rotation speed increases. The maximum power obtained is 3.125 W when the load reaches 15 Watt at 277.92 rpm rotation. But after reaching its maximum power, the generator power will decrease. This is because the generator has reached the maximum limit and the generated power tends to fall so that the curve is parabolic. The characteristics of the generator power to the rotation of the 1:2 double blade are shown in Figure 8. At wind speeds of 5 m/s it can be seen that the minimum power of the generator is 0.405 W when the load reaches 40 W at 148.8 rpm. The generator power will increase as the rotation speed increases. The maximum power obtained is 1.1 Watt when the load reaches 5 W at 192.8 rpm. But after reaching its maximum power, the generator power will decrease. This is because the generator has reached the maximum limit and the generated power tends to fall so that the curve is parabolic. At wind speeds of 12 m/s it can be seen that the minimum power of the generator is 1.8 W when the load reaches 40 W at 238.56 rpm. While the maximum power obtained is 3.36 W when the load reaches 15 W at 288.48 rpm rotation.

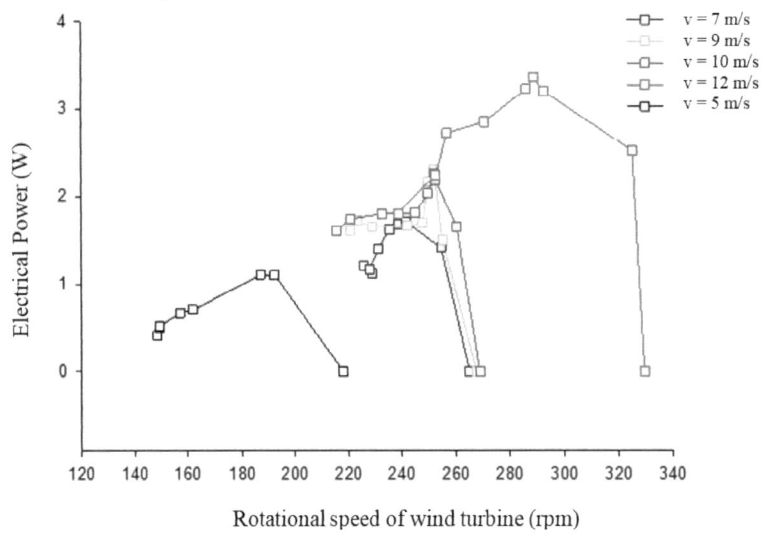

Figure 8. Performance of wind turbine between electrical power versus rotational speed at the ratio double blade 1:2.

3.1. Technological excellence

3.1.1. Innovative excellence. An approach to vertical axis wind turbine technology using concentric type double blade blades to the direction of wind flow at the inlet and outlet side of the blade.

3.1.2. Comparative advantage. To increase the turbine power is done by varying the ratio of wind flow on the inlet and outlet of the blade, so that it will increase the efficiency and power of the turbine produced.

3.2. Application prospect

The design and application of the Savonius double blade turbine will increase the efficiency and power of the turbine as needed and this product can be downstreamed to people in remote areas that have not yet installed the PLN electricity grid.

4. Conclusion

The results of this competitive applied research produce turbine and generator specifications are 3 phase AC generators with additional rectifiers so that the output is DC with 200 W specifications, 12-24 Volts with rotation that can reach 150-300 rpm. Savonius wind turbines are made with galvanized plate with variations of single blade, double blade 1:1, and double blade 1:2. The ratio of the wind flow on the blade influences the speed of the blade output wind. When it reaches the best efficiency, the wind energy used to rotate the blade is more optimal, so that the resulting blade rotation is faster to be able to turn the generator and produce more power. The best system efficiency in Savonius single blade wind turbines is 8.43%, the highest efficiency in Savonius 1: 2 double blade wind turbines is 6.9%. While the best efficiency in Savonius 1: 1 double blade wind turbines is 4.37%.

Acknowledgement

The authors gratefully acknowledge the financial support provided by the Penelitian Terapan Kompetitif Politeknik Negeri Semarang the contract of Polines 0179/PL4.7.2/SK/2019.

References

[1] AWEA 2018 2012 The *American Wind Energy Association* [Online] retrieved from: http://www.awea.org. accesed on 5 Oktober 2012
[2] DESDM (Dinas Energi dan Sumber Daya Mineral) 2018 *Blueprint Pengelolaan* Energi Nasional 2005-2025 [Online] Jakarta. www.esdm.go.id. Diakses pada 12 Januari 2019 pukul 14.00
[3] Herlamban Y D 2008 Unjukkerja Turbin Angin Nibe 3-Sudu Menggunakan Pompa Sudu Luncur untuk Pengambilan Air *Jurnal Eksergi* **IV** 12-16 0216-8685
[4] Hofman H and Harun 1987 *Energi Angin. Penerbit Binacipta* (Jakarta)
[5] Beri H and Yao Y 2011 Effect of Chamber Airfoil on Self Starting of Vertical Axis Wind Turbine *Journal of Environment Science and Technology* **4** 302-312
[6] Cooper P and Kennedy O 2004 Development and Analysis of a Novel Vertical Axis Wind Turbines *Journal of Mechanical and Materials Engineering* Wollongong. Australia
[7] Hau E 2006 *Wind Turbines: Fundamentals, Technology, Application, Economics.Second Ed.* (Germany: Berlin Springer-Verlag)
[8] Leysen E H 1983 *Introduction to Wind Energy*. PO BOX 85/Amersfort/The Netherlands: Steering Committee Wind Energy, Developing Countries
[9] Hansen M O L 2008 *Aerodynamics of Wind Turbines. Second Ed.* Earthscan, USA: New York
[10] Ingram, G. 2009. *Basic Concepts in Turbomachinery. First Ed.,* (Grant Ingram &Ventus Publishing ApS. UK: Durham)
[11] Saha U K and Rajkumar M J 2006 On The Performance Analysis of Savonius Rotor with Twisted Blades *Renewable Energy* **31** 1776-1788
[12] Damak A, Driss Z and Abid M S 2013 Experimental Investigation of Helical Savonius Rotor with a Twist of 180^0 *Renewable Energy* **52** 136-142
[13] Hermawan 2010 Unjukkerja Model Turbin Angin Poros Vertikal Tipe Savonius dengan Variasi Jumlah Sudu dan Variasi Posisi Sudut Turbin *Seminar Nasional Tahunan Teknik Mesin (SNTTM) ke-9*
[14] Soelaiman TA Fauzi, Tandian N P and Rachmatullah R 2012 Optimasi penambahan Selubung (Shroud) pada Turbin Angin Sumbu Vertikal dengan Menggunakan Simulasi CFD 2 Dimensi *Seminar Nasional Tahunan Teknik Mesin (SNTTM) ke-9*
[15] Herlambang Y D and Suwarti 2013 Model Turbin Savonius 1-Tingkat Sebagai Penggerak Mula Pompa Air tanpa Bahan Bakar untuk Pengairan *Eksergi Jurnal Teknik Energi* **9** 18-23
[16] Mahmoud N H, El-Haroun A A, Wahba E and Nasef M H 2010 An experimental study on Improvement of Savonius Rotor Performance *Alexandria Engineering Journal* **51** 19–25

Smart wind turbine three-layers flat blade with control system Arduino-based for blade angle

S Sahid*, S Suwarti, B Prasetyo and M Mulyono

Politenik Negeri Semarang, Semarang, Indonesia

*sahid.polines@gmail.com

Abstract. The objective of the design of a three-layers flat blade smart wind turbine with an Arduino blade angle configuration system is to study experimentally the performance of a wind turbine Three-layersflat blade horizontal guide with various wind speed variations, and apply automatically to blade blade speed average for weather. Tests are carried out to obtain the value of wind turbine power coefficient (Cp). Tests that use variations in wind and blade angle. Variations in wind speed are obtained by variations in refineries from wind turbines. The performance of Three-layersbflat blade wind turbines has the highest Power Coefficiency (Cp) values at certain blade angles on variations in wind speed at a wind speed of 5 m/sec with a blade angle of 15° at 57.84%; at a wind speed of 6 m/sec with a blade angle of 20° of 36.24%; at a wind speed of 7 m/sec with a 25° blade angle of 51.39%; at a wind speed of 8 m / sec with a blade angle of 30° of 39.37%; and at a wind speed of 9 m/s with a blade angle of 35° of 28.24%. After obtaining data about wind speed and angle with the highest Cp value, the wind speed that will be used as the input of the Arduino control system, and the output is a large blade blade that changes to adjust the wind speed.

1. Introduction

Wind is one of the renewable energy potentials that can make a significant contribution to the domestic electricity needs, especially in remote areas [1]. The wind energy generator which is commonly called the Power Plant This wind power is pollution free and the source of energy is that the wind is available anywhere, then this power plant is able to answer environmental problems and the availability of energy sources [2]. Indonesian wind speeds vary greatly. The National Energy General Plan (RUEN) lists 60,647.0 MW for wind speeds of 4 meters per second or more [3]. At low wind speeds using a flat blade with multiple blades (above eight blades) or commonly called Multiblade. The weakness of this turbine has a low performance for high speed. For high wind potential it is suitable to use NACA airfoil blades. weakness of this turbine can not rotate at low wind speeds [4].

With the varied Indonesian wind potential, the application of both types of turbines has times when the turbine works with poor performance [5]. This situation then gave rise to a new idea for designing wind turbine types that can rotate at low and high wind speeds, namely the three blades flat type where each flat is made in layers with bladder angles that can automatically adjust the wind speed to its highest coefficient.

The object of this project is to Create a horizontal axis three-blade flat blade turbine with varying blade angles, Experimental review of the performance of a horizontal three-axis flat axis wind turbine with various variations in wind speed, Testing the performance characteristics of a horizontal three-axis flat blade wind turbine with varying blade angles and obtaining.

1.1. Previous research

Afiffudin, et al. researched about three-layer flat blade wind turbines with Dimensisudu, length: 300 mm; width: 100 mm; thickness: 2 mm. Dimension of lace, diameter: 160 mm; thickness: 35 mm and variations in wind speed between 3-11 m/s with the aim to find out the turbine CP, and torque. The results of the study concluded that the performance of a three-layer flat wind turbine is better than a multiblade wind turbine at a variation of wind speed of 8 m/sec -11 m/sec tested by the highest Coefficient of Power (Cp) obtained at a speed of 8 m/sec with variations in the distance between blades 3 cm with a blade angle of 15° at 13.6042% [6].

The aerodynamic shape of the blade is greatly influenced by the position of the angle of the blade to the direction of the wind, so the position [α] of the blade angle will change. With the change in blade angle, it will affect the angular speed of the wind turbine, whose value is directly proportional to the production of wind energy extracted by the wind turbine through the blade, thus tracking the angular position of the wind turbine blades is important [7].

1.2. Three-layers flat type smart wind turbine

This type of wind turbine has the principle of cooperation with turbines in general. Starting from the use of kinetic energy possessed by the wind, which is then converted by the blade into mechanical energy on the shaft. In this case what distinguishes the type of three-layer flat blade smart wind turbine with three-layer flat blade wind turbine previous research is the existence of an automatic blade angle opening control system depending on the amount of wind speed. Blade can change due to the movement of DC motor back and forth controlled by Arduino based on the input wind speed sensor. Three-Layers Flat Type Smart Wind Turbine shown figure 1.

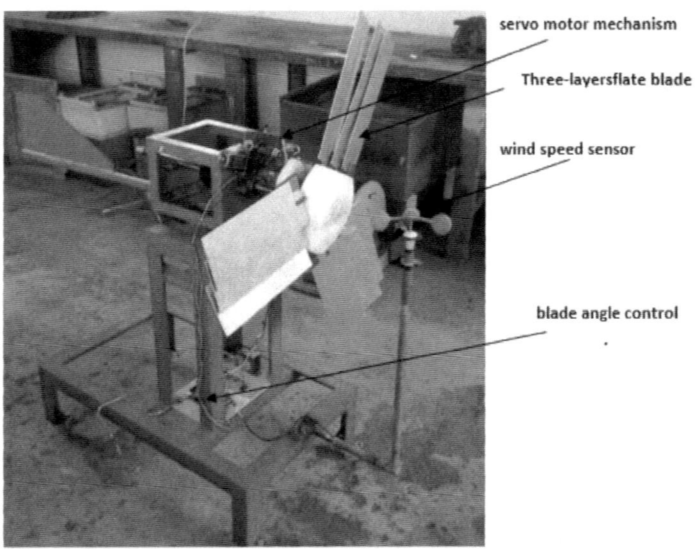

Figure 1. Three-layers flat type smart wind turbine.

Based on the block diagram in figure 2, Data input in the form of wind speed. Datais obtained from the speed sensorwind to be processed on Arduino, so that servo motor moves according to position angles that have been determined so angles blade change. With feedback in the form of data from the wind speed sensor, then if changes in data values, Arduino will reprocess to change the angle blade. The angle value is ranged between angles blade with wind speed with flow diagram as in figure 2.

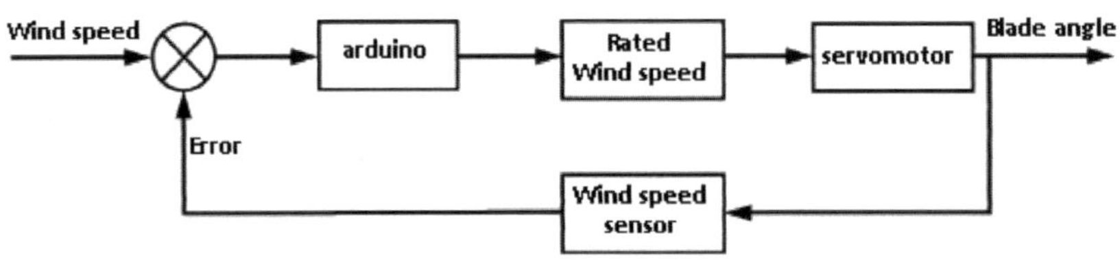

Figure 2. Control block diagram.

2. Methods

The main material in this study is a three-layer flat blade wind turbine equipped with a drive mechanism, sensor and turbine blade control. Variations in this study are wind speed and blade angle. The tools used in this research are Blower, three-layer flat axis horizontal wind turbine, Torsimeter, Digital tachometer, Anemometer, and Load. Before conducting research, preparations need to be made, including making turbine blades whose bladder angles can be varied, preparing Blowers, Digital tachometers, and Anemometers. Data retrieval begins after the flow of air from the blower to the entire wind turbine installation. Wind speed is controlled by looking at the anemometer, where the speed starts from 5 m/s to 9 m/s.

Wind speed is regulated by adjusting the distance of the turbine to the blower. Retrieval of data in this study in the form of a rotation on the turbine shaft using a digital tachometer, torque generated discs with a torsimeter and wind speed in front of the blade with anemometer. Breeding tests are conducted to assess the effect of wind speed on the blade angle. This test will produce the optimum blade angle at a certain wind speed. the angles generated from this test serve as input data on the system settings. The next test is the blade control system test. This test will assess the performance of the blade control system based on changes in wind speed. blade system performance is seen based on the results of the first test.

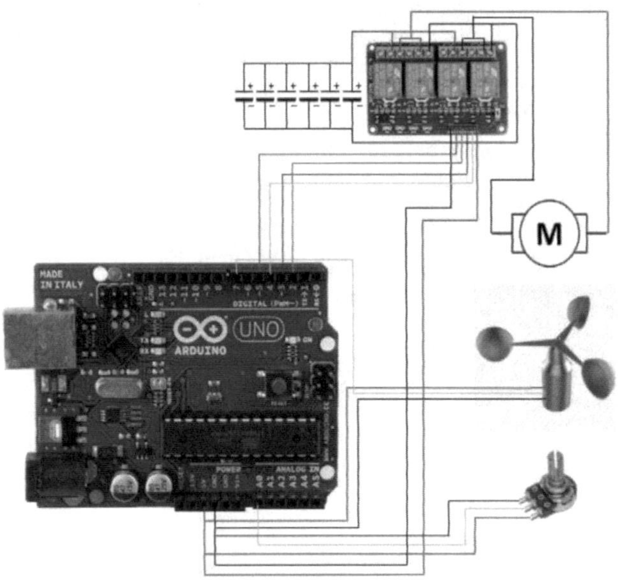

Figure 3. Series of blade control systems.

3. Results and discussion

3.1. Turbine performance testing

Figure 4 is a graph of the characteristics of Cp against TSR. The results of wind turbine testing at speeds of 5 m / s. There are 4 curves in the graph that include three-layer flat wind turbine curves with blade angles of 15°, 20°, 30° and 40°.

The turbine performance testing was started at empty load with the highest rotation at the angle position of the blades of 150, 200, 300, and 400, namely 328.30 Rpm, 308.10 Rpm, 220.80 Rpm, 181.10 Rpm. The next step is to increase the load by braking, increasing the load causes the rotation down and mechanical power increases to reach a maximum value that is at an angle of 150, 200, 300, and 400 is 12.04 Watt, 10.68 Watt, 19.88 Watt, and 20.35 Watt. After reaching the maximum value, the addition of the load is continued until the rotation is down and the mechanical power is down. Based on the characteristic graph that has been made, it can be seen that each blade angle setting has a Coefficient of Power peak. At the 15° Cp blade angle the largest was 57, 84%, the largest 20° Cp was 49.79%, the largest 30° Cp was 29.26%, and the largest 40° Cp was 12.41%. This shows that the highest Cp value at 5 m / s wind speed is at a 15° blade angle.

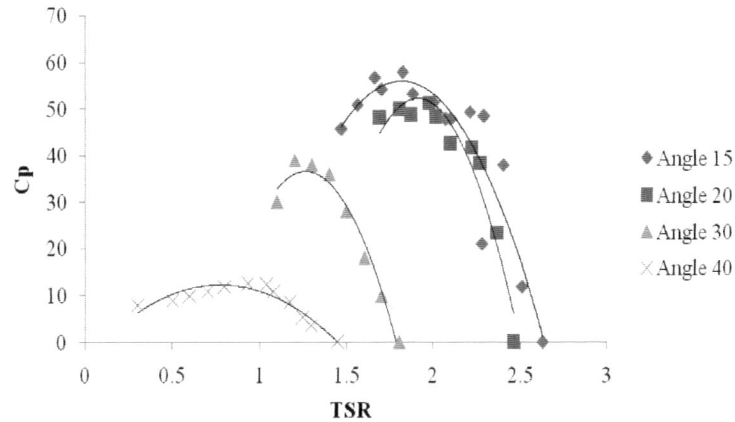

Figure 4. Graph of relationship characteristics between Coefficient of Power (Cp) against tip speed ratio at wind speed of 5 m/s.

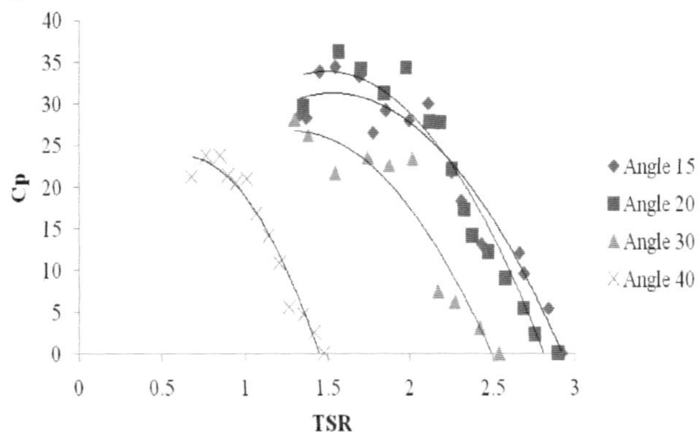

Figure 5. Graph of relationship characteristics between Coefficient of Power (Cp) against tip speed ratio at wind speed of 6 m/s.

Testing and data processing aims to determine the highest Cp value of each wind speed. This highest Cp value will be used as a reference input for the blade angle control system in Arduino. Cp value is influenced by kinetic power (Pk) and mechanical power (Pm). Cp value decreases at greater wind speeds. This is because the Cp value is the result of a comparison between mechanical power (Pm) and kinetic power (Pk). Mechanical power (Pm) is affected by rotation and torque. Torque is inversely proportional to rotation. The smaller the rotation, the greater the torque produced. From the test data it can be seen that the curve arises due to Cp which has reached its optimum point, where the turbine Cp has decreased after reaching the peak point.

Based on the calculation data and graphic figure 4 to figure 8, it can be seen that each wind speed test that is 5 m/s to 9 m/s has the highest Cp value respectively. At a wind speed of 5 m/s the highest Cp value at an angle of 15° is equal to 57.84%. At a wind speed of 6 m / s the highest Cp value at an angle of 20° is 36.24%. At a wind speed of 7 m/s the highest Cp value at an angle of 25° is 51.39%. For the wind speed of 8 m / s the highest Cp value at an angle of 30 ° is 39.37%, and for the wind speed of 9 m/s the highest Cp value at an angle of 35° is 28.24%. From the results obtained, it is known that each wind speed has a blade angle characteristic that shows the highest Cp value, from the data the wind speed is then used as a reference for the input control system or blade control system whose output is the blade angle automatically adjusts the wind speed.

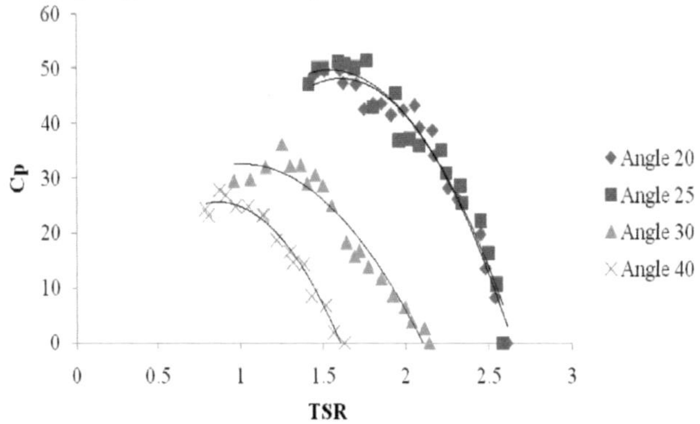

Figure 6. Graph of relationship characteristics between Coefficient of Power (Cp) against tip speed ratio at wind speed of 7 m/s.

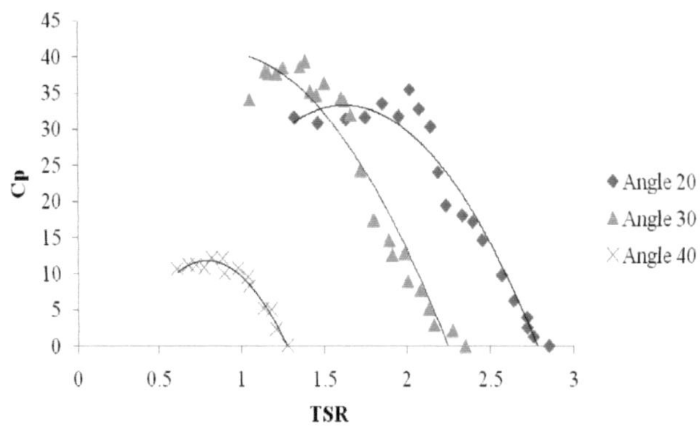

Figure 7. Graph of relationship characteristics between Coefficient of Power (Cp) against tip speed ratio at 8 m/s Wind Speed.

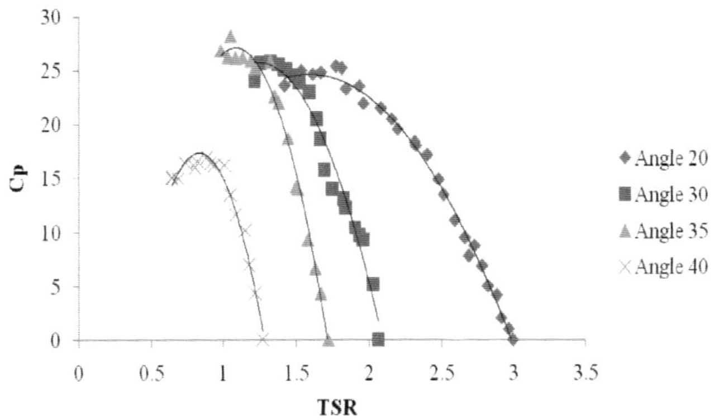

Figure 8. Graph of relationship characteristics between Coefficient of Power (Cp) against tip speed ratio at wind speed of 9 m/s.

3.2. Blade control system test results

Based on the results of the Cp wind turbine input, calculation and testing graphs can be displayed with a graph that can be seen in Figure 9. Based on these graphs it can be seen that the greater the wind speed that hits the turbine blade, the greater the blade opening. At an average wind speed of 5 m/s the position of the blades opening at an angle ranges from 140 to 150. At an average wind speed of 6 m/s the position of the blades opening at an angle ranges from 180 to 200. At an average wind speed 7.5 m/s blades opening position at an average angle ranging from 230 to 320 At an average wind speed of 9 m/s blades opening position at an angle ranging from 320 to 360 At an average wind speed of 10 m/s position of blade opening at angle 370. For control system testing data can be seen in the appendix.

From the data that has been obtained, it can be seen that the results of the wind turbine testing that becomes the control input are in accordance with the desired results. The greater the wind speed, the greater the angle of the blade opening which can be adjusted by the Arduino control system. Although the data results of wind turbine testing without a control system and wind turbine testing with an Arduino control system have a significant difference, these results do not change the results of the whole test that the blade of the wind turbine blade is successfully moved automatically according to the wind speed.

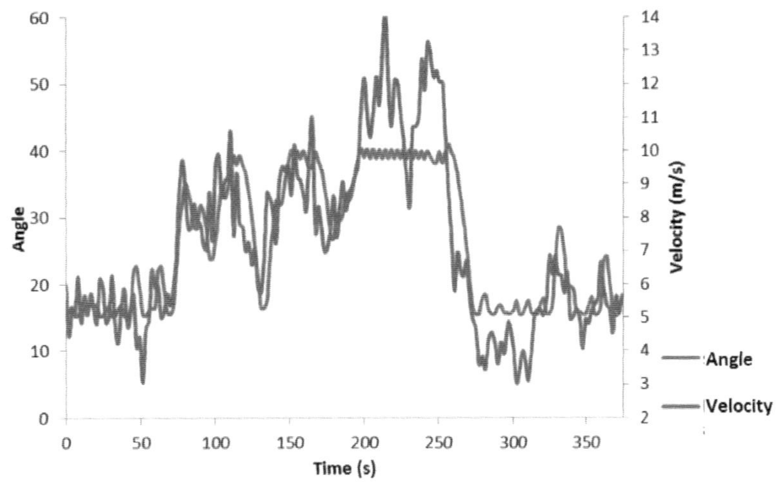

Figure 9. Performance graph of wind turbine angle control system.

4. Conclusions

Based on data from testing and analysis, it can be concluded as follows:

- Three-layer flat blade wind turbine has 3 blades in each layer by using aluminum plate material that can change the angle of the blade automatically. Blade dimension, length: 300 mm; width: 100 mm; thickness: 2 mm. Dimension of blade, diameter: 200 mm; thickness: 40 mm.
- At 5 m/s wind speed and 15° blade angle, the three-layer flat blade wind turbine has the highest Coefficient of Power (Cp) value of 57.84%.
- The performance of a three-layer flat blade wind turbine has the highest Coefficient of Power (Cp) value at a certain blade angle at each variation of wind speed, namely: at a wind speed of 5 m/sec with a blade angle of 15° of 57.84%; at a wind speed of 6 m / sec with a blade angle of 20° at 36.24%; at a wind speed of 7 m/sec with a blade angle of 25° at 51.39%; at a wind speed of 8 m/sec with a blade angle of 30° at 39.37%; and at a wind speed of 9 m/sec with a blade angle of 35° at 28.24%.
- Three-tiered flat blade wind turbines have good performance at low wind speeds and high wind speeds making it suitable for both low wind speeds and high wind speeds.
- The three-layer flat blade wind turbine is equipped with an Arduino-based automatic blade angle adjustment system where there is a DC motor that can move to change the size of the blade angle to automatically adjust the wind speed according to the Coefficient of Power (Cp) value at the best blade angle.

Acknowledgment

Thank you to DRPM, Ministry of Research and Technology for funding this research through the PTUPT scheme.

References

[1] Hau E 2006 *Wind Turbines: Fundamentals. Technologies, Application* (Economics–2nd edition–Springer-Verlag Berlin Heidelberg)

[2] Ginting D 2007 Sistem Energi Angin Skala Kecil Untuk Pedesaan *Jurnal Ilmiah Teknologi Energi. LAPAN: Bogor*

[3] Sekretaris Kabinet 2017 Peraturan Presiden No 22 Tahun 2017 Lampiran 1 [Online] Retrieved from: http://setkab.go.id/potensi-pengembangan-pltb-di-indonesia/

[4] Ramadhan F S 2016 *Pembuatan Turbin Angin Sudu Tipe Flat Berlapis Tiga* (Tugas Akhir. Semarang: Jurusan Teknik Mesin Polines)

[5] Satriani G P *2015 Modifikasi dan Kinerja Turbin Angin Sudu Flat Bingkai Berbasis Trapesium* (Tugas Akhir. Semarang: Jurusan Teknik Mesin Polines)

[6] Afiffudin 2017 *Rancang Bangun Turbin Angin Poros Horizontal Sudu Flat Tiga Lapis Menggunakan Variasi Sudut Blade dan Jarak Antar Lapis Sudu* (Tugas Akhir. Semarang: Jurusan Teknik Mesin Polines)

[7] Musyafa A 2018 Study of SupervisoryControl Implementation in A Small Scale Variable Speed Wind Turbine *Jurnal Ilmiah. Institut Teknologi Sepuluh November*

Performance of single cylinder diesel engine using triple fuel HSD-LPG-Steam

G E Kusuma*, S T Sarena and M Santoso

Marine Engineering Department, Shipbuilding Institute of Polytechnic Surabaya (SHIPS), Surabaya, Indonesia

*kusuma.george@ppns.ac.id

Abstract. Fuel system modification techniques to improve engine efficiency could be the way to slow down the energy crisis caused by the lack of new fuel inventions. The Compression Ignition (CI) diesel engine is one of the most preferred engines for both in-land and marine transportation motors. The advantages of compact dimensions and large generated power make CI diesel engine superior to other engine types. The large number of single CI diesel engines use in small fishing vessel engines and agricultural machinery will contribute substantially to fuel consumption in a cumulative manner. Previous researchs on the modification of fuel system with bi-fuel system High Speed Diesel (HSD)oil – Liquified Petroleum Gas (LPG) system capable of reducing Brake Specific Fuel Consumption (BSFC) up to 24% and exhaust gas opacity up to 68%. Subsequent studies focused on engines with single fuel HSD oil but were injected with superheated steam in 1 atm (abs) 130 ^{0}C by utilizing Heat Recovery Generating System (HRSG) using heat of exhaust gas for a flowrate of 1.5 - 4.5 kg/hour. It is capable of delivering the performance improvement of Specific Fuel Consumption (SFC), which is reduced by 8.92% compared to diesel engines without steam injection. Recent research combines the above two modification methods so that diesel will adopt two types of fuel- oil-LPG with additional superheated steam injection resulting from exhaust gas heat cogeneration system. Experimental results of the diesel engine with LPG injection maintained at 0.4 kg/h and 3.5 kg/h superheated steam with an engine variation loading of 0.3-0.7 kWh provide an average AFR of 30.09 and SFC improvement up to 37.2% lower than SFC engine standard with a single fuel HSD oil.

1. Introduction

Single cylinder diesel engines are used in large quantities for small fishing boats and agricultural machinery due to considerable maintenance and operational costs. The disadvantage of diesel engines is high exhaust emissions that have negative effects on the environment. Fishermen and lower farmers tend to ignore the consideration and awareness of the level of negative effects of exhaust emissions compared to economic considerations. The population of fishermen and farmers of very large diesel engine users will give a cumulative effect of the quantity of exhaust emissions produced. Hence, research to decrease the exhaust emissions of single cylinder diesel engines and simultaneously raise their efficiency will be an important long-term program in the effort to save the environment. The characteristics of high temperature combustion diesel engines generate high power and high side effects of exhaust emissions. The negative effects of exhaust emissions on diesel engines are regulated nationally and internationally and continue to be limited by regulation annually. One of the exhaust emissions generated by combustion at high temperature diesel engines is NOx. There is a wide variety of methods inside and outside the cylinder to reduce NOx emissions. One method of NOx reduction is the injection of water into the combustion chamber. Water injected into the combustion chamber in the emulsion form of oil fuel with water directly or injected directly into the intake manifold (fumigation) on the diesel engine. Based on existing literature studies, the maximum reduction in combustion temperature by water injection will be followed by a decrease in NOx gas emissions [1,2]. Same et al. emphasizes that NOx and soot emissions will be reduced by 20% and 50% respectively when using 10% and 15% water-fuel emulsions. On the other hand, it does not cause a decreasing effect of combustion efficiency with marked increase in SFC [3]. Abu-Zaid et al evaluated the performance of diesel engines in torque, power, effective efficiency, SFC temperature and exhaust chamber when water/fuel ratio increased [4]. Lin and Wang point out that the exhaust, O2, NOx and exhaust gas temperatures drop, while CO2 and CO emissions increase when the three-phase oxygen fuel (water/fuel/water or fuel/water/fuel) is injected into diesel engines and materials three-phase emulsions have higher exhaust chamber gas temperatures and lower CO and NOx emissions compared to two phases (water/fuel) [5].

In the fumigation method, water in the liquid phase is injected into the air intake manifold. Tauzia et al. states that if 60 - 65% of water is injected into a diesel engine air intake manifold with a common rail type injection system, NOx emissions may drop up to 50% and heat losses occur in cylinder walls, which will result in reduced efficiency [6]. Donahue and Ishida et al. investigating the injection of water into a diesel engine will produce NOx, soot emissions in the flue gas and SFC reduction when the diesel engine is given a low loading but otherwise the soot emissions and SFC will rise and the NOx emission decreases when the engine is loaded high [7].

Referring to existing literature reviews, direct injection of water into the combustion chamber such as emulsion or fumigation methods into the intake manifold will be able to reduce NOx emissions [3,5,6,8]. But the exhaust gas parameters for HC, CO and SFC will increase [4]. Additionally, water especially in the fumigation method will damage the lubricant specification and cause increased wear of moving parts of the engine [9]. One method that can be used to reduce NOx emissions is water injection into the intake manifold that previously in the liquid phase is converted into steam phase. Parlak et al explains that NOx emissions will be reduced by 33%, power and effective torque increased by 3% and SFC value decreased to 5% as a result of full load test with electronically controlled steam injection system with optimum steam ratio determined by 20% [10]. In this study, single cylinder diesel engine modification system is focused on the use of two types of fuel, namely HSD (high speed diesel) oil - LPG (liquified petroleum gas) plus steam injection on superheated steam phase to improve power performance and SFC fuel efficiency.

2. Material and methods

2.1. Method of experiment

The experimental method is carried out on a single cylinder diesel engine, where the air injection system is natural, 4 (four) strokes engine and uses water cooling system. This type of engine is chosen because the construction is simple to undergo the process of modification and this type of machine is much and commonly used for engine engine propulsion small fishing vessel and agricultural machinery.

Table 1. Engine diesel –dongfeng specification.

No	Model R180 Specification	
1	Displacement	402
2	Diameter	80x80 mm
3	Compresion Ratio	21 : 1

To generate steam to reach the superheated phase of this system utilizes the engine heat exhaust cogeneration process by using a steam generator unit installed after the exhaust gas flows from the exhaust chamber. Temperature and pressure gauges mounted on the steam to ensure steam quality is in the superheated phase. This steam generator unit is called a Heat Recovery Steam Generator (HRSG)

– Cogenerator that works to reuse the energy that has been wasted on the exhaust chamber. HRSG is self-fabricated by using heat transfer design with stainless steel material as shown in Figure 2 (a). The amount of water mass converted to steam is adjusted by flowmeter. The steam injection system that enters the manifold water is fabricated as shown in Figure 2 (b).

To monitor the exhaust gases coming out of the steam generator, Air Fuel Ratio (AFR) is used to provide information on combustion quality that occurs in the combustion chamber. The single cylinder diesel engine in the experiment is operated on a variety of electrical loads on the 0.3 kWH, 0.45 kWH, 0.6 kWH and 0.75 kWH electrical generators to get the engine performance picture at low to high loads at steady engine rotation at 2500 RPM. To convert the resulting mechanical energy, the diesel engine is coupled with an electric generator using a V-belt. Electric generator has power specification name plate 7.5 KW and 1500 RPM.

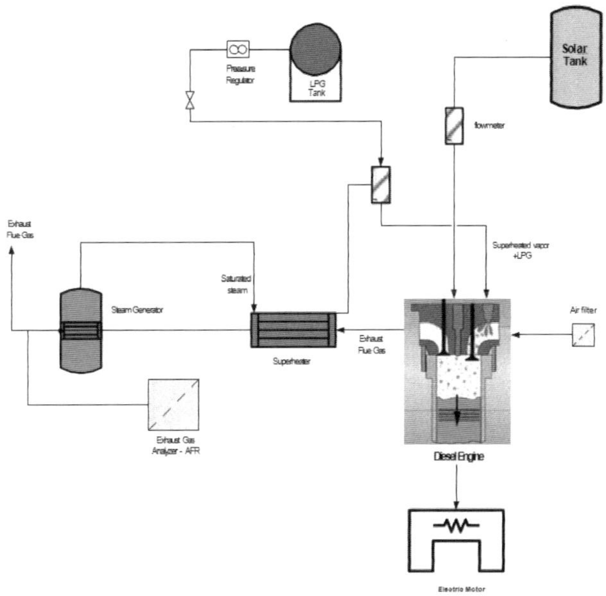

Figure 1. Experiment's set up.

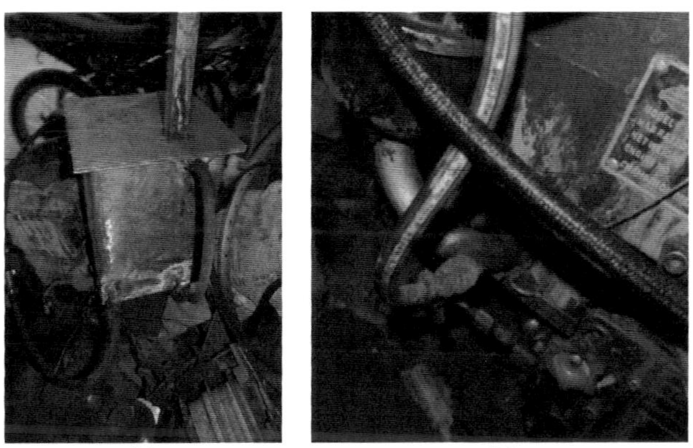

(a) (b)

Figure 2. (a) Steam generator (b) Steam injector on air intake manifold.

As a benchmark standard, single cylinder diesel engines are operated without modifications to the fuel system in order to obtain the performance of the original engine. The data is used as reference benchmark data on the performance of the engine undergoing modification process in its fuel system with steam injection utilizing heat exhaust gas cogeneration.

The steam mass generated by the steam generator is measured by flowmeter to vary from 1.5, 2.5, 3.5, to 4.5 kg/h with a temperature of 130°C with a pressure of 1 (bar) which has reached the superheated steam phase with a specific volume parameter h = 2736 kJ/kg, Vs = 1.841 m^3/kg, Entrophy (s) = 7.517 kJ/kg.C. In the saturated steam conditions, the parameters are h = 2676 kJ/kg, Vs = 1.674 m^3/kg, Entrophy (s) = 7.354 kJ/kg.C [11]. The superheated steam is injected into the combustion chamber mixed with combustion air and LPG gas in the air intake manifold so that steam and LPG enter the combustion chamber through the process of the air intake manifold step mixed with the outside combustion air.

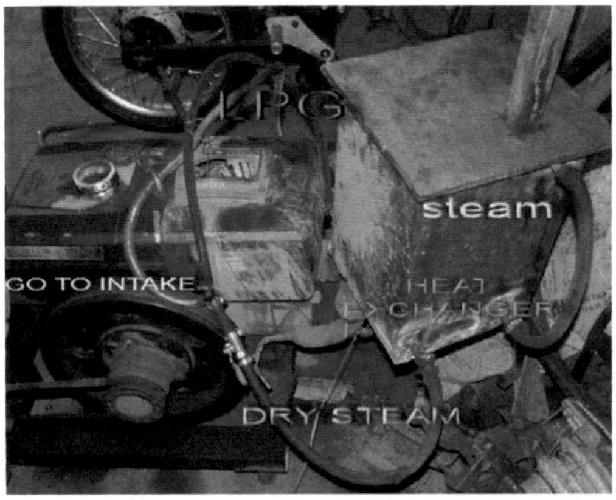

Figure 3. Diesel engine system with HRSG – cogeneration.

Experiments were repeated on each variation of engine load as well as steam and LPG injection to obtain overall performance descriptions from low to high loads. The AFR quality of the combustion process gas is compared with the diesel engine operating conditions when operating under standard conditions without modification.

3. Results and discussion

The effect of the addition of superheated steam and LPG gas to the performance of single cylinder diesel engine was analyzed with SFC, Air-Fuel Ratio (AFR) parameters and exhaust gas temperature. The SFC value shows the ratio of total fuel consumption to electric power that can be generated by an engine-connected alternator to determine the ratio and efficiency of the engine system. In the SFC measurement study, it was carried out at a fixed load for one hour of engine operation to obtain stable condition so an AFR value representing stable engine operating conditions.

AFR shows the ratio of the amount of air mass to the fuel mass in the combustion process, especially the type of internal combustion engine. The AFR value will show the condition of the fuel mixture with the air, the amount of fuel that can be released by the process and how much mass of the pollutant produced by the combustion reaction. Enough air mass combustion will provide a perfect stoichiometric combustion process. A low ratio of stoichiometric values will provide a rich mixture of fuel. The fuel-rich air mix will produce a low- efficiency combustion process but will be capable of producing greater power and lower combustion temperatures. A low mixture of stoichiometric ratio will result in more efficient combustion but will cause engine damage and produce NO_x exhaust emissions in the exhaust gas.

3.1. Performance of standard diesel engine

Diesel engine operation test with original fuel system is done using single fuel HSD oil with variable loading using alternator. the variation of loading conditions on the alternator will lead to an increase in fuel consumption on the engine so that the engine works not only rely on the large inertia moment on the flywheel to be able to achieve stable engine rotation but requires more fuel consumption to be able to achieve stable engine RPM (Revolutions Per Minute) at 2500 RPM reduced by transmission to 1500 RPM on the alternator.

Table 2. Performance of standard diesel engine.

Load (kWH)	SFC (Litre/kWh)	AFR
0,30	2,80	35,5
0,45	2,67	33
0,60	2,33	31
0,75	2,00	26

A decrease in engine AFR indicates that fuel consumption will rise to reach a constant engine RPM while generating an appropriate power supplied by the alternator. The above correlation is shown by table 1 which explains the linear relationship between fuel consumption (SFC) on the engine power generated along with the HSD oil combustion characteristics. The stable load of the engine will cause the AFR of the flue gas combustion to be close to the AFR of stoichiometric burning of HSD oil for 14-15.

The performance of SFC standard diesel engines works on loads raised from 0.3 kWH to 0.75 kWH. Trends in SFC performance improvements occur as the load continues to increase. This indicates that the better combustion efficiency and AFR value in the exhaust gas approach AFR stoichiometry. A good combustion reaction process will result in improved SFC performance so as to decrease HSD oil fuel consumption per minute.

3.2. Performance of diesel engines with the addition of LPG injection (Bi-fuel HSD oil – LPG System)

In this mode of operation, the diesel engine is modified into a dual fuel system that is bi-fuel system HSD oil - LPG. LPG is injected into the combustion process through an air intake manifold with a unit converter kit to avoid extreme machine construction modifications. The regulator converts the LPG liquid phase into a vapor phase. The LPG mass is injected constantly 0.4 kg/hour and the combustion process quality is analyzed using the AFR parameters of the flue gas at each loading variation. LPG is injected to the compressed air from the Bottom Dead Centre (BDC) to the Top Dead Centre (TDC) so that the mixing of air with LPG gas occurs. A high compression ratio of single-cylinder diesel engine is designed 21:1 will not cause premature ignition of LPG + air mixture due to LPG fuel characteristics determined by an octane fuel- ignition characteristic in which the ignition process still requires spark (ignitor). LPG-fueled diesel engines will still require an ignitor so that the design of this machine only allows a combination of HSD oil and LPG fuel for its operation. LPG with higher calorific values has the effect of decreasing HSD oil fuel consumption to achieve the same load condition compared to standard engine without LPG fuel injection [12].

Table 3. Performance of diesel engines with the addition of LPG injection (bi-fuel HSD oil – LPG system).

Load (kWh)	HSD (Litre/Hour)	LPG (kg/hour)	AFR	SFC (Litre/kWh)
0,30	**0,60**	**0,40**	**55**	**2,00**
0,45	**0,88**	**0,40**	**41,5**	**1,96**
0,60	**1,00**	**0,40**	**31,5**	**1,67**
0,75	**1,20**	**0,40**	**29,5**	**1,60**

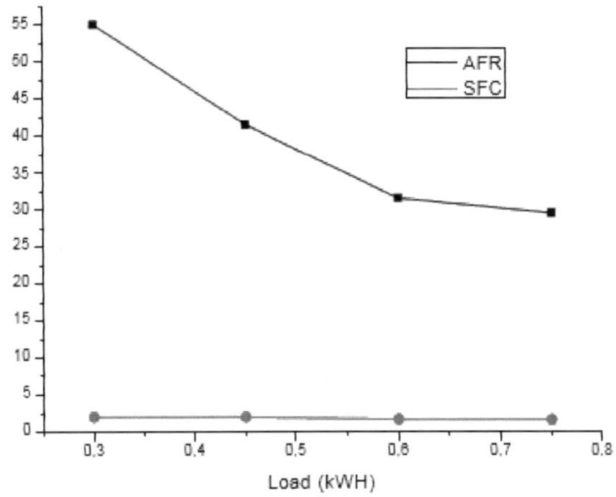

Figure 4. The performance of SFC diesel engines with constant LPG injection of 0.4 kg/hour (Bi-fuel HSD oil + LPG).

3.3. Diesel engine performance with the addition of superheated steam injection from exhaust gas cogeneration system

In this mode of operation, the diesel engine modified the fuel system using HSD oil with the addition of superheated steam. Steam is produced by a steam generator utilizing heat recovery exhaust gas. Engine can decrese the exhaust gas temperature and also regenerate heat from the exhaust gas for converting water into superheated steam. Steam is injected in the air intake manifold and combustion chamber due to the suction stroke process. The study was conducted with variations of superheated steam flowrate injected at 1.5, 2.5, 3.5 to 4.5 kg/hour to provide correlation between engine performance with steam injection value. The combustion quality that occurs in the combustion chamber can be analyzed and shown by the parameters of the AFR value of the flue gas in each loading variation and the mass of the steam injection. Figure 5 shows the AFR engine combustion characteristics at a constant load. it can be seen in the figure that the injection of steam causes a decrease in the AFR value compared to the standard engine AFR. This decrease occurs because part of the space combustion chamber is occupied by steam. The occupancy process causes the percentage of air in the combustion chamber will be reduced. This reduction of combustion air does not reduce the combustion quality of HSD oil in the combustion chamber, as indicated by the engine power generated. The process also causes the engine to consume lower fuel HSD oil at the same load. This significant result can be a reference that the gas cogeneration system for steam generators is able to work well to increase engine power and decrease the consumption of fuel oil HSD oil. The superheated steam injection process will increase the enthalpy value of air entering the engine combustion chamber so that the air in the combustion chamber has higher energy and temperature than standard engine conditions [13].

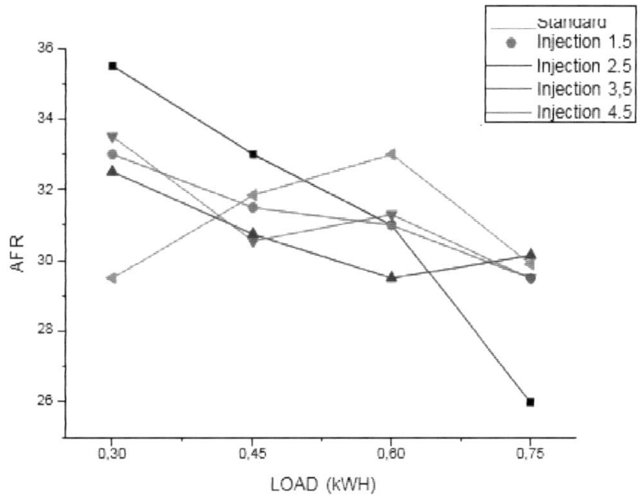

Figure 5. Graph of Comparison between AFR Value of Exhaust Gases with loading variation on Diesel engine with Steam Injection of 1.5, 2.5, 3.5, 4.5 kg/hour.

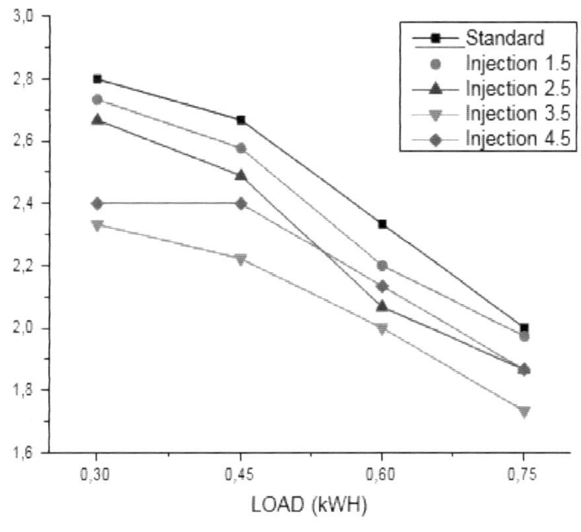

Figure 6. Graph of Comparison between Specific Fuel Consumption (SFC) with load variation on Diesel engine with steam injection of 1.5, 2.5, 3.5, 4.5 kg/hour.

Increased energy and temperature of air in the combustion chamber can speed up the preparation of HSD oil burning at the compression step. These conditions will increase the combustion pressure and generate greater engine torque at the same RPM with lower fuel consumption. Engine performance is analyzed with SFC parameters and exhaust gas temperature shown in Figure 6. Variations of injection flowrate are given to determine the effect of steam injection on engine performance on SFC, AFR and exhaust gas values shown the burning process occurring in the engine cylinder shown in Figure 5 and 6. At low engine loading at 0.3-0.45 kWH superheated steam injection of 1.5 -3.5 kg/hour can decrease the SFC of the engine. However, in addition to steam up to a flowrate of 4.5 kg/hour, the SFC increase again. The increase in SFC is caused by the effect of steam occupancy in the combustion chamber that reduces the air mass coming into the combustion chamber which makes the combustion process more difficult to occur.

The air mass for the combustion process is reduced rapidly as shown by low AFR numbers when the steam injection reaches 4.5 kg/ hour. This causes the combustion process to require more HSD fuel oil consumption to produce the same power compared to operating conditions with steam injection under 4.5 kg/hour and performance of standard engine without steam injection. Rich AFR values show higher fuel consumption to generate per unit of power generated. This condition will cause an increase in the SFC of the combustion process and reduced combustion efficiency.

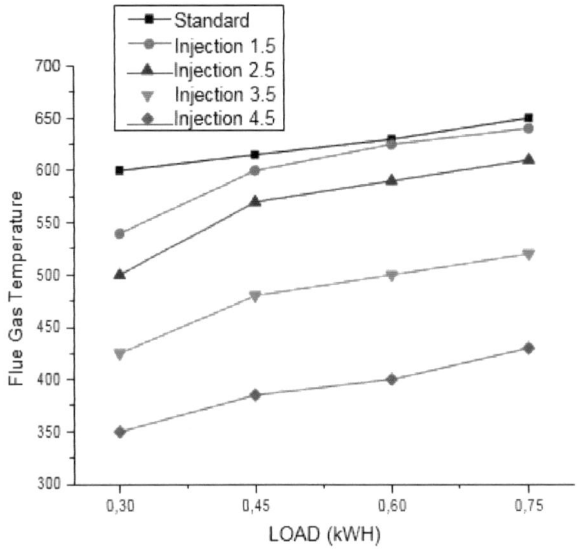

Figure 7. Comparison chart between the exhaust gas temperature and the load variation.

In standard machines without steam injection, conditions where there are low fuel and high amounts of air will result in high combustion temperatures. This condition does not occur in engines with steam injection system because the steam mass in superheated conditions can reduce the temperature of the combustion temperature without decreasing the power due to the increase of air enthalpy value in the combustion chamber. This gives an increase in the energy of the combustion process shown by figure 7 on the comparison gas temperature graph of the flue gas. Literature and previous studies show that combustion in diesel engines with lower combustion temperatures can reduce exhaust emissions released into the air [10,14-16].

3.4. Performance of Diesel engine with triple fuel (HSD oil + LPG) + Superheated steam injection from exhaust gas cogeneration

This mode combines two initial diesel engine modification studies with LPG (Bi-fuel HSD + LPG) and Steam (Bi-fuel HSD + Steam) injection systems at varying loads. Standard single-fuel diesel engine is modified with the addition of LPG injection system and superheated steam injection simultaneously to obtain SFC and AFR characteristics on engine load variation. In this test, LPG injection is kept constant 0.4 kg/hour and steam is maintained at a mass of 3.5 kg/hour on load variations performed from 1200 to 3000 watts. In this condition the machine is able to work normally and well with stable power according to the loading. In this mode, diesel fuel consumption is very small when compared to the standard engine at the same load. So it can be concluded that this system is able to work well with lower fuel consumption.

Table 4. Performance of diesel engines with the addition of triple fuel (hsd oil + lpg) + superheated steam injection from exhaust gas cogeneration.

Load (kWh)	HSD (litre/h)	LPG (kg/h)	STEAM (kg/h)	AFR	SFC (litre/kWh)
0,30	0,45	0,40	3,5	31	1,50
0,45	0,80	0,40	3,5	30,25	1,78
0,60	0,9	0,40	3,5	29,6	1,50
0,75	1,00	0,40	3,5	29,5	1,33

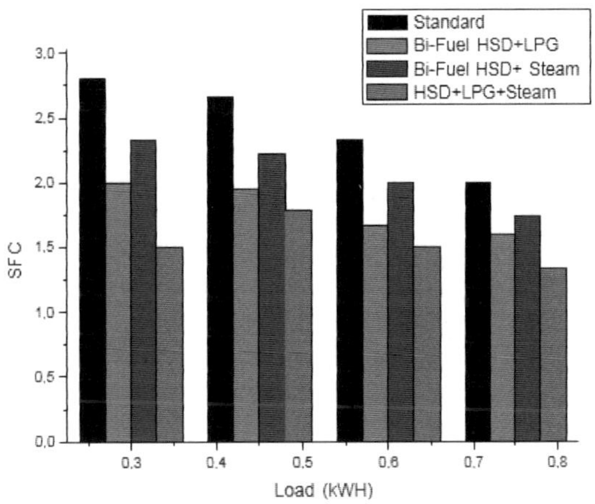

Figure 8. Comparison of SFC performance.

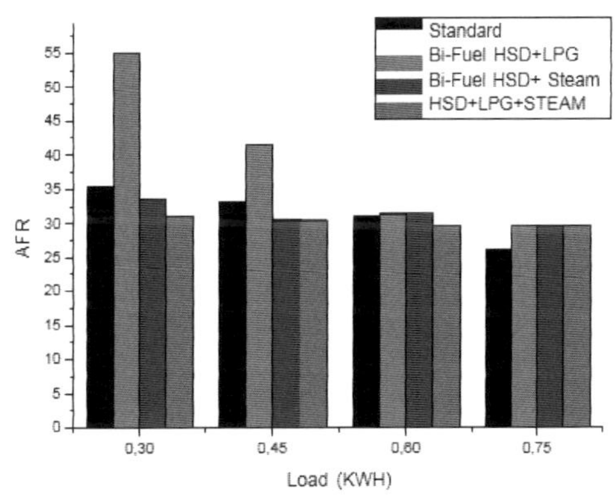

Figure 9. Comparison of AFR parameters.

4. Conclusion

Research on the performance of single cylinder diesel engine with LPG injection and superheated steam 130 °C by utilizing the recovery of wasted energy in an average temperature flue gas of 624 °C with cogeneration system proved to be one of the alternative energy saving breakthroughs. The conclusions that can be drawn from the research results are:

- Maximum performance was achieved at 1200 watt loads due to superheated steam phase and LPG added constantly to the engine. In addition to the load, the engine will automatically add fuel injection HSD oil mass to keep the engine can spend maximum power. The process is seen with the decreasing number of AFR (Air to Fuel Ratio).

- High calorific value of LPG mass injection at Bi-fuel fuel system is able to maintain AFR when load is added to the system so that engine efficiency is maintained.
- Superheated steam injection adds air enthalpy value inserted into the machine. The addition of heat from superheated steam and LPG will increase the temperature of air entering the engine and accelerate the preparation of combustion of diesel fuel. This condition increases the speed of HSD oil combustion so that the combustion pressure increases and improves SFC performance compared to standard engine mode.

Acknowledgment

This research was Supported by Kementerian Riset Teknologi Dan Pendidikan Tinggi Republik Indonesia–Indonesia. We thank our close partner in invention – Mr Himawan Candra Buana from Mashim Maximo – Yogyakarta.

References

[1] Samec N 2002 Numerical and experimental study of water/oil emulsified fuel combustion in a diesel engine *Fuel* 2002
[2] Bedford F, Rutland C, Dittrich P, Raab A and Wirbeleit F 2000 Effects of Direct Water Injection on DI Diesel Engine Combustion *SAE Pap* 2000
[3] Samec N, Kegl B and Dibble R W 2002 Numerical and experimental study of water/oil emulsified fuel combustion in a diesel engine *Fuel* **81** 2035–2044
[4] Abu-Zaid M 2004 Performance of single cylinder, direct injection Diesel engine using water fuel emulsions *Energy Convers Manag*
[5] Lin C-Y and Wang K-H 2004 Diesel engine performance and emission characteristics using three- phase emulsions as fuel *Fuel* **83** 537–545
[6] Kökkülünk G, Akdogan E and Ayhan V 2013 *Prediction of emissions and exhaust temperature for direct injection diesel engine with emulsified fuel using ANN* **21**
[7] Ishida M, Chen Z-L, Ejima K and Ueki H 1997 Prediction of NOx Reduction Rate Due to Port Water Injection in a DI Diesel Engine *Trans Japan Soc Mech Eng Ser B* **63** 3798–3805
[8] Tauzia X, Maiboom A, and Shah S R 2010 Experimental study of inlet manifold water injection on combustion and emissions of an automotive direct injection Diesel engine *Energy* **35** no 3628–3639
[9] Abd-Alla G H, Soliman H A, Badr O A and Abd-Rabbo M F 2001 Effects of diluent admissions and intake air temperature in exhaust gas recirculation on the emissions of an indirect injection dual fuel engine *Energy Convers Manag* **42** 1033– 1045
[10] Parlak, A, Ayhan, V, Üst, Y, Şahin, B, Cesur, I, Boru, B and Kökkülünk G 2012 New method to reduce NOx emissions of diesel engines: electronically controlled steam injection system *Journal of the Energy Institute* **85** 135-139
[11] Cengel Y A and Boles M A 2002 *Thermodyamics an engineering approach* 2002
[12] GE Kusuma M S 2014 Economic And Performance Analysis in Bi-Fuel System HSD Oil-LPG of Single Cylinder Diesel Engine *J Perkapalan* 1– 11
[13] GE Kusuma P S, Emie Santoso and Mardi Santoso 2017 Performansi Mesin CI Diesel Silinder Tunggal Dengan Injeksi Cogenerasi Superheated Steam in *MASTER - Maritim Sain Teknologi Terapan* 142
[14] Aydogan H 2015 Prediction of diesel engine performance, emissions and cylinder pressure obtained using Bioethanol-biodiesel- diesel fuel blends through an artificial neural network *J Energy South Africa* **26** 74–83
[15] Kökkülünk G, Akdoğan E and Ayhan V 2013 Prediction of emissions and exhaust temperature for direct injection diesel engine with emulsified fuel using ANN *Turkish J Electr Eng Comput Sci*

Experimental study of transmission ratio changing mechanism for motorcycle applications

B Supriyo[1,*], S Ariyono[2], S Sihono[3], B Sumiyarso[2] and B Tjahyono[2]

[1] Magister Terapan Teknik, Politeknik Negeri Semarang, Semarang, Indonesia
[2] Program Studi Teknik Mesin, Politeknik Negeri Semarang, Semarang, Indonesia
[3] Program Studi Teknik Elektronika, Politeknik Negeri Semarang, Semarang, Indonesia

*bsupriyo2015@polines.ac.id

Abstract. Transmission ratio change in today's motor cycle rubber belt Continuously Variable Transmission (CVT) are based on centrifugal forces resulted from engine revolution. Compare to manual transmission, this kind of CVT needs higher engine revolution to start engaging the transmision ratio which in turn requiring more fuel consumption. A transmission ratio change concept based on an Electro-mechanical actuator consisting of cam, actuator gears and DC motor has been developed to eliminate the dependancy of transmission ratio on engine revolution. The DC motor rotates the cam through actuator gears. The rotation of the cam causes the primary movable pulley sheave to move in axial direction, which in turn changing the belt-pulley pitch radius and changing the transmission ratio. This research aims to design and implement an electronic measurement system based on Arduino Uno as a data acquisition system and Matlab/Simulink as a controller program to obtain the measurement data of primary pulley and secondary pulley speeds and primary pulley axial position. Based on these data, the transmission ratio can be determined and the relationship between transmission ratio and primary pulley posistion can then be obtained. For future development, these measurement data will be used as important references when developing and testing the transmission ratio controller programs.

1. Introduction

Gear boxes, also called automotive transmissions, play an important role in motorcycles or cars because they regulate both speed and torque on the drive wheels. One form of transmission that is now favored both on motorcyles and cars is Continuously Variable Transmission (CVT). The basic CVT system mainly consists of primary pulley (input), secondary pulley (output) and belt connecting both pulleys to transfer speed and torque from input to output pulley. A common CVT system has one movable pulley sheave in one side and one fixed pulley sheave in the other side, forming V-shape pulley. CVT has a good drivability, smooth gear shifting without jerking and easy to drive. But in general CVT still less efficient than manual transmission, because of some losses inherent in the CVT itself, such as transmission loss [1,2] and speed loss [3].

In the early period many CVTs used rubber belts in the form of simple bracelets and cone systems, such as those developed by the Dutch company, DAF, in 1958 [4]. However, DAF's CVT was only able to be used on 0.6 L engines because heavy problems, noise and rough noise finally dropped its reputation [5]. In 1954, the Honda company introduced a mass production of Scooter to the market under the name Juno, a 0.175 liter engine scooter produced 8.8 kW [6]. Honda continued to produce small motorbikes with V-matic in 1977. Starting 1987 CVT system has been introduced for the first time in the market.

The current concepts of using centrifugal force systems on motorbikes and hydraulic force on sedan vehicles have not been able to adjust the transmission ratios accordingly so that the engine load is not at the optimum line [7-10]. In the centrifugal system, the transmission ratio of the CVT is governed by the centrifugal force caused by the engine rotation, while in the hydraulic system the ratio is regulated by control pattern through the hydraulic system. Another disadvantage in the hydraulic system is that the pulley moves to regulate the ratio of CVT whichs results in belt misalignment [11,12]. Both of these losses cause a decrease in efficiency of CVT hydraulics.

CVT is a transmission that is easy to operate both in a car and on a motorcycle. This transmission system is a very good transmission in improving fuel efficiency [13], since it is able to adjust the engine load on the optimum line if the transmission ratio is set properly. Several studies have been conducted to improve CVT efficiency, such as improving transmission mechanism and belt slip control [14], speed [3] and eliminating belt misalignment [15-17].

The transmission ratio changing mechanism for motorcycle application still uses the single actuated pulley movement concept, because the system still uses a flexible rubber belt, so that the rubber belt misalignment position in the middle of the two pulleys does not cause belt to easily break. This research will focus on the CVT transmission for motorcycle applications. It uses an electro-mechanical system consisting of DC motor as an actuator combined with a gear train set (called actuator gears) to move a cam mechanism attached to the primary moveble pulley sheave. This system is called Electro-Mechanical Single Actuated Rubber Belt CVT (EMSARB-CVT). The use of this cam is intended to axially move the primary movable pulley sheave such that the position of the belt contact point can move up or down on the surface of the pulley, changing the effective belt-pulley radius contact, hence changing the transmission ratio. In addition, the belt is clamped on the secondary pulley surface based on the spring force acting on the secondary movable pulley sheave, so that the belt can transfer speed and torque from the input pulley to the output pulley with no slip, then finally to the driving wheel of the motor cycle via the final gear drive.

This research will utilize Arduino Uno as a data acquisition system and Matlab/Simulink as a controller program to measure both primary and secondary pulley speeds, primary movable pulley position and compute the transmission ratio. Based on this data the relationship between transmission ratio and primary pulley position can be obtained.

2. Research materials and methods

2.1. Research materials

Research materials for this research consists of a mechanical set of electro-mechanical single actuated rubber belt CVT (EMSARB-CVT), primary pulley speed sensor, secondary pulley speed sensor, primary pulley position sensor, H-bridge DC motor driver, DC motor, Arduino Uno, 3 phase AC motor, 3-phase variable speed drive, computer with Matlab/Simulink program.

2.1.1. Mechanical set of EMSARB-CVT. The main parts of this mechanical of EMSARB-CVT consist of primary fixed pulley sheave, primary movable pulley sheave, primary shaft, secondary fixed pulley sheave, secondary movable pulley sheave, secondary shaft, secondary spring, rubber belt and cam mechanism with 4-gear train-set. The belt is placed and clamped inside the V-shape primary and secondary pulleys to transfer the speed and torque from the input (engine) to the output (wheel). The secondary spring makes sure that the belt always tights. For a basic CVT with fixed belt length and no belt-pulley slip, then the tangential velocity of both primary and secondary pulleys and belt are the same, such that the following (1) and (2) holds.

$$\omega p\, Rp = \omega s\, Rs \quad (1)$$

$$\omega p/\omega s = Rs/Rp \quad (2)$$

Where ωp is primary pulley angular speed, ωs secondary pulley angular speed, Rp primary belt-pulley contact radius and Rs secondary belt-pulley contact radius. If the transmission ratio ($rcvt$) is determined based on primary and secondary pulley radii then it can be represented by (3). But if the transmission ratio is determined based on primary and secondary pulley speeds then it can be represented as geometrical speed ratio (*rgs*) by (4).

$$rcvt = Rs/Rp \tag{3}$$

$$rgs = \omega p/\omega s \tag{4}$$

When the primary pulley gap increases, the primary belt-pulley contact radius will decrease, then the secondary spring will push the secondary movable pulley sheave such that the secondary pulley gap will decrease, the secondary belt-pulley contact radius will increase, hence increasing transmission ratio and decreasing the speed of the secondary pulley shaft. Inversely, when the primary pulley gap decreases, the primary belt-pulley contact radius will increase, then the belt on the secondary pulley sheaves will slide inward and push the secondary movable pulley sheave away and push back the secondary spring such that the secondary pulley gap will increase, the secondary belt-pulley contact radius will decrease, hence decreasing transmission ratio and increasing the speed of the secondary pulley shaft.

In this test rig the engine is represented by 3-phase AC motor which drives the primary pulley via a standard belt-pulley mechanism and the secondary shaft (output) does not have any load. The transmission ratio of this CVT can be varied by changing the cam position which in turn changing the effective radius of belt-pulley contact of the primary pulley.

2.1.2. 3-Phase AC motor and Variable speed drive. The 3-phase AC motor represents the engine for this system rig. The shaft of this motor is connected to the primary shaft using a standard belt-pulley mechanism. The speed of this motor is regulated using a variable speed drive. In this experiment, the speed of the primary pulley shaft is mantained to about 1000 rpm.

2.1.3. DC motor and driver. This DC motor is used to rotate the actuator gears which turn the cam mechanism to axially move the primary movable pulley sheave, then changing the belt-pulley contact radius and finally changing the transmission ratio. This DC motor is controlled by PWM signal from a PWM pin Arduino Uno via the H-bridge DC motor driver.

2.1.4. Primary pulley position sensor. The primary pulley posision sensor uses a 10-turn potentiometer to detect the rotational position of cam which also detects the axial position of the movable pulley and indirectly measures the primary belt-pulley effective contact radius. This primary pulley position measurement data is then correlated to the transmission ratio measurements derived from the speed measurements of both primary and secondary pulleys, using geometrical speed ratio equation in (4).

2.1.5. Speed sensors. There are two speed sensors based on 360 pulse/revolution (ppr) incremental encoders, namely the primary and the secondary speed sensors to detect the speeds (in rpm) of the primary and secondary pulley shafts. The number of pulse produced by each encoder for every one second time is converted to its respective angular speed in rpm. Based on these speeds, the real transmission ratio can be derived, in which using geometrical speed ratio equation in (4).

2.1.6. Arduino Uno. Arduino Uno is a very popular microcontroller board with inexpensive price and has good controller performance, especially for less complicated applications. The Arduino Uno is based on ATMega328, with 6 pins configurable for analog inputs and 14 digital input/output pins, in which 6 pins are configurable for pulse width modulation (PWM) outputs. Arduino Uno is used as a data acquisition system controlled by Matlab Simulink program via USB port to read the primary pulley position, to read primary and secondary shaft speeds and to drive the DC motor actuator for controlling the primary pulley position.

2.1.7. Computer. Computer is used as a main controller which is programmed using Matlab/Simulink to command the Arduino Uno via USB port to do the tasks of reading the primary pulley position, reading both primary and secondary shaft speeds and driving the DC motor actuator for controlling the primary pulley position. Data from the readings are then processed and displayed as graphs.

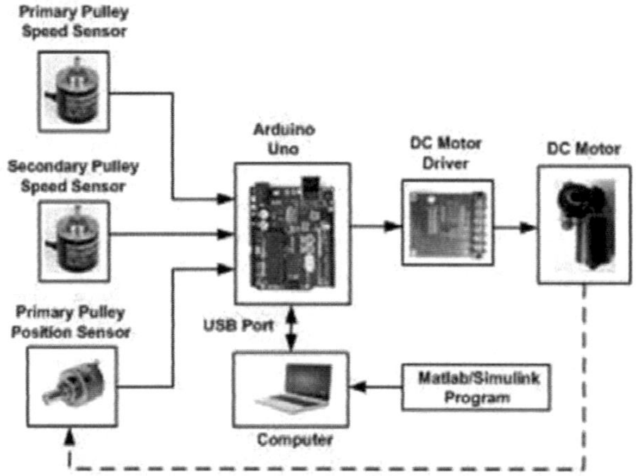

Figure 1. Electronic hardware diagram of EMSARB- CVT.

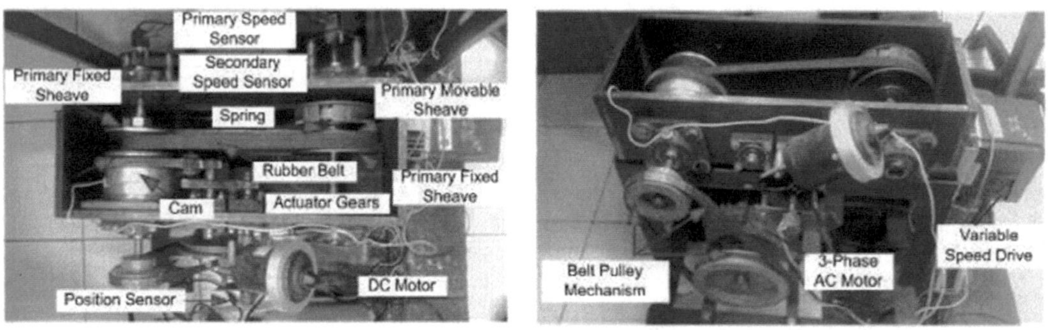

(a) (b)

Figure 2. Test rig (a). Top view (b). Side view.

3. Research methods

The overall research method will cover:

- Literature study related to the basc iconcept of how to control the transmission ratios in CVT.
- Design and develop electronic circuits for speed sensors, position sensor and DC motor Driver.
- Design and develop measurement program using Arduino and Matlab / Simulink programs
- Testing of an electronic circuit consisting of an input pulley and output pulley speed sensors, as well as primary pulley position sensor.
- Software testing involves the Arduino Uno microcontroller hardware, as well as the Matlab / Simulink software installed on the computer to program the measurement process of the transmission ratio.
- Testing the system as a whole to do the simple task of achieving the actual transmission ratio according to the transmission ratio set point.
- Data collection of test results and discuss its performances related to primary and secondary pulley speeds, primary pulley position and transmission ratio.

4. Results and discussion

4.1. Results

The results of these experiments consist of data measurements of primary pulley position, primary and secondary pulley angular speeds.

4.1.2. Relationship between transmission ratio and output voltage of primary pulley position sensor. The number of the DC motor shaft revolution is sensed using a multiturn potensiometer (10-rotation) acting as a position sensor which produces output voltage proportional to that of its respective number of revolutions. This motor shaft rotates the actuator gears which turn the cam and axially shift the movable primary pulley sheave, changing the belt-pulley contact radius hence changing transmission ratio. The actual transmission ratio is calculated based on the ratio between primary and secondary pulley speeds. The measurements of both primary and secondary pulley speeds are done using two rotary encoders attached on both pulley shafts. The graph of transmission rasio as a function of output voltage of position sensor can be seen in Fig. 3.

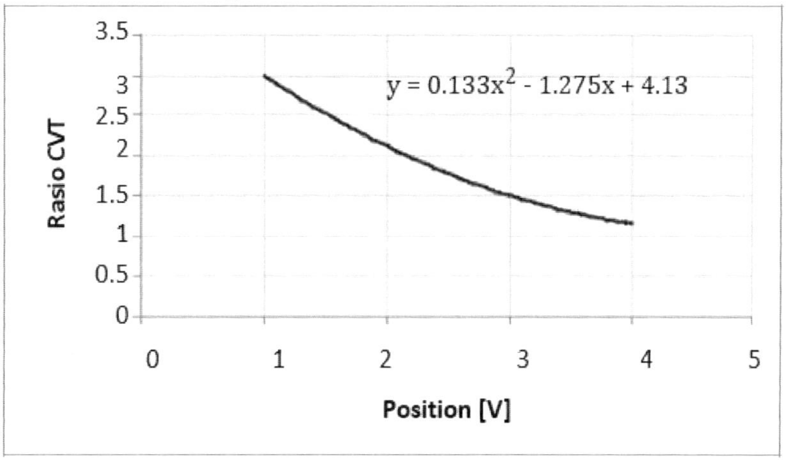

Figure 3. Transmission ratio as a function of position sensor voltage.

4.1.3. Relationship between secondary shaft speed and transmission ratio changes. If the transmission ratio is changed from about 3.0 (under drive) to 1.0 (1-to-1 ratio), as shown in Fig 4.a, then the output speed response from CVT gearbox (rpm out) with engine speed (rpm in) constant about 1000 rpm, can be seen in Fig 4.b.

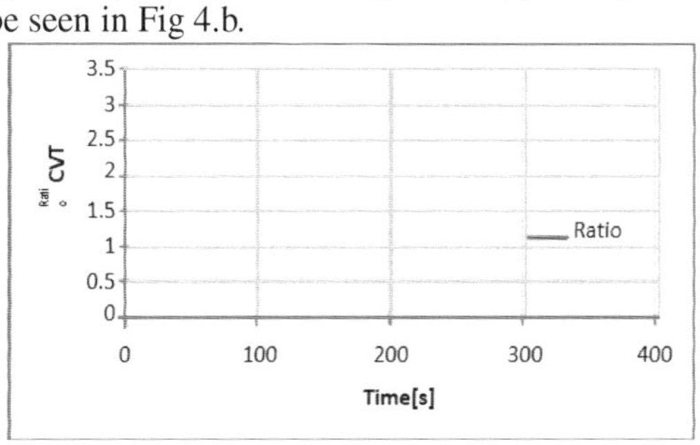

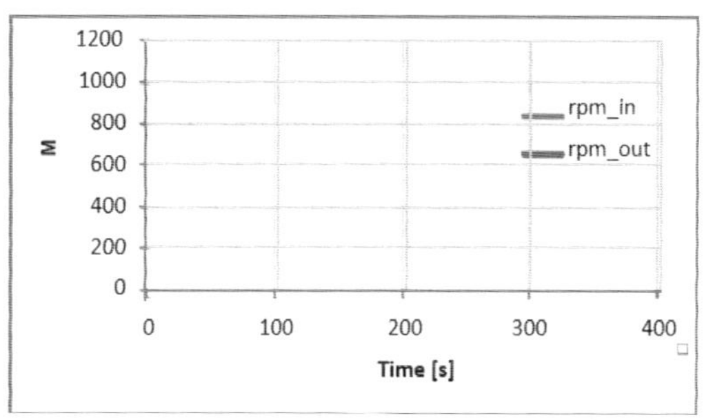

(a) (b)

Figure 4. CVT gearbox response: a) changes in transmission ratio and b) changes in output speed.

4.2. Discussions

Based on the graph in Fig 3, it can be seen that the relationship between the transmission ratio and the position sensor voltage is not linear but tends to be a second order polynomial. By using this polynomial equation, the actual value of the transmission ratio can be obtained by measuring the output voltage of the primary pulley position sensor. By monitoring this output voltage of the position sensor, the transmission ratio controller then can be developed to indirectly control the primary pulley position by actuating the DC motor as cam actuator, accordingly.

By setting a constant input speed of about 1000 rpm, the response speed of the output shaft of the CVT gearbox will change in accordance with the value of the transmission ratio. From Fig. 4. (a) and (b), it can be seen that for the lowest transmission ratio of 3.1, with an engine speed of 968 rpm, the output speed drops to 312 rpm. As for the highest transmission ratio of 1.1, with an engine speed of 935 rpm, the output speed drops to 860 rpm. From Fig. 4. (a), it can also be seen that the input speed (rpm in) which ideally should be constant for all values of the transmission ratio, it turns out that in practice it has gradually decreased slightly from 968 rpm (transmission ratio of 3.1) to 935 rpm (transmission ratio of 1.1). This is due to the increasing loading effect resulted from the increasing spring force exerted on output pulley sheaves, hence increasing the friction between rubber belt and both input and output pulley sheaves, especially when the output speed achieves its biggest value (transmission ratio of 1.1). In this case the AC motor used for power source representing engine does not have enough power to maintain a constant speed of 1000 rpm especially for transmission ratios of 1.1 which produces the highest output speed.

5. Conclusion

Experimental study of transmission ratio changing mechanism based on electro-mechanical rubber belt Continuously Variable Transmission on motorcycle has been carried out properly. The results show that the value of the transmission ratio can be indirectly obtained using a position sensor that uses a multiturn potentiometer (10-turn). The effective transmission ratio that can be detected is between 3.1 (lowest speed) to 1.1 (highest speed). The relationship between the transmission ratio and the position sensor output voltage is not linear, but it follows a second order polynomial equation. This equation is used to determine the value of the transmission ratio based on the output voltage of the primary pulley position sensor. The input speed (rpm in) of the CVT gearbox decreases gradually when the transmission ratio changes from 3.1 to 1.1 due to the increasing spring force exerted on the secondary pulley sheaves to clamp the rubber belt.

References

[1] Ide T 2000 *Effect of belt loss and oil pump loss on the fuel economy of a vehicle with a metal V-belt CVT* (No 2000-05-0130) (SAE Technical Paper)

[2] Akehurst S, Vaughan N D, Parker D A and Simner D 2004 Modelling of loss mechanisms in a pushing metal V-belt continuously variable transmission Part 1: torque losses due to band friction *Proceedings of the Institution of Mechanical Engineers, Part D: Journal of Automobile Engineering* **218** 1269-1281

[3] Balta B, Sonmez F O and Cengiz A 2015 Speed losses in V-ribbed belt drives *Mechanism and Machine Theory* **86** 1-14

[4] Diepstraten J F P B 2007 Modeling of pulley based CVT systems: extension of the CMM model with bands-segment interaction

[5] Birch S 2000 Audi takes CVT from 15th century to 21st century *Automotive Engineering International* **1**

[6] Casadesus-Masanell R and Heilbron J 2016 Decision-Making by Precedent and the Founding of American Honda (1948-1974) *Harvard Business School Strategy Unit Working Paper* (17-016)

[7] Song M, Oh J, Choi S, Kim Y and Kim H 2014 Optimal line pressure control for an automatic transmission-based parallel hybrid electric vehicle considering mode change and gear

shift *Advances in Mechanical Engineering* **6** 216098

[8] Kazemi R and Raf'at M 2014 Nonlinear Optimal Control of Continuously Variable Transmission Powertrain *International Scholarly Research Notices*

[9] Jha S and Gandhi R A 2015 Continuously Variable Transmission Control Strategy Review *International Journal of Engineering Research and Technology (IJERT)* **4** 1030-1034, 2015

[10] Alzuwayer B, Singh A, Muralidharan P and Han Z 2016 Model Based Pressure Control of a Push Belt Continuously Variable Transmission *Modern Mechanical Engineering* **6** 99-112

[11] Tawi K B 1997 *Investigation of belt misalignment effects on metal pushing v-belt continuously variable transmission* (Doctoral dissertation, Cranfield University)

[12] Faye Z and Zhihong W 2009 Control study on the CVT metal V-belt's axial-misalignment of car In *2009 IEEE Intelligent Vehicles Symposium* (pp 1389-1393) IEEE

[13] Tsuji T and Takeoka M 1996 Study of Fuel Consumption Improvement of the Car with the Dry Hybrid Belt CVT Society of Automotive Engineers of Japan *JSAE Review* **17** 381-385

[14] Bonsen B 2006 Efficiency optimization of the push-belt CVT by variator slip control *Dissertation Abstracts International* **68**

[15] Tawi K B, Husain N A, Kob M C, Supriyono B and Ariyono S 2012 Development of electro-mechanical actuation system for continuously variable transmissions in automotive applications In *Prosiding Industrial Research Workshop and National Seminar* **3** 62-68

[16] Supriyo B, Tawi K B and Jamaluddin H 2013 Experimental study of an electro-mechanical CVT ratio controller *International Journal of Automotive Technology* **14** 313-323

[17] Supriyo B, Tawi K B, Jamaluddin H and Hussein M 2014 Experimental Study of Electro-Mechanical Dual Acting Pulley Continuously Variable Transmission Ratio Calibration *Jurnal Teknologi* **71**

IOP Conf. Series: Materials Science and Engineering 1108 (2021) 012045 doi:10.1088/1757-899X/1108/1/012045

Application of CNC machine router 3-Axis for making of Engravired granite or marble

S Suharto[1,*], S Suryanto[1], S Sarana[2] and K Purbono[1]

[1] Mechanical Department, State Polytechnic of Semarang, Semarang, Indonesia
[2] Accounting Departments, State Polytechnic of Semarang, Semarang, Indonesia

*pakharto58@gmail.com

Abstract. The development of a Computer Numerical Controller (CNC) machine specifically for the manufacture of creative product crafts needs to be encouraged to improve the variety and quality of the products. Manual engraving of natural stone (marble, granite) by relying on the skills of people and simple equipment will produce a product with a long time of processing. Competition in the business and industrial world is needed to increase effectiveness and efficiency with short manufacturing time indicators, low prices, more and more interesting design variations. The research objective is to find effective and efficient machining parameters, analyze factors that influence surface roughness, and enrich the variety of creative product designs of marble stone materials. This study uses an experimental method approach which is a study to find the effect of spindle rotation variables, infeed speed, infeed depth on the quality of surface roughness by operating the **CNC Machine Router 3-Axis**. It was found that the following parameters of the machine are cutting speed machining 30 (m / min), spindle rotation 12000 (rpm), and speed / feed rate 2000 (mm / min). Analysis of factors that influence the surface roughness of marble engraver the higher the speed / rate of infeed and the depth of infeed, the higher the surface roughness value of marble. Computer and engineering software applications are able to increase the variety of creative product designs.

1. Introduction

The progress of science and technology today is so advanced, but some people still use the method of making products in a conventional way. Conventional mold making has a lack of productivity, manufacturing time and minimal design variations so that it has not been able to meet consumer demands. The use of 3-axis router CNC (computer numerical controllers) machine is one of the options to meet the increasingly high demands of consumers, both in terms of quality and quantity [1]. Today's era of industrialization has increased the need and use of CNC machines along with the demand for creative products of granite or marble. CNC machines routers are machine tools used for cutting, engraving and marking workpieces with rotating chisels on their axes, and cut surfaces in the form of flat, angular or curved shapes [2,3]. Previous research, namely the manufacture of aluminum batik stamp with CNC Router 3-Axis Machine by applying mastercam software was successfully done to enrich national batik motives [4]. Another study, Ardiansyah [5] in his research resulted in the effect of machining parameters on the surface roughness of teak wood material. Patel K.P. [6] analyzed surface roughness experiments on CNC milling 3-axis machining process with aluminum material. The research objective is to find effective and efficient machining parameters, analyze factors that influence surface roughness, and enrich the variety of creative product designs of marble stone materials.

According to Salam, A. Rohman [7] CNC machines are classified according to shape, spindle position, number of axes and spindle speed. Based on the size of the shape, CNC machines can be classified into 2 types, namely plano and gantry. Based on the position of the spindle, CNC machines can be classified into two types, namely horizontal and vertical types. The axis is a rotating shaft on the spindle and workbench. Based on the number of axes, the number of rotary shaft in CNC machines is divided into several types, namely 2 axis, 3 axis, 4 axis and 5 axis. Based on spindle speed, CNC machines can be classified into two, namely high speed machining and low speed machining [7]. The **CNC Machine Router 3-Axis** can be categorized as CNC machines with high spindle rotation speed or high speed machining.

The research objective is to obtain the optimal setting of CNC router machining parameters, analyzing the factors that influence the surface roughness of granite and marble stone. Optimal conditions for precise and time efficient machining parameters, surface roughness, security for cutting tools to enrich various creative gravired of the stone products in Indonesia. The CNC Machine Router 3-Axis machining parameters consist of cutting speed, spindle rotation, feed rate, and cutting depth. Cutting speed is a rotational speed of cutting tools expressed in meters per minute. The cutting speed depends on the workpiece material to be cut and the material from the cutting tools themselves. Spindle rotation is the ability to rotate the router CNC machine in one minute. The feedrate is the straight-distance feeding motion taken by the cutting tool at a constant rate relative to the workpiece in units of time (mm / min). Cutting depth (depth of cut) is determined based on the difference in thickness of the initial workpiece to the thickness of the final workpiece. The size of the depth of the cut is closely related to the speed of feeding and also the diameter of the cutting tool.

1.1. Software of CNC machine router 3-axis

The selectexd software is an important mean of operating the CNC Machine Router 3-Axis. The program implemented consists of various kinds of instructions from the software to carry out work orders. The things instructed by the software for breakout boards and drivers are the movement of stepper motors, spindles and relay settings, limit switches and emergency stops. The 3-axis CNC machine for making aluminum molds uses several supporting software before being executed by Mach3, namely CorelDraw, Aspire Vectric. Mach3 is a software found by Art Fenerty in 2001 and developed by ArtSoft. This software functions to implement all commands to the driver in running a 3-axis CNC router machine. This can be ordered by the Mach-3 software, namely the movement of the stepper motor (axis movement) in the form of G / M code, spindle rotation and speed, limit switch settings, relay settings and emergency stop settings. These commands are the basis for programming a 3-axis CNC router machine, making Mach3 software the main software in programming 3-axis CNC router machines [8]. CorelDraw is a computer program that performs editing on vector lines. This program was created by Corel, a software company headquartered in Ottawa, Canada. Coreldraw has uses for processing images. Therefore it is widely used in work in the field of publication or printing or work in other fields that require a visualization process. The superiority of CorelDraw mamou produces images produced with vector or vector based. CorelDraw is used as a tool for designing with higher quality results in collaborating between writing and drawing, especially when dealing with curves, lines or angles [9]. Aspire Vectric software is one of the programs used in CAD / CAM to provide creative industrial product manufacturing solutions by highlighting decoration and artistic. This combination of artistic features is often used in making products with 3-axis CNC router machines. The materials used make various kinds of creative products made from aluminum, copper, brass, wood, and acrylic resulting good products [8].

2. Methods

2.1. Location and time of research

The research location was conducted at Mechanical Engineering Laboratory of Semarang State Polytechnic, workshop of CV Retro Kreasi Machine in Semarang, and Erriesa Griya Print Pekalongan from April - October 2018.

2.2. Research materials and machinetools

The main research material: granite and marble obtained at Stone Gallery Semarang with various types of granite and marble available. Machine tool for this research is: CNC Machine Router 3-Axis with the following specifications: machine dimension: 1540x1005x1500 (mm); working area:

600x400x150 (mm); spindle motor: 3 (KW) 18000 (rpm); axis drive: stepper motor X = 8.7 (Nm), Y = 8.7 (Nm), Z =7.7 (Nm), tool holder: ER20 collet, max tool he. 13 (mm); Power: AC220V / 50 Hz, 3.5 (KW); Computer: Dualcore 2GB RAM, 15 "Touchscreen LCD Monitor; software: Mach-3 CNC control (www.machsupport.com). The machine completed with cooling system, clamping system, cutting tools for carbide type Engraving bits of 600 Tip angle of 6 mm diameter, and surface roughness tester. The implementation of granite and marble engraving with CNC router machines to improve the variety of creative products consists of stages, Figure 1:

- Identification of the needs of natural stone engraving (marble / granite) craft.
- The design of marble / granite stone engraving motifs is done using a computer with some coreldraw, aspire vectric, and mach-3 software.
- The granite / marble material is prepared at the CNC router working table clamped to the edges so that it does not move towards the X axis, Y axis, and Z axis.
- The 3-Axis router CNC machine used by local production.
- Machining tests are carried out with granite / marble stones measuring 300x250x10 (mm) with machining parameters to produce the best graphical recommendations.
- Evaluation to obtain information about the best machining parameters (spindle rotation, feed rate, cutting depth, type of cutting tools) is correlated with the quality of marble engraving and the safety of the CNC router equipment and machines.
- Application of machining parameters study for the manufacture of creative marble stone engraving products from the design of the Polines logo, the Higher Education Research and Technology Logo, and the Garuda Logo using materials measuring 300x250x10 (mm). Granite and marble engraving products are painted to show attractive products.

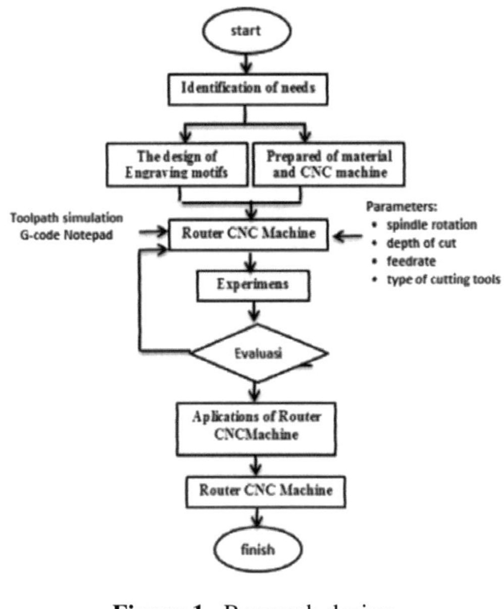

Figure 1. Research design.

This activity includes smoothing sharp corners, and the coating process on the engraved groove with paint. The final stage of the activity of marble stone engraving is compared with conventional methods starting from preparation, execution, to finishing stage. Figure 1 research design of marble stone engraving with CNC router machine.

3. Results and discussion

The experimental data collected in Table 1 was then analyzed to determine the effect of CNC milling machining parameters on the characteristics of optimal quality and condition from a combination of machining parameter levels. The higher the spindle speed, the lower the surface roughness value. Conversely, the lower the spindle speed, the higher the surface roughness value will be. The higher the feed speed / rate, the higher the surface roughness value will be. Conversely, the lower the speed / rate of feeding, the smaller the surface roughness value will result. Experiments were carried out with 300 x 250 x 10 (mm) marble type Ornamental marble using carbide type cutting tools, collant water cooling systems and CNC router machines. Machining parameters for 12000 spindle rotation (rpm), feed speed / rate 2000 (mm / min), plunge 480 (mm / min), pass depth 0.5 (mm), step over 0.12 (mm) produce fine quality marble stone engraving functional and security cutting CNC router machine tools.

The research team would like to thank the Ministry of Research, Technology and Higher Education for their financial support, Director of Semarang State Polytechnic, Head of P3M Polines, Student Team of Mechanical Engineering Department, of Semarang State Polytechnic, Manager of CV Retro Kreasi Machine, Semarang, and Erriesa Griya Print Pekalongan for cooperation so that this Institutional PSN program can run smoothly.

4. Conclusion

The results of the research that have been carried out can be concluded:

- Machining parameters of the CNC Machine Router 3-Axis for engraving granite and marble stones are: cutting speed 30 (m / min), spindle rotation 12000 (rpm), and feedrate / feedrate 2000 (mm / min).
- Analysis of the factors that influence the surface roughness of the first marble engraving is the speed/ rate of infeed, the depth of infeed means that the higher the value, the higher the surface roughness value of marble.
- Security of selected cutting tools of cutting tools of carbide type: Engraving bits of 60^0 tip angle, 0.5 mm diameter capable of finishing the marble logo polines at the time 01:36:22, logo ristekdikti 01:54:53, garuda logo 01:40:26 without replacing the cutting tool (of which a continuous liquid dromus cooler to keep cutting tools working optimally).
- Enrichment of creative marble engraving products can inniciatated with the input design of various photos on the corelDraw page into silhouettes, followed by setting Aspire vectric and Mach3 executions on CNC Machine Router 3-Axis.

References

[1] Suryanto S, Priyo V T, Sarana I H and Suwondo A 2014 Innovation of Making of the Batik Stamp Stand for Development of National Batik In Paper dalam *Seminar International: Engineering International Conference-Education, Concept and Application of Green Technology*

[2] Suharto S, Suryanto S and Sarana S 2016 Pengembangan Rancang Bangun Canting Batik Cap Berkualitas Biaya Murah In *Prosiding Sentrinov (Seminar Nasional Terapan Riset Inovatif)* **2**(1) 344-352

[3] Suryanto S, Priyo V T, Hermawan I and Suwono S A 2015 Pembuatan Canting Batik Cap (Cbc) Menggunakan Mesin Cnc Router Milling Berbasis Software Mastercam In *Sens*

[4] Suharto S, Suryanto S, Priyo V T, Sarana S, Hermawan I and Suwondo A 2014 Bahan Alternatif Pembuatan Canting Batik Cap (CBC) *Prosiding SNST Fakultas Teknik* **1**(1)

[5] Ardiansyah 2011 Pengaruh Parameter Pemesinan pada Kayu terhadap Kekasaran Permukaan. (Semarang: Skripsi UNDIP)

[6] Patel K P 2012 Experimental analysis on surface roughness of CNC end milling process using taguchi design method *International Journal of Engineering Science and Technology*

(IJEST) **4**(2) 540-545
[7] Salam A R 2013 *Proses Pembuatan Part RIB AT STN WFX 365.514 Komponen dari Sayap Pesawat Airbus A380* (Semarang: Laporan Kerja Praktek, Universitas Diponegoro)
[8] Afrilyan N 2015 *Pemodelan Mesin CNC Engraving Untuk Pembuatan PCB* (Fakultas Teknik Program Studi S-1 Teknik Mesin. Universitas Diponegoro. Skripsi)
[9] Irfan A 2014 *Belajar CorelDraw X4 Panduan Belajar dan Praktek* [Online] Modul CorelDraw. MAN 2 Banjarnegara. https://www.scribd.com/doc/89167017/Modul-Corel-Draw-X4.
[10] Getting Started With 3D in Aspire. Introduction to Modelling Ocean Drive House Sign Tutorial . (c) Vectric Ltd The Coach House Upper Skilts Farm Gorcott Hill, Beoley B98 9ET UK. [Online] support@vectric.com. www.vectric.com/support/training-material/aspire.html

Maintenance information system development for Powerhouse Department on diesel engine manufacturer

A S Ardjo[1,*], E D Wardihani[2], A Slamet[1] and M H Habiburrahman[1]

[1] Mechanical Department, State Polytechnic of Semarang, Semarang, Indonesia
[2] Electrical Department, State Polytechnic of Semarang, Semarang, Indonesia

*ardjo.anwars@polines.ac.id

Abstract. The Powerhouse Department on a diesel engine manufacturer has the role of supplying electricity to all departments, especially the production department. The problem that arises is the recording of all maintenance and repair activities carried out in the logbook. The purpose of this research is to develop a maintenance information system that can record all maintenance and repair activities that can be recorded electronically. The method used is Object-Oriented Development (OOD). The results of the research are data of toolst, parts & spare-parts, maintenance history & engine damage, maintenance scheduling documents, and vendor & supplier data. The system was implemented on this manufacturer.

1. Introduction

This research is carried out on an industry that produces diesel engine (tractor) for agricultural tools. This industry in 2016 produced 80,000 diesel engines and in 2017 produced 98,000 diesel engines. The resulting product consists of horizontal and vertical diesel engines with a power of 6.5-14 HP, namely: Corn Sheller, Combine Harvester, Rice Transplater, and Organic Fertilizer Processing Equipment. The running of the production process is also determined by the condition of the machine and tools, because basically the production will not run optimally if the engine condition is not optimal. This will affect production capacity and product quality. Maintenance of production tools must always be carried out as maximum as possible either planned or unplanned.

The Powerhouse Department is part of the Production Department responsible for electricity supply for the entire production process and handling the engine trouble that is in the diesel engine production line, as well as providing spare-parts for existing machines. The maintenance carried out at the Powerhouse Department is mostly applied after damage (Corrective Maintenance). The history record of engine damage and damage to the tools that is still spreadsheet-based has made it difficult to monitor engine damage, tools damage, or component replacement.

In connection with the matters above, basically the management of maintenance in the Powerhouse Department has been arranged according to the maintenance work procedures. But the implementation of maintenance management has not been carried out properly. In order to overcome the problem of maintenance management in the Powerhouse Department, it is necessary to have a maintenance information system that can integrate the work of overcoming engine problems, providing spare-parts, replacing components, inspecting the machine, and tools conditions.

The problem that occurs in the Powerhouse Department is: the history record of machine damage and tools damage is done manually into the recording log in the form of a spreadsheet. However, the recording log is rarely used, preventive maintenance reports are not carried out, technician actions are not reported to the section head, patrols at each line by officers are not carried out, tools inventory and spare-parts are not controlled.

The creation of this maintenance information system aims to improve the maintenance management in the Powerhouse Department. The system will make it easier to manage maintenance scheduling and provide maintenance recommendations on a machine.

2. Literatures review

According to Whitten and Bentley [1], information systems are the settings of people, data, processes and information technology that interact to collect, process, store and provide information needed to support an organization. According to Rochim [2], information systems can be interpreted as a unit of information elements, including how to design, activate, handle, maintain, and utilize information. The needs of maintenance information system will increment efficiency, security, reliability and availability, and will simultaneously reduce operational cost of maintenance too [3]. Maintenance must be carried out to replace, repair and service an identifiable set of manufacturing components. The need of maintenance is to maximize the availability of machinery and tools for production [4]. Good maintenance system must cover: preventive maintenance, corrective maintenance, and maintenance quality control [5]. CRUT model for information maintenance are connected to Control, Reliability, User Participation, and Training to each other. The element of control will lead to efficiency and determine the effectiveness of the systems. The elements of reliability, user participation, and training are included in the proper control of an organization's IS [6]. The goal of maintenance management procedure is the improvement of action based on data analysis from processes and indicators. This will improve overall efficiency in the service of an organization and avoiding costly downtime [7]. Attention to facility management determines competitive, because there is a reduction in costs and added value to the business init. Data from the department of maintenance in the plastic industry incorporated in a computerized maintenance management information system can provide information relevant to workers responsible for repairing machinery. Also management of the company can help them make the right decisions [8]. The management of maintenance is complex and has a significant impact on the profitability of the business, so computer-based supports for achieving successful implementation of these systems [9].

3. Methods and results

3.1. Time and place of research

This research was carried out in September - October 2017 at a joint venture diesel engine company located in Semarang, Central Java, Indonesia

3.2. The method

The method used in making this system is the method proposed by Britton and Doake [7], Object Oriented Development (OOD) which includes stages: Inception, Elaboration, Construction, and Transition. This system built with all of stages that Britton and Doake proposed. The result of each stages are:

3.2.1. Inception stage. This stage includes collecting data from interviews with one of the technicians in the Powerhouse Department. The results of data acquisition are related to machine and tools maintenance scheduling consisting of data: tools, parts & spare parts, engine maintenance & engine damage history, maintenance scheduling documents, Vendors and Suppliers.

3.2.2. Elaboration stage. This stage includes Modeling (UML) in the form of: Use Case Diagram (Figure 1), Class Diagram (Figure 2), Sequence Diagram (Figure 3), Collaboration Diagram (Figure 4), and Activity Diagram (Figure 5).

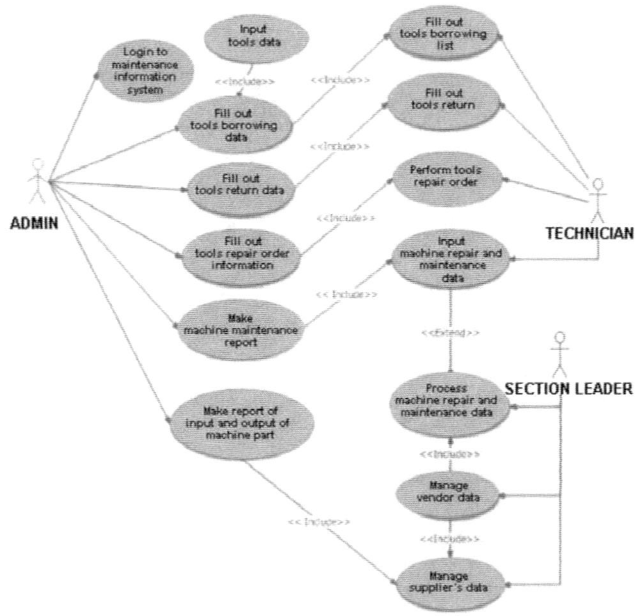

Figure 1. Use case diagram of powerhouse maintenance information system (PMIS).

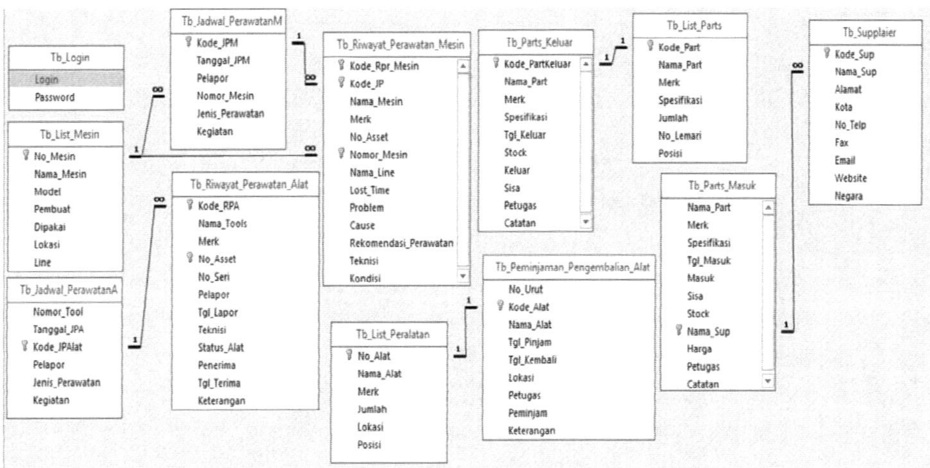

Figure 2. Class diagram of PMIS.

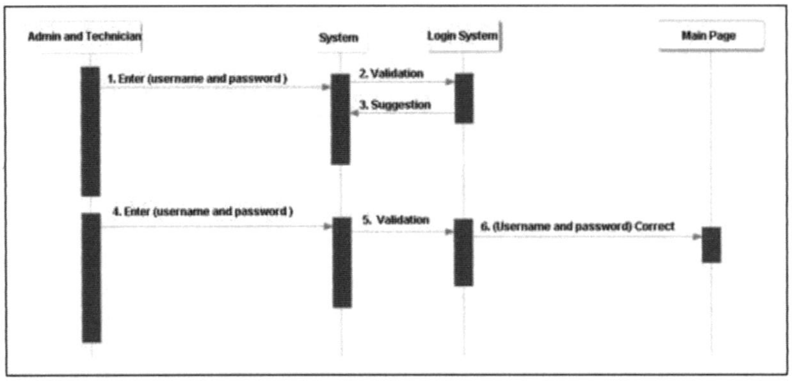

Figure 3. Sequence diagram of PMIS.

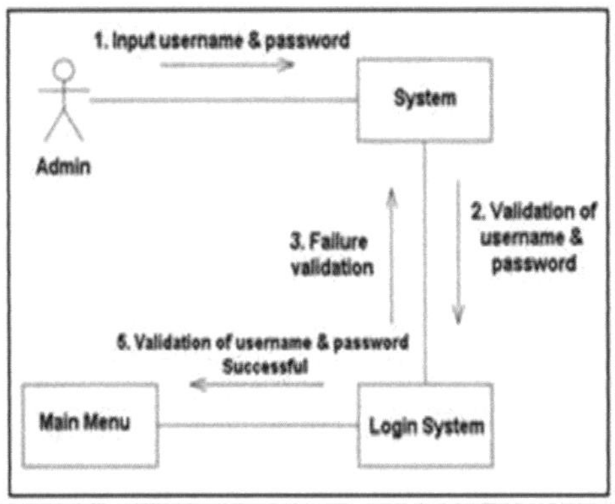

Figure 4. Collaboration diagram of PMIS.

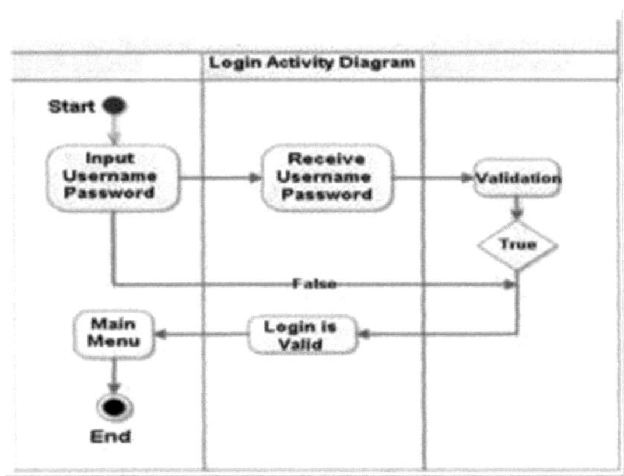

Figure 5. Activity diagram of PMIS.

3.3. Construction

This stage includes; The interface design between the user and the system is done by the user entering the name and password as shown in figure 6. The provided selection button consists of enter and cancel, then the next page will appear.

If the user logs in with a username and password as an admin, it will automatically be directed to process tools data as shown in figure 7, spare-part data as shown in figure 8, tools data as shown in figure 9, and data on tools borrowing and return as shown in figure 10.

If the user logs in with a username and password as a technician, it will automatically be directed to process machine maintenance data and tools repair as shown in figure 11.

If the user logs in with a username and password as a section leader, it will automatically be directed to process vendor data such as shown in Figure 12. and supplier data as shown in figure 13.

The test results from the maintenance information system in the Powerhouse Department on diesel engine manufacturer can be seen in table I.

Figure 6. Initial display design of PMIS.

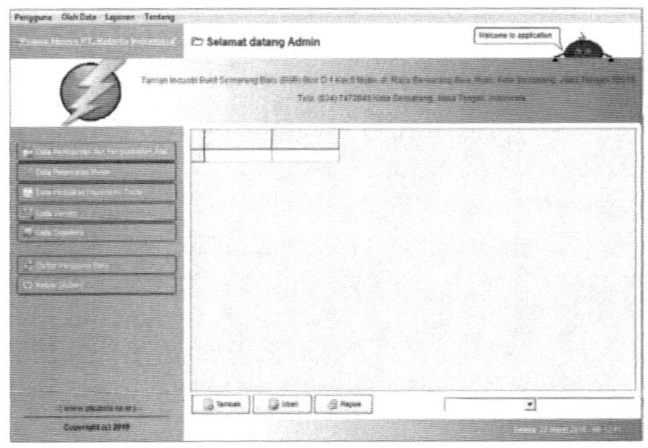

Figure 7. Initial display design of information systems.

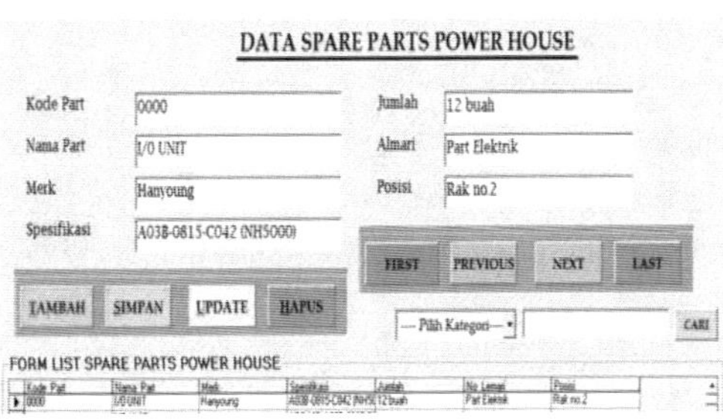

Figure 8. Design of powerhouse spare-parts data form.

Figure 9. Design of powerhouse tools data form.

Figure 10. Data form design for borrowing and returning tools.

Figure 11. Data form design for repairing and machining tools.

Figure 12. Supplier data form design.

Figure 13. Vendor data form design.

Table 1. Test results of powerhouse maintenance information system.

No	User	Test items	The result is corresponding to the design (yes / no)
1.	Admin	Fill out the borrowing data for the tools	Yes
		Fill out the return data for the tools	Yes
		Fill out the order to repair the tools	Yes
		Fill out the maintenance report for the machine	Yes
		Make a receipt and expense report for the tools	Yes
2.	Technician	Fill out the borrowing data for the tools	Yes
		Fill out the return data for the tools	Yes
		Perform tools repair order	Yes
		Enter machine maintenance and repair results data	Yes
3.	Section Leader	Process maintenance and repair data	Yes
		Manage vendor data	Yes
		Manage supplier's data	Yes
4.	All	Printing	Yes

3.4. Transition stage
Transition stage consist of these parts:

3.4.1. Installation. Implementation of installation is easy because the maintenance information system for the powerhouse department is built on a PC based on Microsoft platform.

3.4.2. User training. Implementation training for admins, technicians, and section leaders lasts for 5 days each 2 hours. Training results show that they can operate each function easily. Their task outside of training is to enter data from a spreadsheet.

3.4.3. Evaluation. Observation of daily tasks for 2 weeks shows no significant difficulties and disturbances have been found.

4. Discussion
During building a maintenance information system for the Powerhouse Department, it requires thoroughness in obtaining data. When doing data retrieval requires accuracy because the data is still in the form of information without using database principles, the fields in the table are still in the form of 1NF. Careful observation is needed of the process of recording activities by technicians and service providers and the use of spare parts. Interviews were carried out on the technicians and the section leader

of the maintenance division of the current repair were confirmed. Careful relational database design is needed, especially in determining the attributes of each table so that the separation of data from spreadsheets into normalized tables can be done easily. The process of designing a database table must be supported by query operations performed in front of the technicians and parts to obtain results from those whose query outputs do not match their needs.

Then a discussion of the initial design of the use case diagram was conducted to confirm the accuracy of the function and add functions that they did not know but later would be needed by them. After the use case diagram is agreed upon, we continue to design diagrams as can be seen in figure 3 to figure 13.

In the transition stage, the most time spent is when doing training, because it requires detailed discussion and explanation. Several improvements occurred in the query based on the results of the discussion during the training. The time for installation is relatively short, while the evaluation is carried out simultaneously with the time of death.

Result test of powerouse maintenance information system which includes 13 items in 4 sections all indicate that the output of the system is in accordance with the results of the discussion during all of the design phase. In general, the maintenance information system for the power generation department can be used correctly.

5. Closing

This system was successfully created and has been used as a tool for scheduling maintenance and repair of machine and tools in the Powerhouse Department in the diesel engine industry. The maintenance information system needs to be implemented in the company which includes maintenance or repair, and solving the problems that arise in order to avoid loss time that causes the production process to stop and leads to loss.

Acknowledgement

We would like to convey our gratitude to the Directorate General of Higher Education of the Republic of Indonesia for funding this research according to the letter of agreement no. SP Dipa-042.0.1.2.400997 / 2017 on December 7, 2016.

References

[1] Witten J L and Bentley L D 2007 *Systems Analysis and Design Methods* 7th penyunt (New York: McGraw-Hill)
[2] Rochim T 2002 *Information Systems* (Bandung: ITB)
[3] Mihanović L, Ristov P, Belamarić G 2016 Use of new information technologies in the maintenance of ship systems *Scientific Journal of Maritime Research* **30** 38-44
[4] Ramachandra C G, Srinivas T R and Shruthi T S 2012 A Study on Development and Implementation of a Computerized Maintenance Management Information System for a Process Industry *International Journal of Engineering and Innovative Technology (IJEIT)* **2**(5)
[5] Radoje K, Miodrag B, Darko P and Muharem S 2012 Aviation Maintenance Information System Configuration *Journal of Engineering Studies and Research* **18**(4)
[6] Urhuogo I, Vann V and Chandan H C 2005 Information Systems Maintenance: The Application of Total Quality Management Construct *Journal of Business Studies Quarterly* **3**(3) pp 1-15
[7] Abreu J, Martins P V, Fernandes S and Zacarias M 2013 Business Process Improvement on Maintenance Management: A Case Study. CENTERIS 2013-Conference on ENTERPRISE Information Systems / PRojMAN 2013 - International Conference on Project Management / HCIST 2013 - *International Conference on Health and Social Care Information Systems and Technologies. Procedia Technology* **9** (2013) 320 - 330
[8] Eva S and Kateřina M 2013 The Effect of Implementing a Maintenance Information System on the Efficiency of Production Facilities *Journal of Competitiveness* **5**(3) 60-75
[9] Wienker M, Henderson K and Volkerts J 2015 The Computerized Maintenance Management

System An essential Tool for World Class Maintenance "SYMPHOS 2015" *3rd International Symposium on Innovation and Technology in the Phosphate Industry. Procedia Engineering* **138**(2016) 413-420
[10] Britton C and Doake J 2005 *A Student Guide to Object-Oriented Development* (London: Elsevier Ltd.)

Android based maintenance information system for machining workshop in polytechnic high education

A S Ardjo[1, *], E D Wardihani[2] and P Prayitno[2]

[1] Mechanical Engineering, State Polytechnic of Semarang, Semarang, Indonesia
[2] Electrical Engineering, State Polytechnic of Semarang, Semarang, Indonesia

*ardjo.anwars@polines.ac.id

Abstract. There are six state polytechnics aged 36 years old in Indonesia. The six polytechnics have machining workshops with the same number and machines. The problem that arises in Semarang State Polytechnic is the maintenance system that has not been integrated into a system, even though the electronic maintenance and repair activities have been carried out. The purpose of this study is to develop a maintenance information system that can control all of its activities well. The system is built with an object-oriented approach implemented by android. Research results include: database systems, web application systems, and android application systems that have delivered output correctly.

1. Introduction

Vocational higher education in Indonesia with the names of polytechnics in the cities of Medan, Palembang, Jakarta, Bandung, Semarang and Malang has been running since 1982. One of study programs includes mechanical engineering. This study program has the same type and number of machine tools, such as: lathes, milling machines, surface grinding machines and CNC machines that are used as practical facilities. The use of these machines is more in the form of non-productive practicum with a usage load up to 14 hours/day. The characteristics of the maintenance carried out are routine maintenance in the form of inspection, replacement of lubricants, cleaning, and replacement of spare-part. Characteristics of improvement are incidental, that is, if damage occurs, repairs will be carried out. If the handling of damage is beyond the ability of the existing technical staff, then the work will be tendered to the parties outside the campus. This thing definitely will disturb the student practicum activities. Technical staffs who handle the maintenance of machine tools are not experts, but are constantly getting training.

The machine tools maintenance and repair in Machining Laboratory of Mechanical Engineering Department Semarang State Polytechnic, has the following characteristics:

- There is no machine manual,
- Maintenance and repair are done based on the requests that users upload via the web. The web application is a type of general interference complaints, not specifically for logs maintenance and machine repair.
- There is no database record, the record is in the form of manual notes based on the work statement form that have been completed.
- HR (Human Resource) expertise is limited by training and certification efforts.

The conditions of maintenance and repair activities such as this require improvement to meet the principles of modern maintenance and repair management. Table 1 shows the machine tool data owned by the Machining Laboratory of the Mechanical Engineering Department Semarang State Polytechnic.

Table 1. Types of machines in Polines machining laboratorys.

No	Class of Machine	Brand	Type	Units	Year of Prod/Use	Conditions
1	Center Lathe	Modiale 81	Celtic 14	4	1978/1982	Function
2	Center Lathe	EMCO	Maximat V13	12	1982/1982	Function
3	Vertical Drilling	Wilson	VSG 20	1	20012001	Function
4	Vertical Drilling	Rockwell	420M	2	1981/1982	Function
5	Vertical Drilling	Rockwell	420M	8	2013/2013	Function
6	Vertical Drilling	Arfa	RF-31	2	2003/2003	Function
7	Vertical Drilling	Krisbow	KW1500	2	2014/2014	Function
8	Universal Drilling	Aciera	F3/F4	9	1981/1982	Function
9	Surface Grinding	Prazision Swerkzeug	Compact 600	2	1981/1982	Function

Figure 1. Machining laboratory of Polines.

Based on the information contained in the background, the following problems can be identified:

- There is absolutely no database relating to machine data, maintenance and repair activities, or maintenance schedules.
- No maintenance plan before damage occurs (mean time before failure),
- No estimates and stock of spare-part that are predicted to be often damaged or will soon be damaged.
- Not applying maintenance and repair system management.

Problems set in this study are: What kind of Android-based maintenance and repair information system needed by the Machining Laboratory of Mechanical Engineering Department in polytechnics above?

The purpose of this study is to build an Android-based maintenance and repair information system needed by the Machining Laboratory of Mechanical Engineering Department in polytechnics above.

2. Literature

Choiril et al. create Preventive Maintenance Information for Small and Medium Industries using the SDLC (System Development Life Cycle) method and have tables: Mesin (kode_mesin, nama_mesin, lokasi_mesin), spareparts (nomor_spareparts, nama_spareparts, lokasi_sparepats), pegawai: id_ pegawai, nama_pegawai, jabatan), jadwal (jadwal_inspeksi, cleaning, pelumasan, overhaul) [1].

Husain et al. made Computer-Based Information Systems for the Management of Industrial Facilities in Ship Manufacturing using the SDLC approach. The built system is used in the shipyard industry which has 6 workshops with 162 machines. The built system is equipped with a warning facility, i.e. when the activity schedule reaches the implementation date, a warning code will appear [2].

Lubis et al. examined the Effect of Implementing Office Equipment and Machine Maintenance Information Systems on Efficiency. The Maintenance Information System is built with Visual C Sharp and uses a MySQL database. Information on maintenance of Office Equipment and Machining is stored in maintenance history which includes: date of incident, damage, replaced spare-part, length of maintenance and costs. In addition, the system also provides information on the frequency of machining and office equipment maintenance. The method is SDLC [3].

Melladya et al. developed a Group-Technology-based Machine Maintenance Management Information System Design for coachwork industries in the method of Prototyping. The tables include: data_umum_mesin (depertemen, jenis, merek,tahun), hirarki_komponen (komponen, sparepats), atribut_inventori (lokasi, letak, nomor) [4].

Salim built a Production Machine Maintenance Information System for the steel pipe industry. The system was built using the SDLC method, a system that provides facilities to manage the production of machine maintenance information in a simple and fast manner, reporting breakdown production machine, a function to display the schedule of monthly maintenance activities, meet the need for data maintenance and breakdown in more detail, ease access, system protection, ease of making reports, and easy data search [5].

William built a Computer-Integrated Machine Maintenance System at Crumb Rubber Factory with SDLC approach. The tables are: jadwal_perawatan, data_mesin, data_komponen, perbaikan, daftar_komponen_stok, suppplier, aktivitas [6].

The development of the maintenance information system (Si-Main) uses the approaches: use case / class-object, etc.), programming language (Java), database (Firebesa), based on Android platforms.

3. Materials and research methods

This research was carried out in in the machining workshop of Semarang State Polytechnic using Object Oriented [7,8] method. Developing stage of Maintenance information system (Si-Main) aims to provide an overview of what is done and produced at each stage until it is complete. The data acquisititon were: observing part of maintenance and repair, ensure maintenance and repair part & users are fully covered, compile application specifications, either software or hardware, making UML diagrams, preparing the program code, prototyping, and testing. The installation is done by uploading Si-Main through the Google Play Store and downloading via smartphone. Training is a way to socialize all users. Implementation is the time when Si-Main can be used fully and independently.

3.1. Inception system

The first step taken at the inception stage of the system is to identify problems by observing the repair maintenance unit and machining workshop users. After that, validation is done by ensuring that the needs of the repair maintenance unit and machining workshop users are fully covered. The final step is to compile specifications for both software and hardware. The design of a maintenance information system architecture consists of:

- Online database server that provides data processing capabilities and can be accessed through Android and Web application platforms.
- The android application is needed for technicians because of the flexibility of entering data in each lab. Checker / Technician can enter Maintenance & Repair data.
- The Si-Main web application provides the feature to view data as a whole or report data that has been entered in the Android application.

The design is seen as follows:

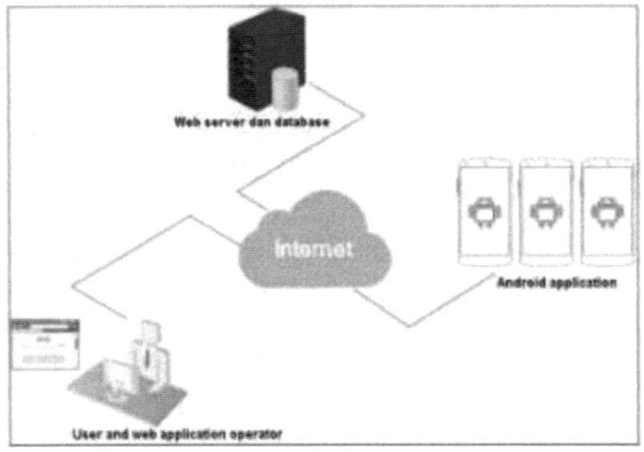

Figure 2. Architecture diagram of SiMain.

3.2. *Elaboration system*

The elaboration stage / design of the maintenance information system that will be built consists of the design *use case diagram, entity relationship diagram (ERD) and Graphical User Interface (GUI).*

3.3. *Use case diagram design*

Figure 4 shows the use case diagram design which consists of 2 actors, namely admin and technician and 3 large use cases (Maintenance Data, Machine Data, Spare-part Data and Report Print)

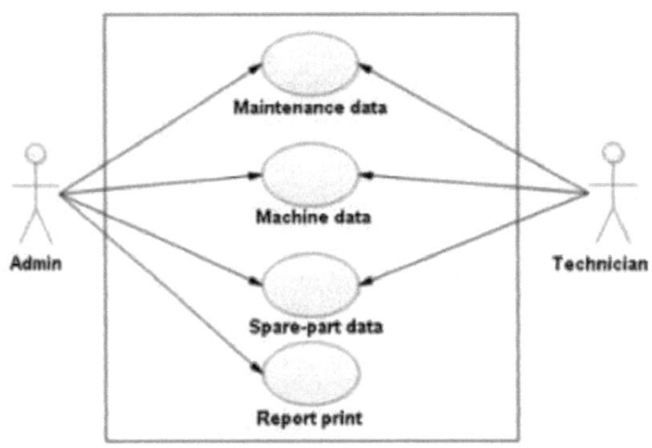

Figure 3. Use case of SiMain.

This design is based on the interaction of actors with the application system made in this study, with descriptions as in the following table.

Table 2. Description of design systems use case diagram.

Use Case	Actor	Specification
Maintenance data	- Admin -Technician	Admin and technician can perform an operation to save, modify, search and delete machine maintenance data. The automatic system gives notification to technicians about the maintenance schedule that will be carried out next.
Machine Data	- Admin -Technician	Admin and technician can perform operations to save, modify, search and delete machine data and there is a maintenance information system application.
Spare-part Data	- Admin - Technician	Admin and technician can perform to save, modify, search and delete spare-part data in the application system.
Report Print	- Admin	In this feature, only admin who can perform report print operation in the form of maintenance, spare-part and machine data. Data report print can only be done through a computer due to the large computer resources.

3.4. Entitiy Relationship Diagram (ERD) design

Before the data in the machining lab is translated into database tables, it is necessary to design ERD in advance, ERD is used to model each the relationship between data and relationship between ERD in the machining lab. Machining lab entity has several machines that contain code information, name, brand, vendor etc. Each machine has several maintenance and repair records such as the repairing officer, maintenance description, treatment date and duration of treatment entered by the technician. When performing maintenance, spare-part information in the form of codes, names, suppliers and spare-part specifications are entered into the system.

Machine

MCode	Name	Brand	Vendor	Spec
Type	SNumber	ProdYear	OpsYear	MTimeToFailure

Maintenance:

Worker	Descrpt	Date,	Duration

Spare-Part

PartCode	PartName	Supplier	PartSpec	Stock

Technician

Username	Password

Database Relation	Relation Cardinality
Machine to Maintenance	1 to Many
Maintenance to Technician	Many to 1
Maintenance to Spare-part	Many to Many

Figure 4. Entity relationship diagram of SiMain.

3.5. System construction

At the construction system stage there are 3 stages of application construction: (1) Construction of the database, (2) Construction of Web Application and (3) Construction of Android Application.

The software database uses MariaDB 10.1.24-MariaDB-cll-lve - MariaDB Server version 10 which is installed on a shared hosting web server. The construction of a maintenance information system website application uses Laravel 5.4 php framework programming language and CSS bootstrap Framework.

The construction of the dashboard web page can be accessed in the url http://simain.silemper.xyz/dashboard or the start page after the user enters a username and password.

The construction can be viewed on http://simain.silemper.xyz/machine or access the machine menu on the left. The machine maintenance data page construction results can be seen when accessing the machine data. On the machine data page there is a menu to access maintenance information according to the machine code listed.

The machine spare-part data web page construction can be found on http://simain.silemper.xyz/sukucadang or access the parts menu on the left. Spare-part page is taken from spare-part table data on the database server.

3.6. Report construction

The results of the report page construction can be seen on the page http://simain.silemper.xyz/ or access the report menu on the left. The code snippet to display parts data can be seen in Figure 5

IOP Conf. Series: Materials Science and Engineering 1108 (2021) 012047 doi:10.1088/1757-899X/1108/1/012047

```
<div class="box">
  <div class="box-body">
    {!! Form::open(['class' => 'form-horizontal']) !!}
    <div class="form-group">
      {!! Form::label('tanggal_perawatan', 'Tanggal Awal', array('class' => 'col-md-2 control-label')) !!}
      <div class="col-md-6">
        {!! Form::date('tanggal_awal', \Carbon\Carbon::now(), array('class' => 'form-control'), '') !!}
      </div>
    </div>
    <div class="form-group">
      {!! Form::label('tanggal_perawatan', 'Tanggal Akhir', array('class' => 'col-md-2 control-label')) !!}
      <div class="col-md-6">
        {!! Form::date('tanggal_akhir', \Carbon\Carbon::now(), array('class' => 'form-control'), '') !!}
      </div>
    </div>
    <div class="form-group">
      <div class="col-md-1 col-md-offset-2">
        {!! Form::button('<i class="fa fa-print"></i> Cetak Perawatan', ['class' => 'btn btn-success', 'type'=>'submit']) !!}
      </div>
    </div>
    {!! Form::close() !!}
  </div>
</div>
```

Figure 5. The system report web page code snippet to display spareparts data.

The report page displays a print menu for the report of machine and its maintenance. Report data is displayed in ready-to-print PDF form.

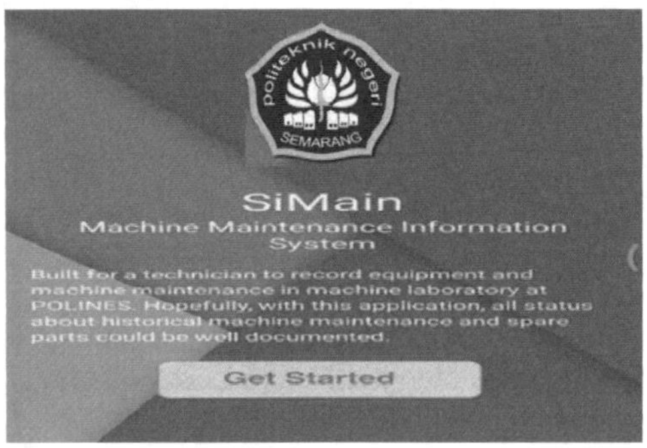

Figure 6. Home page of SiMain Android application.

As shown in Figure 6. the front page of the android application contains information about the SiMain maintenance information system application. The results of the notification of maintenance schedules for each machine and stock of spare-part are displayed on the dashboard page. Android programming implementations use the WebView component to display data on certain web addresses.

3.7. Android application construction

In the application construction stage, the interface design which becomes an Android application program is created by using Android studio software 2.2.2.

The specifications targets are Android Marshmallow SDK 6.0 and Android API 23. The construction of the Android application consists of 4 main pages namely the application splash-screen front page to display information, notification dashboard page, Machine Data and Maintenance Page. The SiMain application can be accessed at the addresshttps://play.google.com/store/apps/details?id=id.web.proditipolines.imimain.

4. Conclusions and suggestions

4.1. Conclusions

The Android application for the Machining Laboratory of the Mechanical Engineering Department Semarang State Polytechnic can be built by using software development method of Object Oriented. Application features are in the form of summarized machine maintenance schedule information, machine and spare-part data in a dashboard that making it easier for technicians to see and control machine maintenance.

4.2. Suggestion
Based on input by mechanical engineering lecturers and technicians of UPT PP (Research and Development Unit) Polines, suggestions that can be used in the development of subsequent research:

- Development of automatic reporting by machines using IoT technology,
- UPT PP (Research and Development) technicians welcome this application and are asked to develop similar applications for different workshops in the scope of Polytechnic.

Table 3. Testing result of siMain.

Maintenance Information System (SiMain)			Output Testing (Y/N)
A. Database System	Processing data on Android and Web plaform		Yes
B. Web Application System	B1. Notification Dashboard	Retrieves data from information board web server and displays television	Yes
	B2. Machine Data	Provides machine maintenance information and is able to add, search, modify, and delete the machine data	Yes
	B3. Machine Maintenance Data	Provides machine maintenance information and is able to add, search, modify, and delete the maintenance data	Yes
	B4. Spareparts Data	Provides machine maintenance information and is able to add, search, modify, and delete spare-part data	Yes
	B5. Report	Displays the machine data report, print feature,prints spare-parts data, and maintenance report	Yes
C. Android on System Application	C1. Notification Dashboard	Display maintenance schedule information and spare-part stock	Yes
	C2. Machine Data	Displays machine data information, adds machine data and prints machine data reports	Yes
	C3. Machine Maintenance Data	Displays machine maintenance data information prints machine maintenance report data	Yes

References

[1] Choiril, Mochamad, Purnomo Budi Santoso and Arif Rahman 2011 Design of Information System Software of Preventive Maintenance for Small and Medium Industries *Proceeding National Seminar on Industrial Engineering & Congress of BKSTI VI 2011* pp vi - 99

[2] Husain, Ulil A. Fida, Sri Rejeki Wahyu Pribadi and Mohammad Sholikhan Arif 2016 Design of Computer Based Information Systems for Maintenance Management of Ship Manufacturing Industry Facilities *Journal of ITS Engineering* **5**(2)

[3] Lubis, Muhammad Amrin, Isteti Murni, Mh. Arfansyah 2010 The Effect of Office Equipment and Machine Maintenance Information Systems on Efficiency *Journal of Edik Informatics in Computer Science and Education Informatics Field Research* V3.i1 (8-17)

[4] Melladya, Raena, Purnomo Budi Santoso, Mochamad Choiri 2010 Design of Group Technology-Based Machine Maintenance Management Information Syste

[5] Salim, Aminuddin 2010 Design of Production Machine Maintenance Information Systems (Study Case in Technical / Maintenance Division PT Indonesia Steel Tube, Ltd, Semarang)

[6] William 2013 Design of Computer-Integrated Machine Maintenance System at Crumb Rubber in PT. Hadi Baru

[7] Britton, Carol and Jill Doake Jill 2005 A Student Guide to Object - Oriented Development, Burlington MA: Elsevier Butterworth-Heinemann. pp.7, 8 & 15

[8] Shoval, Perez 2007 *Functional and Object-Oriented Analysis and Development: An Integrated Methodology* (London: Idea Group Publishing) p 128

IOP Conf. Series: Materials Science and Engineering 1108 (2021) 012048 doi:10.1088/1757-899X/1108/1/012048

Participatory rural appraisal strategy in realizing the standardization of ship components

A Khoryanton[1,*], P Pratikto[2], S Suparman[2] and P B Santoso[2]

[1] Mechanical Engineering Department, Politeknik Negeri Semarang, Semarang, Indonesia
[2] Mechanical Engineering Department, Faculty of Engineering, Universitas Brawijaya, Malang, Indonesia

*ampala.Khoryanton@gmail.com

Abstract. The role of ships for Indonesia is very important as a trigger for economic development and inter-island liaison. In line with the Indonesian government program in realizing the sea toll program and as the ruler of the world maritime axis, the ship needs will increase rapidly then the shipping industry becomes the main foundation in supporting this program. However, the development of the shipping industry in Indonesia is still hampered because ship components are still largely imported. As many as 70% of ship components must be imported from abroad. While many domestic SMEs of ship components have not been standardized, so they have not been able to support the shipping industry. The purpose of this study is to formulate participatory rural appraisal strategy in realizing the standardization of ship components. The results of the research indicate that there are five main factors influencing the success of product standardization: function and performance of Local Working Group Platform, the involvement of strong certification institutions, availability of technical expert and supporting institutions, shipyard in Evaluation Monitoring. The standardization program undertaken by SMEs of ship component is very important to support the shipping industry in Indonesia.

1. Introduction

The world's shipping industry is currently controlled by China, Japan, and Korea, which controls the market for up to 90% of total ship orders worldwide [1]. As a maritime country with the longest coastline in the world, Indonesia should have the potential to compete in the world's shipping industry. But since the monetary crisis hit, the shipyard utilities in Indonesia are declining and the Indonesian shipbuilding industry increasingly difficult to compete in the world shipping industry [2].

There are several major shipyard problems in Indonesia according to the Biro Klasifikasi Indonesia [3]. among others: Capacity and productivity of national shipyard are still low, Facilities and equipment in many shipyards are old, domestic shipbuilding ship price is relatively more expensive 10% to 30%, weak local banking support, about 70% to 80% component and ship material still to be imported. The most dominant main problem is the high dependence on imported materials and components which account for 70 percent of the total cost of shipbuilding [4,5].

The importance of the availability of component and material raw materials made the government make various efforts to encourage the growth and development of ship component industry in the country to be able to supply the needs of the shipyard industry according to the standard [6]. Seeing the large potential of ship component requirement, small and medium of ship component business actors (SMEs of ship component) take the initiative to standardize ship components. Standardization of ship components is an action plan from the community as a ship component business. This standardization action plan was triggered by the problems of market development in supporting the needs of domestic shipyards and facing free trade.

This is in line with Soetrisno's opinion that the effort of standardizing products is very relevant to SMEs in Indonesia, which until now face the problem of competitiveness, especially to face the global market [7].

The purpose of this study is to formulate participatory rural appraisal strategy in realizing the standardization of ship components. The objective used to achieve the objective is to identify existing ship component products and identify factors that influence the success of ship component standardization. Because with standardization can increase market confidence in ship component products [8].

2. Methodology

Standardization will be easier to build based on the wishes of the community as actor of ship component business [9]. Participatory Rural Appraisal (PRA) approach method is very appropriate because SMEs of ship component can participate in realizing the plan, implementation, supervision, and evaluation of product standardization program [10,11]. With the community involved in the program process, it can increase community awareness in running the program. The PRA method is described as a technique that allows the community to participate in realizing the plans [10], implementation, monitoring, and evaluation of policies that have an impact on the life of the community.

To be able to determine which products will be developed as standardization products, it is necessary to identify products that already exist in the market. The product to be surveyed may be more than one product depending on how many products are in the market. Furthermore, training and assistance in the development of product standardization. The steps are illustrated in the Figure 1:

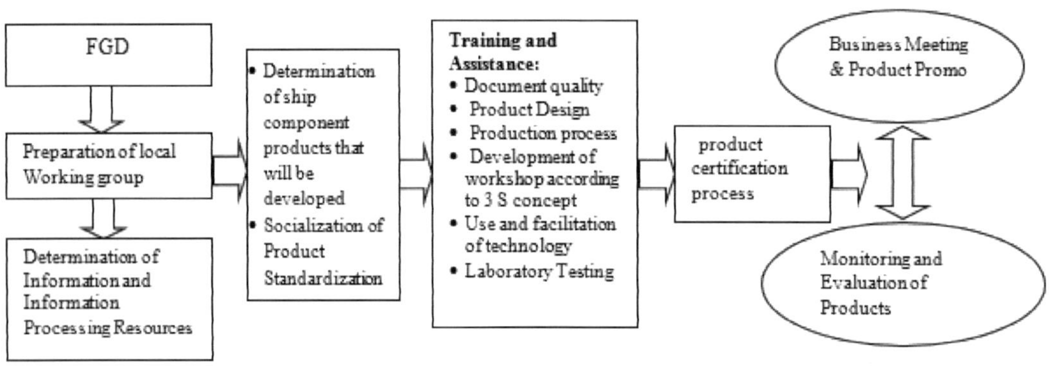

Figure 1. The steps are illustrated.

3. Results and discussions

3.1. Local Working Group (LWG)

Some representatives of ship component business actors need to conduct Focus Group Discussion to formulate action plans. The formulation of an action plan starts with determining the main objective of standardizing ship component products by describing what products to develop and which markets to target. This is done to request support from stakeholders.

Furthermore, to realize this standardization plan, LWG is set up to function as a platform for SMEs of ship component to plan and facilitate the standardization plan and mobilize resources to achieve a challenge of standardization process together. LWG is tasked with designing and scheduling sequentially activities in accordance with the standard certification process and facilitating standardization plan with related institutions. As LWG support data needs to conduct value chain analysis, utilization of ship component product technology, and product standardization socialization.

To determine which component of ship product to be standardized, it is necessary to identify the product including the availability of raw materials, the ability of the ship production process and its market share. The surveyed products of more than one product are Side scuttle, Square window and Weathertight door as seen in Figure 2 with the following considerations:

- Raw materials, i.e. aluminum scrap, aluminum profiles, steel plates, steel profiles and other
- supporting materials such as rubber seals, tempered paint, and glass are all available in the area of SME locations of ship components,
- The ship's component SME has the technology and production process capabilities to the product.
- Has an excellent market potential because it is required in large quantities in a ship.

The results of field identification indicate that the ship component product produced by ship component SMEs can be illustrated in table 1. Product Identification Result of ship component.

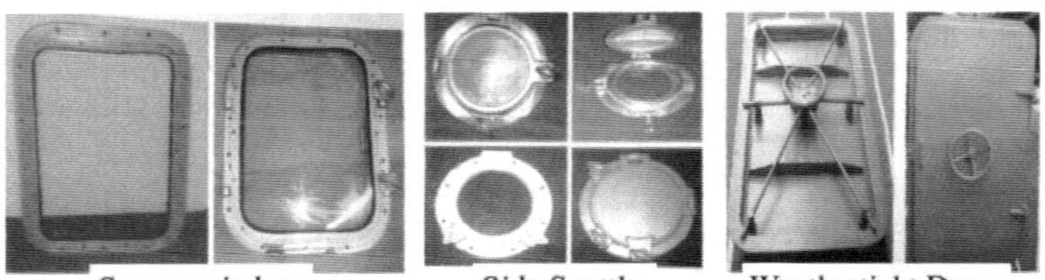

Figure 2. Products of ship components to be certified.

Table 1. Product Identification Result of ship component.

Product	Raw materials	Production process	Description
Side scuttle	Aluminum scrap, rubber seal, tempered glass	Casting, machining, assembling	The composition of casting material is not standardized and not uniform
Square window	Aluminum profile, rubber seal, tempered glass	Machining, assembling	Product dimensions and quality are not uniform
Weathertight door	Steel plates, steel profiles, rubber seals	Cutting, machining, assembling	Product dimensions and quality are not uniform

3.2. Involvement of certification institution

Certification institution such as the Indonesian Bureau of Classification is involved since the standardization planning stage as a source of information through socialization related to the standardization process of ship components. Bureau of Classification Indonesia also plays a role in achieving standardization through the provision of technical guidance and advice related to the manufacture of prototypes and the preparation of documents required for certification. Therefore, when facilitating the industry for certification and development of new markets, agencies with certification authorities should be involved from an early stage.

3.3. Availability of technical experts and supporting institutions

The availability of competent local experts can provide appropriate advice and follow-up services is another key to success. a local expert for metal-working, mobilized by the local government, can make regular visits to SMEs of ship component and provide appropriate advice in the prototype and testing process. The support of government agencies with adequate infrastructures such as Ferro and non-ferrous material testing laboratories, engineering drawing studios, and modern machining workshops also contribute to the successful achievement of ship component quality standards. They have responded to the issues faced by SMEs flexibly. The technical guidance includes the preparation of quality documents, product design, production process, workshop development according to 3 S concept (sort, set in order, and shine), preparation of quality documents, use and facilitation of technology and laboratory testing. As illustrated in Table 2 below:

Table 2. Technical guidance on product standardization.

product	Product design	Production process	Description
Side scuttle Square window Weathertight door	• Design the product as per the quality document • Design Testing of materials and products as per quality documents	• The flow of production processes is simpler according to the 3S concept • SOP as per the quality document	The product is certified by the Indonesian classification bureau

Product development strategy is determined to improve the quality of an existing product. The following results are submitted for quality standards for ship components: non-water tight door (SNI 7361, 2007), Weathertight door for small boat (SNI 7362, 2007), shipbuilding and sea construction ship side window (SNI ISO 1751, 2007), Weathertight one-sided steel door (SNI ISO 6042, 2007), ships Side Scuttles (ISO 1751,1993), shipbuilding and marine structures ships, heated glass panels for ship rectangular window (ISO 3434, 1992), shipbuilding and marine structures ship' ordinary rectangular windows (ISO 3903,1993).

3.4. Strong contribution SMEs of ship component

There is no doubt that the willingness and commitment of ship component SMEs are indispensable in the case of standardization of ship component products. SMEs who participated in this product standardization plan took a strong initiative related to the certification process, namely;

- Development of workshop according to the 3S concept,
- Long-term continuous experiments for prototype manufacturing and material/product tests that meet ship component quality standards,
- The movement of resources to prepare raw materials/labor for the manufacture of prototypes.

3.5. Monitoring and evaluation

The next challenge is how SMEs of ship components enter the global market. The Ministry of Industry facilitates business meetings between ship component SMEs who have successfully obtained product certification with the shipyard. Another promotional tool is to include IKM Components of ships at the international trade fair. While monitoring and evaluation were done by shipyard and Bureau of Classification Indonesia together. When the ship component is installed in the ship unit, it must be evaluated for conformity with the quality standard.

4. Conclusion

The success of facilitation activities of SMEs development of shipping components in the achievement of product quality standards of ship components is determined by four main factors, i.e.:

- The function and performance of the LWG Platform
- Strong involvement of certification institutions
- Availability of technical experts and supporting institutions in the regions
- Contribution and strong willingness of ship component SME to standardize the product
- Monitoring and evaluation

References

[1] Nolan P 2002 China and the global business revolution *Cambridge Journal of Economics* **26**(1) 119-137
[2] Prasetyo T 2018 Analysis of Industrial Developmentof DomesticShip Components *Maritime Technology Scientific Journal WAVE* **10**(2) 39-46
[3] Biro Klasifikasi Indonesia 2015 *Role of BKI in Certification and Standardization of Marble-Based Industry* (Jakarta: Kementerian BUMN)
[4] Windyandari A 2008 Prospect of Domestic Shipbuilding Industry To Face Global Competition *The technique* **29**(1) 73-76
[5] Asmara I P S, Santoso M, Julianto E and Praharsi Y 2016 Development of Innovation Cluster for Improvement of Local Content of Shipping Industry In *Seminar MASTER PPNS* **1**(1)
[6] Hidayat T 2016 *Development of Surabaya Ship Industrial Competitiveness Strategy Using Yardstrat Formulation Model* (Doctoral dissertation, Institut Teknologi Sepuluh Nopember)
[7] Soetrisno N 2005 SME Clustering Strategy In Indonesia: An Integrate Development Support: Improving The Competitiveness of SMEs Through Enhancin Productive Capacity *Proceedings of Four Expert Meetings* 131-137
[8] Rao-Nicholson R and Khan Z 2017 Standardization versus adaptation of global marketing strategies in emerging market cross-border acquisitions *International Marketing Review* **34**(1) 138-158
[9] Hussain A and Khan S 2013 Internationalmarketingstrategy: standardization versus adaptation *Management and Administrative Sciences Review* **2**(4) 353-359
[10] Narayanasamy N 2009 Participatory rural appraisal: Principles, methods, and application *SAGE Publications India*
[11] Campbell J R 2001 Participatory rural appraisal as qualitative research: distinguishing methodological issues from participatory claims *Human Organization* **60**(4) 380-389

IOP Conf. Series: Materials Science and Engineering 1108 (2021) 012049 doi:10.1088/1757-899X/1108/1/012049

Hybrid power plant of the photovoltaic-fuel cell

Y D Herlambang, T Prasetyo, A Roihatin, Y M Safarudin and F Arifin*

Dept. of Mechanical Engineering, Politeknik Negeri Semarang, Semarang, Indonesia

*farifinus@polsri.ac.id

Abstract. The purpose of this research is to study the performance of hydrogen fuel cell to study different fuel flow rate and different concentration to obtain the optimum performance of the cell. First, design a hydrogen fuel cell model, second simulate cell performance toward hydrogen fuel flow rates of 6, 18, and 30 mL/min and fuel concentrations of 50 mol/m^3. Subsequently, we obtained the characteristics of voltage-electric current density and power density-electric current density per stack of fuel cell. The research of this hydrogen fuel cell obtained the power density optimum of 0.0471 mW/cm^2 at a cell current density of 0.135 mA/cm^2 and a cell voltage of 0.35 V. The greater the load the greater the time used by the battery for instant loads. The average time when the load is small is 10 W with a time of 240 hours and the fastest time when a large load is 400 watts with a time of 6 hours. Decreasing battery discharge time to the same load due to the completion of power losses in each device or circuit that produces the power generated by the battery does not reach the maximum load.

1. Introduction

Fuel cells or more commonly known as "fuel cells" are new and renewable energy sources that are being researched and are becoming research topics in the last decade. Fuel cells are an environmentally friendly energy source because they do not cause pollutants and can be used renewably, have a wider application range, are easy to control, have no interference if integrated into the system, and can be combined as a microscale hybrid power generation system [1-3]. Research on proton exchange membrane-type fuel cells (PEMFC: Proton Exchange Membrane Fuel Cells) both using gases and liquid solutions as fuels has been widely carried out [4,5]. This shows that the potential of fuel cell-based power plants has big opportunities and wide applications in various fields, such as medicine, computers, telecommunications, and so on. But research related to micro-scale fuel cells is still very little and has not been done much in Indonesia. The technology of proton-type micro-scale cell fuel membrane proton exchange has been widely patented and researched. However, the average of these patents and research only examines the structure and electrode design and PEMFC performance at certain temperature variations [6-8]. While the parameters of volumetric flow rates, fuel concentrations, generated voltage-currents and numerical simulation models of species transfer and momentum transfer in fuel cells, are rarely studied [9,10]. For this reason, it is necessary to develop a numerical simulation model of these parameters in proton-type membrane fuel cells (PEMFC), with optimum performance characteristics. This research is focused on proton-type membrane cell type (PEMFC) fuel cells oriented to optimum performance, both model design and numerical simulation prior to the micro-scale PEMFC fabrication process, so that PEMFC models will be produced whose components are easily obtained at low cost and easily duplicated.

.

IOP Conf. Series: Materials Science and Engineering 1108 (2021) 012049 doi:10.1088/1757-899X/1108/1/012049

2. Research methods

2.1. Design of PEM fuel cell model

This study carried out through two stages, namely the design of Proton Exchanger Membrane Fuel Cells and numerical simulations of fuel cells. The Proton Exchange Membrane Fuel Cell (PEMFC) Fuel Cell Model has a channel dimensions of 200mm x 10mm x 10mm.

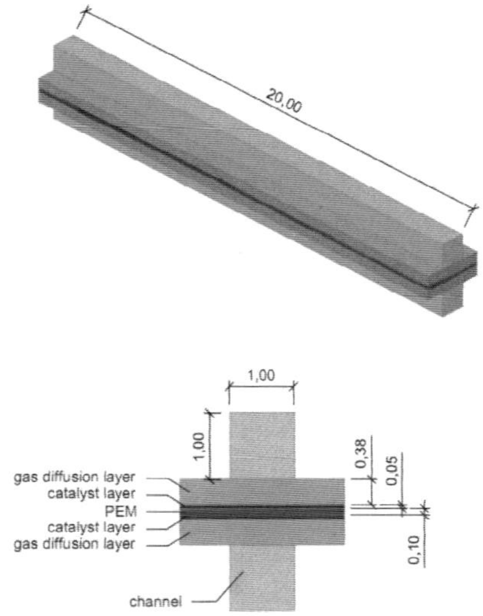

Figure 1. Design of PEM fuel cell stack.

Both the anode and cathode electrodes have dimensions of 2.75 mm x 200 mm, with the distance between the electrodes being 0.1 mm. The two electrodes are placed on the bottom side of the microchannel. Figure 1 illustrates the PEMFC model which consists of seven main parts namely the anode-side micro fuel channel, the gas diffusion porous layer on the anode side, the catalyst layer on the anode side, the polymer thin membrane layer, the cathode-side catalyst layer, the cathode-side porous gas diffusion layer, and the cathode side fuel microchannel. Miniaturization of the fuel cell with the structure and design of the PEMFC unit stack is shown in Figure 1.

2.2. Numerical simulation of the fuel cells

Hydrogen fuel cells enter through the anode inlet side, while oxygen will flow into the microchannel on the cathode inlet channel side. Both gases will react electrochemically to the surface of the anode and cathode catalyst layer. The hydrogen gas ions (H +) will flow through the membrane and move from the anode side to the cathode side, while electrons (e-) will move to flow from the cathode side to the anode side. Electrons from the negative anode pole will flow through the bipolar collector plate through the external circuit to the positive pole of the cathode to generate an electrical load. Fuel cell stack models for micro-scale PEMFCs can be seen in Figure 2 and Figure 3.

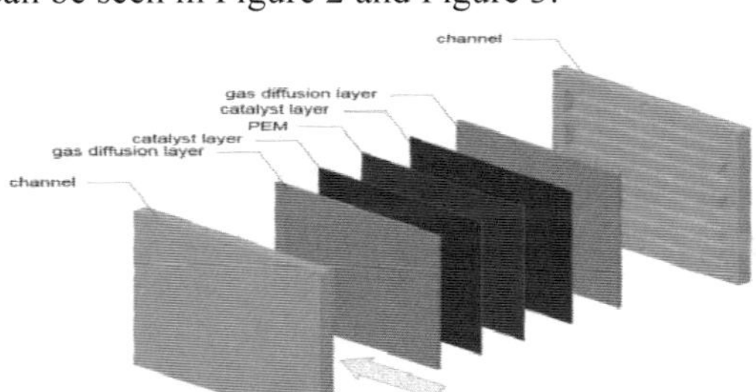

Figure 2. Single cell stack model for micro scale PEMFC.

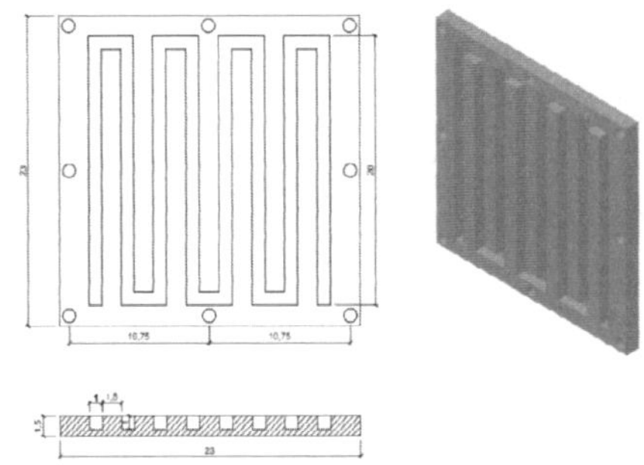

Figure 3. Microchannel serpentine model for micro scale PEMFC.

Numerical simulation based on the 3-dimensional element finite method is implemented in COMSOL Multiphysics 5.1. The 3-dimensional model of the fuel cell is designed with COMSOL to solve the fluid transport phenomenon that is the equation of the flow of gas fluid distribution in the micro channel, the equation of fluid flow passing through the porous media [9], the diffusion and convection equation, the electric charge equation according to the electrochemical reaction that occurs on the surface of porous electrodes and catalysts. This numerical simulation will discuss the performance of proton exchange membrane fuel cells related to the current and voltage polarization curves generated, including overpotential on the electrodes, distribution of fuel and oxidant concentrations, and distribution of fluid velocity on micro channels. In addition, the performance of numerical simulation models is compared with experimental data as a reference to validate that the performance of models is appropriate and accurate.

3. Results and discussions

3.1. Fuel cells calculation analysis

This study to obtain optimal fuel cell performance by varying 3 pieces of fuel flow rates namely 6, 18, and 30 mL/min as shown in Figure 4, Figure 5, and Figure 6. Characteristics of fuel cell voltage and Electric current can be calculated using equation [3,6]:

$$E_{cell} = E_{OCV} - (\eta_{a,a} + \eta_{a,c}) - \eta_r \quad (1)$$

Where E_{ocv} is the open-circuit voltage in the experiment that can be obtained from the characteristic curve IV according to the load flow rate of a given fuel, it can be seen in Figures 4, 5, and 6. The fuel flow rate greatly affects the performance of the fuel cell, this is related with the electro-oxidation reaction rate of the fuel on the anode side and the electro-reduction reaction of the oxygen rate on the cathode side, as given in equation [6,11,12]:

$$S_i = \left(\frac{i_0}{n_i F}\right)\left(\frac{c_i}{c_{i,ref}}\right)^{\beta_i}\left[\exp\left(\frac{\alpha_a F\eta}{RT}\right) - \exp\left(-\frac{\alpha_c F\eta}{RT}\right)\right] \quad (2)$$

Where i_0 is the change in current density in the fuel cell (exchange current density) at the concentration of the fuel $c_{i,ref}$, the order of reaction to the species "i" as a reference step for the charge transfer in the cells α_a and α_c is the charge transfer coefficient at the anode and cathode, η is over-voltage, T is the operating temperature of the cell, F is the Faraday constant, and R is the universal gas constants. Oxygen supply from the surrounding air penetrates through the gas diffusion electrode (GDE) before reaching the catalyst layer at the cathode. Through the boundary conditions approach and entering values into equations (1) and equation (2), the oxygen concentration that passes through the GDE can be determined.

In the fuel cell electrochemical reaction, the electrochemical reaction always takes place on the upper surface side of the electrode, in this case the catalyst layer on the anode side and the catalyst layer on the cathode side. As it is known that, the anode and cathode electrodes are both made of a porous carbon layer which is coated with a thin layer on the side of the catalyst particles. This catalyst layer is tested as a boundary condition, which corresponds to the electrochemical reaction that occurs on the surface side of this boundary condition (catalyst layer). So that from the electrochemical reaction on the surface of the catalyst layer seen at the concentration of species as known as the source (source term), the term source term in the species conservation equation is [6,7],

$$S_i = \nabla \cdot \left(- D_i \nabla c_i + c_i \vec{u}\right) \tag{3}$$

Here S_i is the rate of consumption of "i" species per cubic meter is caused by electochemical reactions along the surface of the catalyst layer on the anode and cathode sides. D_i is the diffusion coefficient of the species "i" for hydrogen gas on the anode side and oxygen on the cathode side according to the porous media at the electrode. Species diffusion coefficient values for the reaction at the anode and cathode are shown as the diffusivity of hydrogen gas and oxygen diffusivity which passes through the gas diffusion electrode (GDE) made of carbon layers.

On the surface along the catalyst layer, only the normal flux of the electrode and fuel surface influences and contributes to the fuel cell's output power as written in equation below [6,7],

$$\vec{n} \cdot \nabla \cdot \left(- D_{ij} \nabla c_i + c_i \vec{u}\right) = S_i \tag{4}$$

The porous structure of the anode and cathode electrodes has an effect on the binary diffusion coefficient used in the electrode as a porous medium to be corrected for electrode porosity (ε) and turtuosity or surface roughness at the electrode (τ) as follows [6,7],

$$D_{ij}^{eff} = D_{ij} \frac{\varepsilon}{\tau} \tag{5}$$

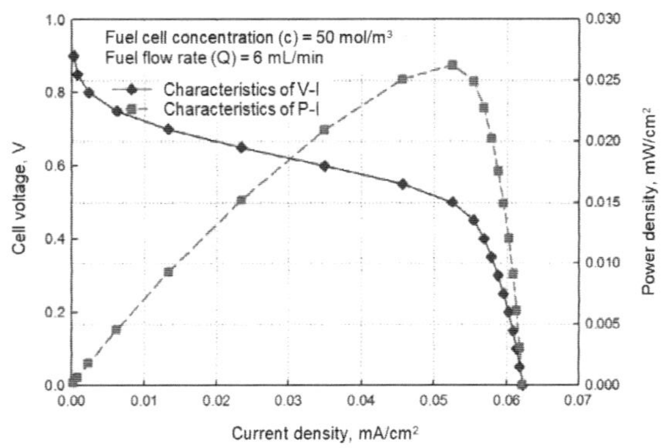

Figure 4. Characteristics of PEMFC with a hydrogen fuel concentration of 50 mol/m^3 and a fuel flow rate of 6 mL/min.

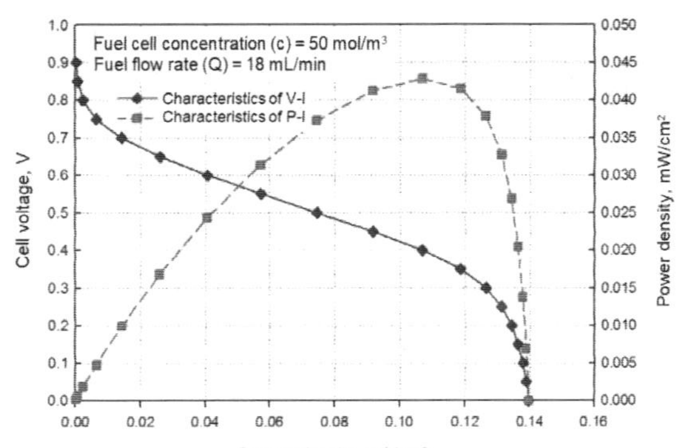

Figure 5. Characteristics of PEMFC with a 50 mL/m^3 hydrogen concentration and a fuel flow rate of 18 mL/min.

Performance simulation of PEMFC models at different fuel flow rates of 6 mL/min, 18 mL/min, and 30 mL/min at 50 mol/m^3 hydrogen fuel concentration, as shown in Figure 4, Figure 5, and Figure 6. Simulation results show that the fuel flow rate at the anode side channel which is getting bigger, will increase the accumulation of electric current produced by the fuel cell. Likewise, there will be an increase in the power of the fuel cell produced. Therefore, one way to obtain the optimum performance of the fuel cell is by knowing the conditions of the fuel flow rate parameters. Although many other parameters that will affect the performance of the fuel cell, among others is to vary the fuel concentration, membrane conductivity value, membrane and electrode permeability value, flow velocity on the inlet channel side, catalyst loading, temperature, and electrolyte concentration.

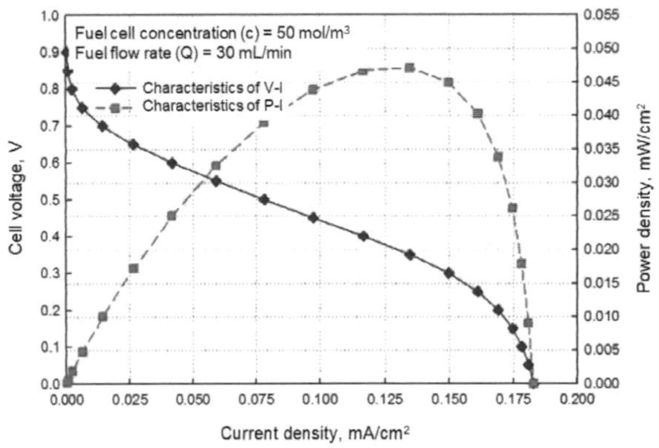

Figure 6. Characteristics of PEMFC with 50 mol/m^3 Hydrogen concentration and 30 mL/min fuel flow rate

3.2. Analysis ofhybrid power PV-fuel cells

3.2.1. Analysis on the battery charging process. The test results show the battery charging process. The battery charging process is influenced by the amount of current used for charging and the battery specifications. In this charging process, the battery uses two car batteries that are installed in series with a total specification of 24V 100Ah.

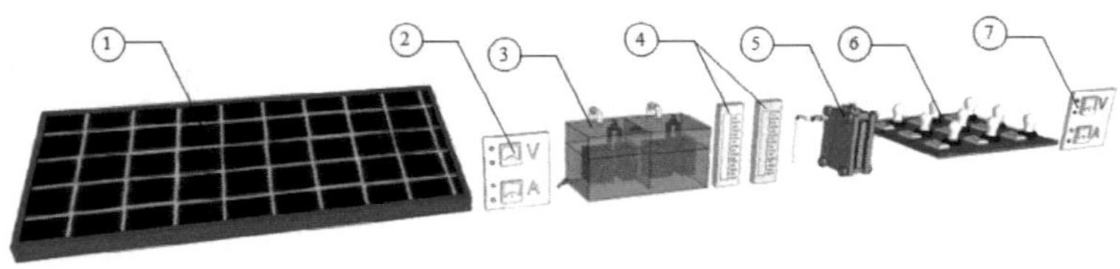

Figure 7. Design of hybrid PV-fuel cell models: 1. power sources (solar cells); 2. voltmeter & ampere meter input; 3. electrolyzer box; 4. flowmeter; 5. fuel cell; 6. load; 7. voltmeter & ampere meter output.

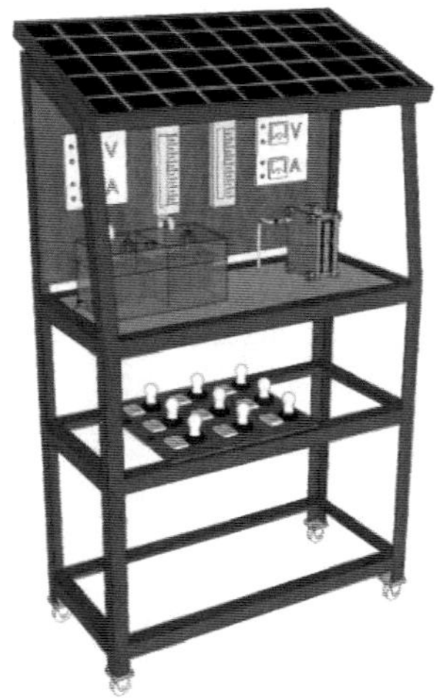

Figure 8. The Hybrid PV-Fuel cell power generation model.

This test is carried out during cloudy weather and starts at 09.00-14.00 with a time scale of 30 minutes. In this test the weather is very influential on the charging system, especially for solar power plants.

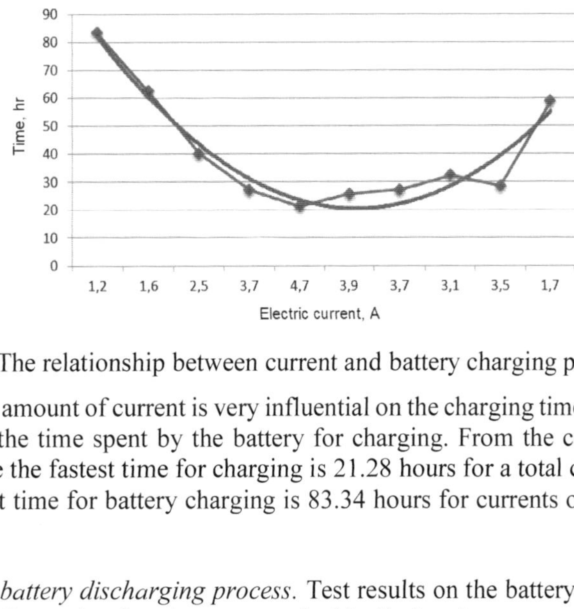

Figure 9. The relationship between current and battery charging process time.

Figure 9 shows that the amount of current is very influential on the charging time. The greater the current produced, the smaller the time spent by the battery for charging. From the calculation of the battery charging process above the fastest time for charging is 21.28 hours for a total current of 4.7 amperes at 11:30 while the longest time for battery charging is 83.34 hours for currents of 1.2 amperes occurring at 09.30.

3.2.2. Analysis on the battery discharging process. Test results on the battery discharging process are shown in Figure 8. Similar to the charging process, in this discharging process the battery uses two car batteries that are installed in series with a total specification of 24V 100Ah and are fully charged. This test uses a light bulb load with a varied power and on a time scale of 10 minutes.

3.2.3. Actual battery discharge analysis. The results of the calculation analysis for the length of time the battery is turned on during the actual battery discharge process are shown in the Figure below.

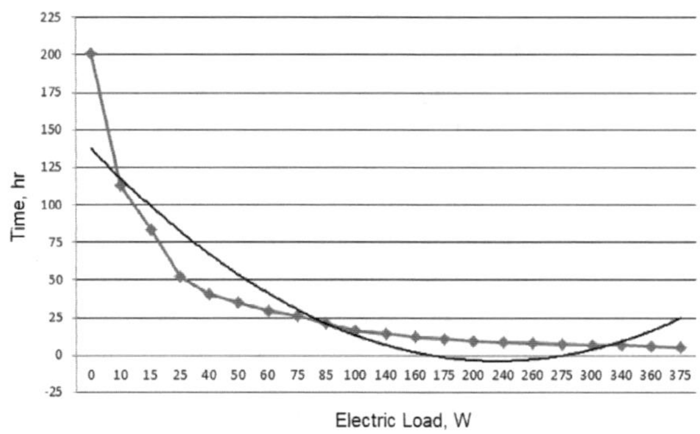

Figure 10. The relationship between load and time of use in the actual battery discharge process.

Figure 10 shows the size of the load is very influential on the discharge time. The greater the load used, the smaller the time used by the battery to turn on the load. From the actual calculation above the highest time taken when a small load of 10 watts with a time of 200.803 hours and vice versa the fastest time taken when a large load of 400 watts with a time of 5.68 hours.

3.3. Theoretical battery discharge analysis

The results of the analysis of calculations for the length of time to start using the battery theoretically in the process of emptying the battery.

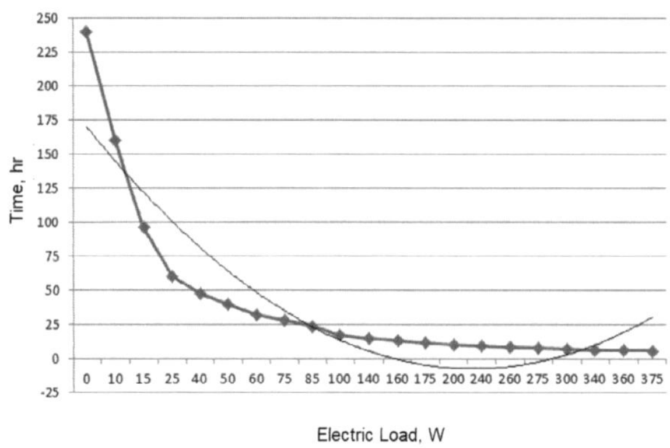

Figure 11. The theoretical relationship between load and time of use in the battery discharge process.

Figure 11 shows the size of the load is very influential on the discharge time. The greater the load used, the smaller the time used by the battery to turn on the load. From the actual calculation above the highest time taken when a small load is 10 watts with 240 hours and vice versa the fastest time taken when a large load is 400 watts with a time of 6 hours.

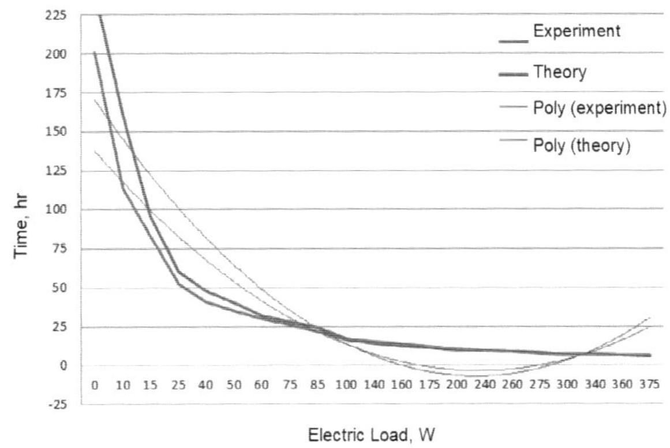

Figure 12. The relationship between the load and time of use in the actual battery discharge process and theoretically.

Figure 12 shows the reduction in battery discharge time for the same load. This is because due to the power losses on each equipment and circuit that results in the power generated by the battery not reaching the maximum load.

4. Conclusion

The results showed that the open circuit voltage in the fuel cell increased slightly with increasing fuel flow rate in the reactants in the micro channel. Open circuit voltage can vary under conditions of varying fuel flow rate which is around 0.35 V. Besides that the surface conditions of the electrodes will affect the change in current. Low open circuit voltage will trigger the formation of air bubbles in the fuel lines. The fuel cell micro channel has a width of 1.0 mm, a depth of 1.0 mm, and a length of 20 mm. The thickness of the MEA (Membrane Electrode Assembly) is 0.53 mm, including a 0.38 mm thick gas diffusion layer, a 0.05 mm thick catalyst layer and a 0.10 mm thick membrane layer. The maximum power density will gradually increase as Reynold's value increases. The maximum power density is achieved at 0.026, 0.042, and 0.047 mW/cm^2, for fuel cells at a flow rate of 6, 18, and 30 mL/min at a fuel concentration of 50 mol/m^3.

Acknowledgment

The author thanks to Politeknik Negeri Semarang for providing financial support with SP DIPA-042.0.1.2.400997/2018 contract number, PNBP no. 537/PL4.7.4.2 / LT / 2018 in accordance with the letter of agreement on the Penelitian Unggulan Doktor.

References

[1] Eikani M H, Eliassi A, Khandan N and Nafisi V R 2012 Design and fabrication of a 300 W PEMFC fuel cell test station Procedia Engineering 42: 406-413 *20th International Conggress of Chemical and Process Engineering CHISA* 2012, 25-29 August 2012, Prague, Czech Republic

[2] Herlambang Y D, Lee S C and Hsu H C 2017 Numerical estimation of photovoltaic-electrolyzer systems performance on the basis of a weather database *International Journal of Green Energy* **14** 575-586

[3] Larmine J and Dicks A 2003 *Fuel Cell Systems Explained Second dition* (West Sussex-UK: John Willey & Sons)

[4] Nasharudin M N, Kamarudin S K, Hasran U A and Masdar MS 2014 Mass transfer and performance of membrane-less micro fuel cell: a review *International Journal of Hydrogen Energy* **39** 1039-1055

[5] Peighambardoust S J, Rowshanzamir S and Amjadi M 2010 Review of the proton exchange

membrane for fuel cell PEMFC application *International Journal of Hydrogen Energy* **35** 9349-9384

[6] Herlambang Y D, Roihatin A, Kurnianingsih, Lee S C and Shyu J C 2020 MEMS-based Microfluidic fuel cell for in situ analysis of the performance on the electrode surface *Journal of Physics: Conference Series* 1444 012044

[7] Herlambang Y D, Roihatin A, Kurnianingsih, Prasetyo T, Lee S C, and Shyu J C 2020 Computation and numerical modeling of fuel concentration distribution and current density on performance of the microfluidic fuel cell *AIP Conference Proceeding* 2197, 090001

[8] Shyu J C, Wang P Y, Lee C L, Chang S C, Sheu T S, Kuo C H, Huang K L and Yang Z Y 2015 Fabrication and test of an air-breathing microfluidic fuel cell *Energies* **8** 2082-2096

[9] Young, D F, Munson, B R, Okiishi, T H and Huebsch, W W 2011 *Fundamentals of Fluid Mechanics Sixth edition* (New York-USA: John Wiley & Sons)

[10] Youssef M E, Amin R S and El-Khatib K M 2015 Development and performance analysis of PEMFC stack based on bipolar plates fabricated employing different designs *Arabian Journal of Chemistry, King Saud University* **5** 1-6

[11] Zhang B, Ye D D, Sui P C,Djilali N and Zhu X 2014 Computational modeling of air-breathing microfluidic fuel cells with flow-over and flow-through anodas *Journal of Power Sources* **259** 15-24

[12] Zhang H, Xuan J, Xu H, Leung M K H, Wang H, Leung D Y C, Zhang L and Lu X 2012 A numerical study on microfluidic fuel cell: Improving fuel utilization and fuel operation concentration *Energy Procedia* **61** 250-253

IOP Conf. Series: Materials Science and Engineering 1108 (2021) 012050 doi:10.1088/1757-899X/1108/1/012050

Wind energy conversion system with capacity of 450 W in order to support teaching and learning process in the laboratory

Y D Herlambang[1], T H Mulud[1], B Bono[2], S Suwarti[1,*] and G Suwoto[1]

[1] Dept. of Mechanical Engineering, Politeknik Negeri Semarang, Semarang, Indonesia
[2] Politeknik Negeri Semarang, Semarang, Indonesia

*suwarti@polines.ac.id

Abstract. This study examines a 5-blade Horizontal Axis Wind Turbine to obtain the best performance at different wind speeds. The dimensions of this wind turbine have a rotor diameter of 1200 mm, blade length of 600 mm, number of blades 3 pieces, and blade material made of fiberglass. The research method uses different wind speed test parameters. The speed test using a blower is 4 m / s, 5.1 m / s, 6.5 m / s, and 7.5 m/s with the load of the lamp is 10 W, 20 W, 30 W, 40 W, 50 W, and 60 W. Subsequently analyze the characteristics of the wind turbine. The results of testing the wind turbine at a speed of 4 m/s obtained a maximum efficiency of 36.4% at 496.3 rpm turbine rotation. Power generator 15.8 W at a voltage of 20V and current 0.79 A.

1. Introduction

The accelerated development program for the new plant aims to increase the electrification ratio from 87.4% in 2017 to around 97% in 2019 [1-3]. The government, in this case, PLN, has a commitment to increase the electrification ratio in rural areas, remote areas and Indonesia's outer islands. The electrification ratio is a comparison of the number of Indonesian people who have enjoyed electricity facilities with a total population of Indonesia. Perpres No. 5 of 2006 concerning the National Energy Policy states that the contribution of the development of renewable energy (EBT) in the national primary energy mix up to 2015 is 17% with 5% geothermal composition, water, solar, wind, biomass, and nuclear at 5%, biofuel (biodiesel and bioethanol) by 5%, and liquefied coal by 2% [4-6]. It is expected that by 2025 the national primary energy mix will increase to 23%. Based on the data above, Indonesia will experience an energy crisis, because all types of Indonesian energy will be exhausted within a period of 90 years [1]. One of the applications of a 75 MW capacity wind turbine in the village of Mattirotasi, Sidrap Regency, South Sulawesi. A total of 30 Wind Turbine Generators (WTG) are able to flow electricity to 70,000 homes. Research related to the use of new renewable energy such as solar power plants, water, microfluidic fuel cells and wind power plants has been conducted by Herlambang. The working principle of the VAWT wind turbine by utilizing the wind coming from the blower then the wind is captured by the blades of the turbine so that mechanical energy arising from the turbine shaft is then converted to electrical energy from the generator [7,8]. The kinetic energy of the mass of air m moving at velocity v [9]:

$$P_{kin} = \frac{1}{2}\,\dot{m}\,.\,v^2 \qquad (1)$$

Referring to the area of a specific cross-section A, which is passed through the airspace at velocity v, the volume V˙ flows past a certain unit of time, called volume flow and the mass flow rate is called $\dot{m}$ [9,10]:

Referring to the area of a specific cross-section A, which is passed through the airspace at velocity v, the volume V˙ flows past a certain unit of time, called volume flow and the mass flow rate is called [9,10]:

$$\dot{m} = \rho \,.\, A_1 \,.\, v_1 \tag{2}$$

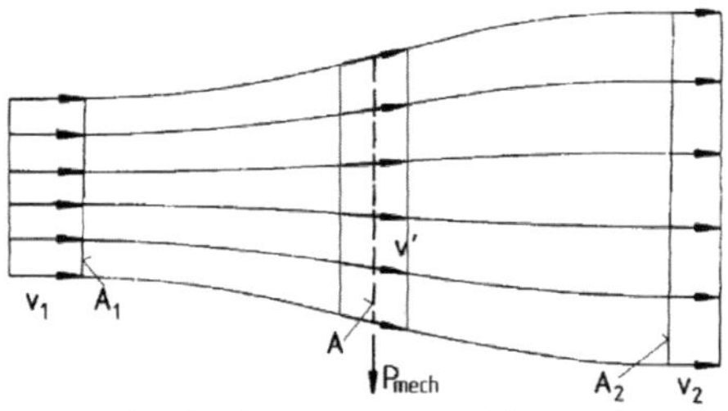

Figure 1. Flow conditions due to mechanical energy expenditure from free flow airflow, referring to the basic theory of momentum.

The mechanical energy which is removed from the disk-shaped converter from the air flow is related to the difference in power from the air flow before and after the converter [10]:

$$P_{kin} = \frac{1}{2}\rho \,.\, A_1 \,.\, v_1^3 - \frac{1}{2}\rho \,.\, A_2 \,.\, v_2^3 = \frac{1}{2}\rho\,(A_1 \,.\, v_1^3 - A_2 \,.\, v_2^3) \tag{3}$$

v_1 : undelayed free-flow velocity, wind speed
v_2 : flow velocity behind the converter
The required continuity equation is:

$$\rho \,.\, A_1 \,.\, v_1 = \rho \,.\, A_2 \,.\, v_2 \tag{4}$$

$$P = \frac{1}{2}\rho \,.\, A_1 \,.\, v_1^3\,(v_1^2 - v_2^2) \tag{5}$$

where:

ρ : Density of air [$^{kg}/_{m^3}$]
A : Rotor swept area [m^2]
v : Wind velocity [$^{m}/_{s}$]
index 1 is the direction of entry; index 2 is the direction of exit.
According to Gupta [8] and Grant [11] explained that the power will be able to reach the maximum if the velocity of v_2 is zero. However, this result cannot be made physically as can be expected, physically the result means it consists of a certain numerical ratio (numerical) v_2 / v_1 where the power that can be released reaches a maximum with v2 / v1 = 16/27 or cp = 0.593.

2. Research methods

The first stage is the selection of fiberglass blade material and the design of tools is carried out to obtain the best design and alternative. This activity is carried out by sketching a picture of the HAWT wind rotor blade and its mechanism so that a construction drawing is produced. The amount of turbine power is determined by the area of the rotor sweep and wind speed after obtained the amount of turbine power is used to determine the amount of electrical capacity needed. The 3-blade HAWT rotor consists of a turbine rotor, a blade holder, and a shaft. The HAWT rotor model made consists of 3 turbine blades made of fiberglass (polyester resin, wool roofing, fiber cloth, glass fiber/mat, pigment coloring, catalyst). The choice of blade material uses a blade mold. The diameter of the 3-blade HAWT rotor is 1200 mm, the blade length is 600 mm. This 3-blade HAWT turbine blade is mounted on a disc as a buffer blade.

Implementation of the test, where at this stage the data will be collected. The test parameters measured in the test are the wind speed (m / s) that will be used to rotate the turbine blade, which are 4 m / s, 5.1 m / s, 6.5 m / s and 7.5 m / s; shaft rotation (rpm), i.e. the rotation generated by the turbine shaft after the turbine blade has been subjected to wind at a predetermined speed; processing and analysis, test results data (wind speed, turbine rotation, voltage, electric current, lamp load) are then processed to obtain generator power, and system efficiency. The processing results are then displayed in a graphical form of wind turbine characteristics. The performance of each turbine is reviewed and analyzed. So it will produce turbines with blade angle variations and simulations that have the best performance.

3. Results and discussion

The process of installing and assembling begins with the manufacture of the supporting frame, rotor assembly (hubs, turbine blades, gears, generators, tailwind direction) as shown below.

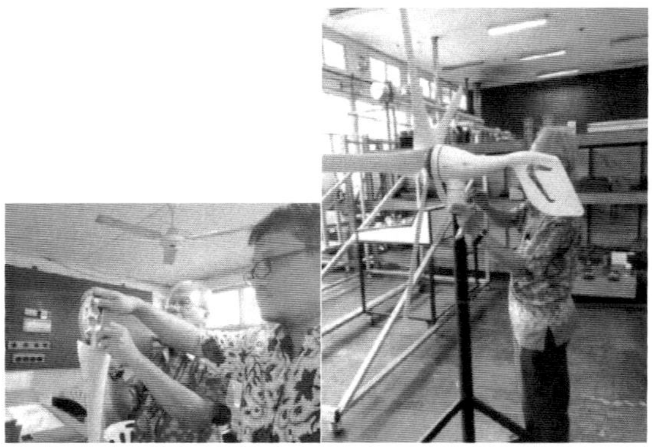

Figure 2. Installing and assembling.

Table 1. Data from tool testing results.

Load (W)	Ampere (A)	Volt (V)	n (rpm)
10	0,79	20	496,3
20	1,4	10	129
30	1,8	8	132
40	2	5	56
50	2,25	2,5	7,4
60	2,2	2	4

In this test wind turbines using wind speeds of 4.5 m / s. Speed measurements are made at several points in front of the turbine to get the desired average results. This test uses a load of 10, 20, 30, 40, 50, 60, watts. Variables that are recorded when making measurements include voltage, current, and rotation speed. Voltage and current are used to measure P-I while calculating kinetic power requires additional data in the form of air density, turbine speed and turbine surface area. Then for efficiency using the formula P-I at the generator divided P_{kin}. The test results show that the greater the load used, the greater the current generated. While the voltage generated is getting smaller, turbine rotation will also decrease. According to the calculation data, the greater the load, the smaller the turbine efficiency, along with the Tsr value (tip speed ratio).

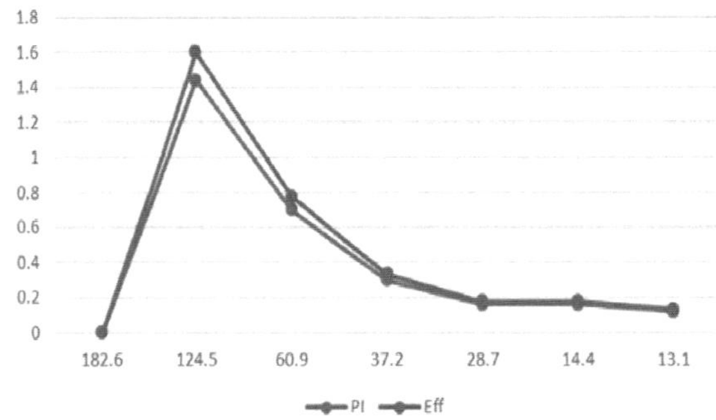

Figure 3. Graph of the relationship between system efficiency, generator power, and turbine shaft rotation at wind speeds of 4.5 m/s.

At Figure 3 that based on the results of data retrieval on wind turbines, it is known that at 0-watt load, the amount of current flowing is not detected because there is no load so that the value of the electric power (P-l) remains 0, as well as the efficiency value. Meanwhile, with increasing load value, the rotation will experience a decrease followed by a decrease in the amount of current and voltage generated by the wind turbine. This has an automatic impact on the amount of Pl so that it affects the value of efficiency which is also declining. The amount of electric power generated due to the rotational speed ultimately determines the working efficiency of the wind turbine, the greater the electric power, the greater the efficiency.

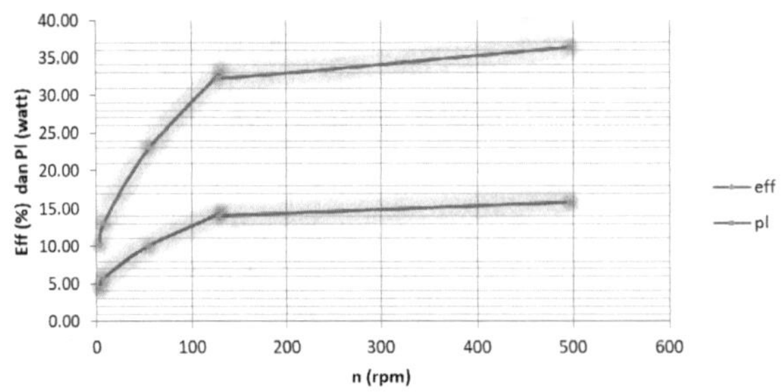

Figure 4. Graph of the relationship between system efficiency, generator power, and turbine shaft rotation at wind speeds of 5.1 m/s.

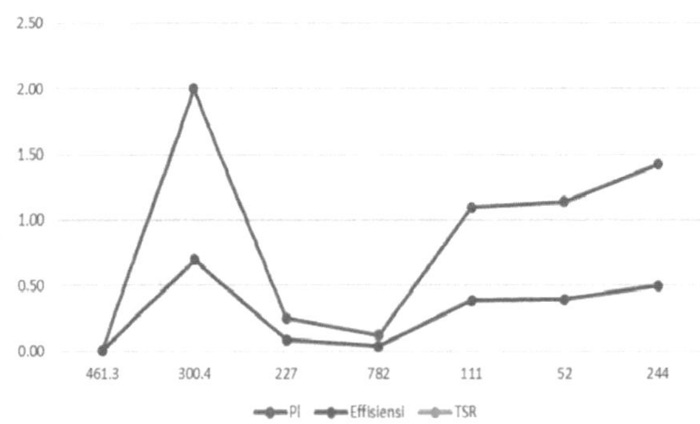

Figure 5. Graph of the relationship between system efficiency, generator power, and turbine shaft rotation at wind speeds of 7.5 m/s.

Based on the results of data retrieval on wind turbines, it is known that at 0-watt load, the amount of current flowing is not detected because there is no load so that the value of the electric power (Pl)

IOP Conf. Series: Materials Science and Engineering 1108 (2021) 012050 doi:10.1088/1757-899X/1108/1/012050

remains 0, as well as the efficiency value. Meanwhile, with increasing load value, the rotation will experience a decrease followed by a decrease in the amount of current and voltage generated by the wind turbine. This has an automatic impact on the amount of PI so that it affects the value of efficiency which is also declining. The amount of electric power generated due to the rotational speed ultimately determines the working efficiency of the wind turbine, the greater the electric power, the greater the efficiency.

3.1. Technological excellence

3.1.1. Innovative excellence

- Is a technological approach using a HAWT turbine blade Aerofoil NACA Laboratory scale.
- If applied to power plants, they can be designed with dimensions which is great for producing power as needed.

3.1.2. Comparative advantage

- If applied for energy generation, it will be able to receive wind energy from all directions and increase the flow of air to hit the turbine, thereby increasing the efficiency and power of the turbine.
- Material that is cheap and easily available on the market will reduce production costs

3.2. Application prospect

- The application of an aerofoil rotor blade will increase turbine efficiency and power wind.
- Support materials that are cheap and easily obtained in the market, will be pressing investment costs for generating plants.
- Turbine design for low wind speeds can be applied throughout regions of Indonesia that have wind potential above 3.5 m/s.

4. Conclusion

The results of this study produce conclusions:

- Tool specifications:
  a. Wind turbine: Turbine height = 1 m - 3 m
     Turbine diameter: 1.2 m
     A number of blades: 3 pieces
  b. Generator, output voltage = 12 volts AC
     Generator power: 500 Watt
     Rotation: 1800 rpm
  c. 12 Volt DC Accumulator, 75 Ah
  d. Rectifier, output voltage = 15 Volt DC
  e. Inverter, input voltage = 12 Volt DC
     Output voltage = 220 Volt AC
- This test uses a load of 10, 20, 30, 40, 50, 60, watts. Variables that are recorded when making measurements include voltage, current, and rotation speed. Voltage and current are used to measure PI while calculating kinetic power (P_{ek}) requires additional data in the form of air density, turbine speed and turbine surface area.
- The test results show that the highest system efficiency is 36.4% with a maximum generator power of 15.8 W. The greater the load used, the greater the current generated. While the voltage generated is getting smaller, turbine rotation will also decrease.
- The greater the load, the smaller the turbine efficiency, along with the decreasing value of the Tsr (tip speed ratio).

Acknowledgment
The author would like to thank the Politeknik Negeri Semarang who has funded the Penelitian Terapan Unggulan Program Studi Teknik Konversi Energi-PNBP with contract number 988/PL4.7.4.2/PT/2019.

References
[1] AWEA 2004 The American Wind Energy Association. http://www.awea.org. BWEA. 2002. The British Wind Energy Association. http://www.bwea.com. diakses tanggal 5 Oktober 2009.
[2] Herlambang Y D, Lee S C, Hsu H C 2017 Numerical Estimation of Photovoltaic–Electrolyzer System Performance on The Basis of a Weather Database *International Journal of Green Energy* **xx** 1-12
[3] Herlambang Y D, Shyu J C and Lee S C 2017 A numerical study on the performance of air-breathing microfluidic fuel cells The 12th *Annual International Conference on Nano/Micro Engineered and Molecular Systems (NEMS)*, 9-12 April 2017, UCLA, California USA
[4] Herlambang Y D 2017 Numerical simulation of the performance of air-breathing direct formic acid microfluidic fuel cells, IET Micro & Nano Letters.
[5] Herlambang Y D, Lee S C and Liu C J 2016 Numerical Study and Modeling of The Solar Radiation Measurement on Tilted Surface for The Local Behavior Database **37** 441-448
[6] Herlambang Y D and Suwoto G 2010 Unjukkerja Turbin Air Mikro Aliran Silang Terhadap Variasi Sudut Sudu Jalan (Runner) Pada Debit Konstan Untuk PLTMH *Prosiding SNST* Fakultas Teknik Unwahas Semarang
[7] Beri H and Yao Y 2011 Effect of Chamber Airfoil on Self Starting of Vertical Axis Wind Turbine *Journal of Environment Science and Technology* **4** 302-312
[8] Grant A and Kelly N 2003 The Development of a Ducted Wind Turbine Simulation Model. Eight Int. *IBPSA Conference Eindhoven*, Netherlands 407-414
[9] Hau, Erich 2006 *Wind Turbines Fundamentals, Technologies, Application, Economics. 2nd Edition* (Germany: Springer-Verlag Berlin)
[10] Leysen E H 1983 Introduction to Wind Energy. PO BOX 85/Amersfort/The Netherlands: Steering Committee Wind Energy, Developing Countries
[11] Gupta R, Roy S, and Biswas A 2010 Computational Fluid Dynamics of Twisted Airfoil Shaped Two Bladed H-Darrieus Rotor Made From Fibreglass Reinforced Plastic (FRP) *International Journal of Energy and Environment* **1** 953-968

AUTHOR INDEX

Abdullah, M I 160
Ahmed, F Y H 160
Alfatih, M F 129
Alfauzi, A S 299
Amalia, R 335
Andromeda, T 109
Antono, D 227
Apriantoro, R 49, 294
Ardjo, A S 299, 379, 387
Arifin, F 400
Ariwibowo, D 335
Ariyono, S 90, 367
Aryanti, A 195
Astor, Y 65
Azizah, A 77, 120
Badruzzaman, Y 227
Bono, B 409
Dewi, R S 195
Dini, O N 246
Endawati, J 12, 24
Giyanto, G 329
Habiburrahman, M H 379
Hadi, T 38
Handayani, S U 335
Handoko, S 255
Hapsari, A A 219
Harijadi, M B 109
Hariyanto, H 305
Harmanto, S 329
Hartati, I 335
Hartono, H 299
Haryanto, H 329
Hasan, A 209, 219, 275
Helmy, H 275, 284
Herlambang, Y D 343, 400, 409
Hestinigsih, I 139
Huboyo, H 57
Ismail, S 152
Iswanto, I 1
Johar, M G M 160
Jumi, J 77, 120, 264, 320
Jum'atun, S 294
Junesco, D 209
Karimah, A I 284
Karnisah, I 65
Kasyanto, H 19
Khoryanton, A 312, 329, 395
Kurniawati, R 152
Kusuma, G E 357
Kusumawardani, R 49
Lehat, N M 234
Mansyur, A W 98
Manullang, O 57
Mardiyono, M 246
Martono, M 49, 202
Mawardi, M 202
Mekongga, I 195
Mufidati, D 174
Mujiman, M 1
Mukhlisin, M 49, 209
Mulud, T H 409
Mulyono, M 350
Mulyono, T 38, 77, 120, 264, 320
Nugraha, D 160
Nursyahid, A 275, 284
Oesman, M 1, 24
Oktaviani, T 12
Pamungkas, N S 38
Pramono, A 312
Pramuditya, F 294
Prasetiyo, B 343
Prasetyo, B 350
Prasetyo, T 400
Pratama, L P 234
Pratikto, P 395
Prayitno, P 387
Purbono, K 374
Puspawaty, K Y 255
Rahayu, E P 189
Riyadi, M A 129
Rochaeti, R 12
Roihatin, A 400
Sadida, A 98
Safarudin, Y M 400
Sahid, S 350
Samadiku, B 57
Santosa, S 202
Santosa, Y P 202
Santoso, A 227
Santoso, M 357
Santoso, P B 395
Sarana, S 98, 374
Sarena, S T 357
Sari, R E 246
Satito, A 305, 329
Saud, S N 234

Sayekti, I 146
Sa'diyah, K 275
Setiawan, I 129
Setiawan, T A 284
Setyawan, T A 275
Sihono, S 90, 367
Slamet, A 379
Somantri, M 109
Suharjono, A 49, 146, 219
Suharto, S 374
Sukamto, S 139
Sulaiman, S 31
Sumiyarso, B 367
Sundara, A 24
Supandi, S 305
Suparman, S 395
Supriyanto, E 209, 219
Supriyatna, S 227
Supriyo, B 90, 367
Supriyo, S 343
Suroso, S 202
Suryanto, S 374
Suryono, S 152, 174, 182, 189
Susanto, A 19
Susilahadi, S 19
Sutrisno, S 335
Suwarti, S 350, 409
Suwondo, A 98
Suwoto, G 409
Syahid, S 227
Tegar, T 329
Tjahyono, B 367
Triyono, L 139, 255
Utami, R 31
Utomo, M B 202
Veronica, S Y 182
Virnawan, D 24
Wahjoedi, W 38
Wardihani, E D 146, 234, 275, 379, 387
Wattimena, R M 299
Widodo, I G 312
Widyawati, M N 174, 182, 189
Yudantoro, T R 139, 294
Yulianti, N P 31
Yulianto, M E 335
Yunanto, B 49
Zaenuddin, A 264, 320